1978

Plant biology

A CONCISE INTRODUCTION

Plant biology
A CONCISE INTRODUCTION

Ross H. Arnett, Jr., Ph.D.

Professor of Biology
Siena College
Loudonville, New York

George F. Bazinet, Jr., Ph.D.

Professor of Biology
Siena College
Loudonville, New York

FOURTH EDITION

with 726 illustrations

The C. V. Mosby Company

Saint Louis   1977

Front cover illustrations

Reading from left to right and top to bottom

Sarracenia purpurea, pitcher-plant (Sarraceniaceae), northern Florida.
Proboscidea parviflora, devil's claw (Martyniaceae), Arizona.
Cypripedium candidum, ladyslipper (Orchidaceae), western New York.

Back cover illustrations

Reading from left to right and top to bottom

Sarracenia leucophylla, pitcher-plant (Sarraceniaceae), northern Florida.
Commelina dianthifolia, blue dayflower (Commelinaceae), Arizona.
Cassia leptocarpa, senna (Leguminosae), Arizona.
Cylindropuntia whipplei, plateau cholla (Cactaceae), northern New Mexico.

FOURTH EDITION

The C. V. Mosby Company
11830 Westline Industrial Drive, St. Louis, Missouri 63141

Library of Congress Cataloging in Publication Data

Arnett, Ross H
    Plant biology: a concise introduction.

    First-2d ed. by D. C. Braungart and R. H. Arnett; and
3d ed. by R. H. Arnett and D. C. Braungart, published
under title: An introduction to plant biology.
    Bibliography: p.
    Includes index.
    1. Botany.   I. Bazinet, George F., 1937-
joint author.   II. Braungart, Dale Carl, 1912-
III.   Title.
QK47.A78 1977   581   76-26531
ISBN 0-8016-0316-1

TS/M/M   9 8 7 6 5 4 3 2 1

To the memory of
Roger Bacon, O.F.M. (1214-1294)
and
Noel A. Fitzpatrick, O.F.M., Ph.D. (1934-1974)

Life on Earth is possible only because of a flow of energy from the sun.

Energy flow is in one direction only, from **plant** to **plant, plant** to **animal, plant** and **animal** to **oblivion.**

Life systems are a complex web of interactions of one organism with another as a result of the flow of energy through the system. These systems exist at three levels as a hierarchy:

- **Cellular level,** wherein chemical combinations operate under the rigorous laws of thermodynamics.
- **Organismic level** of development and reproduction operating under the rigorous laws of heredity and evolution.
- **Ecosystem level,** wherein species remain distinct from their environment, function within the rigorous limits of their environment, and then merge into the environment.

The following pages investigate the truth of these statements. This is the theme of this book.

Preface

Modern trends in biological education seemed for a while to be away from terms, classification, and even organisms. This was first brought about by the rapid development of molecular biology a decade or so ago. Younger teachers were caught up in this rapid swing of the pendulum so that their education often neglected these "old-fashioned" topics. Soon, however, it became apparent that a knowledge of organisms, and even of their classification, was necessary if the new challenges posed by the current pollution of the environment and the destruction of wildlife are to be met. Biological education now requires a broader view of the field than ever before and, fortunately, a much better balance of subject matter. Molecules, cells, organisms, and environment are of equal importance. Students must be well grounded in all of these subjects no matter what their chosen specialty may be. For many years educators have attempted to integrate subject matter, but this has been hampered in the past by an imbalance of available knowledge. No longer is this true. We now are able to treat all topics equally.

As we prepared this edition of our text, we constantly had, as our theme and goal, the hope that all branches of botany could be treated equally. So we set to work moving parts into a more logical order, trimming here, expanding there, and adding new topics as recent developments in the field required. Our efforts are presented on the following pages.

As for each of the previous editions, we are grateful to the many teachers who took the time to write to our publisher sending corrections, suggestions, and constructive comments. These have all been considered and incorporated into the text. We hope our readers will continue to do this because no work can be useful for long without such help.

Several of the chapters have been completely rewritten so that the subject matter could be presented in the best possible form for easy study. This was primarily the work of the new second author. In addition, we are grateful for comments from our colleagues at Siena College, particularly those from Dr. Douglas F. Fraser.

Two very important subjects have been woven into the text throughout this new edition. We have become very aware of the role of botany in the new assault against impending world food shortage and the current world pollution problem. Therefore, we have added much to the chapters on plant genetics and plant ecology. These subjects are summarized in the final chapter, "Plants and Man," where we show the need for a background in botany in order to cope with these problems in everyday life.

Finally, we again hope that when you have finished your study of this text you will fully appreciate the need to preserve in ecological harmony all forms of life and that you will be prepared and be willing to work for the best balance of the environment, wildlife, and man that is possible on this earth.

Ross H. Arnett, Jr.
George F. Bazinet, Jr.

Contents

1 **The plant kingdom,** 1

Plants and animals, 2
Diversity, 2
Features common to plants and animals, 3
Differences between plants and animals, 4
The kingdoms of living organisms, 4
Plant classification, 5
Systematics, 7
Plant names, 8

2 **The science of botany,** 13

Botanical research, 13
Why study botany? 14
Early history of botany, 16
Scientific method, 19
Analogy and homology, 26
Science and modern society, 27

3 **The composition of living matter,** 33

Cellular chemistry, 33
Some basic chemistry, 33
Composition of matter, 33
Atomic structure, 34
Chemical reactions, 36
Chemicals in protoplasm, 36
Inorganic molecules, 37
Organic compounds, 38
Chemical nature of enzymes, 43
Enzyme action, 44
Physical structure of protoplasm, 44
Water, 44
Diffusion, 46
Osmosis, 47

4 **Cell structure and function,** 51

History of cell concept, 51
Methods for studying cells, 54

The cell, 55
Cell membrane, 56
Cytoplasm and cytoplasmic organelles, 57
Matrix, 57
Mitochondrion, 59
Plastids, 59
Organelles associated with protein
   synthesis, 59
Centrioles, 64
Vacuoles, 64
Nucleus, 64
Nucleolus, 66

5 **Viruses, bacteria, and blue-green algae,** 69

Kingdom Prokaryotae, 69

6 **Algae,** 87

Kingdom plantae (Chapters 6 to 11), 87
Features used to separate the division of
   algae, 87
Biological significance of algae, 100

7 **Fungi and lichens,** 103

Some important biological
   interrelationships of fungi, 120

8 **Mosses, liverworts, and hornworts,** 127

9 **Ferns and other primitive vascular plants,** 135

10 **Primitive seed plants,** 147

11 **Flowering plants,** 159

Contents

12 Energy transformations, respiration, and fermentation, 189

Fermentation, 192
Krebs cycle, 194
Oxidative phosphorylation, 195

13 Photosynthesis, 201

Site of photosynthesis, 201
The nature of light, 204
Pigment systems, 204
Light reactions, 206
The dark reaction, 207
Hatch-Slack pathway, 208
Factors influencing the rate of
    photosynthesis, 210

14 Roots, stems, and leaves of flowering plants, 213

Organization of the plant body, 214
The seed and its germination, 215
Differentiation in higher plants, 220
Roots, 222
Root adaptation, 227
Stems, 228
Stem modifications, 239
Leaves, 245
Adaptation of leaves, 247

15 Absorption, translocation, and transpiration, 253

Chemical requirements of growth, 253
Macronutrients, 253
Essential minerals, 253
Micronutrients (trace elements), 257
Absorption, 257
Translocation, 261
Cohesion-tension theory, 264
Translocation in phloem, 264
Mass-flow hypothesis, 265
Transpiration, 267

16 Reproduction by flowers and their fruits and seeds, 275

Basic principles, 275
Asexual reproduction, 276
Sexual reproduction, 276
Sporophytes, 277

Gametophytes, 277
Evolution of the plant body, 278
Flowers, 279
Flower adaptation, 282
Fruits, 287
Seeds, 292

17 Growth, development, and hormones of flowering plants, 297

Mitosis, 297
Regulation of growth by plant
    hormones, 301
Practical uses of plant-growth regulators,
    311
Photoperiodism, 312
Photomorphogenesis, 312
Flowering response, 314
Leaf senescence and abscission, 316
Dormancy, 317

18 Plant genetics, 321

Meiosis, 321
Mendelism, 323
Mendelian heredity, 324
Incomplete dominance, 326
Dihybrid, 326
Modified dihybrid, 328
Converter genes, 330
Modifying genes, 330
Sex determination, 330
Linkage, 331
Crossing-over, 332
Chromosome mapping, 333
Hybrid vigor, 335
Mutation, 336
Lethal genes, 336
Multiple factors (polygenes), 336
Chromosomal variation, 337
Cytoplasmic inheritance, 340
Chemical nature of heredity, 340
Structure of the DNA molecule, 344
DNA replication, 344
Genetic control of protein synthesis, 346
Regulation of gene action, 348
Problems, 350

19 Plant evolution, 353

Early history of evolution theory, 353
Darwin to the twentieth century, 355

Evolutionary process, 356
Speciation, 358
Data used to show the evolution of
    organisms, 359
Species, 361
Coevolution of plants and animals, 363
Phylogeny, 364
Systematics, 365
History of the earth, 366
Evolution of plants, 373

20 Plant ecology, 381

Nature of the environment, 381
Physical environment, 381
Biotic environment, 391
Plant habitats, 395
Plant distribution in relation to
    environment, 397
Niche, 397
Plant forms in relation to environment,
    397
Distribution mechanisms of plants, 399
Classification of biotic communities, 401
Ecology, pollution, and environmental
    science, 410
Plants for conservation and land
    development, 413

21 Plants and man, 419

Origin of cultivated plants, 419
Changes in plants through cultivation,
    421
Food plants, 421
Fiber plants, 423
Lumber trees and forest products, 423
Development of new plants, 424
Insects and insecticides, 425
Wild plants for food, 425
Poisonous plants, 426
Disease-producing plants, 428

Medicinal plants, 429
Plant diseases, 429
Weeds and herbicides, 429
Plants and air pollution, 430
Practical use of botanical knowledge, 430
Botany as a hobby, 431
Preparation of plant specimens, 432
Identification, 433

APPENDIX A
Plant classification, 435

APPENDIX B
Plant life cycles, 447

1—T-even phage, 448
2—*Nostoc* sp., 449
3—*Chlamydomonas* sp., 450
4—*Volvox aureus*, 452
5—*Ulothrix* sp., 454
6—*Chara* sp., 456
7—*Vaucheria* sp., 458
8—*Fucus* sp., 460
9—*Ectocarpus* sp., 462
10—*Nemalion* sp., 464
11—*Saprolegnia* sp., 466
12—*Rhizopus stolonifer*, 468
13—*Schizosaccharomyces* sp., 471
14—*Microsphaera* sp., 472
15—*Puccinia graminis*, 474
16—*Agaricus campestris*, 478
17—*Marchantia* sp., 480
18—*Polytrichum* sp., 483
19—*Selaginella* sp., 485
20—*Equisetum* sp., 489
21—*Polypodium* sp., 492
22—*Pinus* sp., 495
23—*Lilium* sp., 502

Glossary, 510

Fig. 1-1. The plants of this high-rainfall forest in the Pacific Northwest are diverse. Within this macroscopic view are represented all major plant divisions although not all are visible. The ferns on the lower left are representative of this group of plants. The trees and shrubs show the dominance of the flowering plants, while hidden in the ground vegetation is a luxurious growth of moss. Fungi, algae, and lichens are attached to the trunks of the trees. The soil harbors many species of bacteria and fungi, and there is little doubt that some of these leaves are infected with viruses.

CHAPTER 1

The plant kingdom

■ A billion years ago the surface of the earth may well have looked like that of the moon today. Desolate as the surface may have been, the earliest fossils tell us that the seas probably contained **organisms** that were capable of utilizing the sun's energy to build complex molecules out of simple ones like carbon dioxide and water. This process, simple as it may sound, was the one thing that made possible the transition from nonliving to living as we know it now. The process stored energy for future use; that is, it made food. These organisms, the **autotrophs,** * which consisted of only a single cell, are known as **blue-green algae.** They persist to this day in the waters of the earth (see Chapter 5).

The story of the origin of life is still confused because we know so little of the distant past. We can only speculate as to how it might have come about (see Chapter 19). But the fossils we do have suggest that about the time that the oceans teemed with one-celled autotrophs they also contained cells that were unable to manufacture their own food. These, the **heterotrophs,** acquired food from sources outside themselves; that is, they were feeders. Most of the heterotrophs now are animals, or **Animalia,** although in the early seas they all may have been one-celled **bacteria.** Other one-celled organisms living today lack the necessary pigments needed to manufacture their own food. These we call **protozoa.** There is no clear-cut distinction between these organisms and the very simi-

*NOTE: Words in boldface type in the text are defined in the glossary at the end of the book.

lar heterotrophs we call **algae.** Whether bacteria or the blue-green algae came first is actually an unsolved evolutionary question with persuasive arguments supporting both views. Nevertheless, whichever it may have been, this simple life strategy of autotrophs' converting simple compounds into complex foods by means of the sun's energy and then the heterotrophs' feeding upon this food was to be the paradigm for the next 600 million years of plant and animal evolution.

Once life was established, it spread rapidly through the seas, probably from the shallow waters near the shore. Diversity is a remarkable function of the living chemicals, as explained in Chapters 3 and 18. The plants of the ocean became more and more complex because of this ability to diversify. Parallel with this complexity, animals became even more complex because of the many habitats that were created and could be occupied because of the plant life. Complex interrelationships were established and the sea became filled with life.

Plants eventually invaded the rocky shores of the sea. These shores and the land beyond were devoid of soil until occupied by plants. This invasion was not because the sea was crowded but simply because life characteristically pushes into the extremes of every habitat. This is one of the essential features of life—the ability to disperse. Once established on land, even greater diversity of plant life was possible and plants developed eventually into the life forms we know today. This fascinating story is unfolded in greater detail in Chapter 19. As happened in the sea, so again, animal forms followed the plants to the land and occupied all available habitats. Here both groups found a great number of places to occupy because of the extremes of environmental factors that occur on land. So it is that we find the greatest number of species evolving here.

PLANTS AND ANIMALS

The classical problem of how to separate plants and animals is no longer a real scientific problem. Many attempts have been made in the past to make clear-cut distinctions between plants and animals. All have failed; therefore we may conclude that plants and animals are not separate groups of organisms existing in nature. The two groups are separated only as a means of referring to common life forms. We now know all living organisms have common fundamental features, but not all have the same life plan. In this chapter a classification scheme is outlined, based as much as possible on our knowledge of the evolution that has taken place. Many of these groups are described in the early chapters of this book so you can understand the diverse ways basic physiological processes function.

The layman can easily distinguish plants from animals because he concerns himself with the outer appearance of the higher plants and the higher animals, but, as will be discussed in other chapters of this textbook, the scientific dichotomy of certain living organisms is impossible. Although it is applied to the study of those organisms ordinarily called plants, the term **botany** implies the long tradition of separating organisms into plants and animals. This is so firmly established in our colleges, universities, museums, and scientific societies, as well as in our scientific and everyday writing, that it is doubtful if we will discard, or want to, the words botany and **zoology.**

The plant sciences investigate members of the **Prokaryotae** and the **Plantae.** Thus botany includes two of the three kingdoms of organisms now recognized, omitting only the **Animalia.** This includes study of the structure, classification, and history of these organisms. Each of these subjects is considered in detail in the appropriate chapters.

The great variety of plant life is attributable primarily to changes in body form as a result of adaptation to different environments. Tropical rain forests such as the dense forest on the slope of the Andes or in Burma provides a suitable environment for hundreds of different **species** of flowering plants. Fog and rain cover these areas much of the time, yet so great is the diversity of form that hardly any of the species of plants are represented by more than a few individuals in any single area. Some of these forms appear to be much simpler than others. These are referred to as primitive. It is difficult to be sure whether a particular anatomical type is primitive or whether it merely represents a reduced condition in a certain stage because of some specialized function or habit. Modern methods have been developed by students of evolution so that in well-known groups very logical and statistically reliable data can be pieced together in order that the phylogenetic lines, or **phylogeny,** are fairly clear. Often knowledge of the complete life cycle and also of allied species helps to determine whether a particular species is simple or represents a reduced form of a more complex group.

The anatomical structure forms the basis for assigning species to higher groups. It has been shown in some representative forms of these groups that anatomical structure is an indicator of physiological functions and habits. Therefore it is assumed that organisms having similar anatomy also have other features in common, even if this is not known by direct observation. By the use of this principle all the plants and animals are classified.

DIVERSITY

Diversity of plants, as well as other organisms, is possible through heredity, by the functioning of the genes within well-circumscribed limits permitted by the simple codification of the macromolecules—DNA. This variation potentially exceeds numerically many billions of possible combinations. We must consider several levels of variation. The first level of variation is that between individuals. Although it is the individuals we study, we derive concepts by making generalizations about them until we perceive their sameness and discover the gap between one group of individuals and other groups of individuals, gaps that exist at any time period. These groups we call species and these we classify. Variation or diversification of the species is a natural process permitting the orderly operation of many different life systems, that is, particular life cycles. The recognition of this diversity and its classification is the turnstyle of information handling. Only by acquiring a broad prospective of the many kinds of plants can one comprehend this diversity and the relative simplicity of the mechanisms that permit adaptation and evolution.

FEATURES COMMON TO PLANTS AND ANIMALS

The simplest way to define a living organism is to say that it is cellular. In other words, all matter organized in cellular form, either as a simple cell, or a group of cells, is referred to as living. If it is not cellular it is not living, meaning of course, that it has never lived. Thus when we say something is alive, we mean that at least some of its **cells** are functioning. Only cellular material can die. All other matter is simply classified (by biologists) as nonliving, some of which may be the products of cells.

We may conclude from the preceding that the most obvious feature that plants and animals share is their cellular nature. *Both groups of organisms are composed of cells.* Since all cells require basically the same materials to function, it is reasonable to assume that the broad chemical reactions will be the same. Many years of experimentation has shown this to be true. *Both plant cells and animal cells require food in the form of chemicals, water as the basic solute for chemical reaction, and oxygen for respiration.* Details of these processes are exceed-

ingly complex as will be seen in later chapters.

For something to be living, it must be capable of reproduction. Reproduction by itself is not sufficient to define life, however. Machines are able to stamp out millions of identical parts. Reproduction in living organisms involves much more than duplication. *Both plants and animals reproduce their own kind.* This distinctness is not that simple. They produce offspring that are not exactly like the parents. Yet the variation does not and cannot exceed certain limits. Just how they do this is the fascinating tale of speciation and evolution. It is certain, therefore, that *species do exist in all groups of living organisms.*

All species, whether plant or animal, have their own ecological requirements. These requirements are a result of the morphological structure and the physiological processes that are controlled by the genetics of each cell. The end result of these requirements determine where these plants and animals live. Thus, *each species has a distribution pattern that changes from generation to generation as a result of environmental change.*

All organisms react to changes in the environment. These responses include reactions to temperature change, increase or decrease in moisture, pressure changes, gravity, and light. Both also react in varying degrees to chemicals, especially those that are ionized or dissociated. Variations in wavelength other than light and heat may also result in responses. Some of these wavelengths fall within the range of audible sound. Other wavelengths, for example, X rays and gamma rays, cause a response, but usually plants and animals are adversely affected by such wavelengths. So far as is known, neither will respond to unconverted radio or magnetic wavelengths.

All organisms have diverse but characteristic shapes. Plants tend to be asymmetrical and superficially unorganized in shape, whereas animals tend toward definite body plans; however, both also share similar

3

shapes. For example, many marine animal colonies are often mistaken for plants by laymen.

DIFFERENCES BETWEEN PLANTS AND ANIMALS

We have already stated that there is no simple, completely satisfactory way to distinguish plants from animals. Nevertheless, some generalizations may be made that apply to most organisms so that our innate sense that these two groups are different is not really wrong. Just as we were able to tabulate the similarities of both groups, so too can we list certain differences.

A very large number of species of plants are autotrophic; that is, they manufacture their own food either by **photosynthesis** *or by a very similar process.* As we look about us we see that green plants dominate the landscape. Certainly there is no problem recognizing these as plants. The problem develops when we realize that there are many organisms that certainly cannot be called animals, but they are not autotrophic and they are not green. For example, the fungi are ordinarily called plants, and usually the bacteria are thought of more as plants than animals. None of the fungi and very few bacteria carry on photosynthesis. Then there is a further complication. Certain organisms we would ordinarily call animals are capable of carrying on photosynthesis. This occurs in some protozoans, such as *Paramecium*, that have been inoculated with plant cell bodies, the chloroplasts (see Chapter 4). This may occur in some other animals, for example, certain hydras, animals related to the jellyfish (Coelenterata, see p. 90). What this amounts to is that we cannot separate plants and animals just on the basis of their food source.

Most plants store carbohydrate in the form of starch or sugar, whereas most animals store carbohydrate as glycogen. However, the fungi store their carbohydrate as glycogen; so this feature cannot be used to separate the two groups neatly and completely.

Most animals move; most plants are anchored by a root system, or at least passively float. All organisms move at some stage of their lives; so movement alone does not characterize animals. The presence of muscular tissue and a nervous system to coordinate muscles are features that are restricted to animals. Any movement of plant parts is the result of growth or hydrostatic pressure.

Plant cells are usually surrounded by a cell wall composed of cellulose or a similar carbohydrate. Cell wall structure varies considerably from one group of plants to another. If any single feature of plants were selected to distinguish them from animals, it would be this feature. However, once again we see that no single feature can be all inclusive in separating the two groups, for the tunicates, simple sea animals related to primitive fish, have a covering of cellulose produced by their body wall cells. At the same time certain algae lack cellulose in their cell walls, and most fungi are covered with chitin, a product characteristic of animals.

Finally, *animals have genetically fixed limits of growth, whereas plants retain growing tissue throughout their life and may continue to grow.* This accounts for the closeness in shape and usually in size of the members of an animal species. In plants the overall shape and size varies considerably more than in animals. Plant species nonetheless also may be recognized by their shape.

In summary, plants are *usually* nonmotile organisms capable of carrying on photosynthesis. They have cells surrounded by a cell wall, usually composed of cellulose. Animals, in contrast, are *usually* motile organisms dependent on plants and other animals for their food source. Their cells are not surrounded by a wall composed of cellulose except in very restricted cases.

THE KINGDOMS OF LIVING ORGANISMS

Modern **systematists** (one who studies **systematics**) no longer try to place all organisms in one of two kingdoms, plant and animal, but instead they recognize at least three.

Once the step was made in the direction of grouping organisms on the basis of major body features, the two-kingdom, traditional system seemed obviously wrong. Over the past half century several new systems have been proposed with as many as six kingdoms. The most logical and simplist scheme based on current knowledge seems to be the three-kingdom system, and this is used in this text. Two of the three kingdoms are treated here: Prokaryotae and Plantae. The third kingdom, Animalia, is omitted for obvious reasons.

Each kingdom is further divided into either divisions (plants) or phyla (animals). A few organisms are included in two kingdoms reflecting our inability to clearly define the groups as explained previously. These organisms (see Chapter 6) are studied by both botanists and by zoologists, which does no intellectual harm, and serves to emphasize the unity of life.

Certain features of each of the three kingdoms should be pointed out here, and these will be repeated in more detail in the respective chapters to follow.

The kingdom Prokaryotae (sometimes also called Mycota) contain those organisms that lack, essentially, distinct nuclei; that is, their cells are **prokaryotic.** This kingdom is of major importance because it contains the bacteria and viruses, as well as those interesting alga-like organisms, the blue-green algae. There is some debate about the nature of the viruses; some biologists still question their nature. More recently taxonomists have been inclined to include the viruses as a division of the Prokaryotae as we are doing here.

The kingdom Plantae contains all of the **eukaryotic** cell plants, most of which are photosynthetic organisms. This kingdom has many species and includes all of the eukaryotic algae, the fungi, mosses and their relatives, and the vascular plants. The eukaryotic algae are those whose cells contain a distinct nucleus. This is a major difference between the kingdom Plantae and the blue-green algae and others of the Prokaryotae. The fungi are a very distinct group, so much so

that many proposals have been made to place them in a distinct kingdom. Even though they lack **chlorophyll** and are therefore either **saprophytic** or **parasitic,** or sometimes both, they are distinctly plantlike and could never be confused with animals. They have distinct nuclei in their cells; so they are distinct from the Prokaryotae. However, placing this group between the mosses and the algae does not imply that they are an intermediate stage between the two. They probably arose from the algae and went their way separately from the mosses.

You must understand that formal plant classification is used mainly as an information storage and retrieval system for the great many detailed facts that have accumulated about plants and that the reflection of the evolutionary history of plants in such a linear classification list is only secondary to the main classification. The details of the classification system followed here are listed in Appendix I and in the chapters dealing with the various groups.

PLANT CLASSIFICATION

The great variety of plant life and the fact that plants have an evolutionary history have led to a system of classification designed to group similar and probably related plants together (Fig. 1-2). In this way the description of plants can go by steps, from the more generalized statements to the more specific. Thus one can say that most members of the rose family Rosaceae have flower parts arranged in multiples of five, and so on until, through hierarchical steps, the species (plural also "species") is reached. In some cases even the species rank is divided, as is seen in Table 1-1. The distinction between **category** and **taxon** is apparent in the table. When the plant or group of plants is referred to, the name is a taxon, but when its position in the hierarchical scale is considered, the rank is a category.

Several chapters of this text give details about various plant groups. Before you can fully appreciate the general aspects of plant life, it is necessary to know some of the many

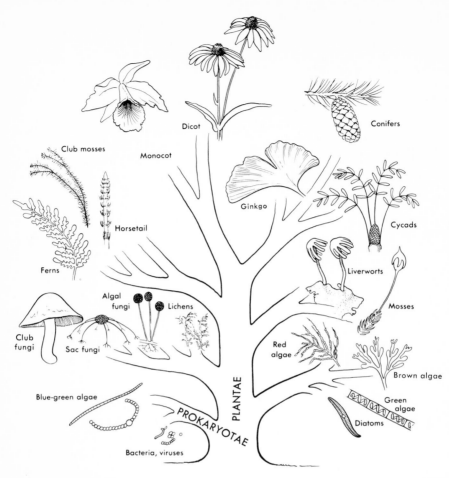

Fig. 1-2. Both divisions, Prokaryotae (Mycota) and Plantae, are diagrammed here, with representative plants grouped as branches of this phylogenetic tree. Arrangements such as these are instructive because they show the probable lines of evolution and at the same time emphasize both the similarities and the differences in the lines of development of major groups.

Table 1-1. Categories of plants and corresponding examples of taxa

Category*	Taxa†	Category*	Taxa†
Kingdom	Mycota, Plantae	Family	Rosaceae, Cornaceae
Subkingdom	Euchlorophyta	Subfamily	Caesalpinioideae
Division	Anthophyta	Tribe	Sophoreae
Class	Angiospermae	Genus	*Crataegus*
Subclass	Dicotyledoneae	Series	Pavifoliae
Order	Rosales, Cornales	Species	*C. intricata*
Suborder	Caesalpinineae	Variety	*C. i.* var. *straminea*

*There are other divisions of these categories, but they are seldom used.
†The examples chosen are taken at random and do not show the classification of any single group of plants.

kinds of plants, especially when considering the history, evolution, and ecology of the plants. At the left is a simplified classification of the common names used in these chapters.

BRIEF CLASSIFICATION OF PLANTLIKE ORGANISMS

Kingdom PROKARYOTAE—Single-celled plants without distinctive nuclei
  Viruses
  Bacteria
  Blue-green algae
Kingdom PLANTAE—The true plants— single-celled, colonial, or multicellular, cells with distinctive nuclei
  Algae (several divisions)
  Fungi (several divisions)
  Mosses, liverworts, and hornworts
  Whisk fern
  True ferns
  Club mosses
  Horsetails
  Cycads
  Ginkgos
  Conifers
  Gnetales
  Flowering plants

The classification of plants has had a long history. You will soon note that there are many classification schemes. Each author prefers one arrangement over another. Some of the older classifications are shown in Table 1-2. Several of these classifications are still used; therefore we believe that these should be compared, briefly, so that when the group names are found in collateral reading, their use will be understood (see also the glossary).

SYSTEMATICS

Darwin's theory of evolution had more immediate effect on other branches of biology than it did on plant taxonomy, where one might suppose it would have its greatest application. In fact, the most immediate reaction took place in social thought instead of in biology. Not all biologists of the period accepted the theory, and many of those who did were reluctant to change from a pigeonhole type of classification to the natural one made possible by Darwin's work. For one reason, the mechanism was not clear. Gregor Mendel's principles were not available to Darwin. For another reason, the idea accepted by Jean Baptiste de Lamarck (1744-1829), that the inheritance of characters acquired by parents was possible, was still prevalent. Lamarck did not actually propose the theory of the inheritance of acquired characters; he simply

Table 1-2. Some former systems of classification of plants

Traditional	Recent	Reproduction*	Morphology
1. Algae	Thallophyta	**Cryptogams**	Nonvascular
2. Fungi			
3. Bryophytes	Bryophyta		
4. Pteridophyta Classes Psilophyta Lepidophyta Calamophyta	Tracheophyta		Vascular
5. Spermatophyta Classes Gymnospermae Angiospermae		**Phanerogams**	

*This terminology is based on the location of the reproductive organs, that is, whether they are hidden or exposed.

did not question the concept that was generally believed during his time. Inasmuch as Darwin did not explain the mechanism of evolution, it took time for the old ideas to disappear from the work of the early post-Darwinian period.

Much of the development of modern evolutionary thought was the result of German zoologists. It was not until after Mendel's work was rediscovered that botanists took an active part in establishing the theory of evolution. The discovery of the gene theory and work with plants as well as with animals rapidly developed the field of genetics, and Wilhelm Ludwig Johannsen (1857-1929), Hugo De Vries, and other botanists took leading roles. By the 1920s the theory of evolution, though generally accepted, was considered in scientific circles to be proved and was no longer of great concern to biologists.

The past 35 or more years have seen a greatly renewed interest in the subject. New ideas of speciation, new definitions of species, and the development of systematic biology as an important fundamental branch of biology, all basic to the theory of evolution, have arisen so that at the present time books on evolution are appearing at a rapid rate. It may be safely said that since 1937, when Theodosius Dobzhansky published *Genetics and the Origin of Species*, more scientific information has been published on the theory than in all the previous books and papers on the subject combined. Both zoologists and botanists have taken an active part. Among the leading botanical works are E. B. Babcock's study of the genus *Crepis*, Edgar S. Anderson's (1897-1969) work on hybridization, and G. Ledyard Stebbin's recent work on flowering plant evolution. A series of books by Clausen, Keck, and Hiesey deals with experimental studies on the behavior of plants when grown in widely different habitats. These experiments opened a whole new field—experimental evolution.

PLANT NAMES

Otto Brunfels (1489-1534) was one of the first workers in modern botany. It was he who introduced the modern concept of the genus as used in botanical nomenclature. Early classifications of plants were based upon utilitarian ideas or on such obvious external characteristics as growth, and these classifications led to the division of plants into herbs, shrubs, and trees. In 1660 John Ray (1627-1705), who was an all-around naturalist, published the first of a series of works on systematic botany. He was able to distinguish between monocots and dicots, using the seeds of these plants in addition to using the flowers, fruits, and leaves, all of which led to a more natural classification of the plants. It was through the efforts of this man that the terminology used by earlier workers passed down to the time of Carl von Linné (Latin, Carolus Linnaeus; 1707-1778). It was he who first consistently used the now famous system of nomenclature, the **binominal system.** This has provided us with a means of cataloging plants (and animals) so that accumulated data may be easily indexed. Over the years it has proved to be a workable scheme of nomenclature that, although not invented by him, used all the best features of preceding schemes, was refined, and resulted in a highly practical system. The system is flexible and capable of expansion and adjustment, which make it possible to include all known organisms as well as those yet to be discovered. For example, only 5,950 species of plants were known to Linnaeus. Today over 300,000 different species of plants are recognized, all of which can be and have been named and included in the Linnaean system.

Since Linnaeus's time a set of rules has been drawn up and universally agreed upon by the taxonomic botanists of the world. Briefly these rules provide for formal Latin names of plants. According to these rules, all plant names must be either in Latin or Latinized Greek words and use the binominal system. The system follows the rules and procedures outlined and described in the *International Code of Botanical Nomenclature.* This code has been developed and refined over a long period of time and is continually studied and modified to serve the taxonomic botanists. The current code was adopted by the Eleventh International Botanical Congress in 1969. The purpose of these rules is to help assure that each taxonomic group will bear only one correct name and that the name will be the earliest name applied, according to the rules, to the group, except in particular cases where particular names are conserved in a special list found in the published edition of the code.

In addition to the Latin name, many well-known plants also have a common name in the language of the country or region where they occur. There are distinct advantages in knowing the **scientific name** as well as the common one. First, the scientific name is recognized as the name of the organism in every part of the world, even though the common name may be more easily remem-

bered. Second, the common name may be different in every language and could even vary from one locality to another within the same country. The adder's tongue (*Erythronium americanum*) is sometimes known as the dogtooth violet. Thus it has two common names, and unless both are given, there is no way to tell that they refer to the same plant. Naming a plant by the use of the botanical code provides a single name for a single kind of organism and results in less confusion, which is one of the aims of the scientist. The scientific name of corn is *Zea mays*. (Note how all **generic** and **specific names** are placed in italics, but other scientific names are not. The reason is that they are foreign words and it has always been the custom to indicate them in this manner.) The first name indicates the genus to which the species is as-

CAROLI LINNÆI

S:æ R:giæ M:tis Sveciæ Archiatri; Medic. & Botan. Profess. Upsal; Equitis aur. de Stella Polari; nec non Acad. Imper. Monspel. Berol. Tolos. Upsal. Stockh. Soc. & Paris. Coresf.

SPECIES PLANTARUM,

EXHIBENTES

PLANTAS RITE COGNITAS,

AD

GENERA RELATAS,

CUM

Differentiis Specificis, Nominibus Trivialibus, Synonymis Selectis, Locis Natalibus, Secundum

SYSTEMA SEXUALE

DIGESTAS.

Tomus I.

Cum Privilegio S. R. M:tis Sueciæ & S. R. M:tis Polonicæ ac Electoris Saxon.

HOLMIÆ, Impensis LAURENTII SALVII. 1753.

Fig. 1-3. The title page of the first edition of Linnaeus's book describing the plants known to him in 1753. This was the first consistent use of the system of binominal nomenclature devised by him and soon thereafter generally accepted by the botanists of the day. Many features of his early classification are still used, as are the names of the species.

signed and is the generic name. This word is always capitalized, and it is always a noun. The second name, the specific epithet, is not capitalized, and lower-case type is recommended and preferred even for the name of a person or a place, although in some cases this kind of specific name is capitalized. The word is either a noun in the Latin possessive case or an adjective modifying the generic name, a noun, and agrees with it in gender. The third name, when used, is the author of the original species description, which in the example given here is Linnaeus. The author's name is not placed in italics; it serves as a very abbreviated bibliographical reference. Sometimes this name is placed in parentheses and is followed by another author's name to indicate that the species has been moved, for botanical reasons, to another genus other than the one in which it was originally placed when the species was described. Names in Latin are used because of their universality. Latin, now a dead language, was formerly the language of educated persons and the language of textbooks and scholarly writings. About all that remains today is the rule that the description of the new taxa (singular, taxon) must be written in Latin. These descriptions may be followed by less formal and usually more detailed descriptions in the language of the author.

By using the first validly described name of a plant any ambiguity is almost entirely overcome. All other Latin names applied to the same species of plant are synonyms and are not used except in catalogs that show the history of the names. Synonymy occurs because the older literature does not clearly define the plant, because older names have been overlooked, or because two or more workers from different geographical areas described the same plant species. As soon as these facts are discovered, the more recent names are placed in synonymy. Sometimes it develops that a very well-known plant has two names and that the older name seldom has been used. To avoid confusion, the more recent name is conserved by placing it on the official list of conserved family and generic names. Priority is established by accepting the first edition of *Species Plantarum* of Carolus Linnaeus, published in 1753 (Fig. 1-3), the starting point for most plant nomenclature.

New plants are being described all the time. These descriptions and keys for their identification are published in botanical journals and monographs. As new plants are discovered and added to the flora list, the manuals for the identification of plants have to be revised and brought up to date.

QUESTIONS AND PROBLEMS

1. Why is it necessary to consider diversity in the study of the physiological processes that operate in plants?
2. List and discuss at least four features all living organisms have in common.
3. Do the same for four features that separate plants and animals.
4. Why is plant classification still changing?
5. What are two advantages of using Latin names for plants?
6. What is the greatest disadvantage of using common names for plants?
7. What is the *International Code of Botanical Nomenclature?* Who uses it?
8. What is the relationship, if any, between classification and evolution?
9. What is a division, a category, or a taxon? Give an example of each.

10. What is the difference between a division and a phylum?

DISCUSSION FOR LEARNING

1. Discuss the advantages of a system of classification that used more than the "standard" two kingdoms; also, the disadvantages.
2. Discuss the pros and cons of a classification based on convenience of handling information and one that attempts to be phylogenetic.
3. Discuss the theory of classification in relation to the theory of information storage and retrieval (see Chapter 2).
4. Considering the scientific method and our present state of knowledge of plants, discuss how we know that all living things have certain features in common.

ADDITIONAL READING

Heslop, I. R. P., and Harrison, J. 1964. New concepts in flowering-plant taxonomy. Harvard University Press, Cambridge, Mass. (Explains in simple terms the nature of taxonomy, diversity, and classification.)

Margulis, Lynn. 1972. Five-kingdom classification and the origin and evolution of cells. In Dobzhansky, T., et al. Evolutionary biology, Appleton-Century-Crofts, New York, vol. 7, pp. 45-78. (Reviews the various ways living things are divided into kingdoms and emphasizes the differences between prokaryotic and eukaryotic cells.)

Ross, Herbert H. 1974. Biological systematics. Addison-Wesley Publishing Co., Reading, Mass. (A brief discussion of plant and animal systematics, including phylogeny and brief rules of nomenclature.)

Solbrig, Otto T. 1970. Principles and methods of plant biosystematics. The Macmillan Co., New York. (A textbook with an in-depth account of the speciation process.)

Stafleu, F. A. (editor). 1972. International code of botanical nomenclature. Utrecht, Netherlands. (The official code as adopted by the Eleventh International Botanical Congress, Seattle, August 1969.)

Fig. 2-1. The pitcher plant, *Sarracenia minor*, common below the fall line from North Carolina to central Florida, combines autotrophic and heterotrophic modes of food getting. Botanists and zoologists alike are interested in this and other complex interrelationships between plants and animals.

CHAPTER 2

The science of botany

■ Plant sciences, except for certain applied branches, merge with other life sciences because they deal with living organisms and share the same principles. The only logical reason for separate study is the different study techniques used by the botanists. The dichotomy of the biological science is most evident in studies relating to **taxonomy** and classification and least evident in investigations of the life systems. The modern botanical scientist is most concerned with molecular processes within the plant cell, with the **ultrastructure** of the cell, with the process of genetic change in plant species, and with the developmental mechanisms in plants. He makes use of the electron microscope and is able to photograph the complex structure of cell parts. The plant taxonomist makes use of the results of this work as a means of demonstrating the unity of the plant world, and he is concerned with the evolution of these organisms as well. It is the taxonomist who must organize and process the available information about plant **species.**

BOTANICAL RESEARCH

Other significant areas of investigation of plant life include **genetics, ecology, physiology,** and life cycle or life history studies. These topics are by themselves large areas of specialization.

Research, using plants as specimens, has both theoretical and practical application—

SUBJECTS INCLUDED IN PLANT SCIENCE

Plants and plantlike organisms
Genetics and populations
Physiology and ecology
Reproduction, development, and evolution

theoretical as the life processes are explored and practical as discoveries are made that relate to agriculture and industry. Universities, government agencies, and private organizations, each with somewhat different objectives, provide the place and equipment for these investigations.

Many coordinated conditions must be present before research of any kind may be conducted. The proper education of the research staff is first. An adequate library with current information easily and promptly available is required for the beginning of any research project. Some research can be done with very simple equipment and by a single individual. Other research areas require sophisticated equipment that is often expensive. The more complex the research, the more likely a team of research scientists will be necessary, and since World War II this has been the trend in scientific research.

REQUIREMENTS FOR RESEARCH
PROJECTS

Educated personnel
Adequate libraries
Proper equipment
Worthy subject matter
Financial backing

Most scientists will agree that one of the greatest difficulties in doing research is the selection of a worthy problem. All aspects of the approaches to the investigation must be carefully considered and harmonized. The problem should be one that can be handled by the investigator with the facilities available and within a reasonable length of time. Most important of all, the project should be concerned with a topic that occupies a real place in the scheme of nature. Or, stated in another way, the phenomenon being investigated should be a real one, be unknown, and, when completed, logically fill a gap in

our knowledge of the universe. Some of the specialized divisions of botany are listed in the box below. These and other areas of plant study form special fields of concentration for the professional botanist.

WHY STUDY BOTANY?

Beginning biology students have a tendency to think of plants as less dynamic than animals simply because they do not have the ability to run and hide, or to attack and kill. But these biological processes are present in plants too, but in different ways—ways that should be understood and appreciated by all biologists. Instead of running and hiding, plants develop other protective devices, such as spines and thorns. Instead of attacking and killing with a rush, they slowly strangle their victims (Fig. 2-2) by growing over the top of their victims and killing them with shade and, in a few cases, actually trapping animals and digesting them.

Many people find great pleasure in watching plants grow. The "green thumb" is not an accidental gift; it is knowledge gained through study and observation of the physio-

SPECIALIZED DIVISIONS OF BOTANY

Plant physiology—the study of functional processes in plants (Chapters 3, 5, 12, 13, 15, and 17)
Plant **morphology**—the study of the structure of plants, including **cytology** and **histology** (Chapters 4, 14, and 16)
Plant ecology—the study of the relationships of plants to their environment (Chapter 20)
Plant taxonomy—the identification and classification of plants, which with other sciences forms the study of systematic botany (Chapters 2 and 5 to 11)
Plant genetics—the study of the genes in the cells of plants and their effect (**heredity**) on changes of species (Chapter 18)
Paleobotany—the study of fossil plants and their evolutionary history (Chapter 19)
Horticulture—the science and art of cultivating garden, orchard, or nursery plants, landscape design, turf management, and so forth (Chapter 21)
Pharmacology—the science of the preparation and use of drugs, including medicinal plants (Chapter 21)
Range management—the science of grasslands as it pertains to food plants for grazing animals (Chapter 21)
Forestry—the science of planting and management of forests (Chapter 21)

Fig. 2-2. A gigantic strangling fig, one of the largest in the world, growing in the outback west of Carins, Australia, has destroyed its host and taken over the space left in this dense subtropical forest. This form of predation in the plant world is active competition for light among the plants of the area.

logical needs of different plant species. This is of interest to botanists both amateur and professional. With this knowledge comes the desire to learn the names of plants. By this means, the door is opened to a wealth of literature. The names of the plants are the key words to the many publications describing the characteristics of plants, their habitats, and their distributions.

As soon as one becomes interested in the great variety of wild plants, it becomes apparent that they are not randomly distributed. Each kind of plant has its own habitat. The natural question to ask then is, Why is this? To find the answer leads one along many different paths. One path is the study of plant ecology, which in turn leads to other fields with practical applications such as range management, forestry, and water resources studies. Another path might lead to

15

studies of genetics and evolution. Plant geneticists, of course, have played a very necessary role in the improvement of agriculturally important plants and in developing horticultural varieties.

The plant systematist finds great satisfaction in understanding the evolutionary process, in identifying the many species of plants, and in assigning names to them. Thus the path leads back to our starting point. Now we can somewhat better appreciate how one science supports another, and each is interdependent.

We have intentionally omitted perhaps the most important reason for studying plants. This is the need for every student of biology to study the flow of energy through living systems, the subject of much of this text. The economy of plant life versus animal life, and its cost in energy might well be considered the price of life itself.

EARLY HISTORY OF BOTANY

Primitive humans, forced to act either positively or negatively to their observations of environmental changes and the conclusions they drew, had little time to ask why. Once they learned to shelter themselves, to plant, and to store, they began to contemplate the many aspects of their environment. Thus, in some long forgotten cave, there began on an unrecorded day the rough **scientific method,** the basic technique we still use. At the same time the animal became man, the rational being.

Hippocrates (460?-377? B.C.) is remembered as an ancient who practiced the art of healing by use of plant drugs. Today Hippocrates is considered to be the father of medicine, and young physicians take what is called the Hippocratic oath upon their graduation from medical school. Hippocrates also taught that there were four elements: earth, air, water, and fire. He also believed in four humors: blood, yellow bile, black bile, and phlegm (Fig. 2-3). Hippocrates taught that health depended on the proper balance of these humors. From him we get the term "good humor" and other expressions.

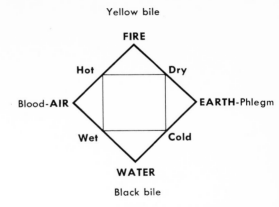

Fig. 2-3. Relationships of the elements according to Empedocles and the humors according to Hippocrates, ancient philosophical and scientific ideas.

Aristotle (384-322 B.C.) was a natural philosopher whose careful observations formed the basis of his many biological writings, most of which, however, were devoted to animals. It was he who first noted that sunlight was necessary for the formation of the green coloring matter (chlorophyll) in plants. He recognized a hierarchy of being.

Hundreds of years were needed to build the foundation upon which modern biology rests. Botany, and all of science, must progress in steps. Each branch of the field progresses to a certain stage; then it must pause awaiting further results in another branch. One specialized area contributes to another to their mutual benefit.

THE FOUR MAIN ADVANCEMENTS
IN THE HISTORY OF
PLANT SCIENCES

1. Discovery of the nature of cells
2. Development of a classification system for plants
3. Experiments that led to the knowledge of heredity, genetics, and the nature of DNA
4. Proposal and refinement of the idea of the evolutionary process

Extremely slow and gradual changes took place in the intellectual evolution of man. About 35,000 years ago Cro-Magnon men of France began to live in caves and paint magic images on their walls. Thus began human culture, that mass of behavior human beings in any society learn from their elders and pass on to the young generation. As man progressed, he transformed himself from a rare animal into one of the most numerous. It is difficult to say just how agriculture started, but it seems certain that it was the women who first fenced off a good patch of greens to protect them from herbivores. Progress toward cultivation varied from place to place. In one place, plants were cultivated before animals were domesticated; in another, the reverse. The domestication of plants and animals developed in uncertain steps until recent times. Nevertheless, with each step came more freedom from work. With time to contemplate, plans could be made and crude experiments attempted, and eventually science developed.

The need to recognize plants as sources of food, drugs, building materials, and fabrics led to the classification of plants as one of the early developments of botany and the establishment of plant taxonomy as a science. With the thousands of plant species that are now under cultivation on farms and about our houses it is natural that they should be named and classified. Hardly a house is without an ornamental plant of some kind, making horticulture the most popular recreation activity in today's busy world. Through science and industry, farming with horses or mules is no longer known. Today it is a business akin to manufacturing; tomorrow the actual manufacturing of food from the raw material of the sea will ensure man's future.

Formal botany had its origin in the efforts of Theophrastus (370?-287 B.C.), who was an apt pupil of Aristotle. Much of the scientific data collected during the campaigns of Alexander the Great were used by Theophrastus to learn about plant functions. He also described the plants brought back by Alexander and attempted to classify them. Theophras-

tus distinguished bulbs, tubers, and rhizomes from true roots and, in addition, had an understanding of the sexual reproduction of the higher plants. All the information Theophrastus accumulated was lost to the world until it was recovered by Andrea Cesalpino (1519-1603) during the Renaissance. Around 30 B.C. the art of agriculture was described in Vergil's Georgics. Another book, by Varro (116-27 B.C.), appeared, in which he described the growth of plants and also suggested that contagious diseases were due to microscopic plants. Finally, about the middle of the first century A.D., Dioscorides (ca. 50-100), who was both a botanist and a military physician, wrote a treatise on plants and pharmacy that lists some 600 plants and their medical properties. Botany therefore had its beginnings in the observations and curiosities of the ancient philosophers.

During the second half of the first century learning was revised mainly through the efforts of Pliny the Elder (A.D. 23-79), a Roman. His 37 volumes on natural history comprising over 20,000 articles written by 475 authors were the encyclopedias of science of the period and included the knowledge and beliefs of the forgotten writers of Greece and Rome. This encyclopedia included a discussion of the general theory of the universe, of the stars in space, and of the earth and its contents. He preserved for all time the superstitions of his era and recounted in good faith the practices and uses of all forms of magic.

Unfortunately, Pliny, a poor compiler, not an original researcher, passed on much misinformation, some of which persists even today. For instance, it is he who is responsible for the idea that every plant and animal must have some use directly or indirectly for man. This idea is frequently expressed by the nonbiologist when confronted by some particularly strange and objectionable organism. Nevertheless, his work remained popular into the eighteenth century.

During the Dark Ages, from A.D. 400 to 1000, few new ideas were added. This was, biologically, the age of the herbalists. The

Arabs reintroduced the work of Aristotle to the western world about A.D. 1000, and his work remained unquestioned through the Middle Ages. Otherwise little remained except the herbals. These were massive books illustrating plants of particular interest as drugs and were the source of information for the alchemists and doctors of the time.

The Age of Scholasticism gives us only two really outstanding men of science who contributed to the field of biology. Roger Bacon (1214?-1294) was an English Franciscan monk who was frequently in trouble for his advanced thinking. His outstanding contribution to biology was the invention of the magnifying glass, the most valuable tool of the biologist. Albertus Magnus (1206?-1280), a Dominican from Germany, was also far ahead of his time. He wrote many accurate descriptions of plant anatomy.

The use of plants in the treatment of disease led to an awakening of interest in the study of botany, a science that was traditionally practiced in the monastery and convent gardens. The shape of a leaf or the color of a flower was regarded as a symbol of its use as designed for the plant by the Creator.

The invention of the printing press in the 1450s permitted the beginning of modern biology, just as it permitted the development of all fields of knowledge. The works of Aristotle were among the first books to be published.

The detailed history of biology after the era of printing began is too involved for anything more here than an account of a few highlights. Mainly during these later periods, tools such as the microscope were brought into use so that great masses of data that formed for the first time the factual basis of biology were gathered.

The history of the microscope spans some 350 years, but the exact time and circumstances of its inception are still uncertain. The first microscope is credited by some to Zacharias Janssen (1590) and by others to Galilei Galileo (1564-1642), who built such an instrument in 1610. More than likely both can be given the credit. Anton van Leeuwen-

Fig. 2-4. The Culpeper microscope, about 1740. (Courtesy Carl Zeiss, Inc., New York.)

hoek (1632-1723) refined the microscope and described for the first time microscopic plants and animals.

When we take a look at historical microscopes, such as those of Edward Culpeper (1660-1740) (Fig. 2-4), not considering the optical design of these instruments, we notice immediately how much they are an expression of the style and spirit of the time. In the seventeenth century microscopes, the tubes, which were generally cardboard cylinders, are often richly ornamented. The surfaces of the objective and eyepiece mounts are extremely elaborate and ornately adorned with an almost playful touch. The tripod, which was used for a time as a support for the tube, is seldom straight or strictly functional. It rather displays a flair for the flamboyant. This intimately artistic and, at the same time characteristic form of the microscope, is most alien to modern man's conception of a scientific instrument. It is rather an expression of the artistic tendencies of the day, which began to disappear late in the eighteenth century. Eventually more and more metal was used in the mechanical parts, and for about 150 years optical instruments, and microscopes in particular, were finished in a polished, lacquered brass. By the latter half of the eighteenth century, more emphasis was placed on the improvement of the mechanical components of the microscopes. Greater ease of operation became a prime consideration. We see the first attempt at focusing by means of a rack and pinion and we increasingly encounter draw tubes and movable illuminating mirrors. Finally, the general trend toward simplicity left its imprint also on the development of the microscope. The light-hearted design and ornamentation gave way to more functional lines.

The ordinary light microscope, similar to those used for classroom study, is sufficient to identify cell types and gross cellular structure. Their limit of resolution, the minimum distance between two points that allows them to be discriminated with the light microscope is about 0.25μ. Obtaining increased contrast with a phase microscope, a device that through the interference of light rays by special condensers causes a transparent object to appear in shades of gray depending on its thickness and its refractive index, is a system particularly useful in studying living cells and tissues because they need not be harmed by contrasting dyes. A similar instrument, the interference microscope, has the advantage of giving quantitative data. In the dark-field microscope the ordinary condenser is replaced by one that illuminates the object obliquely so that no direct light enters the objective. The object appears bright because of the scattered light, and the background remains dark. With this type of microscope, objects that are smaller than those detected with the ordinary microscope can be seen. A polarizing microscope is still another instrument that increases resolution by increasing contrast in an object by polarization of the light passed through it. The electron microscope (Fig. 2-9) permits the study of the cell ultrastructure. The even newer scanning electron microscope, though it lacks the resolving power of the electron microscope, has the advantage of high contrast so that shape and dimension of extremely minute structures can be photographed.

Modern light microscopes are typified by the photomicroscope (Fig. 2-5), which is an example of this development. It is an automatic "notebook" for scientists and laboratory technicians. After the specimen has been brought into sharp focus by the simple pressing of a button, a properly exposed photo is taken and the film automatically advanced for the next exposure.

Further history of specific branches of the field is included in the chapter involved.

SCIENTIFIC METHOD

The scientific method is both simple and sure. The scientist must objectively gather **information.** An examination of these facts leads to the formation of a **hypothesis,** which is then tested. The results of this **experimentation,** or test, are used to form a conclusion, usually stated to be a **theory.** Theories that have repeatedly been demonstrated to hold when applied to many different problems are then termed "laws." But how does a scientist really work? Certainly he cannot follow this simple pattern. A closer look at the scientific process shows us that there are many sources for scientific **data** and many ways of treating these data.

Sources of scientific information

The natural scientist depends on observations as the source for all the information he

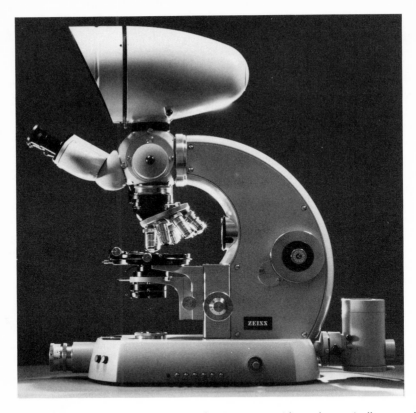

Fig. 2-5. Photomicroscope, a combination of a research microscope with an electronically controlled automatic 35 mm camera. (Courtesy Carl Zeiss, Inc., New York.)

uses. From these observations made directly upon the subject, or indirectly with the aid of instruments, he is able to form concepts and discover the organization of the matter he is investigating. Science, however, does not consist of **research** alone, and scientists are not just researchers. Rather, science is a way of thinking based on logic, and scientists, like other people, must think as well as observe. These apparently simple statements will be developed in detail.

INFORMATION SOURCES

Observation and experimentation
Specimens and samples
Literature and related stored records

The modern scientist enters gradually into his particular field. As a student he begins to restrict his professional interest and delve deeper into a branch that attracts his attention. Through advanced studies and association with teachers and research workers he begins to master a subject. This requires extensive reading as well as practical experience in the laboratory and in the field. At the proper time he becomes a specialist in a particular subject, supported by a broad knowledge of science in general. So equipped, he may enter into a life of research, development of process or product, teaching, or a combination of these activities.

The ultimate and therefore the most important source of information for a scientist is observation. The matter of observation and experimentation is too complex to be treated at once; we shall return to this later. Still one must keep in mind that observations are recorded from specimens and samples and that the published records are the **literature** and related documents that form the storehouse of human knowledge.

The biological sciences utilize as natural study materials, specimens, either as whole organisms, living or preserved, or as samples of materials taken from these organisms. The botanical sciences use, as specimens, plants collected in natural growth areas, grown in the greenhouse, or preserved in **herbarium** collections. All these materials are gathered and carefully selected for the accurate and detailed information they will reveal.

Printed information distributed to libraries and individuals is referred to as literature. Photographs, movies, slides, and charts, some of which may be available in duplicate copies and some only as originals, are also information sources. Modern **data-processing** equipment uses punch cards, tapes, and iron cores to store data for easy retrieval. The last method, information storage by data-processing equipment, has developed very rapidly and is now widely used. Because the term "literature" has a restricted meaning, the term **"record"** is now used to refer to any durable packet of information used in the communication of information. Of great concern is an efficient system for the storage and retrieval of records. Records suitable for storage that utilize automatic data-processing equipment are referred to as source documents. These are converted to punch cards, electronic tape, or iron core for storage and later retrieval. Such a system is known as an **information-retrieval system (IRS).**

Kinds of data used in science

Scientific progress, practically speaking, is the result of (1) the organization of the available information and (2) the active search for supplementary data to complete the information necessary to demonstrate a fact. It is not by chance that many of the great discoveries of the past have been made by several persons almost simultaneously. Once a certain amount of information becomes available, it is natural that the educated minds of several scientists will piece together the separate facts and test these ideas, often in different ways but at about the same time, by gathering additional facts to support their ideas until enough information is available to write a report of their discovery. Such was the pattern of discovery of the evolutionary processes. This, the scientific method, is considered in further detail in the section on the technique of data organization.

KINDS OF DATA USED IN PLANT SCIENCE

Observational
Mensural
Experimental

Observation data are records of information gathered by human senses, generally unaided or with the help of optical instruments. Light microscopes and cameras are generally the aids used for gathering these data. For the plant scientist, information about the shape, size, structure, and color of plants, where they live, and what other kinds of plants and animals live with them constitutes the recorded data.

Measurements of weight, length, and proportion add to the information about plants. Sometimes complex measurements are necessary when one makes physiological studies. Oscilloscopes and other electronic instruments record mensural data. **Computers** may be used for the rapid processing of the great amount of mensural information that may be gathered.

Measurements must be made so that the data is meaningful and can be compared with the literature. To do so, the scientist usually follows certain criteria. First, he measures a biologically important character that is common to and can be measured in an average representation of the specimens being studied. Obviously it is necessary that the measurements be related to the problem. Second, sufficient data should be collected in order to do the proper statistical analyses. Finally, the data must be standardized and have well defined limits.

Almost always the data collected is numerical in nature. When numbers are used that have many zeros either before or after the decimal point, it is convenient to use **scientific notation** using powers of ten (see below).

SCIENTIFIC NOTATION

$$10^4 = 10 \times 10 \times 10 \times 10 = 10{,}000$$
$$10^3 = 10 \times 10 \times 10 \quad\quad = 1{,}000$$
$$10^2 = 10 \times 10 \quad\quad\quad\quad = 100$$
$$10^1 = 10$$
$$10^0 = 1$$
$$10^{-1} = 0.1$$
$$10^{-2} = 0.01$$
$$10^{-3} = 0.001$$
$$10^{-4} = 0.0001$$

Once enough observational and mensural information about a plant or a process is available, it may be possible to conduct experiments leading to the discovery and proof of many facts. Experimental data are gathered to demonstrate growth processes, biochemical actions, reproduction, and genetic change. The validity of observational and experimental facts is demonstrated if, after repetition under the same conditions, the results are the same.

Observation and experimentation

Even though basic studies usually begin with observational data, there is no real distinction between these and data derived through experimentation. Observations may meet all the requirements of an experiment. This may at first seem to be an obscure statement, but as generally understood, an experiment is essentially the measurement of phenomena resulting after the stabilization of most of the factors involved and after allow-

ing the variation of one or a few of the others for comparison. Direct observation is a tabulation or recording of phenomena as they occur unchanged in the field or in the laboratory. Refined experimentation is possible first by observations made with or without instrumentation and second upon comparison with other similar observations. The degree of sensitivity of the instruments does not define the difference between the two procedures. Laboratory and most field experimentation involves a **closed system** and a means of obtaining comparative data by the use of an unchanging control. As the system functions, it is compared with the control. It follows that if a scientist makes comparative observations in the field after careful advance planning, he is experimenting. By the selection of two closely related phenomena he is able to make comparisons, and he is therefore an experimenter. Viewed in this light, the distinction between observation and experimentation fades. The essentials for scientific work are careful planning and detailed testing. It is clear that the only difference between ordinary observation and the usual experiment is the lack of a comparative **controlled system** in the first and its presence in the latter.

Data processing

The need for more knowledge is obvious, and equally obvious is the great amount of complex information gathered and recorded by humans during the approximately 5000 years of accumulation of surviving records. So vast are these records that they are rapidly burying us in a sea of paper—ironically, a plant product. To efficiently gather more knowledge, without repeating what is already known, we must know what we now have. As a result, our systems of organization of data are changing from simple indexing systems of days gone by to modern methods of data processing.

Technique of data organization

The classical description of the scientific method does not adequately describe the process of organization of factual knowledge as it is practiced in the modern, computer-benefited world of science. The classical description of the scientific method has all the elements of today's technique, but it lacks working detail, and most important it neglects the idea of feedback controls, now a part of the continuing research practiced in today's rapid pace of living.

Information retrieval is one of the most pressing problems of the modern world. Vast amounts of information have become available through the increased pace of research since World War II. Stored information, unless it can be quickly and thoroughly retrieved, is useless. If the coding system is not adequate, **source documents** are lost. Therefore output is dependent on an adequate retrieval system, of which there are many.

Books and journal articles have to be processed into order to be used. Titles and indexes are no longer adequate for retrieval of information. Unless they are further processed much information is lost. So each title, with its author and publisher, is treated as a document. These documents are then coded and stored for later retrieval. Once retrieved by any of the several means available, they may be studied, copied, or loaned. Many large libraries have converted to a punch card, tape, or iron core system of information retrieval, using a computer (Fig. 2-6) to search for the documents. Once the information is coded, the computer, properly programmed, will search for the documents. By requesting information using the proper code words, one can obtain a complete bibliography of the current information on the subject, which becomes available very rapidly.

Because of the great amount of published information, abstracts of research have become an important means of keeping up-to-date with research progress. In the United States the most readily available abstracting journal for the botanical sciences is *Biological Abstracts*, published by Biosis (Biological Science Information Service) (Fig. 2-6). This publication and the various aids for information retrieval associated with the publication

Fig. 2-6. Information processing using electronic data-processing equipment at BIOSIS. (Courtesy Biological Abstracts, Philadelphia, Pa.)

are available in every university and most college libraries.

Most recently, information processing, by use of the computer machinery described earlier is being developed as a means of searching for data to use in research projects. It seems obvious that the procedure described as the technique for the organization of data is a cycle that yields more information at each revolution. Ultimately this information, in one way or another, affects modern society.

Experimental design

Both the method and the history of science begin with the recognition of problems and finish with a scientific report. These usually contain a review of the literature already available on the subject, followed by a statement of the problem to be solved, the method used to discover new facts about the subject, the data gathered, and the results. A discussion points out the significance of the discovery made in relation to previous knowledge. The theory of this method and examples of broad problems needing study have been discussed. The actual selection of a specific research problem, as pointed out here, is possible only after considerable experience.

Experimentation usually involves rather sophisticated equipment. **Chromatography,** a widely used laboratory technique for separating similar materials in a mixture, is used in amino acid analysis for many types of research projects. **Isotope tracers** are used by biologists to follow the movement of ions from place to place within a plant or to determine the chemical pathway of a reaction taking place in a living system. Common tracers are carbon 14 or cobalt 60. For use in enzyme and growth studies, one can obtain many different kinds of biochemical compounds labeled with any of several tracer isotopes. Even the particular atom in the compound may be tagged so that the progress of the particular radical may be traced. The following are some instruments used in botanical research:

Measurement
  Wheatstone bridge—a device using matched resistors for the measurement of electrical quantities. Any change that can be expressed electrically, including temperature, pH, light, and energy, can be measured with this device.
  Oscilloscope—an electronic measuring instrument used to show and measure wave patterns.
Analysis
  Centrifuge—a machine used to separate materials of different densities by using centrifugal force.
  Spectrophotometer—any material that absorbs radiant energy within the range of light may be identified and the rate of its reaction measured with this instrument.
  Manometer—a U-shaped tube used to measure changes in gas pressure, which is indirectly a measurement of the rate of respiration or photosynthesis within living systems.
  Scintillation counter—an instrument used to count radioactivity in studies involving isotope tracers.

Experimental design is the application of the scientific method by advance planning of the procedure to be used to investigate a scientific problem. Such experiments are likely to yield more significant data than those otherwise performed because they can be made to meet the requirements of statistically significant data gathering. Not only are controls important in experimental investigations, but also randomization is necessary to ensure that the data are not biased by the experimenter. The careful selection of the instruments used, the test organisms, and the control of the environmental conditions during the test all contribute to the successful gathering of significant data for analysis.

Modern botanical science, like all science, uses three sources of information—the previous literature, samples as stored data, and observed or sensed information. The following chapters report the findings of botanists over the past several hundred years, but most of the information discussed has been discovered only in the past several decades.

24

Gathering botanical data

The general aspects of data organization or data processing explained in the previous section have specific application to the study of plants. Specimens mounted on standard-sized sheets of paper and stored in cabinets as herbarium collections are an important source of botanical information. The classification, distribution, genetics, and evolution of plants are studied by using herbarium collections and the data associated with the specimens. As might be expected, the major collections of plants are to be found in the large cities of the Western world. Some of the most important collections are maintained by the Royal Botanic Gardens, Kew (Fig. 2-7); British Museum (Natural History), London; Museum of Natural History, Paris; United States National Herbarium, Wash-ington; New York Botanical Garden; Gray Herbarium and other herbaria of Harvard University, Cambridge; Missouri Botanical Garden, St. Louis (Fig. 2-8); Field Museum of Natural History, Chicago; and the Academy of Natural Science, Philadelphia. These institutions maintain large botanical libraries associated with the collections, and several taxonomic botanists are employed to care for and do research on the collections.

Other kinds of research, for example, physiological, genetic, ultrastructural, and horticultural, are conducted in national and state laboratories and in the laboratories of many universities. The state agricultural experiment stations are frequently a part of a state university and are either physically located on the campus or usually not a great distance away. The staffs of these stations

Fig. 2-7. Royal Botanic Gardens, Kew, near London, England. The large greenhouse contains growing palms from many parts of the world and represents one of the finest collections of its kind.

Fig. 2-8. The modern building housing the large collection of plants and the library at the Missouri Botanical Garden, St. Louis, Missouri.

often hold joint appointments at the university, and graduate students may conduct research under the direction of their research staff.

Analyzing botanical data

Data gathered through research are used for the preparation of scientific papers for publication in journals and similar research outlets. This forms the literature resources available in our libraries and laboratories and is used, in turn, as an aid for the identification of plants through descriptions and identification keys. When a new project is undertaken, the literature must be thoroughly reviewed for pertinent background information, thus the reason for the information-retrieval procedures.

Literature is available, as previously mentioned, in forms other than journal articles.

By understanding the nature of these publications, one can better judge the reliability of a document. There is no guarantee that published information is accurate. In fact, as a rule, it seems that the farther the information is removed from the original publication, the less likely it is to be accurate. One may call a report of original research a primary publication. All other publications bear upon these reports; therefore they are secondary publications. Examples of these are manuals for identification of particular plant groups, textbooks, and popularized accounts in magazines, newspapers, popular books, or radio and television programs.

ANALOGY AND HOMOLOGY

Throughout the study of biology is woven the unifying theme of evolution. Therefore it is important that in analyzing data one should realize that various parts are similar in function, but are not of the same origin, thus analogous parts. Whereas, other structures are believed to have developed through evolution so that they are homologous. Homologs, or homologous parts, may have the same function, or they may have a different function. The leaves of green vascular plants are all homologous structures. Also homologous with these leaves are the brightly colored petals of flowers—same origin, but a different function. Attention to these concepts is important for an understanding of the results of evolution and classification.

SCIENCE AND MODERN SOCIETY

The impact of **science** on man's welfare through technological advance is obvious no matter where we look about us. Evidence of this may be found even in the most remote areas of the earth, if in no other way than as a discarded can, an unsightly reminder of our triumph over bacterial spoilage and molding.

To detail any of the many advances in recent years in the plant sciences would be beyond the scope of this text, but an understanding of the broad areas of progress and of the way the plant sciences fit into this should be discussed.

Progress in transportation affects botany in several ways, both practically and in research. Plant scientists are able to travel widely and to collect specimens in areas previously difficult to reach. This has greatly expanded knowledge of distribution and floristics. As a result, additional information has been brought into the classroom so that the student of today is not so provincial as in the past.

The study of genetics and **biochemistry** has led to remarkable benefits in crop production all over the world, through greater yields of wheat, corn, and other grains. Biochemical research revealed the type of proteins that give the greatest food value to the grain, and research in the genetics of the plant showed how, through selection and cross-breeding, the crop could be improved.

The details of photosynthesis are explained in Chapter 13, but we have emphasized the importance of plant life by pointing out that until recently all sources of power came from plants or from water. Atomic power is now being developed and probably will replace all other power sources except solar power. Petroleum and coal represent the greatest source of energy in our modern world, but we know they are not limitless. These materials were formed by the process of photosynthesis using energy from the sun, fossilized, and stored as oil and coal deposits beneath the surface of the earth. Scientists estimate that the annual oil formation still occurring today exceeds the amount being used. It is doubtful if this eases any of the current alarm over excessive use of these reserves. The atomic power sources eventually may change our current pessimistic view of the danger of running out of power, but it too has its limitation.

Advances in medicine and the virtual elimination of bacterial diseases through the use of plant antibiotics increase longevity. Similar advances in technology increase time for research as well as for leisure. The use of computers and automatic machinery may even eliminate any need for manual work. The sociological problems resulting from this progress are well beyond the areas covered by this text.

Not all problems have been solved by science. In fact, it seems that as soon as one problem is solved, another is created. The role of plant sciences is apparent in the following brief account of our current ills.

The population explosion continues to keep pace or exceed increases in crop production so that poverty and undernourishment affect at least one third of the world's people. The rapid depletion of the soil and other natural resources, mainly through ignorance of the principles of ecology, increases the threat of even wider starvation. Air and water pollution problems follow as a result of this same ignorance and of the added indifference and greed on the part of some who refuse to accept the responsibility for these transgressions of natural balance.

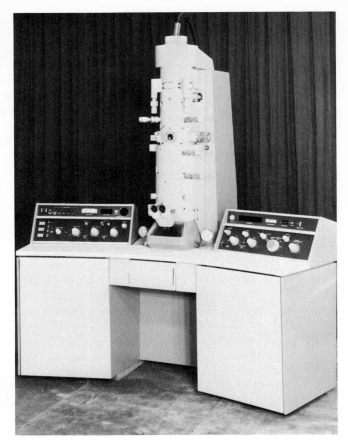

Fig. 2-9. A high-resolution electron microscope capable of viewing macromolecules to 2 Å. (Courtesy Jeolco, Inc., Medford, Mass.)

For example, the multicolored smoke pouring from factory stacks only a few years ago has been trapped by "stack scrubbers." Nevertheless, the same invisible, poisonous hydrocarbons still escape and continue to cause death to plants, disease to man, and destruction of property.

Disease, although now usually not of plant origin, is far from eliminated, and viruses responsible for disease, for example, the common cold, remain to be controlled. One might even add to the list such social diseases as war and riot, as well as psychological unbalance, which are at least partly caused by technological advances and are indirectly related to plant science through the lack of food in the tension areas.

We may learn through the conquest of space that plant life is not restricted to our planet. But for botanists, one remaining frontier is the ocean. Knowledge of marine plant life and of its balance with animal life promises additional food sources to help ease population pressures.

The great problems needing research attention in the plant sciences and related areas are both practical and theoretical. The invention of the electron microscope (Fig. 2-9), which has a magnification power sufficient to bring into view objects as small as 2 **Ångström (Å)** units, resulted in a great interest in **molecular biology** and ultrastructure. Although progress in these studies has been rapid, much remains to be accomplished.

Development and growth of plant life have not received the attention given to parallel studies of animals, so that this remains a potential area of study. Finally, studies that will lead to a better understanding of how life originated, how it evolved, how speciation takes place, and eventually even how higher life forms are created are important modern research areas.

In the minds of some, science is the magic solution to all problems. Even with the remarkable advances possible through science, there are still dangers to man and his world. So far in this discussion we have not attempted to define science. Indeed, such is hardly necessary because our description is itself a definition. But perhaps some distinctions are necessary for a better realization of the possible dangers of science. In a restricted sense, science deals only with facts and their arrangement into a system. So defined, it differs little from technology, which is restricted to the practical use of scientific systems for the development of a machine, product, or process. More broadly defined, science deals with the orderly arrangement of any body of knowledge. Thus the treatment of history, philosophy, and religion may be included in science. Probably the middle ground is to be preferred, so that science is restricted to the knowledge of natural objects, associated phenomena, and theories or concepts about these objects. The systematic arrangement of this knowledge depends on logic. Modern thinkers have developed a system of symbolic logic that is actually a mathematical treatment of thought processes. This has greatly improved the accuracy of scientific thought and the lucid exposition of scientific knowledge.

Perhaps the first danger of science, as it is often presented to the public, is the large-scale oversimplification of complex matters. People are led to believe that the discovery of a theory will immediately eliminate one or another kind of problem. For example, the biological control of plant pests is theoretically possible, and the many dangers of insecticides can be prevented. Pest population management is exceedingly complex, and no simple controls can be expected. Even after scientific discoveries are made, as in the use of atomic energy, their practical application may be long in developing.

A second danger is the acceptance with little question of statements made from scientific laboratories. Acting with insufficient data has led to disappointing results and sometimes disaster. One example is the overpopulation of deer in certain areas through the elimination of natural enemies and the overrestriction of hunting. Failure to properly test some drugs before release to the public has resulted in deformities and even death, although these are certainly rare exceptions. The greatest danger of public acceptance of some of the scientific announcements is the false security that may result.

Returning to the dangers of pesticides, probably the most important scientific dangers directly related to plant science are shown by the public indignation that recently erupted mainly through the publication of Rachel Carson's *Silent Spring*. Laboratory developments in pest control are far advanced over public education in their use. The scientist cannot be blamed entirely for the failure of the public to understand that scientific discovery may not be handled properly by the irresponsible. It is obvious that the use of dangerous insecticides, like drugs, should be restricted to those who are qualified to use them. Fortunately, the overreaction to these problems has resulted in further study, correction, and advances in biological control.

The conflict of interests between conservationists and industrialists is hardly fitting in today's world of scientific achievement. Yet, like war, the illogical situation still exists, and correction comes only at a snail's pace. Unless water and air pollution and soil and forest depletion cease almost immediately, the prospect for future generations seems grim. Without intending to proselytize, we cannot fail to point out that only through wide public and professional understanding

29

of the principles of **ecosystems** can balance be reestablished at a more rapid rate. One way to bring about this understanding is through the study of the plant sciences at the college level, where large numbers of young people can discuss in some depth the facts behind the problem.

Widespread alarm and indignation have been expressed over the increased use of computers in daily life. Where at first they were used to solve lengthy and complex mathematical problems with speed far beyond any previous ability, they now enter the daily lives of each of us, not as a research tool, but as a machine seriously robbing us of our human freedom. This newest danger of science does not differ in principle from any of the others. The solution rests in the acceptance of responsibility. Manual labor is not replaced only by skilled labor. Responsible people who know the need for accuracy are required as well. No computer yet invented will assume the responsibility of accurate input, and none are capable of making decisions requiring moral judgment. Only the operator can be held responsible for the output.

Science of the future will differ from that of today only in detail, refinement, and perhaps speed of discovery. People are inclined to complain when science seems too slow in solving every problem they pose. So many discoveries have been made so rapidly that they have flooded the market, and we accept this with the weary nonchalance of overindulged children. But this is only in its technological advances. The computerized world of the future, with its space travel, controlled environment, disease-free populations, and almost complete leisure time, will be boring indeed to minds. Overindulgence, too, will bring its problems. Hopefully the mind will turn more to pure intellectual pursuits. The heroes of battle will be forgotten, and in their place will be heroes of theoretical discovery.

Scientists differ from people with other interests and professions only in the zeal for organizing knowledge into systems. Certain qualities seem to dominate in men of science

as in many with other professional interests. A high intelligence is noted in all academic pursuits, coupled with a passion for accuracy. Most scientists seem to enjoy what they do, and early retirement and change of work does not seem to be their hope. Instead they prefer to work long hours to complete a project and find their reward in the acceptance of their discoveries in the sympathetic understanding of their colleagues. This has resulted in the public view of the scientist as a man aloof to social contact and unaware of today's social problems. Lack of communication between scientists and the public has sometimes caused great problems that can be corrected only through education.

QUESTIONS AND PROBLEMS

1. Why is botany more than a study of the kinds, classification, and cultivation of plants?
2. Discuss the nature of scientific research. Include the conditions needed for research and what ways scientific research differs from, and is similar to, research in the arts.
3. Name the several branches of botany and tell what topics are the special interest of each.
4. The work of several ancient scientists is briefly discussed in this chapter. Explain how progress in science was in steps, such that the results of one step was needed before the next step could be taken. Give examples.
5. Contrast observation and experimentation. What is meant by a controlled experiment? How could this be applied to observation?
6. Some of the scientific experiments, such as those performed in chemistry, have predictable results. Is this likely to be true of experiments in the life sciences? Explain.
7. What are the similarities between modern data processing and the scientific method? What are the differences?
8. Make a list of the several types of information sources for the botanical sciences. Which of these are available to you?
9. How would you design an experiment to show the following: (a) whether plants use water at night, (b) the proper amount of fertilizer to use for optimum growth of a common house plant, (c) the effect of the length of day on the flowering of a plant, (d) whether a plant is self-pollinated or cross-pollinated, (e) whether the absence of a particular kind of plant in your

area is attributable to environmental causes or simply to not having been distributed to your area from its natural range.

10. What are the aims of science? Of scientists?
11. What are the limits of science? The dangers?

DISCUSSION FOR LEARNING

1. Discuss several applications of the scientific method; for example whether the juice of a certain plant is beneficial as a drug, the use of the system to solve a murder mystery, the use of the system as a way to study for an examination, the use of the system to determine whether or not plants have emotions.
2. List and discuss some of the problems you believe have been caused by technological advancement through scientific discovery.
3. Considering the rapid discovery of the past 35 years, list and discuss some of the advances you feel will be made by the year 2000.

ADDITIONAL READING

Breck, A. D., and W. Yourgrau (editors). 1972. Biology, history, and natural philosophy. Plenum Publishing Co., New York. (This work relates biological science and the humanities for a comprehensive understanding of nature.)

Flanagan, Dennis (editor). Sept. 1966. Information. Sci. Am. **215**:64-260 (Useful as an introduction to the theory of handling information.)

Gardner, E. J. 1972. History of life sciences, 3rd ed. Burgess Publishing Co., Minneapolis. (Useful summary of biological science adapted for use by the biology student.)

Holman, H. H. 1962. Biological research methods. Hafner Publishing Co., Inc., New York. (Helpful suggestions for conducting biological research.)

Nagel, E. 1961. The structure of science. Harcourt, Brace & World, Inc., New York. (The theory of scientific method is explained.)

Taylor, G. R. 1963. The science of life. McGraw-Hill Book Co., New York. (A picture history of biology, interesting and informative.)

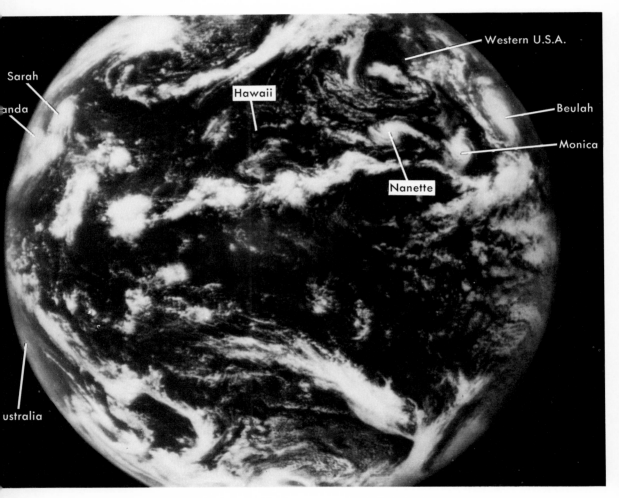

Fig. 3-1. The forces of nature are evident in this dramatic view of five hurricanes (named after women) in this photograph taken on September 19, 1967. This is a reminder that biological life is possible only because basic ingredients of living materials follow the same laws of physics as these violent forces of nature. The complex molecules that have evolved on our earth have been moderated by these forces into the controlled processes we call biochemical pathways. (NASA.)

CHAPTER 3

The composition of living matter

CELLULAR CHEMISTRY

In the cell beyond the resolution of the most powerful electron microscope lies an amazing array of molecular structures. All plants and animals, including the smallest bacteria, are widely diverse organisms that differ functionally and morphologically but have a very similar chemical composition. All cells contain water, minerals, carbohydrates, lipids, proteins, and nucleic acids. All biological phenomena are ultimately based on biochemical processes; it is important that we understand the nature of the biochemical reactions that occur in the cell. Such important processes as photosynthesis, nitrogen fixation, cellular energy transformations, and the genetic control of cell structure and function can best be explained at the molecular level. To adequately understand many of these biological processes, we need to learn or review some basic chemical concepts including molecular and atomic structure.

SOME BASIC CHEMISTRY

According to the cosmic theory, the two basic components of our universe are **matter** and **energy.** Matter is anything that occupies space and has mass. It can assume the form of a solid (ice), a liquid (water), or a gas (water vapor). Whatever form it assumes, it will always occupy space and have mass. Energy is defined as the property of matter that enables it to do work. When something has energy, it is capable of exerting a force on something else and performing work on it. When something has work performed on it, energy is added to it. There are two forms of energy that are at least partially convertible. Potential energy is that possessed by a substance or object because of its position. For example, a bowling ball at rest at the top of the stairs has potential energy by virtue of its position. Kinetic energy is present in a body in motion, as when the ball rolls down the stairs. Other kinds of energy are also found in nature. Electrical and magnetic energy are used to turn motors and light homes. Heat energy from burning gas and oil is used as a source of heat. Radiant energy from the sun is sometimes used as a source of heat, light, and electricity. The chemical energy of gasoline is used to drive automobiles, and the chemical energy of food enables the bodies of living organisms to perform work. Biological systems most often use stored chemical bond energy in order to grow, reproduce, and move.

COMPOSITION OF MATTER

A great deal of experimental evidence supports the concept that all matter is composed of submicroscopic discrete particles called **molecules.** These molecules are in constant motion. Molecules in turn are composed of **elements.** Elements are substances that cannot be broken down into simpler substances by ordinary chemical means. The smallest unit of an element that retains the chemical properties of that element is called an **atom.** Scientists represent an atom of an element by using an abbreviation of the element's name. For example, an atom of oxygen is represented by the symbol O, an atom of hydrogen is H, and an atom of chlorine is Cl. Using ab-

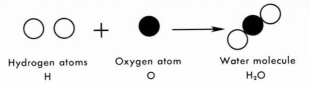

Fig. 3-2. The reaction of two atoms of hydrogen with one atom of oxygen to form one molecule of water.

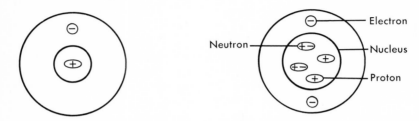

Fig. 3-3. Two examples of atoms. The hydrogen atom consists of a single proton in the nucleus and an orbiting electron. The helium atom consists of 2 protons and 2 neutrons in the nucleus with 2 orbiting electrons.

breviations in this manner, scientists write formulas to represent molecules. A molecule of water is written H_2O. This indicates that water molecules contain two atoms of hydrogen and one atom of oxygen (Fig. 3-2).

ATOMIC STRUCTURE

Atoms are tiny particles composed of a nucleus with a positive charge and an orbiting electron cloud with a negative charge (Fig. 3-3). The nucleus is composed of **protons,** which have a positive charge, and **neutrons,** which are uncharged or neutral. The **electrons** are some distance away from the nucleus and move in orbit. As one can see from this concept, the atom is mostly space. The proton and neutron are about the same size, with the neutron being very slightly heavier. Both are about 1837 times heavier than the electron. Ordinarily, we think of an atom as neutral, since positive charges (protons) are balanced by negative charges (electrons). The addition or subtraction of an electron from an atom creates a charged particle, which is called an **ion.** All atoms of one element have the same number of protons in the nucleus, and the proton number is different for different elements (Table 3-1). The

Table 3-1. Comparison of atomic weights

	Protons	Neutrons	Electrons	Atomic weight
Hydrogen	1	0	1	1
Chlorine	17	18	17	35
Carbon	6	6	6	12

Table 3-2. Atomic weights of common elements found in biological substances

Element	Symbol	Atomic weight*
Carbon	C	12.0111
Hydrogen	H	1.0079
Oxygen	O	15.9994
Nitrogen	N	14.0067
Sulfur	S	32.0640
Phosphorus	P	30.9738
Potassium	K	39.1020
Iron	Fe	55.8470
Copper	Cu	63.5400
Zinc	Zn	65.3700
Iodine	I	126.9044
Manganese	Mn	54.9380
Chlorine	Cl	35.4530
Magnesium	Mg	24.3120

*When using atomic weights, one customarily rounds off these numbers to the nearest whole.

number of protons in the atomic nuclei of the atoms of an element is called the atomic number of that element. Hydrogen has an atomic number of 1, and oxygen has an atomic number of 8. The **atomic weight** of an element, a relative measure based on assigning a weight of 12 to normal carbon, is computed by the summation of the protons and neutrons in the atomic nucleus (Tables 3-1 and 3-2). However, all atoms of an element do not necessarily have the same atomic weight, but they do have the same number of protons (atomic number). This small variation found in the atomic weight is caused by a difference in the number of neutrons present. Some atoms of an element have either more or less than the normal number of neutrons. Atoms with the same atomic numbers but different atomic weights are called **isotopes.** Some isotopes emit radiation and are called radioactive isotopes. It is important to remember that isotopes maintain the chemical identity of the element because the number of protons remains the same. The first isotopes discovered were those of heavy oxygen, ^{17}O and ^{18}O. The radioactive isotope of carbon, ^{14}C, has been most useful in many biological investigations, particularly photosynthesis.

Atoms have properties known as the laws of chemical combination. For instance, when two or more elements combine to form a particular molecule, they unite in a constant and definite way, depending on the **valence,** or combining power of the element. The simplest example of this is water, H_2O. Two atoms of hydrogen always combine with one of oxygen to form water in a ratio of 2:1. Carbon dioxide is formed from one atom of carbon and two of oxygen, CO_2, but methane is formed from one atom of carbon and four of hydrogen, CH_4. Therefore, the ratio in carbon dioxide is 1:2, and in methane, 1:4. According to the above combinations hydrogen has a valence of 1, oxygen 2, and carbon 4. The valence of an element is the number of electrons each of its atoms has gained, lost, or

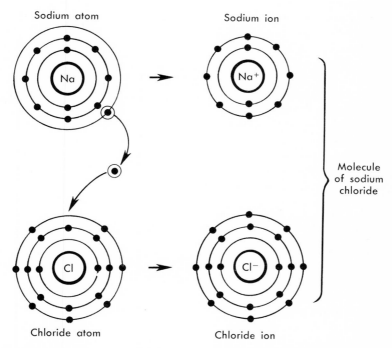

Fig. 3-4. The chemical union between an atom of sodium and an atom of chlorine to form a molecule of sodium chloride.

35

shared in a combination with other atoms. Valence numbers can be positive or negative (Fig. 3-4). The **molecular weight** of a molecule is calculated as the sum of the atomic weights of the atoms that comprise the molecule, each multiplied by the number of times that it appears. Water for example has a molecular weight of 18.

$$2H = 2 \times 1 = 2$$
$$1O = 16 = 16$$
$$\overline{18}$$

Molecular weights are used to derive the concept of the **mole**, which is fundamental to chemical calculations. A mole of a compound is that amount of it whose weight is equal to its molecular weight expressed in grams. A mole of sodium chloride, for example, weighs 58.44 grams.

CHEMICAL REACTIONS

Chemical reactions are the processes by which atoms combine to form molecules or molecules interact to form new molecules or atoms. Equations for chemical reactions can be written by use of molecular formulas, for example, the reaction of sodium hydroxide (NaOH) with hydrochloric acid (HCl):

$$NaOH + HCl \rightarrow NaCl + H_2O + heat$$

The components on the left side of the equation are the reactants, whereas those on the right are the products. The arrow indicates the direction of the reaction, which may be reversible. The above reaction occurs spontaneously, releasing heat in the process, and it is not reversible. One can say that sodium hydroxide reacts spontaneously with hydrochloric acid producing sodium chloride plus water and heat. The heat represents a release of energy that was initially stored in the reactant molecules. In this case the energy level of the reactants is greater than that of the products. To reverse the reactions, one would have to raise the energy level of the products. Obviously the heat generated is lost and cannot be used to reverse the reaction. Substances that lower the energy level needed for a reaction and thus speed up a

Covalent bond

Fig. 3-5. Two atoms of hydrogen and one atom of oxygen form a covalent bond.

chemical reaction are called **catalysts.** Almost all reactions in biological systems are mediated by catalysts. Biological catalysts are different from nonbiological ones and are known as **enzymes.** Enzymes are proteins and will be discussed later in the chapter.

A molecule is a stable union of two atoms. When chemical reactions occur, it is the electrons in the outermost orbit of the atom that react. If electrons are donated from one atom to another, the resulting chemical bond is ionic, an association based on the electrical attraction of the two now oppositely charged ions. A covalent bond is formed when electrons are shared between atoms (Fig. 3-5). In biological systems, we will often be concerned with **oxidation-reduction** reactions, which are actually electron transfers. When an atom loses an electron, it is oxidized; when it gains one, it is reduced.

CHEMICALS IN PROTOPLASM

All cells contain protoplasm, which is a very general term that applies to the various materials found in a living cell. Note later that not all cells found in mature plants are living.

Certain generalizations that will be true for all kinds of living organisms may be stated about protoplasm. Protoplasm is a complex mixture of organic and inorganic molecules. Approximately 95% by weight of protoplasm is composed of four elements—oxygen, 62%; carbon, 20%; hydrogen, 10%; and nitrogen, 3%. It so happens that all of them are present in some form or other in the atmosphere, the oceans, and the fresh waters. The remaining 5% by weight of protoplasm is made up of approximately 30 elements, some in relatively large amounts, such as calcium, and others only in trace amounts.

Most of these elements are present in a combined form, so that it is possible to distinguish both classes of compounds—inorganic and organic. Organic compounds, with the exception of the artificially synthesized ones, are always found in living things or their products and always contain carbon. Inorganic compounds are abundant in both living and nonliving matter, very rarely contain carbon, and are combinations of mineral substances.

INORGANIC MOLECULES

Water is the most prevalent **inorganic** compound in protoplasm. The protoplasm of organisms rarely contains less than 60% water, and it may be present in amounts as high as 99%. Water is the solvent for all inorganic constituents of protoplasm and most of the organic molecules. It is also a dispersal medium for the various particles that are not in solution. The solvent properties of water are very important because water provides a medium within which chemical reactions can occur. Some molecules when dissolved in water have the ability to break apart and form ions, which are charged atoms that become involved in chemical reactions. When molecules break apart to form charged atoms, this is called dissociation. Sodium chloride dissociates as follows:

$$NaCl \rightleftharpoons Na^+ + Cl^-$$

The thermal properties of water are also important for an organism. Water has a high heat content so that when a sudden temperature change occurs in the environment a living organism changes its temperature gradually because of its high water content. Water is also used to dissipate heat from the cell in the form of water vapor and thus overheating is prevented. Water dissociates to an extremely small degree to form hydrogen (H^+) ions and hydroxyl (OH^-) ions, which participate in many cellular reactions.

Both hydrogen and hydroxyl ions are very active and determine the acidity or basicity of the protoplasm. The acidity of a solution is measured as the **pH,** which is equal to the negative logarithm of the hydrogen-ion concentration (Fig. 3-6). Compounds that resist a change in the pH of a solution are called **buffers.** It is absolutely essential that the pH of protoplasm be maintained within the limits of 6.8 to 7.6; therefore protoplasm contains many buffers that are crucial to its structure.

With the exception of gases, the remaining inorganic constituents in protoplasm can be classified as acids, bases, or salts. **Acids** are compounds that produce hydrogen ions when they dissociate; **bases** produce hydroxyl ions. **Salts** are combinations of a positively charged ion with a negatively charged one, when neither ion is the hydrogen or hydroxyl ion (Fig. 3-7). Examples of salts are the phosphates, carbonates, chlorides, and sulfates of calcium, sodium, potassium, and magnesium. Although most of these salts may be in solution in water, some make up the more solid parts of the organism. In addition, such gases as oxygen and carbon dioxide may be present.

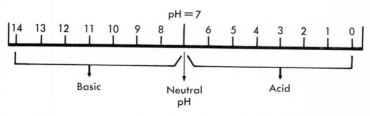

Fig. 3-6. The pH scale with which we measure the acidity or basicity of a solution; pH = 7 is neutral; numbers above 7 are basic, and those below 7 are acid.

Acid
$$HCl \rightarrow [H]^+ + Cl^-$$

Base
$$NaOH \rightarrow Na^+ + OH^-$$

Salt
$$NaCl \rightarrow Na^+ + Cl^-$$

Fig. 3-7. Dissociation of (1) HCl, which is an acid and forms H$^+$ ions; (2) NaOH, which is a base and forms OH$^-$ ions; (3) NaCl, which is a salt and does not form either H$^-$ or OH$^-$ ions.

ORGANIC COMPOUNDS

Organic compounds are complex combinations that always include the elements carbon and hydrogen. Carbon has the unique capacity to form a great variety of compounds, because of the structure and behavior of the carbon atom. It can combine with both positively and negatively charged atoms. In addition, carbon has a relatively high valence of 4, which enables it to combine with four univalent atoms at the same time. Carbon dioxide and carbonates are not usually considered organic compounds because of their simple nature and mineral origin. Carbohydrates, proteins, nucleic acids, and lipids are the chief organic compounds found in protoplasm. Generally, they are dissolved or suspended in water, and in some cases they may combine to form various complexes, for example, fat-protein.

The simplest organic compounds are found in the **carbohydrates,** which are composed of carbon, hydrogen, and oxygen. The carbons are usually in a chain, —C—C—C—, and hydrogen and oxygen are present in the same 2:1 ratio as in water. The simplest sugars, such as glucose and fructose, are made up of

6-carbon chains and have the formula $C_6H_{12}O_6$ (**monosaccharide**).

$$
\begin{array}{c}
H \\
| \\
C=O \\
| \\
H-C-OH \\
| \\
HO-C-H \\
| \\
H-C-OH \\
| \\
H-C-OH \\
| \\
H-C-OH \\
| \\
H
\end{array}
$$

Two 6-carbon sugars may combine to form a 12-carbon sugar (**disaccharide**), with the loss of water, as follows:

Glucose + Fructose → Sucrose + Water
($C_{12}H_{22}O_{11}$) (H_2O)

Several disaccharides may combine to form longer chains called **polysaccharides,** such as starch and cellulose, which are chains of about 2000 glucose units.

Carbohydrates decompose into water, carbon dioxide, or fragments of several linked carbon atoms, with the release of energy as illustrated below:

$$C_6H_{12}O_6 \rightarrow 6\ CO_2 + 6\ H_2O + \text{energy}$$
$$C_6H_{12}O_6 \rightarrow 2\ C_2H_6O + 2\ CO_2 + \text{energy}$$

Alcohol

Carbohydrates in protoplasm function as fuels. Their stored energy being released for cellular work.

The **fats,** or **lipids,** like the carbohydrates, are composed of carbon, hydrogen, and oxygen. However, the ratio of oxygen and hydrogen is not 2:1 as it is in the carbohydrates;

Fatty acid: *Palmitic acid*

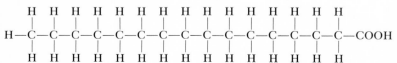

lipids contain far more hydrogen than oxygen atoms. The most common fats are composed of a 3-carbon compound called **glycerol** attached to longer carbon chains, called **fatty acids,** as illustrated at the bottom of p. 38 and below:

Common fat:

$$H-\underset{\displaystyle H}{\overset{\displaystyle H}{\underset{|}{\overset{|}{C}}}}-O\ \ldots\ldots\ \text{Fatty acid}$$

Glycerol

The fatty acids usually consist of even-numbered carbon chains because they are synthesized from two carbon atom units. The carbon chains may be saturated or unsaturated. In a saturated fatty acid, all the inner-chain carbon atoms are saturated with hydrogen atoms.

Saturated fatty acid:

$$H-C-C-C-C-C-C-C-C-COOH$$

In the unsaturated fat, carbon atoms are not totally saturated with hydrogen atoms.

Unsaturated fatty acid:

$$H-C-C=C-C=C-C-C-COOH$$

Long and saturated fatty acids are usually the hard fats, such as tallow. The unsaturated fatty acids have shorter carbon chains and are generally liquids (oils). A wax is a long-chain fatty acid that has combined with some substance other than glycerol. Complex ring compounds known as **sterols,** which form the basic structure of many hormones, are closely related to the fats.

The fats, like the carbohydrates, are important cellular fuels. They release energy through their decomposition. As carbohydrates can be converted to fats, so also can fats be converted to carbohydrates. Fats are one of the fundamental building blocks of protoplasm; for example, they contribute to the framework of membranes, and thereby have an important role in diffusion and osmosis.

The most complex organic compounds found in protoplasm are the **proteins.** In addition to carbon, hydrogen, and oxygen, proteins also contain nitrogen and sulfur. They are large molecules called polymers that have molecular weights ranging from 5000 to many millions. A polymer is a large molecule that is made up by joining together many small molecules. The small molecules are called subunits. The subunits that make up proteins are amino acids (Table 3-3) and are represented by the general formula below:

$$R-\underset{\displaystyle H}{\overset{\displaystyle H-N-H}{\underset{|}{\overset{|}{C}}}}-COOH$$

The NH_2 is the amino group; the R is a ring or chain of carbon atoms having side linkages of H or OH or even of other groups; and fi-

Table 3-3. Table of amino acids

Glycine	Gly
Alanine	Ala
Valine	Val
Leucine	Leu
Isoleucine	Ileu
Serine	Ser
Threonine	Thr
Cysteine	Csh
Methionine	Met
Glutamic acid	Glu
Aspartic acid	Asp
Lysine	Lys
Arginine	Arg
Tyrosine	Tyr
Histidine	His
Proline	Pro
Hydroxyproline	Hyp
Cystine	Cys
Tryptophan	Try
Phenylalanine	Phe

nally, the COOH is the carboxylic acid group, which can release H^+ ions under certain conditions.

Proteins are linear chains of **amino acids;** protein specificity is determined by the kind, number, and arrangement of these substances. The almost unlimited number of arrangements of amino acids in proteins accounts for the great variation of life forms and processes found on earth today and in the past.

So far about 60 amino acids have been isolated or identified in plant tissues, but only about 20 are found in proteins. The nature of a protein is determined by the sequence of amino acids in its chain. The series of amino acids are linked together by **peptide** bonds. The structures resulting from the formation of peptide bonds are termed "peptides," and the individual amino acids of peptides are termed "residues." Thus a dipeptide (one with two amino acid residues) looks like the following:

$$NH_2-\underset{\underset{R}{|}}{\overset{\overset{H}{|}}{C}}-\overset{\overset{O}{\|}}{C}----\underset{\underset{H}{|}}{\overset{\overset{H}{|}}{N}}-\underset{}{\overset{\overset{R}{|}}{C}}-COOH$$

A **polypeptide** (in this case a tetrapeptide) is shown below. The R groups are variable and determine different amino acid residues.

The complexity of the protein molecule is relative to its size. In order that we might better understand and deal with protein structure, we will consider several levels of organization based on the nature of the chemical bond necessary to maintain that level. The **primary structure** of a protein, defined as the sequence of amino acids, is maintained by the peptide bonds. Once this long strand has

been formed, it can assume a variety of configurations. **Secondary structures** are maintained by hydrogen bonding. Hydrogen bonds are formed as a result of the tendency of hydrogen to share electrons with neighboring atoms, especially oxygen and nitrogen. Hydrogen bonds are individually weak but exert a considerable force when found in great numbers over the entire length of a protein molecule. The angles of the C—C and C—N bonds cause a protein chain to form a coil or helix, which may be stabilized by hydrogen bonding within the same molecule. Hydrogen bonding between protein molecules can result in a so-called sheet structure (Fig. 3-8). You can imagine

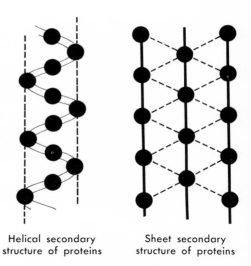

Helical secondary
structure of proteins

Sheet secondary
structure of proteins

Fig. 3-8. A comparison of protein secondary structure. Helical forms and sheet forms are illustrated. *Solid lines* are peptide bonds; *dotted lines* are hydrogen bonds; *black dots* are amino acids.

that the protein in a secondary structure can then fold on itself to assume a number of additional different configurations. Such configuration is known as a tertiary structure. The chemical bonds necessary to maintain a tertiary structure are many, including hydrogen bonds and covalent linkages. A protein **tertiary structure** is almost always associated with its function, for example, the action of a specific enzyme. When we destroy or change a protein's tertiary structure, we denature the protein, and, depending on the extent of denaturation, the protein's function may be altered temporarily or permanently. A good example of protein denaturation is the frying of an egg white. Sometimes fully folded protein molecules form aggregates with one another because of various molecular attractions, and such aggregates constitute an even higher level of organization, the quaternary structure (Fig. 3-9).

Nucleic acids are giant molecules similar to those of proteins. **Nucleotides** are the building blocks of nucleic acids, much like the amino acid is the building block of proteins. The nucleotide consists of a 5-carbon sugar, either ribose or deoxyribose, purine or pyrimidine bases, and phosphoric acid.

There are two kinds of nucleic acids found in cells, **deoxyribonucleic acid (DNA)** and **ribonucleic acid (RNA).** The nucleotide units of DNA contain deoxyribose, whereas the units of RNA contain ribose. The nucleotides are bound together in long chains so that the bases are free to react. The nucleic acids found in cells are sometimes combined with protein to form a nucleoprotein complex. DNA contains the bases adenine, guanine,

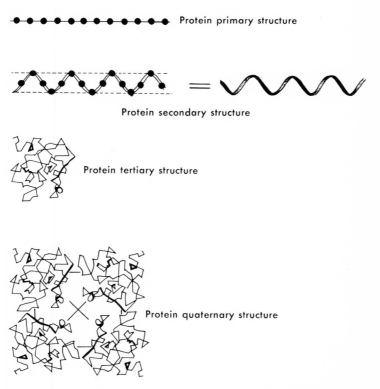

Protein primary structure

Protein secondary structure

Protein tertiary structure

Protein quaternary structure

Fig. 3-9. The primary structure of a protein is its amino acid sequence. The secondary structure is a helix, and it can fold on itself to form globules, which are tertiary structures. The globules can associate to form aggregates called a quaternary structure.

PURINES

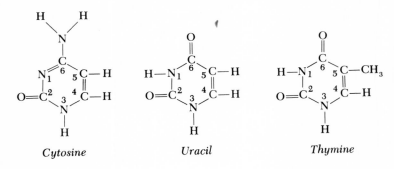

Adenine

Guanine

PYRIMIDINES

Cytosine

Uracil

Thymine

SUGARS

Ribose

Deoxyribose

NUCLEOTIDE: *Adenine monophosphate*

(Base)
Adenine

(Sugar)
Ribose

Phosphate

Table 3-4. Classification of enzymes

No.	Systematic name	Example	Reaction
1.	Oxidoreductases	Glucose oxidase	Oxidation-reduction
2.	Transferases	Glucokinase Ribokinase Triokinase	Group transfer
3.	Hydrolases	α-Amylase Pepsin Trypsin	Hydrolytic
4.	Lyases	Isocitrate lyase Citrate synthetases	Addition of groups to double bonds
5.	Isomerases	Lysine racemase Ribulose phosphate epimerase	Isomerizations
6.	Ligases (synthetases)	GMP synthetase CTP synthetase	Condensation of two molecules in ATP or similar triphosphate pathways

cytosine, and thymine; RNA, however, contains uracil in place of thymine. Nucleic acids function as carriers, translators, and transcribers of the genetic information carried by each cell. We will discuss them in more detail in Chapter 18.

Reactions of a biological nature require rather high activation energy sources. If there were no other means of obtaining these reactions than by the direct application of energy to the chemical systems within the cell, life would be much different from the way we now know it. Perhaps somewhere in the universe such life forms do exist unrecognized. In our system the reactions occur with the help of catalysts. In biological systems the catalysts are termed "enzymes."

Enzymes are organic catalysts produced by a living cell. Enzymes speed up a cellular reaction by lowering the energy level required for activation of the molecules. The molecules with which the enzyme reacts are called the substrates. Enzymes are named either after their substrates or according to the type of reaction that they catalyze. Thus an enzyme catalyzing the hydrolysis of the sugar sucrose is termed "sucrase." There are six general groups of enzymes classified according to the type of reaction that they catalyze (Table 3-4).

CHEMICAL NATURE OF ENZYMES

Enzymes are proteins. They are sensitive to temperature and ion concentration, particularly changes in pH. The tertiary structure, or specific configuration of the protein molecule, is necessary for enzyme activity. Enzyme proteins are usually very large. Although some enzymes act alone, many others need nonprotein factors in order to catalyze a reaction. These nonprotein factors are called "cofactors" and they act in concert with the protein to cause the catalysis of a certain reaction. Many of the vitamin compounds such as niacin, thiamine, and riboflavin are enzyme cofactors or coenzymes. In addition to the organic coenzyme, a metal ion such as Mg^{++} or Mn^{++} may also be needed for activity of the enzyme. Usually then an enzyme consists of a protein portion, a coenzyme (vitamin), and a metallic ion. In such cases, all three must be present for enzyme activity. Enzymes are extremely specific and will react with only one or a few substrates. In-

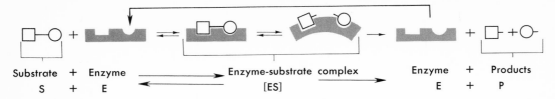

Fig. 3-10. The mechanism of enzyme action: (1) enzyme and substrate react to form an enzyme-substrate complex; (2) the energy of the substrate is raised in the union; thus the chemical bond is made to break; (3) products are released and the enzyme is regenerated.

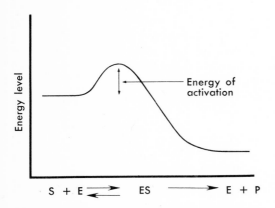

Fig. 3-11. The enzyme-substrate complex causes the substrate to have an increased energy level. The increase in energy of the substrate is the energy of activation.

organic catalysts do not possess this characteristic of specificity.

ENZYME ACTION

A specific spot on an enzyme protein combines with the substrate. This site is called the enzyme's active site; it is here that an enzyme-substrate (ES) complex forms, when the enzyme initially reacts with the substrate. The ES complex then undergoes reaction and dissociates to form the product (P) plus the free enzyme again, which can then react with another substrate molecule. The enzyme has speeded up the reaction somehow as a result of its combination with the substrate and yet remains essentially unchanged (Fig. 3-10). The generalized enzyme reaction can be written as indicated.

$$E + S \rightleftharpoons (ES) \rightleftharpoons E + P$$

When the enzyme combines with the substrate to form the (ES) complex, the energy level of the substrate is lowered so that the reaction can occur (Fig. 3-11). Enzymes are sensitive to inhibitors that may mimic the substrate and combine with the active site, as do some insecticides and fungicides. Other poisons inhibit enzyme activity by disrupting the tertiary configuration.

PHYSICAL STRUCTURE OF PROTOPLASM

Although Huxley aptly terms protoplasm the "physical basis of life," protoplasm is not a single uniform substance, as we have seen from the previous discussions. The functional differences and similarities need to be considered.

WATER

Protoplasm is largely a colloidal system, the liquid phase of which is water containing dissociated ions and small molecules. The solid or dispersed phase consists of large protein molecules and complex fat and carbohydrate particles. Any system that consists of a liquid phase and a solid phase must be either a **crystalloid** or a colloid, depending on the size of the particles (Fig. 3-12). If all the particles are very small, such as inorganic ions or small molecules, the system is a true solution or crystalloid. It is so named because crystals form readily from it. Extremely large particles would merely settle to the bottom by gravity, as in a suspension. On the other hand, if the particles are intermediate in size, they will neither settle out nor form a true solution. Such a system is called a **colloid.**

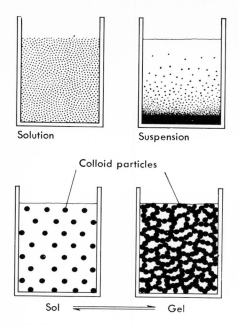

Solution   Suspension

Colloid particles

Sol ⇌ Gel

Fig. 3-12. Diagram of crystalloids, suspensions, and colloids.

Examples of general types of colloids include gas dispersed within either a solid or a liquid; a solid within a gas, liquid, or solid; and a liquid within a gas, solid, or liquid. For example, water droplets in air (fog) and protein and fat particles in a liquid (milk) are examples of nonliving colloids. The dispersion of a liquid within another liquid is commonly called an **emulsion.**

Since protoplasm is a colloid, it is necessary to discuss the characteristics of colloids. The molecules that make up a liquid are constantly in motion, thereby bombarding any particles in the liquid. Large particles would fall to the bottom of a container, but smaller particles are kept suspended. The random movement or agitation of these small particles is called **Brownian movement** and can be easily detected under the microscope in the living protoplasm of an ameba or in the vacuoles of some desmids (unicellular green plants) or even in dust in a sunbeam. Although Brownian movement helps to keep these particles suspended, they are not prevented from settling out by this force alone.

Electrical charges are largely responsible for their dispersion, since colloidal particles are either positively or negatively charged.

As a colloid, protoplasm may undergo a phase reversal transforming from a **sol** to a **gel** and likewise from a gel to a sol. If more colloidal particles are added or if water is withdrawn from the sol phase, the particles will become more closely associated and form a gel. This gel state of the colloid is an intricate network of colloidal particles within which are held droplets of water. Thus there is formed a resilient semisolid characteristic of protein colloids such as gelatin. Should the water phase be increased or colloidal particles be removed, the sol or more fluid state will return. Although some protoplasm may be in a more or less permanent gel state, it is not abnormal for protoplasm to alternate repeatedly between the sol and gel phases in accord with the fluctuations in the water–organic compound ratio. Many physical and chemical conditions such as hydrogen-ion concentration (pH), pressure, temperature, and absorption of energy may influence the sol-gel phase reversal. For example, the temperature tolerance of colloidal protoplasm is confined to a relatively few degrees found in the areas of the earth inhabited by living organisms, that is, from a few degrees below 0° C. to approximately 75° C. above.

When it is freshly formed, any colloid is enveloped by what is known as bound water. This is the molecules of water held to the surface of the particles by electrochemical attraction. With time, the particles lose this capacity to bind water. For example, it is common knowledge that a gelatin or a custard will lose water when it stands for any length of time to the open air. The water lost is bound water, the loss of which causes a significant drying of the colloid.

Colloidal particles, such as molecules concentrated in a certain area, are constantly agitated either as a result of Brownian movement or as a result of heat. In a concentrated region they will collide with each other and be stopped. Those moving toward a less concentrated area will be stopped less frequently through collision since the particles are farther apart. This migratory movement of particles from a region of greater concentration toward one of lesser concentration is called **diffusion.**

When any colloidal system comes into contact with a different medium such as water, air, or even a colloid of a different kind, a membrane will form where the two faces meet. This is known as an interfacial membrane. The plasma membrane of a living cell is more complex than an interfacial membrane. In a protoplasmic membrane it appears that the boundary molecules of water, fat, and protein are oriented in layers or are parallel or are both, since they are subjected to a variety of electromolecular forces. It is obvious that such colloidal membranes in living systems are extremely important, since they are the medium through which all materials enter and leave the cell.

DIFFUSION

The principal process involved in the passage of water and solutes into, within, and out of the cell is diffusion. It is dependent on the movement of ions, molecules, and molecular aggregates as a result of their own kinetic energies. Diffusion occurs only when there is a difference in the concentration of particles in two different regions. The particles will move from a region of greater concentration to one of lesser concentration as a result of their intrinsic energy. The force or pressure developed when the particles move is called the **diffusion pressure.** This pressure is proportional to both the movement and the concentration of particles and is dependent on their concentration per unit volume and temperature, among other things. Therefore diffusion pressure is not only the result of diffusion but also is the cause. There are instances in which particles are moving in all directions simultaneously at the same rate, as when water molecules pass in equal numbers in both directions through a membrane. Such a system is said to be in **dynamic equilibrium,** since there are no differences in concentration nor is there any diffusion pressure

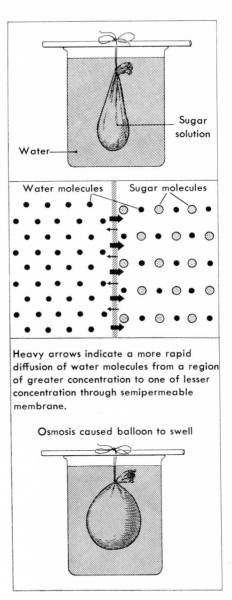

Heavy arrows indicate a more rapid diffusion of water molecules from a region of greater concentration to one of lesser concentration through semipermeable membrane.

Fig. 3-13. Osmosis. *Top,* Balloon made of a semipermeable membrane filled with a sugar solution, the molecules of which are too large to pass through the membrane. *Middle,* Water does pass through, however. *Bottom,* The result is the swollen balloon.

and, therefore, no diffusion. Similarly, when particles are moved by the outside forces (forces other than those of the particles themselves), the process is not diffusion but mass movement. An example of mass movement is smoke carried by the wind from a stack.

A special kind of diffusion known as **imbibition** occurs when water molecules move into the living and nonliving parts of a plant or a cell through adhesive attraction between the imbibing and the diffusing substances. The swelling of gelatin is caused by the imbibition of water; that is, the colloidal particles absorb the water in tightly compressed shells around the particles. Plant materials such as cellulose, starch, and protein are capable of imbibing water, which is a factor responsible for the swelling of seeds in water. Imbibition occurs only when the diffusion pressure of the imbibed water is less than that of the water surrounding the imbibing substance and when there is an adhesive attraction between the water and the imbibing substance.

OSMOSIS

An osmotic system is established when a semipermeable membrane is placed between two solutions. In biological systems **osmosis** is the exchange of water between the protoplasm and the surrounding medium, with the plasma membrane being the semipermeable membrane (Fig. 3-13).

In the living cell both the cell wall and the plasma membrane determine the permeability of the cell, but the latter is the least permeable of the two. The ability of a substance to pass through these barriers is known as the **permeation** of the substance. Since not all substances may pass into or out of the cell, the cell surface is **selectively permeable** or **semipermeable.** The semipermeability varies from cell to cell and under certain conditions. Water molecules alone permeate it freely, but most ions more slowly, and many large molecules and nonelectrolytes, for example, sugar, not at all. Yet it is known that sucrose will permeate some cells. So it appears that

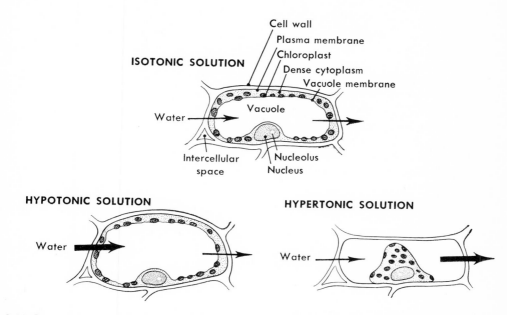

Fig. 3-14. Osmosis in plant cells showing the effect on the cell of different solutions. Normal, or isotonic, solution does not distort the cell. A hypotonic solution, that is, one with a lower amount of salt in the water, causes the cell to swell. The reverse occurs when the solution has a greater amount of salt than does the cell, as can be seen by a shrunken cell in the hypertonic solution. *Arrows,* Relative amounts of water entering and leaving the cell.

47

Fig. 3-15. The Badlands of South Dakota are another reminder of the subtle forces of nature. Here physical action of wind and water create a land nearly uninhabitable by living organisms even though they do conform to the laws of physics and chemistry described in this chapter.

the cell wall will pass a wide range of molecules, which includes all those necessary for the cell's metabolism.

Heretofore, living membranes have been compared to nonliving ones such as cellophane, since that material is also semipermeable. But a living membrane is not like a window screen that allows air to pass through freely, keeping out larger particles because of its pore size. The pore hypothesis is inadequate to explain what occurs in a living membrane, since two particles of the same size do not necessarily enter with equal facility, and one type may not be able to diffuse through the membrane at all. The living membrane is

therefore selective, and an expenditure of energy is involved in the passage from one side to the other. Therefore, instead of visualizing a living membrane as a passive membrane wherein there is merely a diffusion through pores from a region of greater concentration to one of lesser concentration, it appears that the membrane is a dynamic structure in which the passage of almost any substance is accomplished through a complex chemical process. The fatty components of the membrane may be particularly active in this transfer.

Osmotic processes characterize many well-known phenomena, such as those that result when a concentrated salt solution is placed around the leaves of an *Elodea* plant. The sodium chloride cannot pass through the cell membrane but water can pass freely in both directions. As a result the initial concentration of water inside the cell is greater than outside (since the salt concentration is less inside the cell than outside). As a result, the cell becomes dehydrated as the water moves from the protoplasm to the surrounding medium until equilibrium is reached. A solution in which the concentration of salt is greater than that inside the cell is called a **hypertonic** solution and will cause the plant cell to become plasmolyzed. In other words, it loses water and becomes dehydrated through **plasmolysis.** If the concentration of the salt solution outside the cell is less than that inside the cell, a **hypotonic** solution, the water will diffuse into the cell, making it **turgid.** Should the **turgor pressure** become great enough, the cell membrane will burst. When the concentration of salt is the same on both sides of the membrane, the outside solution is said to be **isotonic.** This explains why aquatic plants must remain in water to which they are adapted and cannot be changed from fresh water to salt water or vice versa (Fig. 3-14).

Active secretion takes place when the principal ions in the cell are at higher concentrations than in the solutions outside. This is usually accomplished by substitution of one ion for another This transfer or exchange of ions, such as potassium ions for the sodium ion in salt water, is termed secretion. **Guttation** is more general and involves the excretion of water droplets or very dilute solutions from a highly turgid plant (Fig. 3-15).

QUESTIONS AND PROBLEMS

1. Why is it necessary to study the chemical and physical nature of protoplasm?
2. Define an isotope.
3. What are oxidation-reduction reactions?
4. From what we have just learned about enzymes, what is one of the prerequisites for life on this earth?
5. How is osmotic pressure related to diffusion?

DISCUSSION FOR LEARNING

1. Discuss the organic constituents of protoplasm.
2. Would the pH of a system affect enzyme activity? Explain.
3. What sort of problems would one expect to encounter when trying to reconstruct protoplasm from simple inorganic chemicals?

ADDITIONAL READING

Beiser, A., and A. Krauskopf. 1969. Introduction to physics and chemistry. McGraw-Hill Book Co., New York. (A useful introductory reference book.)

Glasstone, Samuel. 1958. Sourcebook on atomic energy. D. Van Nostrand Co., Inc., Princeton, N.J. (Explains basic concepts of nuclear physics.)

Lehninger, A. 1973. Biochemistry. Worth Book Publishers. (A biochemistry textbook with a good discussion of enzyme mechanisms.)

Mahler, H. R., and E. H. Cordes. 1971. Biological chemistry. Harper & Row, Publishers, New York. (A biochemistry text with a suitable discussion of levels of protein organization.)

Orten, J. M., and O. W. Neuhaus, 1975. Human biochemistry, 9th ed. The C. V. Mosby Co., St. Louis. (A biochemistry text with a detailed discussion of cell chemistry.)

Shortley, G., and D. Williams. 1959. Principles of college physics. Prentice Hall, Inc., Englewood Cliffs, N.J. (A physics textbook that shows energy relationships clearly.)

White, A., and P. Handler. 1973. Principles of biochemistry. McGraw-Hill Book Co., New York. (A biochemistry textbook with recent developments in enzyme chemistry clearly discussed.)

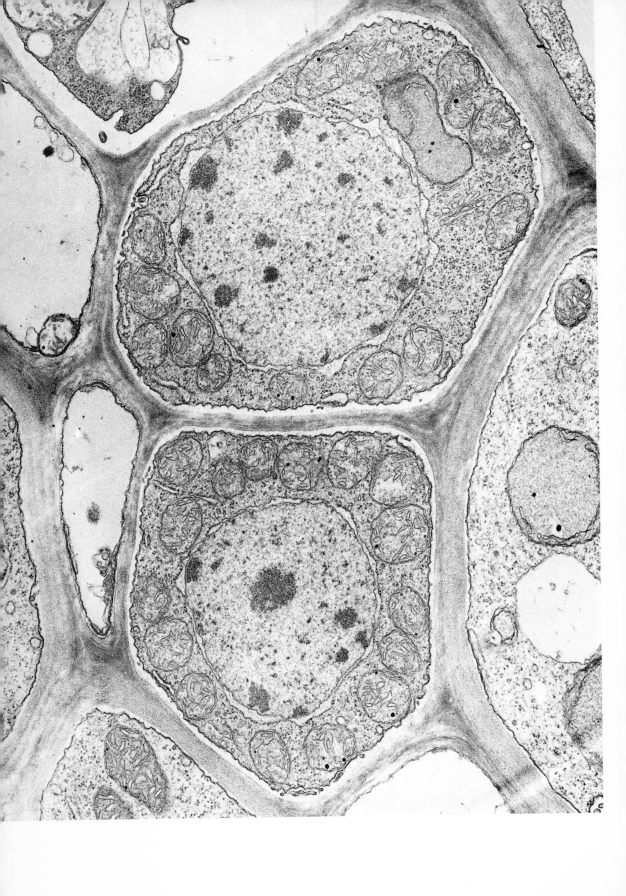

CHAPTER 4

Cell structure and function

■ Basic to all biology, cells are the units of operation as well as the determiners of structure, the carriers of all hereditary traits, and the site of all biochemical processses. As self-evident as this is to us today, the discovery and proof that all organisms are cellular did not come easy.

HISTORY OF CELL CONCEPT

William Harvey (1578-1657) originally proposed the biological maxim "ex ova omnia" (everything from the egg), but that was forgotten for years. It took Redi, Spallanzani, and Pasteur over 200 years to prove empirically what seemed evident to the logical mind of Harvey and what seems so obvious to us today: the proof that all living things come from other living organisms. Yet this represents one of the longest, most involved, and most interesting research projects in the history of science. The theory of spontaneous generation held by men during the Dark Ages was finally disproved in the world of the educated. Still, in the dark corners of uneducated minds, there lurks this old misconception.

Francesco Redi (1626?-1697) tried to disprove spontaneous generation by a series of interesting experiments showing that substances wouldn't spoil at high altitudes if they had been previously protected from contamination. The final proof came when Louis Pasteur (1822-1895) showed that spoilage was caused by microorganisms. By so doing, he established a principle upon which biologists could build a unified concept of disease transmission. At the same time it led to the process we now know as pasteurization. No longer do intelligent people believe that decaying bodies generate flies or that swamps generate disease. Persons who now understand the nature of microorganisms, the production of spores, and the process of decay, find it hard to believe in the generations of new lower forms of life from dust and decay. Because of the former lack of understanding of these simple processes, we can see why the laws of evolution took so long to become established and why some people still refuse to accept them. The things that they cannot understand or that they are warned against by persons in whom they trust are not accepted. Once the concepts of biology are elaborated and unless the minds are completely rigid, the logic of evolution should at least be apparent.

The invention and refinement of the microscope was necessary before the universality of cells could be proved. In 1665 Robert Hooke (1635-1703) used one of the new instruments. He observed a piece of cork and for the first time described the empty spaces in cork tissue, calling them cells. The use of these lenses spread. More and more information was gathered. For instance, Marcello Malpighi (1628-1694) described the details of

Fig. 4-1. An electron micrograph of bermuda grass leaf showing clearly defined nucleus, cytoplasm, and cell walls. (Courtesy H. H. Mollenhauer, College Station, Texas.)

plant vessels. Then in 1682, Nehemiah Grew (1641-1712) made drawings of plants in which he depicted cells, although unaware of their structure and significance. It took almost two centuries for someone to realize that cells were not the empty spaces described by Hooke. In 1835 Félix Dujardin (1801-1860) recognized that each cell contained some liv-

ing material, and he labeled it sarcode. Today we call it protoplasm, a term coined in the early part of the nineteenth century by both Johannes Purkinje (pronounced poor' kin-yay) (1787-1869) in 1839 and Hugo von Mohl (1805-1872) in 1846. Both are given credit for using the term.

Matthias Schleiden (1804-1881) in 1838

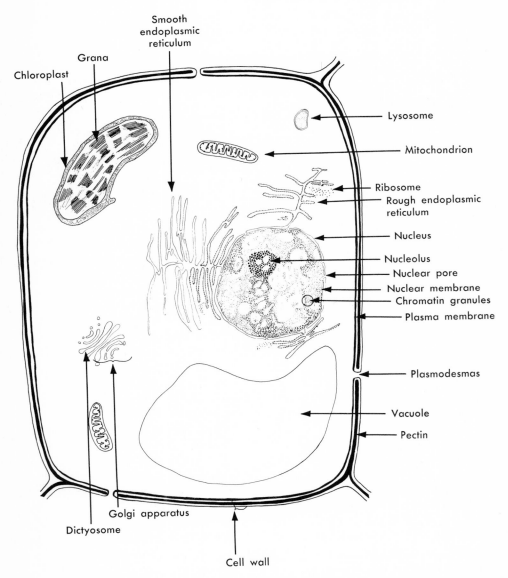

Fig. 4-2. Diagram of a typical plant cell showing chloroplast, mitochondria, Golgi apparatus, lysosomes, endoplasmic reticulum, plasma membrane, cell wall, vacuoles, and nucleus.

and Theodor Schwann (1810-1882) in 1839 independently proposed the cell theory, the theory that maintains that all life comes from cells and is composed of one or more cells as the operational unit. This was probably the first unifying concept of biology.

The world of atoms, molecules, chemical bonds, and energy exchange is transformed into the world of living substances when we focus on the cell. Although the contents of a cell are referred to as **protoplasm,** the term now has relatively little meaning when used to describe the living substance. Cells are never at rest, as can be seen in the photograph above. Even when they are not dividing, they are otherwise active chemically and physically.

Cells have a general structure and size (Fig. 4-2), but their physical appearance varies according to their specific physiological functions. Their chemical composition is much the same in general terms, but it, too, varies from moment to moment. This variation accounts for the great diversity of living organisms, yet it is the cell that is the unit of life.

Cell shape ranges from spherical, in the freely floating cell, to flattened, irregular, cylindrical, cuboid, rectangular, or spindle-shaped. Cell size in plants ranges as much as a million to one, the range in the size of plants is much greater. The smallest flowering plant is approximately 1 mm long (*Wolffia* sp., a floating aquatic plant) whereas the largest (the giant redwood tree of California) measures 360 feet and weighs at least 1000 tons, and some algae (see Chapter 6) are even larger.

Cells are classifed into two major categories based on their internal anatomy. The most primitive cell is the **prokaryon** (*pro-,* 'before'; *karyon,* 'kernel, nucleus'). This type of cell contains only one **cell membrane** system which is derived from the outer covering or plasma membrane. The genetic material (DNA) is not separated from the rest of the cell by a membrane. Examples of the prokaryotes are the bacteria and the blue-green algae (Chapter 5). The **eukaryon** (true nucleus) type of cell is characterized by containing a complex system of membranes that divide the interior of the cell into distinct regions. The most notable division is that of the **nucleus** and the **cytoplasm.** Many other **organelles,** such as **mitochondria** and **chloroplasts,** are also present and delineated by membranes. In addition, the eukaryon has a more complex internal molecular system than the prokaryon.

A cell may be described as a small mass of protoplasm (the cytoplasm and a nucleus) surrounded by a **plasma membrane.** Each cell, at least at the beginning of its life, is a complete unit, carrying on all life functions, but some cells of higher organisms, for example, cells in the stem of woody plants, are incomplete and do not live long after they are formed.

In cooperation with the cytoplasm the nucleus controls and regulates growth, development, and continued existence of the cell. Even though a cell can function without a nucleus for a short time, the denucleated cell will soon cease to function.

The nucleus cannot extend its control over an indefinite amount of cytoplasm, since as a cell enlarges, its volume increases as the cube of the radius of the cell, whereas that of the surface area of the nucleus increases merely as the square. Such a disproportionate increase in cytoplasm would soon upset the metabolism of the cell. However, a nucleus can increase its surface area by changing its shape or by doubling its chromosome number, thereby doubling its volume. Mature cells tend to maintain a constant nuclear-cytoplasmic ratio, and growing cells will divide; thus that ratio is kept below the maximum value.

Surface area also limits the cell's size. The substances required for metabolism can only pass in and out of a cell through its surface membrane. For example, nearly all cells require free oxygen, and if sufficient oxygen is to reach the center of a cell, the concentration outside the cell must be at or above a critical value. The amount of any material in a cell depends on the ease with which it can enter the cell, its rate of utilization within the cell, its diffusibility after it has entered, and the dimensions of the cell. Since the surface area increases by the square of the radius and the volume increases as its cube, any increase in volume is governed by the ability of the cell surface to provide for the entrance or exit of the various gases, water, and foods necessary for the demands of metabolism. These limitations can be overcome by cells when they abandon their spherical shape by flattening, elongating, developing large vacuoles filled with liquids, or otherwise increasing their surface area.

The rate of metabolic activity within the cell is a third factor affecting cell size. Small cells and small organisms metabolize more rapidly than do larger ones. Thus surface exchanges in smaller organisms must be accomplished more rapidly, and the cells must be smaller so that the amount of surface area compared to volume can be kept at a maximum. The cell contents are confined by the cell membrane, which is 0.1μ or less thick, forming an elastic but firm support. Small cells do not make efficient use of this support, whereas large cells could burst the cell membrane as a result of internal pressure. Therefore a balance must be maintained between the need for support and the need for adequate surface-volume relationship.

In summary, three factors govern size: (1) the nuclear-cytoplasmic ratio, (2) the ratio of the cell surface area to the cell volume, and (3) the rate of cellular activity or metabolism.

METHODS FOR STUDYING CELLS

Two approaches are used in cytological studies: (1) direct study using microscopes and photographs and (2) chemical testing of the composition of the cell. Although the discoveries of recent decades have greatly expanded our knowledge of cell structure and function, cytological investigation is still an extremely active area of research.

It is almost impossible at present to study structure and function simultaneously. By using special techniques, we can study the functions of a cell while the component parts

are in motion and chemical reactions are taking place. At the same time the composition and the position of a cell's parts will be undergoing alterations. Hence, it is not possible to study the details of structure at the same time function is studied. When a cell is stained so that its structure can be studied, the cell usually ceases to function. Consequently, at the time it is possible to study the structure, it may not be the same as it was while it still functioned. In other words, structure is not static but is as dynamic as function. Therefore there can be no such thing as the study of the structure of a living cell. Nevertheless, much can be learned of structure by investigating a dead cell at any one instant.

The observation of a living cell under the microscope reveals that protoplasm is a relatively transparent material, the components of which, when unstained, refract light to some and the same degree. One part cannot be distinguished from another in the living state. To overcome this problem, cells are stained with dyes, either in the living state or after the cell has been preserved in as lifelike a state as possible. Because the chemical composition of cellular components varies, differential dyes can be used to demonstrate the cellular components. In addition, a number of other techniques are available today, such as microsurgery, electron microscopy, and phase microscopy—all of which contribute to our knowledge of the cell and its makeup (see Chapter 2).

Remarkable advances have been made in our knowledge of the molecular biology of cells through X-ray diffraction, a technique making use of a beam of X rays brought into parallel lines by means of a defining slit. The beam passes through the molecule to be analyzed, and the rays are recorded on a photographic plate. The diffraction pattern shown is then interpreted in terms of the molecular size and structure. It was this technique that led to the discovery of the double helix structure of the DNA molecule. The chemistry and chemical activity of structures within the cell is determined by the technique of selective staining. The careful selection of stains and their use on specimens, fixed, or killed at various stages or on the intact living cell, will allow the assay of small quantities of material. In addition, the technique of radioautography makes it possible for one to follow a radioactive isotope into the cell and associate it with a specific structure. If a living plant cell is treated with radioactive thymidine (^{3}H), which is a precursor of DNA, the isotope will be associated with the structure that is actively synthesizing DNA. The cells are exposed to the isotope for various time periods and then sectioned. In the dark they are placed against a photographic plate. When the plate is developed, the position of the isotopes in the cell appear as black dots. By comparing this photograph with the cells as they appear under the microscope, one can locate the isotope within a specific structure.

THE CELL

> ### THE CELL
>
> *Cell wall and cell membrane*
> *Cytoplasm*
>   Mitochondrion
>   Plastids
>   Vacuoles
>   Centriole-like bodies
>   Lysosomes
> *Organelles in cytoplasm associated with*
> *protein synthesis*
>   Endoplasmic reticulum
>   Rough (granular)
>   Smooth (agranular)
>   Ribosomes
>   Golgi apparatus
> *Nucleus*
>   Nuclear membrane
>   Chromatin (chromosome)
>   Nucleolus

The items on the checklist of cell structure shown in the preceding study box are considered one by one in the following paragraphs.

The student should not forget that the structure and content of the living cell is constantly changing. These minute factories are sites of great activity and energy exchange difficult to describe in mere words. The best way to understand what is taking place is to imagine oneself as an observer on a tour of one of these factories. As one passes from place to place, he should remember that activity continues throughout the factory at all times.

On the outer surface of the cell is a permanent cell structure in the form of a **cell wall.** This fairly rigid surface is characteristic of most plant cells and one that makes them different from animal cells. The cell wall is secreted by the cell. It is formed in various stages during the cell's development, first as a jelly-like **pectin** compound, the middle lamella, which acts as a cell cement between cells.

The **middle lamella** is composed of pectin, which is a polysaccharide consisting of galacturonic acid units. When a new plant cell is formed, the first material laid down around the plasma membrane is the jelly-like pectin. The cell then synthesizes and secretes a primary cell wall between the plasma membrane and the middle lamella. The primary cell wall consists mostly of a polysaccharide composed of **cellulose** and about 10% protein. It is a thin rigid structure. The cell will then form a secondary cell wall in the same manner as the primary one, forming it between the plasma membrane and the primary cell wall. It may be thick or thin

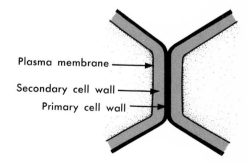

Plasma membrane

Secondary cell wall

Primary cell wall

Fig. 4-3. Diagram showing layers of adjacent cell walls and middle lamella.

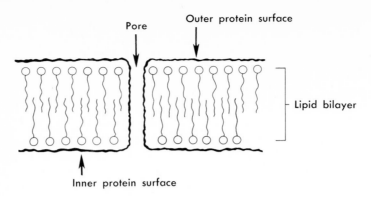

Fig. 4-4. Molecular model of Danielli's unit membrane.

depending on the nature of the plant. If the plant deposits materials such as lignin in the secondary cell wall, it will form a woody cell. Adjacent plant cells form cytoplasmic connections through the cell membrane and cell wall that are called plasmodesmas. The cell wall of fungi is composed of **chitin** and is much simpler structurally than the plant cell wall (Fig. 4-3). The cell wall is freely permeable to all dissolved substances and water. However the plasma membrane is selectively permeable, and it concentrates dissolved substances and ions within the cell. As a result water moves into the cell, causing it to expand, and presses the plasma membrane against the rigid cell wall. If it were not for the cell wall, the cell would burst. It is the osmotic pressure created against the cell wall that causes the growth of the plant cell. Because of its rigidity, the cell wall gives form to the plant cell and serves as a skeletal structure. Plant cells in particular acquire walls that give permanent shape and support to the plant. In addition, in some plants the surface cells accumulate a layer of cutin (waxlike material) that forms a cuticle on the exposed cell surface; for example, the shiny surface of a poison ivy leaflet is a cuticle.

CELL MEMBRANE

The cell, to remain a discrete unit, must have a boundary, but not solely to outline the cell. The cell, or **plasma, membrane** is a functioning structure composed of lipid and pro-

tein molecules. The outer living limit of the plant cell is the plasma membrane. The exact nature and structure of the plasma membrane remains to be elucidated; however there is sufficient experimental evidence to give us a concept of the structure and function of this membrane. Most of the research has been done either on unicellular organisms or the red blood cell membrane. It is composed of lipid and protein with a thickness of about 40 to 60 Ångstroms. Danielli and Davson proposed a model for the membrane based on their own and previous observations (Fig. 4-4). According to their model the membrane is a lipid bilayer composed of protein and lipid. The lipid bilayer is stablized by a layer of protein both at the outer and inner surface. Pores, coated with protein, are found at intervals in the membrane. The pores are extremely small and allow the passage of water but not larger charged molecules. Robertson observed that the membrane was consistently shown in the electron microscope as two dense parallel lines separated by a less dense area. These studies supported the Danielli model, the dense lines being protein and the less dense area being lipid. He also stated that the two surfaces of the membrane are different with respect to their protein content. Robertson said that all biological membranes were basically of this same construction, and he called the concept the **unit membrane.** More recently, Singer has proposed a fluid mosaic

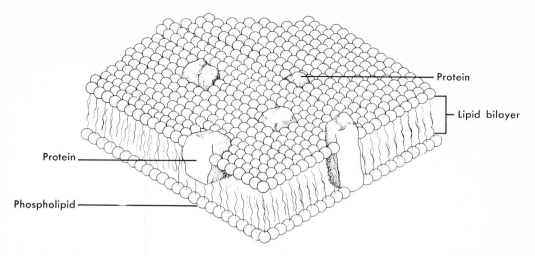

Fig. 4-5. Three-dimensional representation showing mosaic structure of the membrane. Both large and small proteins are embedded in the matrix. (From Singer, S. J., and A. L. Nicholson. 1972. Science **175:**320.)

model for the membrane (Fig. 4-5). According to this model, most of the membrane is in a lipid bilayer form, such that the globular proteins are attached to lipid molecules, with the proteins making the inner and outer surfaces. Studded in this lipoprotein complex are proteins or glycoproteins projecting out on either side. These proteins may move laterally but not vertically in the membrane. Over the past few years the concept of the membrane has gradually changed from that of a relatively solid, rigid structure containing the contents of the cell to that of a liquid or melted fat containing proteins that float and bob in it.

The proteins floating in the lipid bilayer give the membrane its specific functions. First, it functions by regulating the flow of molecules across it. It is semipermeable and some molecules pass through the membrane unimpeded but others cannot pass through at all or only very slowly. The membrane in addition contains systems that actively transport substances either into or out of the cell. The most studied active-transport system is that of the sodium-potassium pump. Second, it functions by regulating all of the interactions of each cell with other cells and with the environment.

CYTOPLASM AND CYTOPLASMIC ORGANELLES

Once inside the cell membrane, we see that the cytoplasm is the metabolic center of the cell, where its functins are executed. We emphasize that parts of the cytoplasm such as mitochondrion and chloroplasts also replicate. Therefore both the nucleus and the cytoplasm execute the twofold purpose of carrying out metabolic functions and self-perpetuation. Thus they are both interdependent; one cannot usually survive very long without the other.

The cytoplasm contains many organelles and inclusions that interact to give the cell its properties.

MATRIX

Various descriptions of the nature of cytoplasm have been given in the past. It was once believed to be a mixture suspended in the cell membrane, either with a changing structure or an unknown structure. Our present idea, because of the great resolution of the electron microscope, is that the cytoplasmic matrix is the basic molecular fabric of the cell and is composed of many small molecules, macromolecules (proteins and nucleic acids), and a complex ultrastructure. In

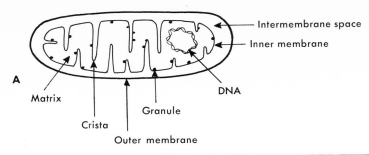

A

Intermembrane space
Inner membrane
DNA
Matrix
Crista
Granule
Outer membrane

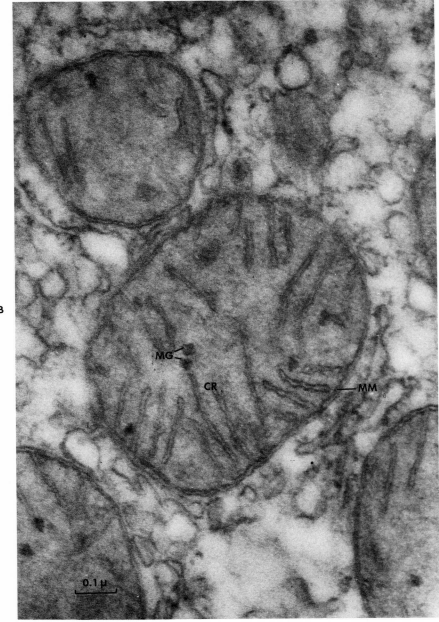

B

MG
CR
MM

0.1 µ

Fig. 4-6. A, Diagram of a mitochondrion showing membranes, cristae, and spaces. **B,** Electron micrograph of a mitochondrion.

the eukaryotic cells there are, in addition, many intracellular membranes (endoplasmic reticulum and Golgi apparatus) and cytoplasmic organelles such as **mitochondria, plastids,** and **ribosomes.** The matrix is colloidal in nature (see Chapter 3), and it is these macromolecules that carry out the biosynthetic functions and energy transfers of the cell. The movements of the cytoplasm during photosynthesis, cell cleavage, and the formation of fibers are all the result of matrix activity.

MITOCHONDRION

The mitochondrion is a spherical or boat-shaped organelle that is about the same size of a bacterium (Fig. 4-6). They are roughly 1 to 2 μ in length and about 0.5 μ in diameter, barely within the limits of resolution of the light microscope. The number of mitochondria per cell may vary from 1 to 100,000 depending on the function of the cell. Through the work of Palade and Sjö-strand with the electron microscope it has been demonstrated that it consists of two separate membrane systems. The outer membrane is continuous and it separates the interior of the organelle from the cytoplasm. The inner membrane is not connected to the outer one, and it forms many infoldings, called **cristae,** which appear as flattened discs in electron microscope studies. The space between the two membranes is called the intermembrane space, whereas the space enclosed by the inner membrane is referred to as the matrix. The intermembrane space appears structureless under the electron microscope. The matrix however appears granular and has many inclusions, including microtubules, crystals, fibrils and large granules. Also found within the matrix are fibrils of DNA very similar to bacterial DNA. The outer membrane is freely permeable to most small molecules and ions, whereas the inner membrane is impermeable to many small ions or molecules. The membranes contain specific transport systems to regulate the passage of such molecules as ADP, and long-chain fatty acids. The mitochondrion is a very dynamic flexible organelle capable of grow-

ing, branching, dividing, and coalescing in a time period of less than 1 minute. Functionally the mitochondrion is the powerhouse of the cell, producing the cell's energy in the form of ATP. It is the organelle of energy production. The systems responsible for respiration are found in the intermembrane space, whereas the systems causing carbohydrate and fatty acid oxidation are found in the matrix.

PLASTIDS

Plastids are globular bodies present in the cytoplasm of plant cells. The most important of these plastids (chloroplasts) (Fig. 4-7) contain **chlorophyll** and are the seat of photosynthesis in the green cell. The ultrastructure of the chloroplast is discussed in Chapter 13 with respect to photosynthesis.

Etioplasts are found in seedlings grown in the dark, and they can differentiate into chloroplasts. In addition they exhibit the internal membrane structure characteristic of the chloroplast. Other plastids are packed with materials other than the photosynthetic membranes and have a variety of names. **Leukoplasts** are colorless and when they store starch they are called **amyloplasts.** **Chromoplasts** are plastids that contain red or yellow carotenoid pigments and are usually seen in the fall. Chromoplasts can develop from the chloroplasts or independently of them.

ORGANELLES ASSOCIATED WITH PROTEIN SYNTHESIS

Within the cytoplasm there is an extensive system of membranes that appear to form a continuum with the nuclear envelope and the plasma membrane. This membrane system is called the **endoplasmic reticulum** (ER). The membrane is found in the form of tubules, vesicles, and large flattened sacs that extend throughout the cytoplasm. It does not appear to be as thick or as complex as the plasma membrane. The endoplasmic reticulum (ER) is found in two forms, rough and smooth (Fig. 4-8). The outer membrane of the rough ER is studded at regular inter-

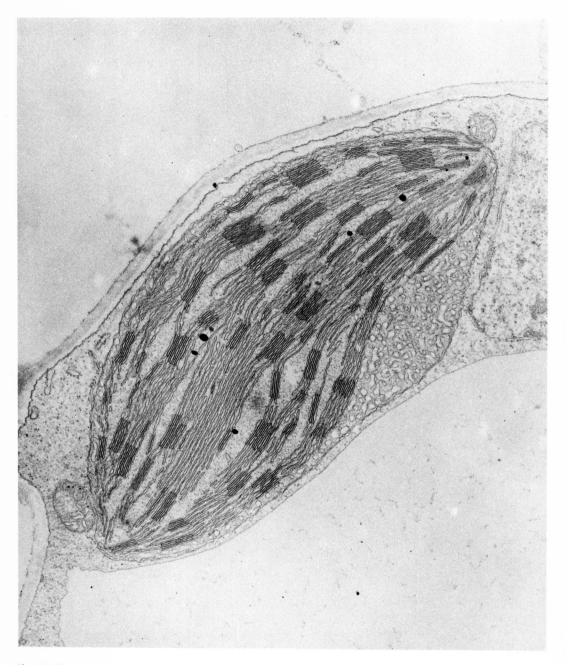

Fig. 4-7. Electron micrograph of a chloroplast in bermuda grass leaf showing stromal lamellae and grana. (Courtesy H. H. Mollenhauer.)

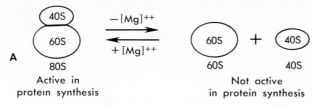

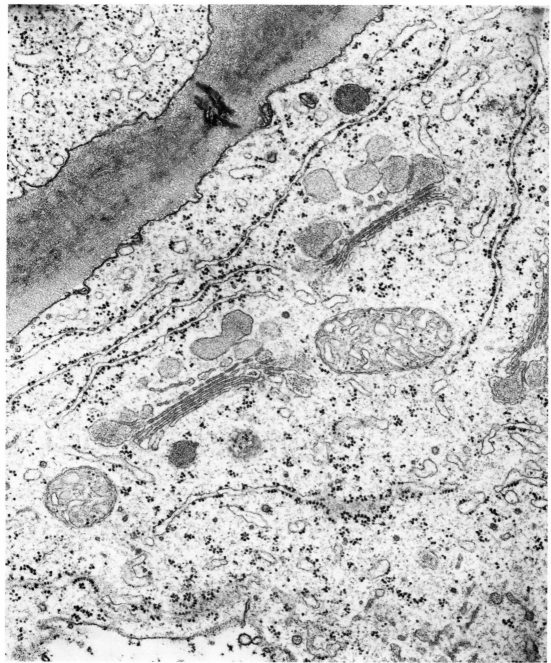

Fig. 4-8. A, Diagram of the dissociation of an 80S ribosome into 60S and 40S subunits. **B,** Electron micrograph of maize root-cap ribosomes on the membrane surfaces of the rough endoplasmic reticulum. (Courtesy H. H. Mollenhauer.)

vals with small granules of ribosomes, and this form is most often associated with cells actively engaged in protein synthesis. The smooth ER appears to be continuous with the rough, but it is smaller and tubular rather than vesicular. The smooth ER functions as a tube for the transport of newly synthesized proteins, and it has also been identified as

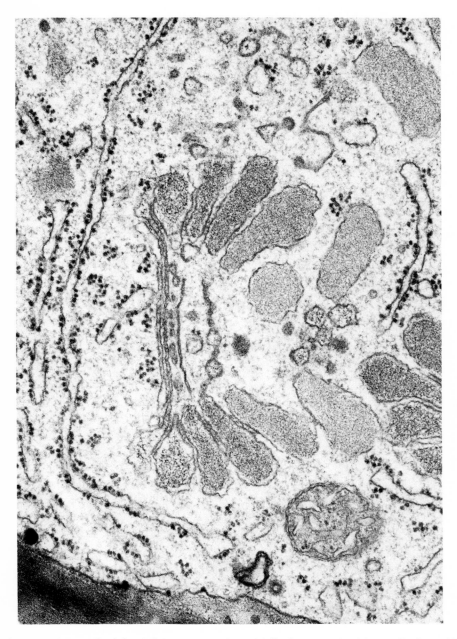

Fig. 4-9. Electron micrograph of the Golgi apparatus. (From Mollenhauer, H. H., Mouve, D. C., and Toten, C.: Protoplasma **79:**333-336, 1974.)

the site for lipid and steroid metabolism. The ribosome is the organelle directly associated with protein synthesis. It is composed of RNA and protein and is approximately 200 Å in diameter. It is classified according to its S value (**sedimentation-velocity coefficient**) in the ultracentrifuge. The S value is a measure of the size and density of a particle. In general ribosomes can be dissociated into a large subunit and a small one, neither of which are active in protein synthesis. Prokaryotic cells contain a 70S ribosome, which dissociates into 50S and 30S subunits, whereas eukaryotic cells contain an 80S ribosome, which dissociates into a larger subunit of 60S and a smaller one of 40S. Ribosomes in the cytoplasm may be free or attached to the endoplasmic reticulum. The free ribosomes are usually found associated in clusters. The proportion of ribosomes attached to the ER varies with time and is usually related to periods of active protein synthesis (Fig. 4-8).

The **Golgi apparatus** is a structure found in plant cells that resembles the smooth ER but is structurally quite different from it (Fig. 4-9). It is composed of a central core of slightly flattened, slightly curved discs, and individually called a **dictyosome.** Each disc is enclosed in a continuous membrane. At the periphery and on the concave side of the discs, many circular vesicles that have arisen by budding off of the Golgi membrane are found. The function of the Golgi apparatus is a secretory one. The structure receives newly synthesized protein from the smooth ER through small vesicles, called transition elements, that bud off the smooth ER and fuse with the Golgi apparatus (Fig. 4-10). It stores and secretes glycoproteins and mucopolysaccharides.

The **lysosome** is a vesicle formed by budding off of the Golgi-complex membrane. It is found only in a few plants, one example being the root cap cells of corn. The lysosome

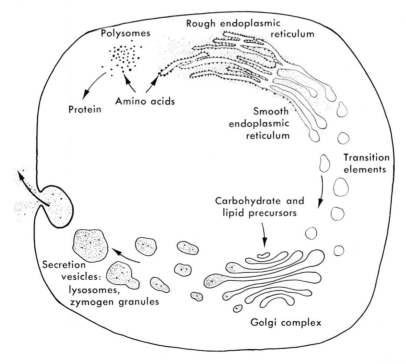

Fig. 4-10. Diagram of the integration of cytoplasmic elements in the synthesis of protein and the formation of secretion vesicles. (From Wolfe, S. L. 1972. Biology of the cell. Belmont, Calif.; reprinted with permission of Wadsworth Publishing Co.)

contains hydrolytic enzymes that degrade larger particles and molecules when they fuse with vacuoles containing such. The enzymes that it contains are capable of destroying the cell when they are released into the cytoplasm.

The mitochondrion and chloroplasts contain specific ribosomes within their structure and are apparently capable of carrying out protein synthesis independent of the cytoplasmic system.

CENTRIOLES

Centrioles are stuctures found in pairs in the animal cell cytoplasm. Each centriole is a small cylinder composed of a set of nine microtubules. The individual microtubules in turn consist of three tiny cylinders. Centrioles are always found in cells that possess either flagella or cilia and apparently generate the formation of these structures.

However, centriole-like structures are found in some primitive plant cells that produce motile spermatozoids and in the primitive gymnosperms such as *Zamina* and *Ginkgo*. These structures are functionally related to the development of flagella for the motile sperm. As the plants evolve from the motile sperm to the pollen grain carrying the simple nucleus, they no longer possess anything similar to the animal cell centriole. The animal centriole has a high degree of organization and ultrastructure, but the centriole-like body in plant cells has not been so elucidated.

VACUOLES

Most plant cells depending on age, size, and environment have **vacuoles** in the cytoplasm. Cells that are metabolically active and have thin cell walls have the greatest number and the largest size of vacuoles. The vacuole membrane, or **tonoplast,** is structurally very similar to that of the endoplasmic reticulum. It is therefore reasonable to assume that the vacuole develops from the endoplasmic reticulum. Vacuoles contain sugars, salts, proteins, and carbohydrates and are hypertonic to the cell, thus causing water to move into the cell and maintaining turgidity. Some plants utilize vacuoles to store plant pigments and waste materials.

NUCLEUS

The nucleus can best be described as the control center of the cell. It is usually spherical, appears densely granular, and is bounded by a double-layered membrane (Fig. 4-11). The outer layer is continuous with the endoplasmic reticulum and contains pores. The protoplasm in the nucleus is called nucleoplasm to distinguish it from the cytoplasm.

The granular material in the nucleus is called chromatin. When the cell divides, the **chromatin** condenses and forms discrete structures called **chromosomes.** The chromosomes then are the most highly organized form of chromatin. In the cells of eukaryotes the chromatin is composed of DNA, RNA, and protein. There are only trace amounts of RNA and about twice as much protein as DNA. The RNA is probably functioning in the transcription of a specific gene. The DNA carries all the information for the development, maturing, and functioning of the cell. The proteins are classified into two categories, **histones** and **nonhistones.** The histone proteins can best be defined by their net positive charge, because of the presence of the amino acids arginine and lysine. DNA has a net negative charge; so the histones can bind readily to the DNA. There is as much histone in the nucleus as there is DNA. The histones are synthesized in the cytoplasm and transported to the nucleus for combination with DNA. There are slightly more nonhistone proteins in the cell than histones. Compared to the histones, the nonhistone proteins are found to be of a much greater variety. They are associated with DNA in a rather dynamic fashion when compared to the histones, which have a rather stable association. Nonhistone proteins are also synthesized in the cytoplasm and transported to the nucleus. It has become apparent that histones function by preventing the transcribing of DNA into mRNA and also by determining

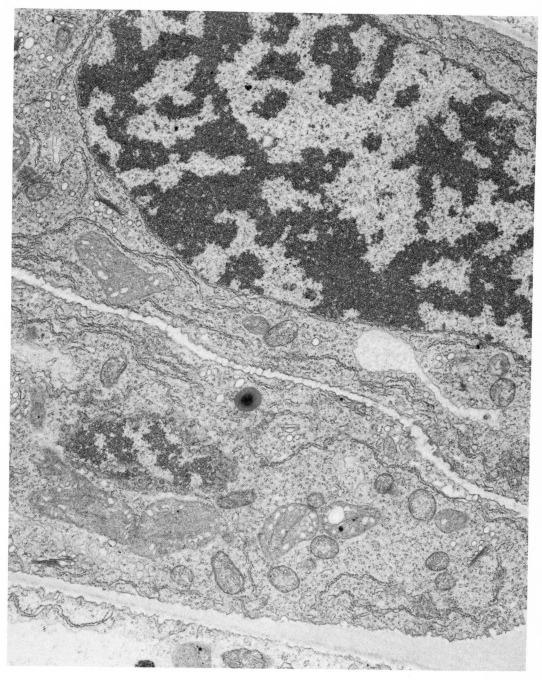

Fig. 4-11. Electron micrograph of a young leaf guard cell of *Vicia faba* showing chromatin granules and the nuclear envelope. (Courtesy H. H. Mollenhauer, College Station, Texas.)

the structure of the chromosome by not allowing the DNA double helix to coil so tight that it could not be transcribed. According to recent research the nonhistone proteins apparently function by regulating the transcription of certain genes.

NUCLEOLUS

The **nucleolus** is a spherical body found within the nucleus that is one fifth to one sixth the size of the nucleus. It is not bound by a membrane. A nucleus may contain more than one nucleolus, and these nucleoli are associated with specific areas of certain chromosomes. The nucleolus is absent in prokaryotes and is found in about the same form in both plant and animal cells. In the light microscope it appears to be divided into two distinct regions—a central core region, which appears structureless and is called the **pars amorpha,** and a filamentous zone surrounding this central core, called the nucleolonema. However under the electron microscope a greater degree of ultrastructure was elucidated. It can be divided into four distinct regions. First there is a granular region consisting of small granules of RNA slightly smaller than ribosomes linked together by a filament and giving the appearance of beads on a string. This region is composed of RNA and protein. The second structural region is a filamentous one consisting of many long fibers that are somewhat indistinct and are composed of RNA and protein. The third distinct structure in the nucleolus is the filaments of chromatin that are intermingled throughout and are much more evident at the periphery. These filaments are somewhat less electron dense and are composed of nuclear DNA. The forth structural area is the background matrix, which is composed of protein and appears to be structureless. This area is somewhat more electron dense than the surrounding nucleoplasm.

Nucleoli are formed only in cells that contain a **nucleolar organizing region** on one or more of their chromosomes. It is clear that the nucleoli are the precursors of ribosomes that are found in the cytoplasm and bound to membranes. A cell that does not have the capacity to form ribosomes certainly cannot survive for any length of time because it cannot synthesize proteins. It has been observed that cells lacking the capacity to form nucleoli do not survive.

QUESTIONS AND PROBLEMS

1. Distinguish the primary from the secondary cell walls of plant cells.
2. Is the concept of a unit membrane compatible with the fluid mosaic model?
3. Define chromatin, chromosome.
4. What is the function of the mitochondrion and how does it relate to the chloroplast?
5. If a mitochondrion were stripped of its outer membrane, would it still be able to carry out its functions?
6. Explain the possible function of the histone and

nonhistone proteins found associated with chromosomes.

7. What is the relationship of the nucleolus to protein synthesis?
8. Discuss the function of vacuoles.
9. Discuss the function of the endoplasmic reticulum, Golgi apparatus, and lysosomes.

ADDITIONAL READING

Albersheim, P. April 1975. The walls of growing plant cells. Sci. Am. **232**(4):80. (Explains the relationship of chemical structure to some of the properties of the cell wall.)

Brown, W. V., and E. Bertke. 1974. Textbook of cytology, 2nd ed. The C. V. Mosby Co., St. Louis. (A discussion of cell fine structure, relating structure to function, an in-depth textbook on the subject.)

Dupraw, E. J. 1970. DNA and chromosomes. Holt, Rinehart & Winston, Inc., New York. (An excellent discussion of the relationship of DNA and its operation in the chromosomes.)

Loewy, A. G., and P. Siekevitz. 1969. Cell structure and function, 2nd ed. Holt, Rinehart & Winston, Inc., New York. (A good, easy-to-understand treatment of cell structure related to function.)

Stein, G. S., J. S. Stein, and L. J. Kleinsmith. Feb. 1975. Chromosomal proteins and gene regulation. Sci. Am. **232**(2):46. (An excellent discussion of the role of histone and nonhistone proteins in the regulation of gene repression.)

Stryer, L. 1975. Biochemistry. W. H. Freeman & Co., Publishers, San Francisco. (A textbook with an excellent discussion of ribosomes.)

Watson, J. D. 1968. The double helix. Atheneum Publishers, New York. (A best seller about the politics of the discovery of the structure of DNA.)

Wolfe, S. L. 1972. Biology of the cell. Wadsworth Publishing Co., Belmont, Calif. (A textbook of the structure and function of cells.)

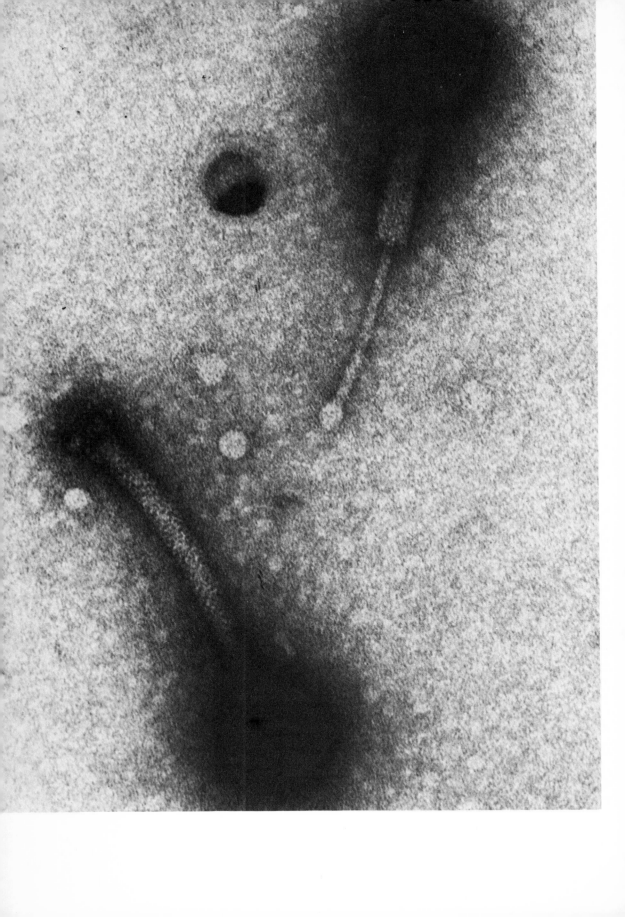

CHAPTER 5

Viruses, bacteria, and blue-green algae

■ The simplest forms of life now known to us have prokaryotic cells. These organisms are believed to be similar to the earliest forms of life on earth. The only living organisms showing this primitive form of cell structure are the bacteria and the blue-green algae. They do not usually reproduce by means of sexual reproduction, but there is some exchange of nuclear material in certain groups of bacteria. In addition to the bacteria and the blue-green algae, the **viruses** are discussed here. The exact placing of these submicroscopic structures remains undecided. Some biologists believe that viruses are living organisms operating as degenerate parasites, whereas other biologists view them as superchemicals. We will treat them in this chapter for convenience sake, as if they were a part of the kingdom Prokaryotae.

Traditionally the bacteria were included with the fungi and the blue-green algae with the algae. At the same time the viruses and the bacteria are studied together because of their great importance as disease-producing organisms. This is the major portion of the science of **microbiology.** Therefore, we are starting our review of the plants with this important group. They will be referred to many times in later chapters because many of the most significant biological processes depend on these forms.

KINGDOM PROKARYOTAE

The kingdom Prokaryotae may be defined as solitary, unicellular, colonial-unicellular, or filamentous organisms, rarely forming a mycelium, without nuclear membranes, usually without plastids or mitochondria, and sometimes with flagella. They are saprophytes or parasites, a few (blue-green algae and some bacteria) are **photosynthetic,** and some bacteria are chemosynthetic. They reproduce either asexually by fission, by budding, or by a simple prosexual process without differentiated sex cells. These organisms are usually nonmotile, but some have flagella, and others move relatively fast, compared to their size, by gliding.

DIVISION 1. Virulenta—viruses

Our inability to satisfactorily classify the viruses certainly does not make them of less interest. The division name Virulenta has been proposed as a formal term for use in a classification system. This has been adopted here with the full realization that it is very unlikely that the viruses represent only a single group. Until more is learned about their possible evolution no more may be said about their classification.

The agents assigned to the group are very simple, hardly much more than complex macromolecules. They vary from 10 to 200

Fig. 5-1. The AS1 type of cyanophage. The sheath of the tail contracts (upper specimen) exposing the core, which injects DNA into the host cell where it reproduces and then causes a disintegration of host cells of the blue-green algae. (×21,000.)

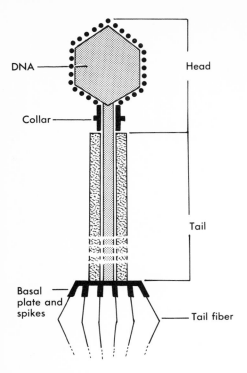

DNA

Head

Collar

Tail

Basal
plate and
spikes

Tail fiber

Fig. 5-2. Diagram of a phage drawn from one shown in Fig. 5-5. The head measures about 950 Å, the tail about 3000 Å, and the tail fibers 1500 Å. Note the protein coat, *large dots*, surrounding the head.

mμ in diameter (mμ = 1/1000 μ, or 1 nm); they are usually so small that they readily pass through a very fine porcelain filter. It is for this reason they were once known as "filterable" viruses. They are now well illustrated by numerous electron micrographs, but for a long time they were known only from their effects on tissues or bacterial cultures. Viruses are obligate cellular parasites; that is, in order to function they must be inside a living cell (Fig. 5-1). This property is not distinctive to them as a group. They are capable of infecting and destroying bacteria, fungi, plants, and animals. The infectious principle of the virus is attributable to a molecular complex of nucleic acid, either DNA (Fig. 5-2) or RNA (Fig. 5-3), and protein coat (capsid). The virus alternates its life cycle between an extracellular phase and an intracellular phase. In the extracellular phase it exists as an inert infectious particle called a viron. In the intracellular phase it consists of a replicating nucleic acid. The nucleic acid causes the formation of **capsid** subunits called **capsomeres,** which eventually will form the protein coat.

Several forms are no longer considered to

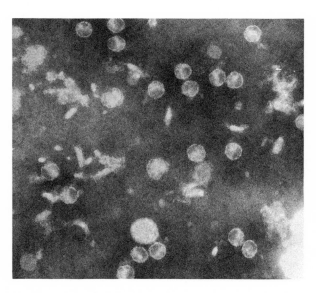

Fig. 5-3. RNA phase MS2. (×350,000.) (Courtesy Jeolco, Inc., Medford, Mass.)

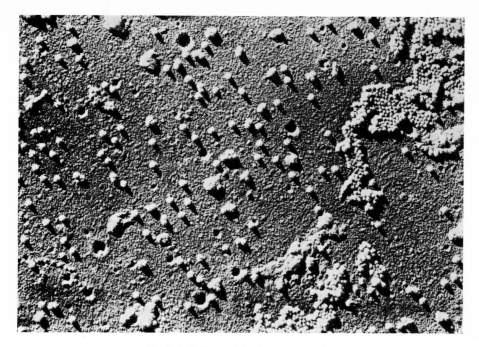

Fig. 5-4. Poliomyelitis virus. (×77,000.)

be viruses. Those removed and returned to the bacteria are the psittacosis agents (causative agents of parrot fever), Eaton agents (a **pleuropneumonia-like organism, PPLO**), the trachoma agents, and the rickettsias. These are all extremely small bacteria and therefore are confused with the viruses.

The viruses are divided into classes, orders, and families according to a new classification proposed in 1965 by the Provisional Committee for the Nomenclature of Viruses.

Viruses or bacteriophages deserve some further special mention because of the adverse effects of some and the indirectly beneficial effects of others. All known viruses are pathogenic because they require a living cell in which to live. Such well-known diseases as mumps, poliomyelitis (Fig. 5-4), and smallpox are caused by viruses. In this respect they are unique. However, both chemically and physiologically they resemble nuclei, which gives room for some interesting speculations. It has stimulated researchers in cancer to investigate the role of viruses in the abnormal cell division of malignancies.

Viruses causing disease in both plants and animals are barely more than complex nucleoprotein molecules. Some are even capable of crystallization and can be stored in jars indefinitely. Upon contact with a suitable cell, they will "come alive" and multiply.

The viruses that infect bacteria, and there are many, are termed **bacteriophages,** or **phages** (Fig. 5-5). Some viruses need to be "helped" by other viruses before they can replicate within the cell. These **defective viruses** seem to arise as mutants, and each has its satelite or **helper virus.** Tobacco necrosis virus is an example of this. Two sizes of particles are present, one, the larger, is the virus that lacks sufficient information to code all the proteins necessary for their replication. The helper virus, the smaller particle, seems to supply the code needed to complete the life cycle.

Bacteriophages were first discovered in 1896 when it was found that the water of the Ganges River contained organisms that would check the growth of bacteria. The electron microscope shows that there are

71

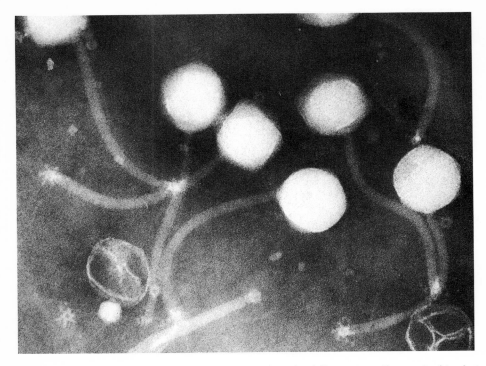

Fig. 5-5. D32 mycobacteriophages. (×400,000.) The head, tail, and tail fibers are easily seen in this photograph. (Courtesy Jeolco, Inc., Medford, Mass.)

minute structures resembling a coccus with flagella. Phages infect only bacteria, within which they reproduce and which they eventually destroy. When first discovered, it seemed likely that phages could be used to combat pathogens. Except in the treatment of cholera, all attempts in this direction have been unsuccessful.

A bacterium host cell for a phage is the cell whose metabolism is used for the growth and reproduction of a phage. One condition, termed a **lysogenic bacterium,** contains a prophage, a latent virus whose DNA is actually combined with that of the bacterium so that its genes are transferred with those of the bacterium at each cell division. These "strains" of bacteria eventually cause infection in others, and lysis takes place in the culture.

The multiplication of bacteriophages (Fig. 5-1) leads to the **lysis,** or breaking down, of the host cells, eventually causing **plaques,** round clear areas in the colony of bacteria. The lysis kills the contiguous cells through several cycles of virus growth (see Appendix B, life cycle 1), and the plaques are formed. Unfortunately, little practical application of this effect has been discovered. Bacterial disease treatment still depends on antibiotics.

The life cycle of a phage (Fig. 1 of Appendix B) is a good indication that these very small organisms are not so simple as was once believed. Two types of infections may take place, single or mixed. If a single **virion** attaches to a bacterium (Fig. 5-10), a simple growth cycle takes place. If more than one attaches, under certain circumstances, a "prosexual" or genetic recombination cycle may occur.

Certain viruses (Fig. 5-1) become important biological agents for controlling undesirable species of blue-green algae. Some species of these plants form an unpleasant scum on the surface of water. This decreases the oxygen content of the water, causing fish

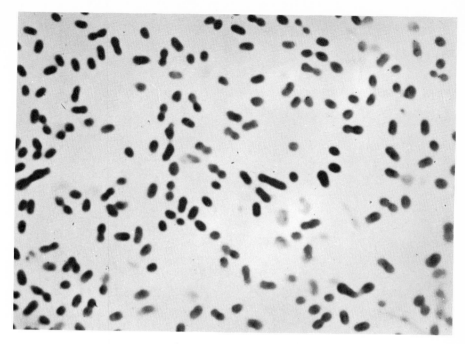

Fig. 5-6. Although the pneumococcus *(Diplococcus pneumoniae)* is the common cause of pneumonia, it is only one of several bacterial agents responsible for the disease. (Courtesy Chas. Pfizer & Co., Inc., New York, N.Y.)

kills as well as spoiling recreational water use. The viruses, called cyanophages or phycoviruses, were discovered in 1963. Attempts are being made now to use their dependence on, and host specificity for, the troublesome blue-green algae as a means to control or eliminate these pest plants.

With these few facts in mind we again ask, "Is a virus a living organism?" The recombination phase of mixed infections has been likened to sexual reproduction, and the growth phase to asexual reproduction.

DIVISION 2. Schizophyta—bacteria

The organisms assigned to the Schizophyta are all microscopic or submicroscopic in size. They vary from 0.1 to 20 μ in diameter. Most are very simple in structure (Fig. 5-6) always without readily demonstrable or distinctive nuclei. They feed on organic materials and are usually incapable of synthesis of organic materials from inorganic substances. Usually

they are not dependent on other cells for their life processes.

Because of the small size of bacteria, the structure of the cell appears to be very simple, without specialized organelles. The cell wall is a rigid layer composed of fatty substances, proteins, and carbohydrates chiefly polysaccharide in nature. Other combinations of these substances are used in the cell walls of the group. Sometimes the outer wall forms a thick, mucilaginous capsule that protects the cell from drying and makes aerial dispersal possible. The chromatin material is distributed throughout the cell in the cytoplasm. This material divides when the cell divides. In this manner it acts as a nucleus. When special preparations are made, some bacteria can be demonstrated to have a circular mass of chromatin. An exchange of chromatin material by conjugation has been observed in a few species. This is the genetic material, and in this respect it seems likely

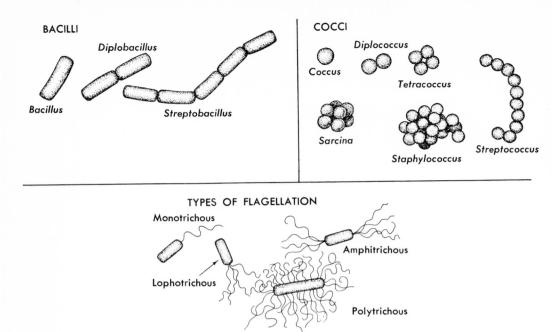

Fig. 5-7. Bacteria are differentiated by their shapes, colony characteristics, and appendages, as well as by their physiological reactions. Some shapes are diagrammed here.

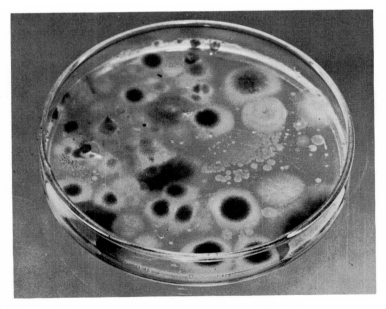

Fig. 5-8. Microorganisms of many kinds are grown in pure or mixed culture (as above) on agar plates. (Courtesy Chas. Pfizer & Co., Inc., New York, N.Y.)

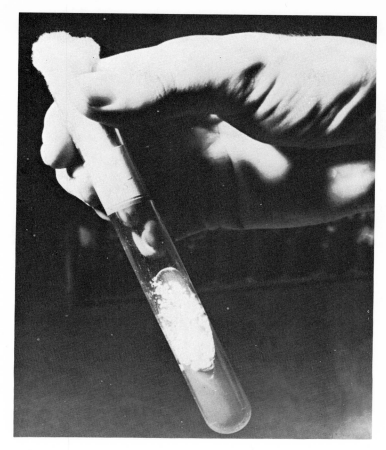

Fig. 5-9. Bacteria may be isolated from an agar plate (Fig. 5-8) and grown in pure culture on agar slants as is the case here. (Courtesy Chas. Pfizer & Co., Inc., New York, N.Y.)

that some form of sexual reproduction takes place perhaps in all or most species. Other details of cytoplasmic structure show that obscure vacuoles are sometimes present, and a few inclusions have been observed. Many bacteria have a strong affinity for basic stains, which further indicates the presence of nuclear materials.

The external anatomy differs in such a way as to be characteristic of groups of species. Some species possess flagella and are therefore motile. Others are motile only by the effect of Brownian movement on their small size. On the basis of their general shape, bacteria are classified as **bacillus** if rod shaped, **coccus** if a sphere, or **spirillum** if a spirally twisted rod (Fig. 5-7). Other details of struc-

ture vary within these groups. Sometimes cocci form chains or cubes in small colonies.

Reproduction. The usual means of reproduction in the bacteria is simple cell division or binary fission. This division is rapid in a suitable medium providing optimum conditions (Figs. 5-8 and 5-9). In 24 hours under optimum conditions a single individual would produce 281,472,976,710,656 descendants. Fortunately, conditions are never perfect, and this number is never reached. As just alluded to, certain forms have been observed to carry on conjugation. To this extent, they show sexual reproduction. When proper conditions are present, some species of bacteria are capable of forming spores in the following manner. The cell loses water

75

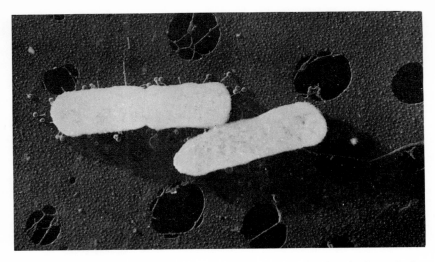

Fig. 5-10. Electron photomicrograph of two bacteria in conjugation. Note also the bacteriophages attached. (Courtesy Thomas F. Anderson, Philadelphia, Pa.; from Wollman, E. L., and F. Jacob. 1956. Sci. Am. **195:**109, July.)

and the protoplasm becomes separated from the wall, causing the entire cell to shrink. This is followed by the secretion of a very impervious cell wall, and often a change in cell shape occurs. Except in a few cases, spore formation is confined to the family Bacillaceae. These spores are resistant to great extremes of temperature, to many strong chemicals, and to drying. For example, anthrax bacillus can withstand boiling water for over an hour. Others will germinate after being immersed in liquid helium ($-269°$ C.) for several hours. Spores in the dry state have been known to remain viable for 30 years and more, awaiting the return of or the chance contact with a suitable environment. Under certain conditions bacteria are known to conjugate (Fig. 5-10), a process permitting the exchange of genetic material.

Colony formation. After division a daughter cell may remain in contact with the parent to form colonies of characteristic shapes. Bacilli often remain attached in pairs or form long chains. Cocci form pairs, chains, four-celled plates, cubes of eight, or large masses resembling a cluster of grapes (the most common type).

Physiology. Bacteria may be saprophytic,

chemosynthetic, or photosynthetic, but most forms are either saprophytes or parasites. Those types feeding on nonliving organic material (**saprophytes**) are responsible for the process known as decay. This may take place by means of a succession of species that are responsible for various stages of decay. Sometimes the result of this process yields useful by-products such as alcohol and cheese or nitrogen in a form available to higher plants. The parasitic species live on or within living plants and animals, where they may cause little noticeable damage or, in varying degree, several pathogenic conditions known as disease. Not a few species are symbiotic with other organisms; that is, they contribute something useful to the host organism while receiving protection and food material from the host. Examples of this include the normal flora of bacteria inhabiting the human digestive tract. They aid in the digestion and processing of food material by breaking down cell walls through the secretion of various enzymes. In return, the bacteria thrive on the otherwise unusable food material.

Some species are **chemosynthetic,** for example, the iron and sulfur bacteria that are capable of reducing compounds of these elements into useful materials for their metabolism. Thus they are independent of organic substances and probably represent primitive forms capable of living on the earth in early geological ages when sunlight was obscured by dense cloud layers. The sulfur bacteria play a prominent part in the purification of sewage and are found in all natural waters. The purple and green bacteria perform photosynthesis different from that of higher plants by the use of unique pigment systems found in the cell membrane.

Several families comprising part 1, the Phototropic Bacteria (see p. 79) carry on primitive photosynthesis, making use of a variety of pigments.

Bacteria in general have remarkable powers of food synthesis, which means that few substances are indispensable to them. In a general way their metabolic requirements may be calssified as follows:

1. Bacteria that can use ammonia as a source of nitrogen for protein synthesis, and simple carboxylic acids as a source of carbon for proteins, carbohydrates, and fats, the nitrogen-fixing bacteria (Fig. 5-11). These include common soil species.

2. Bacteria that require amino acids as a source of nitrogen. A high percentage of the species fall in this classification. Tryptophan seems to be the only amino acid indispensable to them. They can convert all other amino acids from higher plants and animals into those suitable for their own needs.

3. Bacteria dependent on a complex of nutritional needs, including many vitamins (vital amino acids). These are represented by a few specialized parasitic species.

Besides these special features of metabolism, bacteria have the same requirements as other forms of life, that is, water, oxygen, and carbon compounds. Not all bacteria require free oxygen, however. Some species known as **anaerobes** are capable of reducing other materials to obtain energy necessary for their

Fig. 5-11. Nodules on red clover caused by nitrogen-fixing bacteria. (USDA photograph.)

metabolism. In fact, such bacteria cannot live in the presence of free oxygen and are found only in the spore stage when not deep within tissues or other dense media. Other bacteria, dependent on free oxygen for their energy source (for respiration), are known as **aerobes.**

Waste products from metabolic activities are produced by the bacteria, as in all living organisms. Often these substances are poisonous to the host and are therefore known as **toxins.** Waste material is always poisonous in varying degrees to bacteria. If these products are not removed somehow, the bacteria form spores until the conditions change.

All physiological processes of bacteria are possible by diffusion within the cell and by osmosis through the cell wall, as in other organisms.

Ecology. Bacteria are universally distributed. Two features have permitted them to inhabit every place occupied by other forms of life: (1) extremely simple structure requiring exactly the same nutritional materials as other organisms and (2) spore formation for the preservation of the individual organism. They are found in soil as deep as 16 feet, in the upper atmosphere, and on and in all other organisms, even though movements by their own efforts are very restricted. Movement measurements record their speed at 7 inches per hour if these movements were sustained. Transposed to the size of man, these speeds would be magnified to the astounding rate of 90 miles per hour.

Classification. A natural classification of bacteria is extremely difficult to formulate because of their lack of morphological diversity and their apparently poorly organized hereditary mechanisms. Lack of stability of species to the degree commonly observed in higher organisms is perhaps the main reason for classification difficulties. Biologists are used to a reasonable amount of readily observable stability of species. Therefore attempts to define bacteria species in the same terms used for higher organisms is virtually impossible. This is not to imply that species do not exist in the bacteria. It simply means that they have a wider range of adaptability and therefore a more rapid evolution. Mutations occur frequently, and because of the simple nature of the bacteria, they are more likely to survive than are higher plants. However, it is well to think of the bacteria as a group with a fluid "gene pool"; that is, temporal species may be short lived, constantly changing entities. There is little doubt that they have a genealogical lineage like that of other organisms. It is equally certain that any possible attempts to demonstrate this phylogeny will be highly subjective because of the lack of any direct evidence.

It is no surprise, therefore, that the current classification systems are rapidly changing and do not meet with universal ac-

ceptance. The classification proposed in the eighth edition of *Bergey's Manual of Determinative Bacteriology* is followed in this text. This manual describes and classifies all known species of bacteria and rickettsias (and omits phages and viruses).

The rickettsias. The rickettsias are small bacteria and, like viruses, are dependent on living cells for their growth. Most rickettsias are transmitted by arthropods. Several diseases, including typhus and Rocky Mountain spotted fever, are caused by them.

A discussion of the details of the bacteria would be of little service here, since it would involve details beyond the scope of this text. However, in order that relationships between the various forms of common bacteria do not remain limited to a miscellaneous assemblage of disease organisms, we have made an abridged and annotated classification that should enable one to organize the bacteria in his mind somewhat (see opposite page). In addition, some particularly distinct groups are discussed below.

The gliding bacteria. This small group, sometimes called the slim bacteria, are separated from the true bacteria. The cells are rod shaped, flexible, and grouped into variously shaped fruiting bodies that attain macroscopic size. Most species occur in the soil; some are aquatic; and a few are parasitic. Because they move by gliding, they are classified as a distinct class. In many ways they resemble the Cyanophyta.

The mycelial bacteria. The members of the actinomycetes form filaments and some are stalked or branched. Some are pathogenic, but most are soil forms. However, this group includes the pathogens causing tuberculosis and leprosy.

The spirochetes. The spirochetes (Fig. 5-12) are thin, usually flexible, spiral, filaments without flagella, but with an undulating membrane attached along the body. All species are motile. One group of spirochetes is found in stagnant fresh or salt water and in the intestinal tract of mollusks. The other group is mostly parasitic. This group includes *Treponema pallidum*, the causative agent of

CLASSIFICATION OF KINGDOM PROKARYOTAE

Division 1. **VIRULENTA**—viruses

Division 2. **SCHIZOPHYTA**—bacteria (the total number of genera assigned to each part is given in each case, although some genera in some parts are only tentatively assigned)

Part 1. **Phototrophic bacteria** (1 order, 3 families, 18 genera)—sulfur, photosynthetic, and nitrogen-fixing bacteria

Part 2. **Gliding bacteria** (see text) (2 orders, 8 families, 27 genera)—cellulose-decomposing, algicidal, and certain fermentation bacteria

Part 3. **Sheathed bacteria** (7 genera)—iron- and manganese-oxizing bacteria

Part 4. **Budding and appendaged bacteria** (17 genera)—methane-oxidizing and denitrifying bacteria

Part 5. **Spirochetes** (see text) (1 order, 1 family, 5 genera)—causitive agents of the diseases syphilis and relapsing fever

Part 6. **Spiral and curved bacteria** (1 family, 6 genera)—aquatic bacteria that prefer a low oxygen concentration

Part 7. **Gram-negative aerobic rods and cocci** (5 families, 20 genera)—bacteria that are important in the mineralization of organic matter

Part 8. **Gram-negative facultatively anaerobic rods** (2 families, 26 genera)—includes *Escherichia coli* (colon bacteria) and *Yersinia pestis* (bubonic plague)

Part 9. **Gram-negative anaerobic bacteria** (1 family, 9 genera)—intestinal bacteria causing fermentation

Part 10. **Gram-negative cocci and coccobacilli** (1 family, 6 genera)—includes *Neisseria gonorrhoeae* (gonorrhea) and other disease bacteria

Part 11. **Gram-negative anaerobic cocci** (1 family, 3 genera)—parasites of man, pigs, and rodents, particularly of alimentary tract

Part 12. **Gram-negative chemolithotrophic bacteria** (2 families, 17 genera)—nitrifying, sulfur (some), iron, and hydrogen bacteria

Part 13. **Methane-producing bacteria** (1 family, 3 genera)—includes some colon bacteria and anaerobic bacteria found in swamps, lake sediments, and other anaerobic environments

Part 14. **Gram-positive cocci** (3 families, 12 genera)—anaerobic cocci living in the intestinal tract of animals and causing fermentation

Part 15. **Endospore-forming rods and cocci** (1 family, 6 genera)—includes *Bacillus anthracis* (anthrax) and other disease bacteria and *Clostridium botulinum* (botulism, food poisoning)

Part 16. **Gram-positive asporogenous rod-shaped bacteria** (1 family, 6 genera)—bacteria found in the human intestinal tract

Part 17. **Actinomycetes and related organisms** (see text; miscellaneous groups including 1 order, 9 families, and 37 genera in diverse arrangements)—lactic acid bacteria and many disease pathogens

Part 18. **Rickettsias** (see text; 2 orders, 4 families, 18 genera)—organisms that cause Rocky Mountain spotted fever and similar diseases of man, and many others

Part 19. **Mycoplasmas** (1 class, 1 order, 2 families, 4 genera)—pleuropneumonia-like organisms (PPLOs), which cause pneumonia, etc.

Division 3. **CYANOPHYTA**—blue-green algae

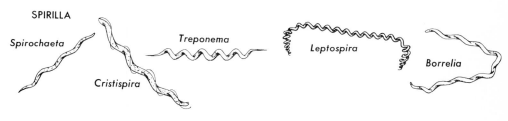

Fig. 5-12. Examples of spirilla.

syphilis. *Borrelia* spp. are the causative agents of relapsing fever and other diseases.

Some principles illustrated by bacteria. Several important biological principles are illustrated by the bacteria, most of which are found in other groups as well. One reason these principles are illustrated by bacteria in formal course work is the relative ease of culturing and experimenting with them. Standard methods have been developed so that testing can be done easily. When all or nearly all environmental factors can be controlled, it is then relatively simple to alter one of these factors and observe the results. Of all the living organisms studied today, this empirical method is best illustrated by the bacteria. In this respect, then, bacteriology is one of the few branches of biology close to becoming an exact science. However, this is relative because there are many perplexing problems in the field, not the least of which is the rapid mutation of species immune to antibiotics. Incidentally, as a result of these new drugs, a flourishing industry has been developed.

Germ theory of disease. Robert Koch (1843-1910) used bacteria to demonstrate the fact that germs cause **disease.** He first illustrated in 1876 that *Bacillus anthracis* was the cause of anthrax, a disease that was epizootic at the time in cattle and sheep and also occurred in man **(epidemic).** Koch isolated *Bacillus anthracis* from infected animals and succeded in growing the organism in a series of cultures made of serum. After many generations of growing in the serum, the organisms still had the capacity to cause disease when new animals were infected with it. The four steps followed by Koch have since become known as Koch's postulates:

1. Find the suspected organism in all cases of the disease and demonstrate its absence in healthy individuals.
2. Isolate and grow the organism in pure culture.
3. Reproduce the disease in suitable animals, using the organism from pure culture.
4. Reisolate the same organism from the reinfected animals.

Principle of pathogenicity. Disease-producing bacteria are certainly in the minority as far as the total number of bacteria is concerned. When a particular species of bacteria lives within another organism and causes sufficient damage to impair its normal function, the bacteria species is a **pathogen.** The damage to the host is usually the result of noncompatible waste material, toxin, given off by the bacteria. It may be either caustic in effect or a protein to which the body reacts. Various degrees of pathogenicity are expressed, dependent on several factors. The bacteria multiply and spread from organism to organism by various means.

Principle of disease. Two causes of disease are known: (1) metabolic or physiological disease, a malfunction of the body not directly caused by living organisms and (2) pathological disease caused by the invasion of the body by some living organism, which in turn does damage to the cells and tissues of the body and otherwise affects its proper function. Most disease is caused by the toxins of bacteria, that is, their metabolic wastes. These substances may be extremely poisonous to the body. The toxin of tetanus, for instance, is 200 times more poisonous than arsenic, 150 times more poisonous that strychnine, and 40 times more poisonous than the venom of rattlesnakes.

Principle of immunity. As a result of bacterial infection or the invasion of the body by a foreign protein (an **antigen**), an innate defense mechanism (an **antibody**) against the invading organism is brought into play. For each antigen introduced a specific antibody is developed that, by chemical union with the antigen, renders it ineffective. It is in this manner that immunity to pathogens is either present or built up in the animal body. When the antigen is a poison, it is called a toxin, and antibodies to it are called **antitoxins.** With the use of this principle, antitoxins can be induced to develop in certain animals, and the substance can be extracted and injected into human beings to provide an artifically built-up immunity. This is done in the prevention of diphtheria by growing an antitoxin in horses for later use in man. Vaccination is much the same, except that small doses of the organisms induce the manufacture of antibodies without the severe clinical symptoms of the disease itself, as is done in smallpox (a virus) vaccination.

Principle of parasitism. A parasite and a pathogen are the same. A distinction is made on the basis of the type of organism and its effect. The organism invading the body is the parasite, and it is parasitic. The result of this parasitism is pathogenicity when the effects become symptomatic. The process whereby the parasite enters into the relationship with the host is infection. A parasite causing pathogenicity is, therefore, a pathogen.

There are other characteristics illustrated by this group of organisms, for example, asexual reproduction and spore formation. However, the ones just described are those that are best illustrated by the bacteria.

Bacteria are frequently distributed in the spore stage and are able to multiply very rapidly once they find suitable conditions. This characterizes both the nonpathogenic and a few pathogenic species.

Disease-causing bacteria, fortunately, because of various bodily defense mechanisms and sanitary measures, often cannot find the proper conditions in which to multiply in large enough numbers to overcome the natural resistance of the host. Once they do, however, and the disease condition is established, they are dispersed from the host in large quantities either orally or by defecation; unless proper measures are taken, an epidemic is in the making.

Beneficial bacteria. Most bacteria species are beneficial to man either directly or indirectly. Their use in industry, particularly in the production of fermentation products, is well known. They also have a part in the making of cheese and butter. Curing and ripening of tobacco, tanning of skins into leather, and disposal of sewage are also dependent on bacteria.

From a biologist's point of view, the bacteria that cause decay play an important and necessary role in the cyclic changes of organisms. Nearly all decay is caused by bacteria or fungi or both. Dead plants and animals are immediately attacked by bacteria, since there are no longer any resistant elements to oppose their action. These bacteria live as saprophytes on the dead bodies. By a succession of species, eventually the entire body, plant or animal, is decomposed into the simplest inorganic and organic compounds. The effect of weather as well as scavengers is minor compared to the action of bacteria. If it were not for the effect of bacteria, dead plant and animal material would accumulate, making a shortage of the raw materials necessary for new growth. Thus the concept of "from dust to dust" is not only a fact but is also necessary for the continuance of life.

The well-known effect of nitrogen fixation by some of the primitive bacteria has a very beneficial effect in agricultural practices by building up the available useful nitrogen in the soil, one of the most necessary raw materials for growth. Thus the planting of legumes, which harbor nitrogen-fixing bacteria in a symbiotic relationship, is a routine procedure for the rejuvenation of the soil by nitrification. Bacteria also add to the fertility of the soil by their decay action. Old roots and similar organic materials are broken down into useful plant food.

DIVISION 3. Cyanophyta—blue-green algae

Blue-green algae are unicellular or filamentous organisms with cells having no well-defined nucleus. They occur mainly in the sea as a part of the plankton, but many species occur in freshwater pools, ditches, and stagnant water, and a few species are found in lakes or water reservoirs. Many are located in damp situations such as stream banks and in hot springs. The species found in hot springs can withstand temperatures as high as 85° C. These algae cause the trass color around the famous geyser in Yellowstone National Park, Old Faithful (Fig. 5-13). They are very abundant in soil, including desert soils. At least 20 species are capable of fixing atmospheric nitrogen, on land and in the ocean.

Fig. 5-13. Blue-green algae are abundant in the runoff from hot springs such as this one in Yellowstone National Park, Wyoming. Species of many genera may be represented, including such genera as *Oscillatoria*, *Chroococcus*, and *Microcystis*. The temperature of this water may be as high as 85° C.

In addition to these characteristics the cells and, in filamentous types, the thallus are always enclosed in a gelatinous sheath. Besides chlorophyll, the cells contain **phycocyanin** (blue) and carotene (orange) pigments, which often give them a blue-green color, from which they get their common name. All species do not show this feature however, but many species vary in color from blue to brown, or even red. These photosynthetic pigments are not located in definite plastids. The cell wall is very thin cellulose, but the surrounding gelatinous sheath may be very thick. Several cells or filaments may be surrounded by this sheath, binding the whole mass of cells into a single colony-like organism (Fig. 5-14). No species

of this group forms flagellated cells or gametes.

Their cell structure is very simple. There are no vacuoles or organelles of any type, except for granules of chromatin. Usually these are grouped near the center of the cell in an area of colorless protoplasm. The granules divide prior to cell division and represent the genetic material of the cell. Mitosis is atypical. The pigment material is distributed in the outer portion of the cell. Also present are oil droplets, small granules of glycogen, and other substances.

Reproduction. Reproduction is by simple cell division in the unicellular species. The filamentous species break into small segments of cells periodically. These are set free

Fig. 5-14. Colonies of *Nostoc* sp. form floating, jellylike balls. (Courtesy General Biological Supply House, Inc., Chicago, Ill.)

Fig. 5-15. *Nostoc* sp. filament showing hormogones.

Fig. 5-16. The vegetative cells in this filament of *Oscillatoria* sp. include several dead cells that are responsible for breaking it up into smaller segments (diagrammatic).

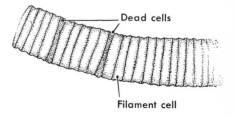

Dead cells

Filament cell

from the sheath. The new organism is capable of slow movement, but it soon settles down to the substrate to form a new filament. During adverse conditions hormogones (Fig. 5-15) are formed (Appendix B, life cycle 2, *Nostoc* sp.).

There are about 125 genera and 1200 species. Many genera are commonly encountered in the field. These all exhibit similar structures and may be readily recognized as members of this group. In the more primitive genera such as *Gloeocapsa* and *Chroococcus*, both nonfilamentous, the resemblance to bacteria is apparent. The more advanced genera such as *Lyngbya* and *Oscillatoria* (Fig. 5-16) show their relationship to the green, filamentous algae of the Plantae. *Nostoc* sp. forms complex colonies and has a primitive but definite reproductive cycle (Appendix B, life cycle 2). Such genera as *Hyella* and *Chamaesiphon* resemble some of the members of the kingdom Fungi. *Hyella* sp. grows within the shells of mussels, and *Chamaesiphon* sp. occurs on old filamentous *Oedogonium* sp., a green alga. The *Oedogonium* sp. occurs as a tangled branching mass of filaments; the *Chamaesiphon* sp. occurs as epiphytic cells producing endospores.

Economic importance. It is not infrequent that some species of this group of plants grow so abundantly in water reservoirs that they must be eradicated because the decaying plants cause diagreeable tastes and odors (see p. 72). Occasionally, lakes "blooming" with these algae cause the death of fish or render the water injurious to the livestock that drink from it, but this is not common.

Of the five genera of blue-green algae that are commonly involved in algal blooms and that are fairly well distributed, worldwide, only two—*Microcystis* and *Nostoc*—are reported to have species attacked by these viruses. Recently, Soviet scientists claimed experimental success in employing a virus isolated from the blue-green alga, *Plectonema boryanum*, to clear a *Microcystis* bloom in a large reservoir in the Ukraine.

QUESTIONS AND PROBLEMS

1. Why do we hesitate to call viruses living organisms? What was our definition of life in previous chapters, and do the viruses meet the requirements?
2. What two recent discoveries have enabled researchers to learn more about viruses?
3. How does the virus, like the common cold virus, cause disease?
4. How are viruses transmitted?
5. Name some common diseases caused by bacteria. Name at least one that is not.
6. Why are bacterial diseases, in general, not so serious as they used to be?
7. How does a disease-producing bacterium cause a disease? Describe the disease cycle.
8. List at least two features of the blue-green algae that are similar to the bacteria, and two that are different.
9. What great evolutionary principle was established by Pasteur using bacteria as his study organisms?
10. Why can we not consider bacteria as plants?

DISCUSSION FOR LEARNING

1. Discuss the reasons why you believe a virus is or is not a primitive organism. Include in your discussion reasons why you do or do not believe that a virus is a degenerate bacterium and therefore a living organism.
2. Some bacteria have become resistant to ordinary disinfectants. Discuss possible ways that this might have come about.
3. Bacteria are classifed principally by means of their physiological reactions, and the plants are classified by means of morphological characters. Discuss the reasons why both means of classification are in reality the same.
4. Discuss some ordinary health practices in relation to the previous discussion of these microorganisms, the bacteria. Consider also food preservatives in relation to the bacterial flora of the digestive tract.

ADDITIONAL READING

Breed, R. S., E. G. D. Murray, and N. R. Smith. 1974. Bergey's manual of determinative bacteriology. 8th ed. The Williams & Wilkins Co., Baltimore. (A complete classification of all bacteria.)

Clayton, R. K., and M. Delbruck. Nov. 1951. Purple bacteria. Sci. Am. **185**(5):68. (A basic explanation of the purple bacteria.)

Jacob, F., and E. L. Wollman. June 1961. Viruses and genes. Sci. Am. (A good discussion of the bacteriophage.)

Morowitz, H. J., and M. E. Tourtellotte. March 1962. The smallest living cells. Sci. Am. **206**:117. (A discussion of pleuropneumonia-like organisms.)

Pelzar, M., and R. Reid. 1972. Microbiology. McGraw-Hill Book Co., Inc., New York. (Introductory text.)

Stanier, R., Doudoroff, M., and E. Adelberg. 1970. The microbial world. Prentice-Hall, Inc., Englewood Cliffs, N.J. (An advanced in-depth textbook)

Wollman, E. L. July 1956. Sexuality in bacteria. Sci. Am. **195**(1):109. (A discussion of bacterial mating.)

CHAPTER 6

Algae

KINGDOM PLANTAE (Chapters 6 to 11)

The plants commonly referred to as algae may be easily recognized as a group even though they are diverse and represent six different divisions of the kingdom Plantae. This kingdom has been defined in Chapter 1. All of the plant divisions are described in this and Chapters 7 through 11.

The Plantae have a higher level of cellular organization than do the Prokaryotae. Plantae have eukaryotic cells; Prokaryotae have prokaryotic cells. There are many unicellular species, but most species are multicellular. Reproduction is both asexual and sexual. **Alternation of generations** is almost universal.

The algae have various photosynthetic pigments in several combinations, differing according to the division to which they belong. Only a few species lack these pigments and are heterotrophic. These species are unicellular and are classed with the animal phylum Protozoa in addition to their respective plant groups.

The size of the algae ranges from 0.5μ to 100 feet or more in some of the marine groups. In fact there are records of specimens up to 700 feet in length, making these the largest species of organisms. The plant body of the multicellular algae may be unicellular, filamentous, colonial, or a flattened grasslike or lettucelike blade. When it is multicellular, it is termed a **thallus** because there

are no true roots, stems, leaves, or complex reproductive organs. Leaflike organs may be present, but no true **vascular tissue** is present even in very large species. No true tissues of any kind are present, but the thallus may be complex in form even if the cells are all very similar. The sex cell–producing organs are always one-celled. The vegetative cells or growth cells are haploid. Therefore, the sexual stage is formed without the reduction of the chromosome number (**meiosis**). Instead, when the sex cells fuse to form the **zygote,** there follows immediately the meiotic or reduction division. Therefore, there is a **sporophyte** and a **gametophyte** stage meaning there is true alternation of generations. These stages are always distinct.

FEATURES USED TO SEPARATE THE DIVISIONS OF ALGAE

As stated previously, the type of photosynthetic pigments vary from division to division. This results in an obvious feature of each division, a distinctive color. The chemical nature of the reserve food products differ one from another. Carbohydrates are stored as starch, leucosin, laminarin, paramylon, and other forms. Flagella may be present. They are all of the 9 + 2 strand type; that is, in cross section there are two central strands surrounded by a circle of nine strands. Their number, length, and insertion varies from

---

Fig. 6-1. Payne Lake in Alabama with floating aquatic vegetation, including many kinds of freshwater algae, is an excellent example of a thriving aquatic community. The algae form an important part of the food chains in such an ecosystem. (U.S. Forest Service.)

group to group. The cell wall is often cellulose, but in many species it is pectic compounds or even protein. The composition of the cellulose differs in the various divisions as does its thickness and shape. This accounts for differences in body form and size. Specific differences are noted in the life cycles and forms of reproduction. The habitats of each division is noted in the following accounts. About 24,000 species of algae have been described.

DIVISION 4. Chlorophyta— green algae

The green algae have plastids containing primarily chlorophyll, with small amounts of **carotenes** and **xanthophylls** giving a bright green color. They produce motile reproductive cells with flagella. Reproductive cells are always borne in one-celled organs. The plant body may be one or many cells, and if many celled, it is filamentous.

Distribution. The green algae are aquatic, usually living in freshwater, but a few are marine. They are most abundant in freshwater ponds, lakes, and streams. A few species are attached to roots or debris, but some float and form so much plankton that they color the water. They are also found in brine lakes and as green snow in snow fields.

Structure and reproduction. Sexual reproduction occurs in most species. The gametes are formed in **antheridia** (male) and in **oogonia** (female) cells. Like all algae, these structures are always single cells, producing either a single female cell or sometimes many male cells. These may be released into the water where fertilization takes place, or the male cells, being motile, may go to the oogonium where fertilization takes place. In the asexual process small motile bodies termed **zoospores** are formed. They are released into the water, after which they swim around for a short time and then settle down for vegetative growth.

Classification. These algae are the most common forms encountered in freshwater habitats. More than 6900 species are known. The following list gives the classification of

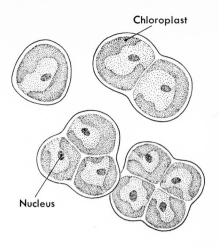

Fig. 6-2. *Protococcus viridis* is a green alga found growing usually on the north side of trees. Colonies of one, two, three, and four cells are shown.

the common forms and some examples, several of which are described in more detail in Appendix B.

Classification	Some genera
Order Volvocales	*Chlamydomonas, Pandorina, Eudorina,* and *Volvox*
Order Tetrasporales	*Tetraspora*
Order Ulotrichales	*Ulothrix* and *Protococcus* (Fig. 6-2)
Order Ulvales	*Ulva*
Order Schizogoniales	*Prasiola*
Order Cladophorales	*Cladophora*
Order Oedogoniales	*Oedogonium*
Order Zygnematales	*Spirogyra* and desmids
Order Chlorococcales	*Scenedesmus* (Fig. 6-3), *Hydrodictyon* (Fig. 6-4), *Chlorella* (Fig. 6-5), and *Pediastrum* (Fig. 6-6)
Order Siphonales	*Vaucheria*
Order Siphonocladiales	*Valonia*
Order Dasycladales	*Acetabularia*
Order Charales	*Chara*

Progressive development. Progressive development of the Chlorophyta is shown by primitive forms consisting of single-celled motile organisms that progress to single nonmotile cells. Others are united end to end to form a platelike mass, and the most advanced forms are filaments.

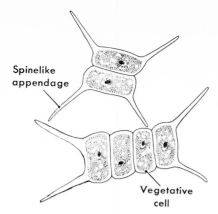

Fig. 6-3. Colonies of *Scenedesmus* sp., a green alga, are common in almost all standing water; this alga often appears in practically pure culture in aquariums and jars of water standing in a laboratory for some time.

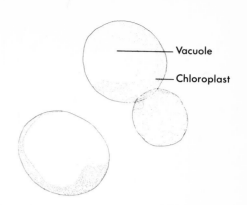

Fig. 6-5. *Chlorella* sp. A colonial alga widely cultivated for food as well as experimentation.

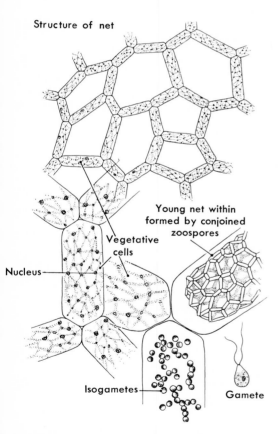

Fig. 6-4. The structure of *Hydrodictyon* sp. is netlike in appearance and often traps small fish in an aquarium. It reproduces by zoospores and isogametes.

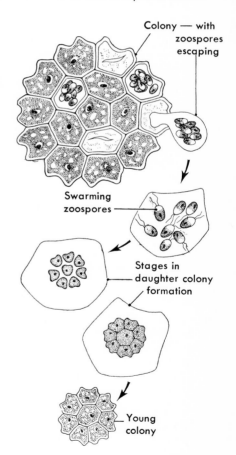

Fig. 6-6. *Pediastrum* sp. is a widely distributed colonial alga that grows free floating in ditches and in the plankton of lakes.

Examples of single-celled green form. *Chlamydomonas* sp., which is described in Appendix B, (life cycle 3), is a good example of this type of single-celled organism. These plants are very similar to the Protozoa of the animal kingdom, with which they may be classified as the family Chlamydomonadidae of the class Mastigophora. A colonial alga, *Chlorella* sp. (Fig. 6-5), is extensively used in the study of physiology of photosynthesis and respiration. These plants are nonmotile and have a very simple organization. They may be cultured like bacteria, and attempts have been made, particularly by the Japanese, to grow them commercially as a cheap source of food. One of the most interesting features of *Chlorella* sp. is that these plants may be induced to live in the bodies of the protozoa *Paramecium* sp. and *Stentor* sp. and of the coelenterate *Hydra* sp. By so doing, they live **symbiotically** and by virtue of the ability to carry on photosynthesis are able to manufac-

ture their own food. Thereafter, in *Paramecium* sp. at least, these plants are transmitted from generation to generation by cell division as a part of the animal.

Examples of motile colonies. Two common pond forms, *Pandorina* sp. (Fig. 6-7) and *Volvox* sp. (Fig. 6-8 and life cycle 4), illustrate motile colonies. These genera consist of species with groups of 16 cells (*Pandorina* sp.) or many cells (up to 20,000 in *Volvox* sp.). They differ primarily in their complex-

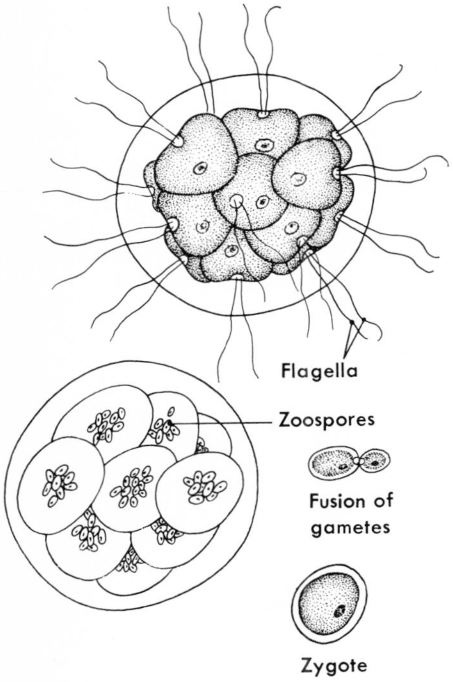

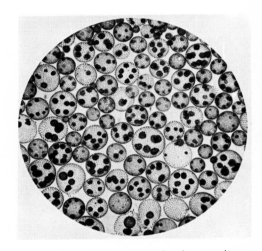

Fig. 6-8. *Volvox* sp. is an example of a complex motile colony. (Courtesy General Biological Supply House, Inc., Chicago, Ill.)

Flagella

Zoospores

Fusion of gametes

Zygote

Fig. 6-7. *Pandorina* sp. is an example of a simple motile colony.

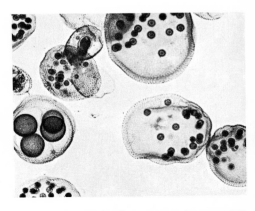

Fig. 6-9. Colonies of *Volvox aureus* show specialization of cell types. (Courtesy Carolina Biological Supply Co., Burlington, N.C.)

ity. *Volvox aureus* shows some cell specialization (Fig. 6-9) and a certain amount of food exchange, although in both of these exchanges, the individual cells morphologically resemble *Chlamydomonas* sp. and generally carry on independent physiological functions.

The cells in *Pandorina* sp. are somewhat pyramidal in shape because they are tightly packed in the colony so that their inwardly projecting portions are thus compressed. The whole colony is surrounded by a mucous covering, out of which project the two long flagella borne by each cell. Even in this simple association there is slight cell specialization. The colony has a definite anterior end, and the eyespots of the cells in this area are larger and better developed.

These genera belong to the plant family Volvocaceae, but they are also assigned to the Protozoa family Volvocidae of the class Mastigophora.

Examples of filamentous types. Many familiar pond species of this class are filamentous and are best represented by the order Ulotrichales. The filaments may be simple or branched. The individual cells are without flagella, except as reproductive forms.

Branching filaments and other chlorophytes. Many marine species of the chlorophytes grow rather complicated **thalli**. These plants (for example, *Canlerpa* sp. and *Bryopsis* sp.) are **coenocytes**—single, large multinucleated cells. The cell is differentiated into a main axis from which arise branches or pinnae. The lower portion of the axis forms a basal **rhizome** of branches that serves to anchor the plant. The plant body is one continuous vacuole joined by cytoplasm containing numerous, round chloroplasts and nuclei. Vegetative reproduction may take place by the breaking off of small sections of the plant. Sexual reproduction is effected by flagellated gamete formation in a gametangium formed from one of the small branches.

The familiar *Oedogonium* sp. and *Spirogyra* sp. are similar to *Ulothrix* sp. (life cycle 5) but are somewhat more highly developed. *Oedogonium* sp. produces **heterogametes**

and, after formation of the zygote (Fig. 6-10), **"swarm spores."** The process consists of rapid cell division of the zygote in which are produced flagellated cells that swim some distance from the parent filaments and serve to disperse the species. *Spirogyra* sp. (Fig. 6-11) carries on sexual reproduction by a process known as **conjugation** (Fig. 6-12). This consists of the passing of the contents of one cell through a conjugation canal formed between two cells in juxtaposition to unite with the neighboring cell. Both cells are identical, but the cell contents that move are considered to be the male. Movement is effected by the pumping action of many con-

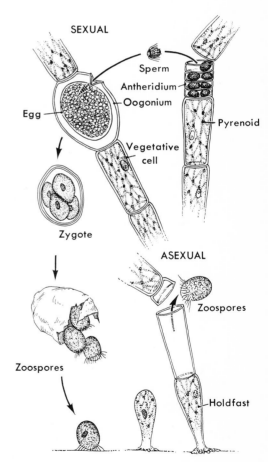

Fig. 6-10. Life cycle of *Oedogonium* sp. illustrates the initial development of heterogametes and "swarm spores."

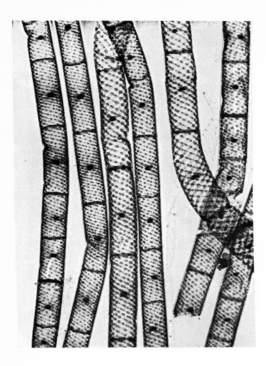

Fig. 6-11. *Spirogyra* sp., a familiar pond and laboratory plant, is characterized by spiral chloroplasts. (Courtesy General Biological Supply House, Inc., Chicago, Ill.)

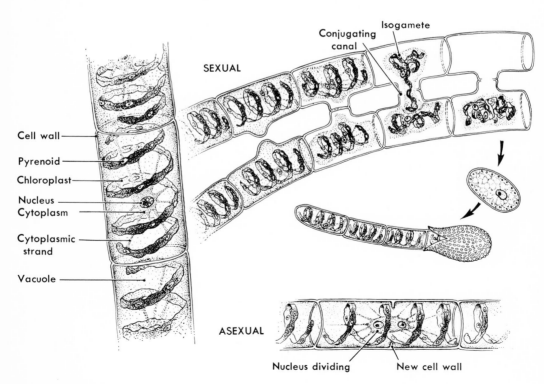

Fig. 6-12. Sexual and asexual reproduction in *Spirogyra* sp. diagrammatically shown.

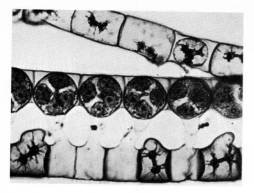

Fig. 6-13. Scalariform conjugation in *Zygnema* sp. (Courtesy Carolina Biological Supply Co., Burlington, N.C.)

tractile vacuoles. A closely related genus, *Zygnema* sp. (Fig. 6-13), reproduces like *Spirogyra* sp., by **scalariform conjugation,** which is conjugation between cells of different filaments. Conjugation between adjacent cells of the same filament is called **lateral conjugation.**

The most complex of the green algae are commonly called stoneworts because of the heavy calcium deposits in the cells. The habitus of these plants is so similar to certain aquatic vascular plants that they are often mistaken for the latter by the layman. The thallus consists of an erect, branched body differentiated into a regular succession of **nodes** and **internodes.** The latter branches at the nodes resemble small needlelike leaves. The cells contain chlorophyll and carry on independent metabolic process, with only slight food exchange. The leaves or branches are capable of limited growth, but the stem has unlimited growth and may also give rise to branches with unlimited growth. Each stem is organized in similar fashion, with nodes and internodes. The plants may be 1 to 3 feet or more in length and are anchored to the bottom of the pond or lake in which they are growing. The holdfast cells are colorless and appear to be rhizoids. They may form extensive subsurface meadows in favorable areas.

The common stonewort *Chara* sp., grows in freshwater pools, ponds, and small lakes.

Some species grow in brackish water. Asexual reproduction is affected by the production of vegetative structures of several types but not by zoospores. These structures are called **amylum stars, bulbils,** and **protonema**-like outgrowths. Amylum stars are star-shaped aggregaes of cells developed from lower nodes filled with starch. These break away and generate into a new vegetative plant. Bulbils are small nodules developed on the rootlike basal extremities of the plant, which eventually may break away and form new plants. Protonema-like outgrowths sometimes appear at a node and resemble small filaments of algae. When they are freed, they will mature into new vegetative plants.

Sexual reproduction is by the formation of gametes in multicellular antheridia and oogonia of fairly complex structure. Fertilization takes place in the oogonium. The resulting zygote forms an oospore, which is shed. This eventually germinates to form a small protonema and finally the young plant (life cycle 6).

The order Charales is a small one with about 6 genera and 250 living species. Many of these plants are also found as fossils, having occurred as long ago as the Devonian period, as evidenced by the calcium content in their cells. All extant species are assigned to the family Characeae, but three fossil families are also described in the botanical literature.

DIVISION 5. Chrysophyta—yellow-green and golden brown algae and diatoms

The plastids of this division contain carotenes and xanthophylls, giving them a variety of colors from yellow-green to brown. There are about 300 genera and 6150 species divided among three classes, each of which are often treated separately. They are Xanthophyceae, the yellow-green algae; Chrysophyceae, the golden brown algae; and Bacillariophyceae, the diatoms.

Class Xanthophyceae—yellow-green algae

The plant body of the yellow-green algae is either unicellular or multicellular. Unicellular forms are flagellated, whereas the multicellular forms are filamentous and often multinucleated. All species have yellowish green chromatophores, which give them their characteristic yellow-green color.

The class is small, with only about 400 species, most of which are aquatic, occurring in freshwater. However, some are aerial, growing on moist tree trunks, damp cliffs, or with mosses, and a few occur in the soil.

In many species, structure of the cell wall is similar to that in the diatoms, that is, overlapping halves that fit together like a petri dish. Asexual reproduction is usually by means of zoospores. Sexual reproduction is known only in a few genera in which it is effected by means of free **zoogametes** of similar form (life cycle 7, *Vaucheria* sp.).

Class Chrysophyceae—golden brown algae

The plastids of the cells of the golden brown algae contain carotenes and xanthophylls, giving the plants a golden brown color. The cells may be uniflagellate or biflagellate but rarely triflagellate. Most species are unicellular, but some form small, loosely held together colonies or loose filaments. They rarely appear as compact, branching filaments. Sexual reproduction has not been conclusively demonstrated.

The class contains about 70 genera and 250 species divided into several orders and families. Most of the species occur in cool, freshwater that is low in calcium. The motile species often form a large part of the plankton of lakes. Others occur in cold springs and brooks.

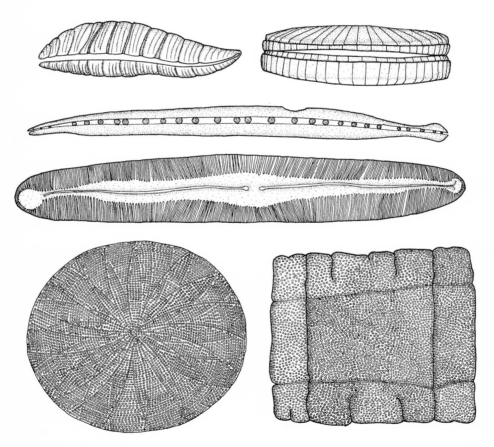

Fig. 6-14. A variety of diatoms from a sample of diatomaceous earth, some in girdle and others in valve view.

Reproduction is asexual by either simple cell division of the unicellular species or by fragmentation of the motile colonies. Non-motile genera produce zoospores formed singly within a cell. **Statospores** similar to those formed by the diatoms may be produced in some genera. These unique spores have silicified walls that form two overlapping halves and a pore closed by a conspicuous gelatinous plug, which enable the species to survive unfavorable conditions. Because of this type of spore the class is often placed in the same group as the diatoms.

Class Bacillariophyceae—diatoms

The diatoms are an ancient group of plants that form important deposits, called **diatomaceous earth** (Fig. 6-14), used in the manufacture of abrasive products, filters, and insultation.

The Bacillariophyceae are unicellular and rarely colonial, and the cells are surrounded by a two-part wall resembling a petri dish (Fig. 6-15). They are microscopic in size. The cell contents are difficult to see, but their geometric shapes make them easily recognizable. The cell wall is composed of silicon, which accounts for their presence in great quantities in fossil formations. The chromatophores are yellow or brown. Food is stored as fat or as a proteinlike substance, not as

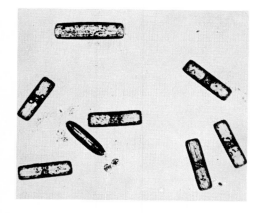

Fig. 6-15. Girdle and valve views of living diatoms. (Courtesy Carolina Biological Supply Co., Burlington, N.C.)

starch. There are about 5500 species, most of which are both living and fossil. A few are known only as fossils.

Asexual reproduction is by a cell division complicated by the necessity to secrete a new cell wall to match the one lost to the separated half of the cell. The new cell wall is secreted inside the old one so that with each successive generation the diatom grows slightly smaller. This small size is counteracted by the formation of sex cells.

Sexual reproduction is affected by a casting off of the cell wall. The cells then divide and form haploid ameba-like gametes. When they fuse, the zygote forms new cell wall halves, preserving the symmetry of the parent cell; thus they regain their original size. Sexual reproduction also occurs in other ways, such as by the formation of auxospores, reproductive cells formed by the fusion of two diatom cells. These may be produced by processes similar to conjugation in *Paramecium* sp., a protozoan.

DIVISION 6. Euglenophyta— euglenoid algae

The euglenoid algae are often treated as protozoa. The zoologist classifies them as the family Euglenidae of the class Mastigophora. Some are colorless and therefore dependent on organic food for energy and growth. Others, like *Euglena* sp. (Fig. 6-16), are equipped with chlorophyll and can therefore manufacture their own food. All members of this division are flagellated, and most of the species are motile. Some, however, form sessile colonies and thus are nonmotile. They are all unicellular and simple in structure. It is impossible to distinguish such organisms as either plants or animals. Undoubtedly this division and the animal phylum Zoomastigina have common ancestors. If all members of the group lacked chlorophyll, they would be treated as protozoa, but they are more advanced forms of algae and are therefore treated in this position.

Sexual reproduction so far is unknown in the group, although dubious cases of gametic union of vegetative cells have been reported.

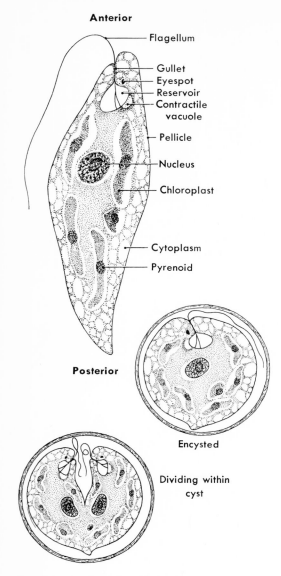

Fig. 6-16. *Euglena* sp. *Above,* Motile green alga stage. *Below,* Encysted stage. Note cell division for asexual reproduction.

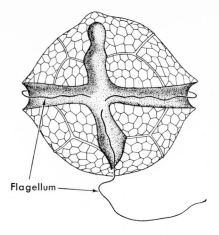

Fig. 6-17. *Peridinium* sp. is a chlorophyll-bearing unicellular alga having a platelike covering.

Asexual reproduction takes place by simple cell division; cyst stages are formed. Meiosis occurs apparently after the asexual fission of nuclei in a process termed **autogamy,** but this is rare.

These organisms are very common in small, stagnant, freshwater pools rich in organic matter. Their common use in biologi-

cal laboratories testifies to their abundance and rearing ease.

DIVISION 7. Pyrrophyta— flagellated algae

Members of this division are mostly motile, flagellated forms. A few are filamentous. They are characterized by a transverse groove or girdle within which one of the flagella is placed in repose. The other flagellum extends backward from the girdle, as in *Peridinium* sp. (Fig. 6-17). The chromatophores contain a brownish pigment in most species, but some lack photosynthetic pigments and are saprophytic or **holozoic** like the Protozoa.

There are only about 300 species. They occur in both marine and freshwater habitats but are rarely abundant. Occasionally *Gymnodinium* sp. becomes exceedingly abundant, causing a "red tide" and a severe disease of fish.

The cryptomonads are small, oval flagellates generally classed as protozoans and are found in foul water.

DIVISION 8. Phaeophyta—brown algae

The brown algae contain, in addition to chlorophyll, a brown pigment, fucoxanthin, which almost completely masks their green color. Unicellular, filamentous, and thalli-

Fig. 6-18. *Fucus* sp. sporophyte. Note air bladders and terminal receptacles. (From Russell, N. H. 1958. An introduction to the plant kingdom. The C. V. Mosby Co., St. Louis.)

Fig. 6-19. The brown alga, *Dictyosiphon* sp., occurs in the cold waters of the northern hemisphere. It grows to about a meter in height.

form species occur. The majority of species form the complex, relatively thick plant body commonly referred to as seaweed. This thallus is the most elaborate and complex of any group of algae. All but 3 of the 1000 or so known species are marine. The freshwater forms are very rare.

The common marine *Fucus* sp. (Fig. 6-18) or rockweed can be seen at low tide along the coast of the United States. The **kelps**

(*Laminaria* spp.) are familiar forms on bathing beaches. Another branching form, *Dictyosiphon* (Fig. 6-19) with its several species, is common in cold ocean water of the northern hemisphere. *Sargassum* sp. played a prominent part in the discovery of America. This seaweed forms such an extensive mat in the middle Atlantic ocean that it is known as the Sargasso Sea. Passing through this mat on his first voyage to the New World, Columbus convinced his sailors that they were nearing land, which gave them courage to continue the voyage.

Classification. The relationship of the various orders of this division is well known except for the order Fucales, which contains the common rockweed *Fucus* sp. The life cycle (life cycle 8) of this group does not include the free-living haploid stage common in the other members of the division. There are three orders, with over 200 genera and several hundred species.

Laminaria sp. is a still more advanced type of brown algae, commonly known as a kelp. These large algae are confined to cold marine waters where they often grow to giant sizes, as long as 30 to 50 meters. These large plants are very highly developed and have some indication of a vascular system in the form of **trumpet hyphae,** filaments with large cells inflated at the ends. There is a well-defined epidermis one or two cells thick, with numerous small cuboid **chromatophores.**

Unilocular sporangia are borne in **sori,** covering the leaf surface at certain times of the year. Haploid zoospores are formed and discharged during high tide. The zoospores, after a period of swarming, develop into small thalluslike gametophytes. Half of the zoospores develop into male gametophytes and half into female ones. Under proper conditions they form oogonia or antheridia. Each oogonium produces a single egg. Fertilization takes place in the oogonium, and the result is a zygote. The plant develops without leaving the oogonium, and the cycle continues (also life cycle 9, *Ectocarpus* sp.).

The fucus type of life cycle seems to represent the end of the algal line of development because of the near elimination of the gametophyte. This condition is not present again on the evolutionary scale until the flowering plants are reached.

DIVISION 9. Rhodophyta—red algae

The red algae are chiefly marine forms characterized by the dominance of the red pigment **phycoerythrin,** which masks the other pigments that are present. However, because of varying amounts of the blue pigment phycocyanin, many shades of red, pink, brown, purple, violet, and olive green occur.

Definition. The most characteristic feature of this group is their passive, nonflagellated sperm cells. Most species are multicellular and form a branching filament or a thallus of moderate size. Most of them are small, but a few reach a meter in length. The cells are usually uninucleated, with a large central vacuole. The cell wall is composed of cellulose and pectic compounds, from which is obtained commercial agar used medicinally or as culture media for bacteria. There are about 2500 species of red algae.

Vegetative reproduction by fragmentation is employed by many species, and in some of the primitive species cell division is the only known form of reproduction. Asexual reproduction is effected by the formation of various kinds of spores. Sexual reproduction is unique in this group, and special terminology must be used. The life cycles of many red algae are incompletely known. The life history may be rather complicated, involving the alternation of as many as three separate generations. The first, the gametophytes, are the large vegetative plants. These produce, on separate plants, **carpogonia,** and **antheridia,** which in turn produce the egg and the spermatium. The latter are released into the water to eventually float into contact with a carpogonium. The carpogonium has an elongate terminal projection, the **trichogyne.** When the spermatium makes contact with this structure, a sperm nucleus penetrates the cell wall and migrates to the egg nucleus. The resulting zygote divides to form the second generation, a parasitic diploid **car-**

posporophyte, which buds **carpospores** on tiny branches. They are liberated, float away, and form a third type of thallus, the third generation, the **tetrasporophyte.** When mature, these plants resemble the gametophytes. They produce, in special internal cells in groups of four, spores known as **tetraspores,** which are formed by meiosis and are therefore haploid. When these are released, they germinate and develop into gametophyte plants.

Several other types of life cycles are known among the species of the red algae. Variation is mainly evidenced by the elimination of the tetrasporophyte and carposporophyte stages.

Some of the genera, such as *Porphyra* sp. have no distinct alternation of generations. The vegetative plant or gametophyte produces gametes, which fuse to form the zygote. Meiosis takes place in the zygote stage, in which the zygote matures into a new gametophyte, so that the sporophyte is represented only by the zygote.

Nemalion sp. (life cycle 10) lacks the tetrasporophyte generation. The sporophyte is represented by the carposporophyte, parasitic on the gametophyte. However, it is haploid because of the formation of this generation after a series of divisions, one of which is meiotic, in the zygote. Therefore the diploid stage is represented again by the next zygote.

The life cycle of *Polysiphonia* sp. involves the three generations just mentioned. All three plant stages, gametophyte, carposporophyte, and tetrasporophyte, are macroscopic. The **gametangia** of the gametophyte are located on separate plants. In these the zygote divides to form the carposporophyte, but these divisions do not involve meiosis; therefore, the sporophyte is diploid. This plant, like *Nemalion* sp., develops from a cystocarp, a seedlike structure containing diploid carpospores. These eventually break away from the parent thallus and form new plants upon germination. The diploid tetrasporophytes resemble the haploid gametophytes. When mature, these plants produce tetraspores after reduction division

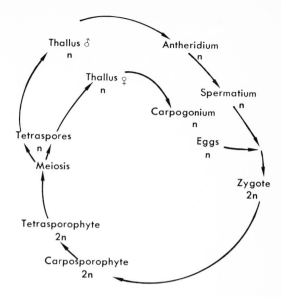

Fig. 6-20. Three alternate generations of the complex life cycle of the red algae.

has taken place. These, then, are haploid, and they in turn produce new haploid gametophytes. This type of life cycle is common in this phylum.

The three alternate generations of a complex life cycle of the red algae are shown in Fig. 6-20. Each generation represents a separate plant: (1) the gametophyte generation, with dioecious plants (male and female plants are separate), (2) the carposporophyte, the sporophyte that is parasitic on the carpogonium of the gametophyte, and (3) the tetrasporophyte, which is a free-living plant resembling the gametophytes.

The economic importance of the red algae is increasing because some species, for example, species of *Porphyra*, are used as food. The Japanese are consumers of millions of tons of this plant, grown in bamboo bundles in shallow water near the shore. The plants are used either dried or fresh and are cooked or made into soup. Others, such as species of *Chondrus* and *Rhodymenia*, are similarly eaten. *Gelidium* sp. is a source of agar, which is used in the manufacture of jellies and in a variety of desserts, as well as in the familiar agar plates used by bacteriologists.

BIOLOGICAL SIGNIFICANCE OF ALGAE

Although the algae are of significant economic importance, they are not currently an important part of the economy of man. They could conceivably become an important food source in the event that the land becomes overcrowded and not suitable for cultivation. They are extremely important as food for aquatic animals and are a part of many food chains, involving all kinds of marine and freshwater animals used as food by man. It goes without saying that they are likewise a part of the natural balance of all aquatic communities (Fig. 6-1).

From the viewpoint of the biologist they also occupy an important place because of diversity of form and the functions they exhibit. This single group illustrates so many lines of development that recent systematists have divided them into many divisions, and included them in two separate kingdoms. The ranking of the groups is not important to students of general botany.

The biological significance of these divisions may be summarized as follows:

1. The unicellular algae and the several animal phyla comprising the protozoa cannot be distinguished in a natural way. It is currently believed that the latter were an offshoot of the algae, and that these in turn gave rise to the remainder of the animal kingdom. The vascular plants are a direct line of development from the algae, with the fungi and bryophytes as sideline developments.

2. The various life cycles beginning with simple asexual growth and ending in the complex alternation of three generations suggest all of the reproductive processes found in the higher plants, plus many special forms unique to the algae. The essential features of these reproductive processes are (a) the formation of gametes having haploid nuclei, (b) the union of gametes by various mechanisms to ensure their fusion, (c) the development of spores for rapid vegetative multiplication and dispersal, and (d) motile stages for gamete dispersal to ensure a mixing of genes throughout the breeding population.

3. The development of the plant body from a simple single-celled plant to a large plant with specialized cells and tissues occurs in this kingdom and leads to the development of a vascular system. A larger size is possible by the secretion of gelatin-like or cuticle-like materials that can occur only in aquatic situations. The thickness of the thallus is never great, since it lacks an efficient vascular system.

4. Unlike most species in the plant kingdom, the gametophyte is dominant.

QUESTIONS AND PROBLEMS

1. What features do the divisions of algae have in common?
2. Where are most of the species found?
3. What pigments are present in the plastids of the algae?
4. Why does the term "algae" no longer have any classification significance?
5. What are the major characteristics used to separate the divisions of organisms formerly known as algae?

6. What is the role of these various divisions (as discussed in this chapter) in their biota?
7. What is seaweed?
8. Of what economic importance are algae?

DISCUSSION FOR LEARNING

1. Discuss the logic of the linear sequence for the "algae."
2. Review the various types of life cycles found in the "algae" and discuss which ones are the most successful and which ones are the least likely to give rise to new groups. Why?
3. Discuss the pros and cons of man's ancestors being "seaweeds."

ADDITIONAL READING

Boney, A. D. 1966. A biology of marine algae. Hillary House Publishers, Atlantic Highlands, N.J. (An account of the ecology and physiology of seaweeds.)

Leedale, G. F. 1967. Euglenoid flagellates. Prentice-Hall, Inc., Englewood Cliffs, N.J. (A careful consideration of the biology, structure, and classification of a significant group.)

Prescott, G. W. 1964. How to know the freshwater algae. William C. Brown Co., Publishers, Dubuque, Iowa. (A useful identification tool.)

Prescott, G. W. 1968. The algae: a review. Houghton Mifflin Co., Boston. (A comprehensive summary of the algae.)

Smith, G. M. 1933. The fresh-water algae of the United States. McGraw-Hill Book Co., New York. (See below.)

Smith, G. M. 1955. Cryptogamic botany. Vol. 1. Algae and fungi. 2nd ed. McGraw-Hill Book Co., New York. (The volumes by Smith are the standard references for the algae.)

Taylor, W. R. 1957. Marine algae of the northeastern coast of North America. 2nd ed. University of Michigan Press, Ann Arbor. (A useful guide for the identification of marine algae.)

Tiffany, L. H. 1968. Algae: the grass of many waters. 2nd ed. Charles C Thomas, Publisher, Springfield, Ill. (A nontechnical account of the algae.)

Fig. 7-1. A common woodland sight. Rapid decay of dead trees is aided by saprophytic fungi. The fruiting portion of this bracket fungus is here visible, but this represents only a very small part of the total plant because the mycelia are hidden from view, deeply imbedded in the wood of this tree.

CHAPTER 7

Fungi and lichens

■ The fungi are at the same time so universally distributed, yet so perfectly integrated into the maze of biological interrelationships, that the nonbiologist rarely senses their presence. Farmers, physicians, and ecologists know the importance of the group because of the plant and human diseases they cause, as well as their role in decay, soil formation, and food production. Except in more or less unusual circumstances (Fig. 7-21), the lichens are seldom dominant plants, yet they are important food sources for arctic animals, play a part in soil formation, and more recently have been shown to be indicators of air pollution. They cannot survive in hydrocarbon-polluted air.

Fungi have many important roles in every kind of ecosystem. Most species are hidden from our sight except when their fruiting bodies appear above the surface of the soil or decaying wood (Fig. 7-1) to cast their spores. Some also grow as epiphytes on leaves and stems of herbaceous plants and leaves, branches, and trunks of trees. The presence of **chitin** in the cell walls of fungi is a curious similarity to certain animal forms. This same substance, chitin, has enabled the insects to become the most numerous animals in species as well as numbers; perhaps this is also true of the fungi. All known species of fungi are saprophytic, parasitic, or in some cases, may start as parasites, kill their host, and continue their existence as saprophytes.

The fungi most familiar to everyone are those that cause food to mold. This fascinating group of organisms is a good example of a saprophyte. Several fungal diseases, particularly athlete's foot and ringworm, are also familar forms and examples of parasites of animals. As we will see, the fungi also are important in our lives as beneficial organisms. The yeasts cause our bread to rise and our grains and fruits to ferment. Other species, for example, *Penicillium*, help cure disease, and still others provide us with cheese, mushroom, and gourmet delights such as truffles.

The specialization for direct absorption of nutrients permits this radically different group of plants to maintain themselves in all aquatic and terrestrial habitats. As a group they are specialized, however, and are not really a fundamental advance over the algae because they have no structural complexities that cannot also be found in at least some of the algae.

Fungi may be defined as a group of plant divisions all with eukaryotic cells surrounded by cell walls. They have tissue differentiation, but they lack organ systems, meaning that there is no grouping of tissues to form roots, stems, leaves, and complex reproductive organs. The cells always lack photosynthetic pigments; thus they need to directly absorb nutrients. Their nutritional requirements include nitrogen and an array of minerals, all of which they obtain from living or dead hosts. Their body is usually composed of branched filaments; these are sometimes interwoven to form a complex body of large size; for example, some of the bracket fungi attain very large proportions. A few species are unicellular.

In the fungi the branching, filamentous thallus is known as a **mycelium,** a single branch of which is a **hypha.** Hyphae in the substrate upon which the fungi are feeding are called **rhizoids,** whereas those that connect one set of rhizoids with another are called **stolons.** The mycelium may be a multinucleated coenocyte with no transverse septa; when transverse walls are present, it is called **septate.**

Fungi have some characteristics they share with animals: the presence of chitin and the storage of "animal starch," glycogen, a carbohydrate typical of animals rather than plants. Fungi cell walls are composed of two main materials. First a polysaccharide, **glucans,** is laid down, and then another polysaccharide, chitin, is secreted farther behind and outside of the growing tip. Thus there is a flexible, penetrating tip of glucans covered by a rigid supporting substance, microfibrils of chitin. This structure enables fungi to penetrate even the hardest of woods and is the means whereby they are able to assume their several important roles in the ecosystem as saprophytes, parasites, and **symbionts.**

Fungi are rather difficult and challenging. It is necessary to grow the fungi so that all stages are available for study. Often critical microscopic examination is necessary before the species can be identified, which partly accounts for the fact that the group is poorly known, except for the economic species.

Classification. The six plant divisions grouped together comprising the fungi have over 40,000 described species. The origin of the group is still a matter of speculation. Few are known definitely as fossils, and their rather radically different mode of life obscures the evidence that shows relationships. The unique relationship peculiar to the lichens, that of two taxonomically distinct organisms, representing two different divisions, arranged into an obligate association with distinctive body form, makes their assignment in a classification system very difficult. However, they may be treated with the fungi, provided that one realizes that in each case the species of organisms involved in the association are at the same time classified in their respective groups. The deuteromycetes or imperfect fungi are also an unnatural group. All the species are temporarily assigned to a single group pending the discovery of further knowledge of their reproductive processes.

At one time the bacteria were considered to be fungi, but it is now clear that they are prokaryotes so that they are now treated as a separate group. It is now believed that the two groups originated independently.

Reproduction. The great variety of reproductive mechanisms in the fungi allows little more than a few general statements. Asexual reproduction is by means of spores, either produced from unspecialized hyphae or in specialized structures, the sporangia. All but the deuteromycetes reproduce sexually by the production of and union of gametes. The gametes may be motile; however, more often, the sex organs themselves are contiguous, and copulation of the gametes is effected by the growth of hyphae. In some cases the gametes depend on insects or water for fertilization. Because of the many structural differences of the sex organs and the complications in the life cycle, many special terms are used in describing the life cycles of fungi. These are related to the standardized system used throughout this book in the following sections and in Appendix B.

DIVISION 10. Myxomycota— slime molds

Slime molds are funguslike plants that pass through a vegetative stage called a **plasmodium,** a multinucleated mass of protoplasm having an indefinite, **ameboid** form instead of the thallus typical of fungi. Cell walls are completely lacking in this stage. However, reproductive stages are very different. Fruiting bodies and **spores** are produced like those of the true fungi. The definite cell walls covering the spores are the only convincing evidence that slime molds are plants. Even so, they are sometimes classified as animals. These organisms very likely arose from flagellated algae.

Classification of slime molds. Four classes of organisms are generally known as slime molds.

Class Myxomycetes — Plasmodial slime molds
Class Acrasiomycetes — Cellular slime molds
Class Labyrinthulomycetes — Cell-net slime molds
Class Plasmodiophoromycetes — Endoparasitic slime molds

Most of the 450 species live as saprophytes on moist, rotting wood or on moist organic soil. A few species parasitize seed plants and cause plant diseases. The group is undoubtedly very ancient in origin and is now widespread throughout the world. Many are brightly colored, the fruiting bodies particularly, but none have photosynthetic pigments.

Plasmodial slime molds: Myxomycetes

Members of this class are so named because they form a multinucleated growing body, the plasmodium. This is a flat, slimy body capable of slow locomotion, with a thick, veinlike arrangement that at first glance may be mistaken for branching filaments. The edges of the vegetative body produce pseudopods like those of amebas, and nutrition is accomplished in the same manner. The organic material is surrounded by the pseudopods and incorporated into the cytoplasm, where digestion takes place. All waste products are left behind as the plasmodium moves on.

Reproduction. Under suitable conditions, usually at the end of a rainy period, the entire nature of the plasmodium changes. When food supply or moisture is lessened, the plasmodium becomes round, thick and less transparent, and the color is intensified. If the plant has been living inside a log, it creeps out to an exposed surface, after which the entire body then rounds up into a shape characteristic for the species. Division of the plasmodium into distinct cells, each with single nucleus, takes place. When this transformation has been completed, the plasmodium changes to one or several sporangia, or fruiting bodies, which may be stalked or sessile. The capsule portion is a network of filaments, the **capillitium** (Fig. 7-2). The shape of the entire sporangium is characteristic of the species, in contrast to the uniform, nonspecific nature of the plasmodium. Fruiting bodies may be variously colored, and the surface of the capsule as well as the capillitium may be sculptured with a pattern peculiar to the species.

Spores are produced by the capillitium and are released as soon as they are mature. They

Fig. 7-2. Capsule portion of the slime mold is a network of filaments known as a capillitium. (Courtesy General Biological Supply House, Inc., Chicago, Ill.)

pass through the network of capillitia and are dependent on the wind for distribution. Dry conditions facilitate this dispersal, and their microscopic size permits them to reach even the upper atmosphere so that they are blown for great distances, all of which accounts for the widespread distribution of the species.

The spores may remain in a dormant condition for either a short or long period of time until they reach a suitable place for germination. On germinating, they give rise to naked swarm spores, which move about by means of two flagella, or in some species spores develop into small nonflagellated ameboid cells. Both types can swim about in the water of the soil or on moist wood. They behave as gametes, two of opposite strains fusing to form a zygote, which usually develops directly into a plasmodium, with only nuclear division instead of cell division. The developing zygotes may annex others to form a single, large plasmodium composed of several individuals. The plasmodium presumably is diploid; therefore, meiosis probably takes place at the time of spore formation.

The species that parasitize seed plants do not produce fruiting bodies but instead produce a mass of spores within the host. These in turn form zoospores, some developing directly into a new plasmodium and others, acting like gametes, fusing to form a zygote and then a new plasmodium.

Cellular slime molds: Acrasiomycetes

These organisms have been called "social amebas" because they start as independent, ameba-like cells in a swarm. They increase in size, divide, and continue to multiply for some time, probably feeding on soil bacteria. When conditions are right, an aggregation stage takes place. The amebas give off an evanescent substance that acts as an attractant, identified as cyclic adenosine monophosphate. They form into an easily visible mass. The mass of cells crawls about and orients toward a light source. Soon a central, cylindrical stalk forms, stiffened by cellulose secreted by the ameba-like cells, is formed by some of the cells. The remaining cells, still

acting independently, stream up the stalk and transform into small spores. These are eventually dispersed, and when they reach a suitable habitat, germinate, and release ameba-like independent cells. The cycle is repeated one or more times each season.

DIVISION 11. Oomycota— oosphere fungi

This is a group of water molds, white rusts, and downy mildews, some of which are serious plant diseases. They are characterized by a biflagellate zoospore, with one flagellum directed forward and one backward, borne in sporangia.

Classification. The division contains three classes as follows: Oomycetes, Chytridiomycetes, and Hypochytridiomycetes.

Class Oomycetes—oosphere fungi

The fungus body may be a unicellular thallus or a filament, some of which are profusely branched. Asexual reproduction consists of the formation of zoospores, which are, in the higher genera, retained within the sporangium, and it, in turn, functions as a spore and germinates to form the mycelium.

Sexual reproduction is usually by means of gametes of two types produced in two different kinds of gametangia. The formation of oospores give the group its name. These originate in an oogonium. The sexual cycles of many species remain unknown, but in one, the *Saprolegnia* sp. or water mold, it has been followed through (life cycle 11). The oospheres of this group are fertilized by nuclei of the antheridium formed from the same somatic hyphae. The zygote thus formed is termed an oospore. These eventually germinate to produce new mycelia.

Class Chytridiomycetes—true chytrids and related fungi

This group is characterized by the production of motile cells with a single posterior whiplash flagellum. They are found generally in water or soil, rarely as parasites on algae and some economic plants, or as saprophytes on dead organic material. Their life cycle is similar to that of the true molds.

**Class Hypochytridiomycetes—
hypochytrids**

Members of this division are freshwater or marine fungi. The presence of motile cells with an anteriorly placed flagellum is unique for fungi. They are parasitic on algae and other fungi, or some live on plant and insect debris in the water in which they are found. There are only about 15 species.

Economic importance. Several species are of great economic importance. Their role in preventing crop yields is of great concern and has resulted in costly control. The potato famine of Ireland in 1845 was the direct result of crop failure caused by a member of this class, *Phytophthora infestans*. Another species, *Plasmopara viticola*, threatened the wine industry of France by causing downy mildew of grapes. There are many other examples. The recent development of organic fungicides has resulted in more effective control measures for these plant diseases.

DIVISION 12. Zygomycota—
conjugation fungi

Most members of the division, the most common member of which is bread mold, have coenocytic hyphae that are filamentous, branched, and multinucleated. The sporangia produce a large, indefinite number of spores. The sporangia may be borne on special reproductive hyphae, where they produce either motile or nonmotile spores.

Classification. Two classes of Zygomycota are recognized. Most species belong to the class Zygomycetes. A small number of species are placed in a separate class, the Trichomycetes, the commensal fungi. These live in the digestive tract of insects and nematodes, where they are believed to be **commensals.**

Reproduction. Two generations occur in these plants: the asexual generation, which produces spores, and the sexual generation, which produces gametes. One more or less typical life cycle is described in Appendix B

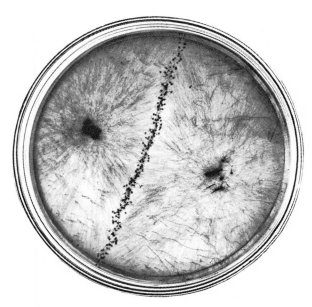

Fig. 7-3. *Rhizopus stolonifer* illustrates a fundamental aspect of sexual reproduction. It shows two sexually compatible strains of the black bread mold. Growing close to each other, they form gametes, and sexual union takes place. Such associations are responsible for the variety of life cycles. (Courtesy Carolina Biological Supply Co., Burlington, N.C.)

(life cycle 12, *Rhizopus stolonifer,* and Fig. 7-3). The spores are produced on special hyphae, the **sporangiophores,** as a branch of the mycelium or thallus of the sporophyte. Two types of spores are produced by members of this class: zoospores, which are uniflagellated or biflagellated, and **aplanospores,** which are nonmotile. Both are discharged into either the water or the air. All aquatic species and many terrestrial species have flagellated spores. Therefore these terrestrial species depend on rain for their dispersal. Spores germinate under suitable conditions and produce new sporophytes.

Sexual reproduction takes place by the production of unicellular gametangia (Fig. 7-4) characteristic of all algae and fungi. Some species discharge free-swimming gametes that unite in water to form motile zygotes. In others the gametes are retained in the gametangium. Fertilization takes place by union of the two gametes of opposite sex. True sexual differentiation does not occur in the group. The differences are physiological and are distinguished as plus and minus strains. The zygote remains dormant for some time. If it is uninucleated, it is then called a resting spore; if it is multinucleated, it is a resting sporangium. Meiosis takes place within the

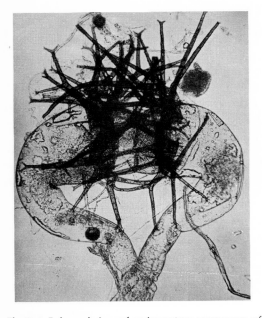

Fig. 7-4. Enlarged view of an immature zygospore of *Phycomyces* sp. showing the suspensions with branched outgrowths and fused gametangia. (Courtesy Carolina Biological Supply Co., Burlington, N.C.)

Fig. 7-5. The search for new cures for disease goes on. In this case the scientists are looking for antipathogens that may be found in the soil of the Sahara desert region. (Courtesy Chas. Pfizer & Co., Inc., New York, N.Y.)

developing zygote, so that the vegetative plant is haploid.

Biology. The primitive members of the division are saprophytes living on dung and decaying organic matter. Bread mold is among this group. Others are parasites on stored fruits and potatoes, some on plants and animals. The fly fungi are parasites of insects, and one of this same group is the cause of a serious disease of humans. At least one, *Mucor* sp., causes asthma and rhinitis (see Chapter 21). Finally, still others parasitize protozoans and nematodes.

DIVISION 13. Ascomycota—sac fungi

The ascomycetes are both saprophytic and parasitic. All species produce spores in sacs called **asci,** which will contain two, four, or eight **ascospores,** usually eight. Although a true mycelium is absent in yeasts and similar forms, when present, the mycelium is divided into sections by walls or septa. Motile cells are not produced by any of the species.

Among the familiar species in this class are the yeasts used in breadmaking, the black and green molds, the powdery mildews, the cup fungi, and the edible morels and truffles that are a real delicacy. Many species are of great economic importance, causing a number of plant diseases, whereas others are of great benefit in curing human disease (Fig. 7-5), for example, the green mold, *Penicillium* sp., which is responsible for the flavors of Roquefort and Camembert cheeses as well.

Classification. The ascomycetes comprise one of the largest groups of fungi, with about 30,000 species. The division is divided into two major groups: those that form ascogenous hyphae and those that lack these structures. The latter group is smaller, with only about 250 species, and includes the yeasts. The other subdivision has many orders.

Vegetative structure. The yeast and related groups form round or oblong nucleated cell masses that may be held together in poorly organized chains. The rest of the sac fungi form mycelia such as those of the phycomycetes and others. The mycelia are septate; that is, multinucleated portions are divided by crosswalls, which give the mycelium the appearance of being cellular. However, there is a pore in each of the septa large enough to permit a steady flow of cytoplasm throughout the entire plant.

Reproductive structures. The reproductive organs of the ascomycetes are varied and complicated. Special terminology is applied to the parts because they deviate from the normal for other plants. Asexual reproduction is by both simple and complex processes. Some yeasts reproduce asexually by a process of fission or equal division of the protoplasm. A septum is formed between the two more or less equal parts. Budding, also characteristic of yeast, is the constriction of a small portion of protoplasm from the mother protoplasm. This part may stay attached to form a small chain, but usually the buds become detached. Fragmentation may occur either naturally or artificially and results in the formation of new individual plants. When fragmentation is natural, **oidia** are produced. This is the separation of a hypha into individual, multinucleated cells. If the cells develop a thick wall before they separate, they are termed **chlamydospores.** However, the production of **conidia** is the principal method of spore formation. Conidia are produced in chainlike fashion at the end of special branching hyphae, the **conidiophores.**

Sexual reproduction is by the development of asci on the same mycelium that produces the conidia and also involves the association of two compatible nuclei brought together in the same cell by one of several methods. However, they do not fuse but instead remain in close association to undergo several divisions forming **dikaryotic** cells, with two closely associated cells that do not fuse. Nuclear fusion eventually takes place before the production of ascospores in the ascus. The process of asci development is very similar to that of sporangia or in this case conidiophores, except that there is also the close association of two strains of plants. The process involves bringing the antheridial nuclei to the ascogonial nuclei. The ascogonium produces a trichogyne through which the antheridial nuclei pass to pair with the ascogonial nuclei. Various modifications of these structures aid copulation in the different species. After the two nuclei are brought together, ascogenous hyphae grow out of the **ascogonium.** A pair of associated nuclei enter one of the hyphae, nuclear division by mitosis takes place, and an ascus mother cell is formed by the growth of a crosswall that separates the apical portion of the hypha from the ascogonium. Fusion of the two nuclei takes place to form a zygote. This is followed by meiosis, so that eventually eight haploid ascospores are formed. Meanwhile the adjacent sterile haploid hyphae have been stimulated to develop a mass of sterile hyphae that enclose the ascogenous hyphae and their asci. These fruiting bodies or **ascocarps** are of three kinds: the cleistothecium, a closed type, the perithecium, an ascocarp with an ostiolate opening, and the **apothecium,** an open, saucer-shaped ascocarp. All these vary with the species and are used in the description and classification of sac fungi.

Life cycle. The mycelium of a sac fungus develops from either an ascospore or a conidium. Only the latter are multinucleated. The mycelium develops conidiophores, which produce numerous conidia (Fig. 7-6) throughout the summer months, resulting in large numbers of asexual generations (life cycles 13, *Schizosaccharomyces* sp., and 14, *Microsphaera* sp.)

Later in the growing season the same mycelia that produce conidia will develop ascogonia. Nearby, an antheridium develops on the end of a special hypha, either on the same mycelium or on a separate one. Compatible nuclei are brought to the ascogonia, the formation of asci takes place, and ascogenous hyphae develop. The zygote is formed by the eventual union of the nuclei in the ascus mother cell. Meiosis follows to result in four haploid nuclei, each of which divides by mitosis to form the eight nuclei that eventually become ascospores. The as-

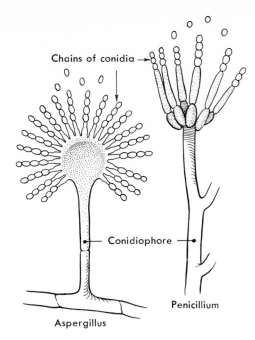

Fig. 7-6. Blue and green molds. Both *Aspergillus* sp. and *Penicillium* sp. produce conidia (spores) on upright structures called conidiophores.

cospores are formed by the development of a wall surrounding a portion of the cytoplasm near each of the nuclei.

Economic importance. The fermentation of sugar to produce alcohol is dependent on yeast. The entire wine and liquor business is built around this simple plant process. Baking makes use of yeast for the generation of the carbon dioxide that causes bread and cake to rise. It is probable that the sac fungi have been domesticated along with the grains and grapes needed for these foods and beverages. The economic importance of these relationships is obvious. Other yeasts occur on plants, in the soil, and in the body cavities of some insects, where they enter into a symbiotic relationship. A few are pathogenic, causing diseases of the skin and lungs.

The blue and green molds are cosmopolitan in their distribution. A few are parasitic and cause plant disease, but most of them are saprophytic, and several are very beneficial

Fig. 7-7. The green mold shown here is *Penicillium chrysogenum,* a mutant form from which almost all the world's supply of penicillin is obtained. (Courtesy Chas. Pfizer & Co., Inc., New York, N.Y.)

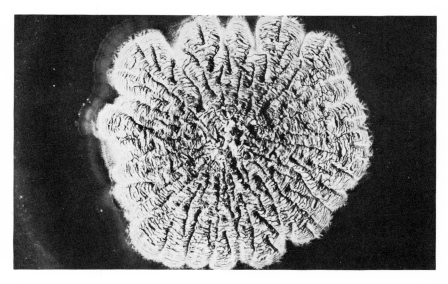

Fig. 7-8. A colony of *Streptomyces rimosum* such as this is the source of Terramycin, discovered and developed for treatment of disease in 1949. (Courtesy Chas. Pfizer & Co., Inc., New York, N.Y.)

Fig. 7-9. Masked technicians in sterile clothing operate freshly sterilized machinery to fill bottles with liquid antibiotics. The room in which they are working has specially filtered air and special lighting to maintain sterile conditions. (Courtesy Chas. Pfizer & Co., Inc., New York, N.Y.)

in the production of food and in medicine. As already mentioned, *Penicillium* sp., by its enzymatic hydrolysis of dairy products, is used in the production of cheese and for the production of high-grade citric acid. Members of this same genus cause storage rot in some fruits. Since World War II a new series of plants has come into use, the fungus plants such as *Penicillium* sp. (Fig. 7-7), and *Streptomyces* sp. (Fig. 7-8), the organisms producing terramycin and others. These plants are grown by drug companies, using special culture methods (Fig. 7-9). The antibiotic action of penicillin, now well known, was accidentally discovered by Fleming in 1928, but not until 1940 was it used as a medicine. The inhibition of enzymes that prevent the cell from forming a wall accounts for this remarkable action. At the same time penicillin is nontoxic to human beings. Production of penicillin and related drugs has become a large industry in the United States and other countries. A closely related fungus, *Aspergillus* sp., also has antibiotic properties, but the substances it produces are pathogenic to man. Several are the causative agents of diseases of the ear and lung.

Powdery mildews such as *Erysiphe gra-* *minis* affect various grains, and *Erysiphe cichoriacearum* affects some fruits and vegetables to cause great loss to farmers by damaging these products and bringing about low production. The ergot fungus *Claviceps purpurea* and others are parasitic on cereals, causing crop loss. In times past, grains used from infected plants caused **ergotism** in man and livestock because of the presence of alkaloids. Holy fire, epidemics of which are described in ancient literature, was presumably a disease caused by these fungi. The drug ergot, obtained from infected plants, is sometimes used in obstetrics to produce contractions of the uterus.

The genus *Neurospora*, which belongs to an order related to the powdery mildews, has become well known as an experimental organism in genetic research (Chapter 18). Actually it is a bakery mold known as red bread mold, which infests bakeries and causes considerable damage.

DIVISION 14. Basidiomycota— club fungi

The spores of club fungi are produced on the outside of special spore-producing bodies, the **basidia,** and are called **basidio-**

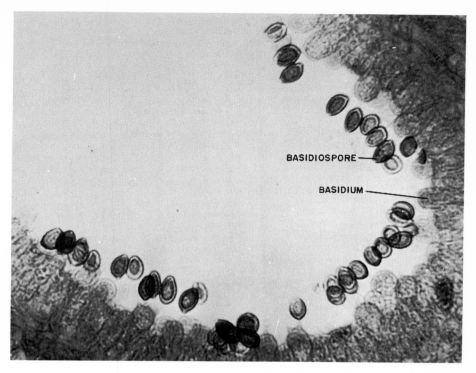

Fig. 7-10. Cross section of the fruiting body of a club fungus showing the basidia with sterigmas and basidiospores. (×530.) (Courtesy George H. Conant, Triarch Products, Ripon, Wis.)

spores (Fig. 7-10). They are uninucleated and haploid and are definite in number, usually four per basidium. The club fungi are macroscopic, forming a mycelial mass having a characteristic shape and growth on plants.

The club fungi are commonly called smuts (Fig. 7-11), rusts, mushrooms, deadly amanita, bracket fungi (Figs. 7-12 and 7-13), puffballs (Fig. 7-14), jelly fungi, and stinkhorns. This group, like the previous one, is of economic importance because of its beneficial nature, as well as its harmful nature. Mushrooms (Fig. 7-15) form a part of our diet; growers produce millions of dollars worth each year. Beneficial as these fungi may be, others, such as the smuts and rusts, cause crop loss valued at millions of dollars annually, and still others attack wood and wood products, especially in tropical and subtropical areas, necessitating a substantial annual outlay for preservatives to prevent this damage. Needless to say, the group is generally distributed throughout the world.

Classification. Club fungi are divided into two groups, one typified by the rusts and jelly fungi and the other by the bracket fungi, mushrooms, puffballs, truffles, and similar forms. Each group is divided into several orders and many families. There are about 25,000 species.

Vegetative structures. The mycelium of these plants is septate; that is, complete crosswalls are present, separating (in mature growth) binucleated and uninucleated cells to form the plant. The rhizoids penetrate the substratum to absorb nourishment and form masses in decaying, moist wood, on leaves, or in rich soil, where they spread out in fan-shaped fashion and are noticeable because of their white, yellow, or orange color. Development from the basidiospore is complicated

113

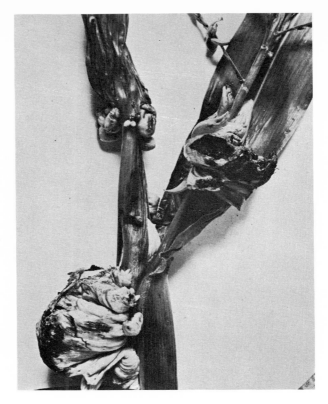

Fig. 7-11. Common smut on stalks, sheaths, and leaf of corn. (USDA photograph.)

Fig. 7-12. Bracket fungi.

Fig. 7-13. Tinderbox fungus, used by the early settlers of eastern North America as a means of keeping glowing embers with which to light the morning fire.

Fig. 7-14. Common puffball pictured here shows only the developing spore-producing body. Much of the plant is hidden as hyphae in the deadwood below. It is their saprophytic means of food getting that makes these organisms an important part of the biota.

115

Fig. 7-15. Fruiting bodies and stalks of the common edible cultivated mushroom. (USDA photograph.)

and goes through three stages, with characteristic features for each stage.

The sporophyte produces fruiting hyphae or basidiophores, with apical **basidia**. By a complicated process the basidium produces basidiospores as a part of the asexual cycle. The sexual cycle involves the conjugation of hyphae of different strains, as in the other classes of fungi. The higher members of the class produce the basidia in a large, complex fruiting body, the **basidiocarp**. In these plants the main body goes unnoticed, whereas the basidiocarp has the characteristic shape of the plant.

Reproductive structures. Sexual reproduction is so reduced that most species have no sex organs. Regular somatic hyphae, uninucleated basidiospores, and oidia serve in sexual reproduction, which usually involves the conjugation of separate strains. Formation of the sex cells is complicated, however, by the genetic factors determining sexual compatibility. The sexual cycle involves the formation of binucleated cells by the union of the cytoplasm of two uninucleated cells of the primary mycelia or by the formation of binucleated sperm from oidia. The latter unite with somatic hyphae. The union of the two nuclei forms a zygote-like structure that is comparable to the diploid zygote found in other fungi. It develops into a basidium, by means of which meiosis again produces basidiospores. The sexual phase of most basidiospores is unknown, but it is presumed to be the same in all species.

Life cycle. The vegetative plant begins by one of three methods: (1) a haploid basidiospore (the spore) germinates into a primary uninucleated mycelium, one of the cells of which fuses with another uninucleated spore, (2) the hyphae of different strains produced by two basidiospores conjugate, or (3) two uninucleated basidiospores of different strains fuse. This in effect is the sexual cycle. The union of two uninucleated cells by one of these processes produces binucleated hyphae. These grow into a larger vegetative plant that develops basidia and eventually basidiospores; thus the cycle is completed. The production of basidiospores involves the fusion of nuclei and meiosis in the following manner. The basidium may be septate or nonseptate, depending on the subclass to which the species belongs. The septate basidium as found in smuts and rusts is formed from the germination of a binucleated basidiospore. On germination the two nuclei fuse, after which meiosis follows and four haploid nuclei are formed. They migrate into a short club-shaped hypha that grows from the spore. Each of the nuclei are then delimited by a septum that cuts them off from the remainder of the hypha. They often transform into conidia. The nonseptate basidium characteristic of the gill fungi are binucleated, with nuclei of opposite strains. As

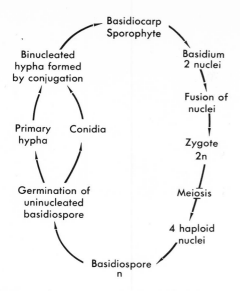

Fig. 7-16. Life cycle of a hypothetical basidiomycete.

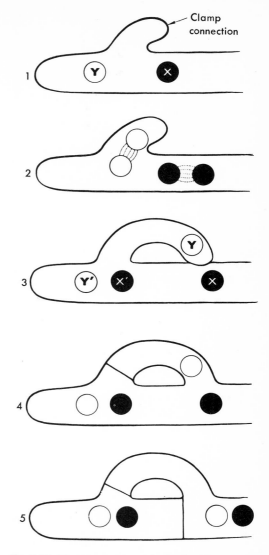

Fig. 7-17. Clamp connectors in the basidiomycetes.

the basidium develops, the two nuclei fuse and then divide meiotically to form four haploid nuclei. These migrate into four little projections at the tip of the basidium where they develop into basidiospores to be shed. Thus it is apparent that there is little similarity between this life cycle and the relatively simple alternation of generations found in other plants. Yet the basic elements of that fundamental plan are there. Fig. 7-16 will serve to illustrate this. This life cycle is hypothetical and rather simplified. Two actual cycles are illustrated in Appendix B (life cycles 15, *Puccinia graminis*, and 16, *Agaricus campestris*). The many variations of life cycles found in other species will be omitted from this introductory discussion. Note that the binucleated condition in these plants is the result of the fusion of two nucleated cells without the fusion of their nuclei and that this condition is retained for some time during the growth of the plant. Thus we may find in a given plant two types of hyphae, those with uninucleated cells and those with binucleated cells. An interesting mechanism has developed, called **clamp connections,** which ensures that the sister nuclei arising from the division of the nuclei of binucleated

cells become separated in the two daughter cells during cell division. The clamp connection is a special structure formed during nuclear division as a short branch arising between the two nuclei, x and y, as shown in Fig. 7-17. Nucleus y migrates into the clamp connection and divides at the same time x divides; x remains in place, whereas x' migrates toward the former location of y. The clamp connection grows toward the position

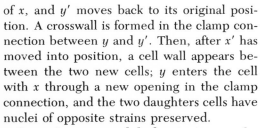

Fig. 7-18. *Aecidium* on the lower surface of a bar-berry leaf, the rust spots. (USDA photograph.)

Fig. 7-19. Spores of wheat rust being collected from a stem. (USDA photograph.)

of *x*, and *y'* moves back to its original position. A crosswall is formed in the clamp connection between *y* and *y'*. Then, after *x'* has moved into position, a cell wall appears between the two new cells; *y* enters the cell with *x* through a new opening in the clamp connection, and the two daughters cells have nuclei of opposite strains preserved.

Some important club fungi. Among the many kinds of club fungi, the following deserve special mention either because of their economic importance or because of their academic interest. No attempt will be made to give complete life cycles of these species, but some of the interesting features of their life processes will be mentioned.

The jelly fungi are so named because their fruiting bodies have the consistency of jelly. Many of the species are white, bright yellow, or orange and live on the branches and trunks of living or dead trees. They are most abundant in the fall, winter, or spring. Some species are extensively eaten as food in the

Orient. Other species in the group are parasitic on flowers and some on scale insects. An interesting host relationship is noted in the latter species. The fungus builds a hyphal mat over the scale colony and penetrates the insect's body. They do not kill the insects, but render them sterile.

The rusts, order Uredinales, form a large group, causing great loss to cultivated plants such as pine and many cereals. The life cycle of wheat rust (Appendix B, life cycle 15) is typical of the group. Two hosts, barberry (Fig. 7-18) and wheat (Fig. 7-19), are involved, through either of which the fungus can be controlled biologically. The elimination of the noncultivated, but equally essential, host plant will prevent the infection of the cultivated plant. A related species causes the cedar apples or galls of cedar trees, and another causes white-pine blister rust, resulting in the destruction of important lumber trees. The primary host of this fungus is found on the leaves of currants and gooseber-

Fig. 7-20. "Fairy ring" of toadstools.

ries. However, since these plants are of less value than lumber, it is they that are eradicated for effective and permanent control.

The smuts are related to the rusts. They get their name from the black, sooty spores they produce. Corn smut, wheat smut, loose smut of oats, and onion smut are a few examples of economically important species. Unlike the rusts, these species are not obligate parasites.

The large basidiomycetes include a wide variety of plants; many are well known and several are edible. Some cause plant disease and others produce galls. Many are brightly colored and are often pictured in full color in natural history books.

The bracket fungi are distinguished by their large basidiocarps growing on the sides of trees. When they are young, they are soft, but as they mature, they become hard and woody in nature. It is this group of plants that causes wood rot and necessitates large expenditures for wood preservation.

Agaricaceae is a large family including the mushrooms and toadstools. A few are edible, but many are poisonous. There is no positive way to distinguish the edible ones from the poisonous ones, except with the help of an experienced mycologist. Therefore it is best to exercise extreme caution in picking wild mushrooms to eat. Many species may be eaten, however, and a serious student can learn to identify those that can be used. In this same family is the famous and deadly amanita. Toadstools and other Agaricaceae often grow in such a way as to form what is termed a "fairy ring." This is a ring of darker green grass caused by the increase of available nitrogen in the soil due to the growth and death of fungal mycelia. The succeeding seasonal growth of the fungus is approximately equal and radiates out from the original starting site. This eventually forms a ring. The fruiting bodies appear around this ring in the fall (Fig. 7-20).

The puffballs (Fig. 7-14) form an interesting group, and as far as is known their round, fruiting bodies are all edible; no poisonous species are known. Basidiocarps sometimes grow to basket-sized proportions. Unfortunately, they are difficult to market; otherwise, their steaklike flavor would soon make them a popular delicacy.

By contrast there are the Phallales or stinkhorns, a near relative. These plants emit a fetid odor that attracts flies. Their mouth parts in turn pick up sticky spores that are disseminated after passing unharmed through the alimentary tract.

The bird's-nest fungi have a most remarkable mechanism for the dissemination of **peridioles** containing the basidiospores. The fruiting body of this fungus is cupshaped and resembles a bird's nest. The cup is called a splash cup—when it rains, drops of water strike the cup and splash out the peridioles. These complicated structures have a cord and an adhesive **hapteron** at the end of the cord. When the raindrops splash out the peridioles, the cord and sticky hapteron are extended. The peridiole flies through the air, strikes a leaf or stem, adheres to it by the hapteron, and ties itself fast, the momentum of the peridiole winding the cord around the stem.

DIVISION 15. Deuteromycota— imperfect fungi

The Deuteromycota include those fungi known to be or believed to be the monidial stages of Ascomycota or, rarely, of Basidio- mycota. They are called imperfect fungi because no sexual stage is known. These stages are perhaps undiscovered or may no longer exist. At any rate, the group is be- lieved to be an unnatural one, merely repre- senting a dumping ground for those fungi that lack the necessary stages for placement in the proper classes of the division. As many as 10,000 species remain in this group. Many more were once placed here, but they have since been removed upon discovery of the so-called perfect or sexual cycle. The only known means of reproduction in the imper- fect fungi is by conidia. It is probable, therefore, that most species are really as- comycetes.

Despite the artificial nature of the group, many of the species are of importance to man because of their disease-producing effects. *Streptomyces* sp., *Alternaria* sp., and *Hor- modendrum* sp. have been mentioned as causing asthma and rhinitis. Ringworm infec- tions and athlete's foot, as well as other types of skin infections, are caused by these plants. Others cause low-grade infection of the heart, lungs, brain, meninges, and bone. A few of these diseases are serious in nature, and all are difficult to treat.

Morphology. Conidia are usually devel- oped on conidiophores generally distributed on the hyphae, or some may be grouped into small fruiting bodies. A variety of arrange- ments occur in the many species of the group. Pathogenic species produce other types of spores to which several terms have been applied, depending on their method of formation. The hyphae of deuteromycetes are septate. The mycelium may be extensive in such subcutaneous forms as those causing ringworm. Some species are known only by their sterile hyphae; no form of reproduction has ever been observed.

Classification. There are four orders. Sphaerossidales produce conidia in pycnidia, globose or flask-shaped bodies on short coni- diophores. The Melanconiales produce coni- dia in acervuli, flat open heads of generally short conidiophores growing side by side and arising from a mass of hyphae. Moniliales reproduce conidia by different means from those found in the other orders, or they re- produce by oidia or by budding. The order Mycelia-Sterilia has no known means of re- production.

SOME IMPORTANT BIOLOGICAL INTERRELATIONSHIPS OF FUNGI

The fungi as a group contribute little new to the overall general phenomena manifested by organisms because they represent a closed sideline. However, their complicated life cycles and involvement with other organisms as saprophytes and parasites demonstrate how complicated protoplasm can be even when appearing simply organized. The in- herent properties of the nucleus and the suc- cessive morphogenesis of a generation of fungi under control of these nuclei amply ex- emplify the potential of complex, highly organized proteins and of the enzymes they secrete. The mechanisms involved are poorly understood and embody the very nature of life itself. How the fungi came to have such a complicated life cycle as that of wheat rust, for instance, will not be understood for some time. It does show the paths organisms may take as changing conditions drive them to further and further specialization in their struggle against extinction. It is well known that spore-bearing plants are poorly adapted for present conditions on the earth. Once, when the seasons were uniform the world over, the plants reproducing entirely by spores were dominant, and perhaps the fungi were free-living green plants—terrestrial representatives of their algae counterparts. Like many groups that struggle to survive after changed conditions render them no longer fit as dominant organisms, they have reverted to a saprophytic or parasitic life, making use of more advanced organisms to aid them in their propagation. By so doing, they have regained, at least in part, their former position, but as a result, they are so changed that their existence is entirely dependent on the success of their secondary role as subordinate organisms in the great cyclic changes typical of all life processes, among the most important of which is the decay cycle.

Saprophytic fungi

The role of saprophytes in the ecosystem is of critical importance. Without these organisms, some nutrient materials would be unavailable for recycling because they would be retained by the dead bodies of plants and animals. The recycling of nitrogen is absolutely necessary for the continuation of life. This process is entirely dependent on bacteria and fungi.

As stated in the introduction to this chapter, some fungi are able to live first as parasites, then later, after they have killed their host, as saprophytes, switching from one type of nutrition to another according to circumstances. Various nutrients may be utilized. For example, many different sugars are absorbed and used directly by various species. Others are able to use more complex carbohydrates. However, cellulose, the most abundant carbohydrate on earth, cannot be used by most species and only slowly hydrolyzed by others. The degrading of cellulose involves a complex of enzymes. In addition to the carbohydrate, almost all organic compounds may be used as a food source by some fungus. All saprophytes require nitrogen and minerals. Nitrogen sources are not so general. Many can use nitrates, whereas others are unable to do so. This may be ecologically important. Almost any amino acid provides a nitrogen source, but no fungi are known to fix atmospheric nitrogen as do some Prokaryotae.

Parasitic fungi

Parasitism is the way of life of many species of fungi in each of the divisions. Their hosts include almost all forms of life. Some are ectoparasitic, growing on the outside of the host, and others are endoparasites, living internally in the tissues of the host. When damage occurs to the host, it is said to be diseased. In many cases, however, parasitic fungi may be grown in artificial media, an ability that indicates that these species are faculative parasites and shows that they are ecologically parasites, but biologically saprophytes. Other species are obligate parasites. Details of fungal parasitism of plants is a special branch of biology, plant pathology. Details of fungal parasites of animals is part of the field of medicine and veterinary medicine.

Mycorrhiza

The roots of terrestrial plants depend on many biological processes taking place in the soil (see Chapter 15). One of these activities involves an association with **mycorrhizae,** specific kinds of fungi that invade rootlets during periods of active root growth. This association is believed to be symbiotic. Mycorrhizal symbiosis is found in the roots of all but a few families of vascular plants. In fact, so general is this phenomenon (as many as 80% of the species) that is is probable that the absence of these fungi would greatly change the appearance of terrestrial plant communities. Forest trees cannot survive, or at least grow poorly, without them. This knowledge is applied in forest-tree cultivation especially when pine is being grown outside of its natural habitat. The soil must be infected with mycorrhizae to achieve proper growth. Or-

Fig. 7-21. Matlike mycorrhizae of pine. This fungus suppresses the development of root hairs and takes over their function. (Courtesy E. Hacskaylo.)

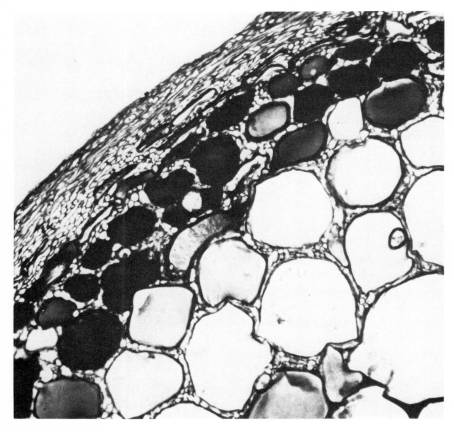

Fig. 7-22. Portion of a transverse section of pine mycorrhiza. Note the compact hypal mantle covering the surface and the hyphae extending between the cell wall into the cortex of the root. (Courtesy E. Hacskaylo.)

chid seedlings cannot grow without being infected by their special mycorrhizae.

The mycorrhizal associations are of three types: mycorrhizae that are outside the root and form a matlike covering (Fig. 7-21) with intercellular hyphae penetrating the root (Fig. 7-22); mycorrhizae that have hyphae penetrating the root cells; and mycorrhizae that have hyphae both intercellular and intracellular. Plants with intercellular symbionts have few root hairs. The roots are short and branched. Those with intracellular mycorrhizae do not have the matlike covering over the root.

Four classes of fungi form mycorrhizal associates. Most mycorrhizae belong to the Basidiomycota, including many of those whose fruiting bodies are edible, such as many wild mushrooms (not the common mushroom, however) and truffles. A few kinds of Ascomycota are involved, some Zygomycota, and a very few Phycomycota.

The physiological effects of mycorrhizal fungi is not entirely known. Tests using radioactive materials show that there is a transfer of nutrient materials. Minerals are absorbed by the fungus and transferred to the host, which in turn provides the fungus with carbohydrates. The association is not that simple, however. Apparently the fungus secretes auxin, the growth hormone, which stimulates the growth of the host.

Mycorrhizal associations are evident in fossil remains; so it is apparent that this relationship has been in operation for millions of years.

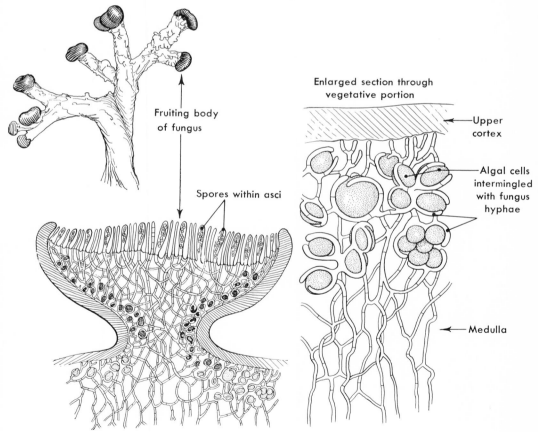

Fruiting body of fungus

Spores within asci

Enlarged section through vegetative portion

Upper cortex

Algal cells intermingled with fungus hyphae

Medulla

Fig. 7-23. Details of a lichen showing the relationship of the fungus and alga.

Lichens

The lichens are not taxonomically a separate group. They represent symbiotic associations of two plants, an alga and a fungus, both of which are otherwise classified in their respective groups. They are, therefore, combination plants and can only be classified artificially. However, because of their importance in ecological associations, soil building, and as food for certain animals, it is convenient to group them together for general discussion.

Component plants. Species of two groups of algae, the Chlorophyta and the Cyanophyta, form the algal component. The fungal elements are mostly ascomycetes, with a few species of basidiomycetes represented in one group. It is by the fungal species that the lichens are artificially classified. All species are named according to the fungus elements in the association.

Nature of association. Considerable controversy still exists among botanists about the nature of the association of these plants, which indicates that further study of the group is necessary. It is generally taught that the fungus provides moisture retention and protection for the alga and, in turn, receives nourishment from the chlorophyllic alga in a perfect symbiotic arrangement. However, there is some evidence that the alga is held captive by the fungus, which may well be cultivating the former for its own purpose, so that the relationship is merely one of host and parasite. The flow of carbohydrate from the alga to the fungus has been demonstrated by the use of carbon 14.

Reproduction. Vegetative multiplication may take place by any portion of the body developing into a new lichen or by granules, called **soredia,** composed of one or more algal cells surrounded by fungal hyphae breaking off and forming a new plant.

Spore formation also takes place (Fig. 7-23). The fungus element forms spores in a special cup-shaped structure. In ascomycetes these spores are in sacs or asci. The sacs rupture and spores are discharged. If the spore settles down in contact with the proper alga, it develops into a lichen; if not, it dies. The

Fig. 7-24. Foliose lichen.

Fig. 7-25. Fruticose lichen. (Courtesy Carolina Biological Supply Co., Burlington, N.C.)

algal element reproduces independently and is spread by rain.

Importance. The lichens are important elements in the formation of soil, dissolving and disintegrating rocks. Lichens form an important food source for caribou, musk ox, and reindeer. Lichens on barren plains and mountains of western Asia and northern Africa formed the manna of the Israelites and are still known as bread of heaven. Some of the lichens have medicinal properties; others are a natural source of many dyes such as litmus; still another, the oak moss, is used in southern Europe as a fixative for perfumes.

Classification by types. There are three types of lichens: (1) crustose lichens, which form a crust on trees, rocks, and soil and are the most abundant; (2) foliose lichens, with leaflike thalli, the upper and lower surfaces of which are different, and (3) fruticose lichens, with the major portion pendant or erect.

In the crustose lichens the upper hard layer is formed of fungal hyphae. The lower layer is composed of algal cells intermixed with hyphae. Finally, there is a loosely woven mass of hyphae forming the substrate. A familiar example is the common *Graphica scripta* on tree bark or on stones.

The foliose lichens (Fig. 7-24) form flat lobes, some of which adhere to the substrate by strands of hyphae. Beneath are pores that allow free passage of air into the algal layer.

These are common forms on stones in exposed places.

The fruticose lichens (Fig. 7-25) develop from a primary thallus, which may be crustose or foliose. They form branches of woven hyphae, with algae within the branches. Among the familiar examples are old-man's-beard and reindeer moss.

QUESTIONS AND PROBLEMS

1. List three reasons why the fungi might be considered to be a separate kingdom.
2. Why are the bacteria not included with the fungi?
3. Describe the food-getting process as found in the fungi.
4. Why is the fungus body often compared to an iceberg?

DISCUSSION FOR LEARNING

1. Discuss the reasons why some of the fungi are beneficial, but some are the cause of very serious diseases.
2. Discuss a way you think the lichens may have evolved.
3. Discuss the problem the fungi have with a binucleated cell. How might have this come about? Do you think it indicates a "wrong way" of development or evolution?

ADDITIONAL READING

Ahmadjian, V. 1967. The lichen symbiosis. Blaisdell Publishing Co. New York. (A monographic treatment of the kinds of lichens and their biology.)

Alexopoulos, C. J. 1962. Introductory mycology. 2nd ed. John Wiley & Sons, Inc., New York. (The standard textbook on fungi.)

Bessey, E. A. 1950. Morphology and taxonomy of fungi. Hafner Publishing Co., Inc., New York. (The basic work on the classification of fungi.)

Christensen, C. M. 1951. The molds and man. University of Minnesota Press, Minneapolis, Minn. (A semi-popular account of the molds and related fungi.)

Clements, F. E., and C. L. Shear. 1931. The genera of fungi. Hafner Publishing Co., New York. (Still the only comprehensive manual for the identification of fungi.

Gray, W. D. P. 1959. The relation of fungi to human affairs. Henry Holt & Co., New York. (A general account of fungi as they serve or hamper man.)

Miller, O. K., Jr. 1974. Mushrooms of North America. E. P. Dutton & Co., New York. (A guide for the identification of mushrooms and related fungi illustrated with many full-color illustrations; especially useful to those who gather mushrooms to eat.)

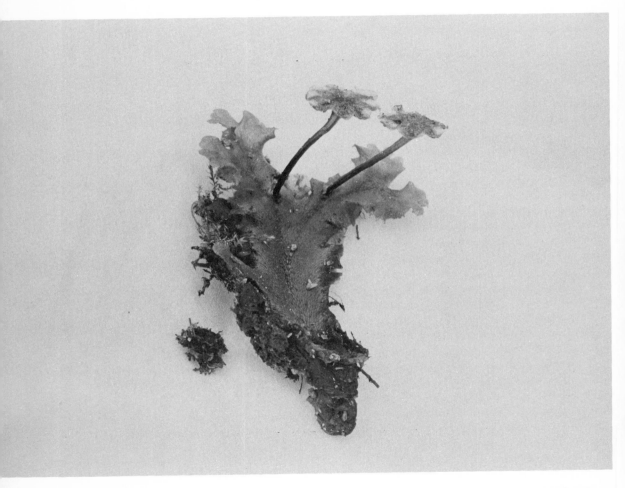

Fig. 8-1. Male gametophyte of the liverwort *Marchantia* sp. (Courtesy Carolina Biological Supply Company, Burlington, N.C.)

CHAPTER 8

Mosses, liverworts, and hornworts

■ The two divisions included in this chapter are collectively referred to as the bryophytes, the amphibia of the plants. They probably evolved from some group of algae as the first true land plants. However, their members are still dependent on moisture for sexual reproduction, and it is for this reason that they are likened to the animal Amphibia. The first clear-cut example of alternation of generations by land plants is found in the groups included here, the other land plants being certain algae that formed terrestrial mats but reproduced while submerged in water. There are distinct sporophytes and gametophytes (Fig. 8-1) with sexual reproduction taking place on separate plants.

There are about 23,000 species in these two divisions, but most of them are mosses. The two divisions are Hepatophyta, with the Hepatopsida (liverworts) and the Anthoceropsida (hornworts), and the Bryophyta, with the single class, Musci (mosses). The Hepatophyta produce gametophytes that are dorsiventral and differentiated into leaves and stems. The sex organs of the Hepatopsida of the Hepatophyta are arranged on stalks on the dorsal side of the plant and are separate, frequently on different plants. The sporophyte is small and retained in the archegonium. The Anthoceropsida of the Hepatophyta have their sex organs embedded in the gametophyte. The sporophyte is prominent.

The bryophyte gametophyte is erect and has stems and leaves, the apices of which develop the sex organs. The sporophyte, which grows from the archegonium as in the other groups, may or may not be prominent.

DIVISION 16. Hepatophyta— liverworts and hornworts

The two classes that belong to this division differ considerably and are treated separately in the following discussions.

Class Hepatopsida

Most members of this class grow in bogs or on the soil near running water. A few are found floating on water. The life cycle of *Marchantia* sp. (female, Figs. 8-2, *A*, and 8-3, and male, Fig. 8-2, *B*), a typical representative of the group (life cycle 17), involves an alternation of generation typical of the division.

The liverworts are characterized by thalluslike or straplike plant bodies without prominent stems. These are prostrate, having rhizoids that anchor the thallus to the soil. The rhizoids are unicellular and unbranched. The cells of the thallus are several layers thick; the upper epidermal layer is usually covered with a sturdy cuticle. **Stomas** open into the plant body and allow the transfer of gases in the spongy layer of chlorophyllic cells. The plastids of these cells are small and numerous. Photosynthesis is carried on by all species. The protonema is very small and transitory or absent. Elaters are present in the capsules and serve to disseminate spores.

Classification. Liverworts are found as fossils, and they were probably more abundant in past geological periods than they are today. Extant species are assigned to four orders and several families. There are about 8500 species.

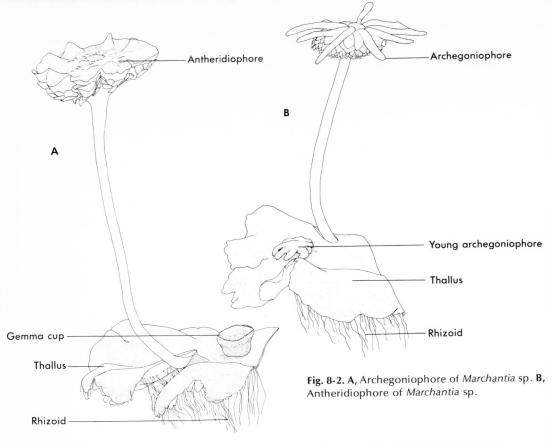

Antheridiophore

Archegoniophore

B

A

Young archegoniophore

Thallus

Rhizoid

Gemma cup

Thallus

Rhizoid

Fig. 8-2. A, Archegoniophore of *Marchantia* sp. **B,** Antheridiophore of *Marchantia* sp.

Fig. 8-3. *Marchantia* sp. with archegoniophores (female gametophytes). (Courtesy Carolina Biological Supply Company, Burlington, N.C.)

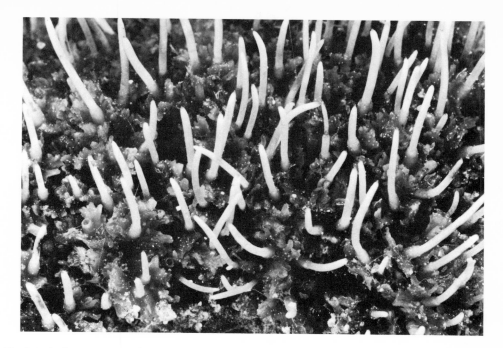

Fig. 8-4. *Anthocerus* sp. with sporophytes. (Courtesy Biological Supply Company, Burlington, N.C.)

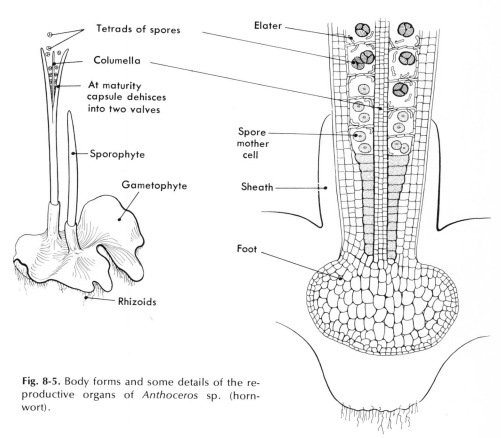

Tetrads of spores

Columella

At maturity capsule dehisces into two valves

Sporophyte

Gametophyte

Rhizoids

Elater

Spore mother cell

Sheath

Foot

Fig. 8-5. Body forms and some details of the reproductive organs of *Anthoceros* sp. (hornwort).

Class Anthoceropsida

The hornworts (Fig. 8-4) have a similar structure, but appear somewhat different because of a more prominent sporophyte that extends above the thallus in stemlike fashion. The plastids in the cells of the hornwort are larger, and pyrenoids are also present. The body form and some details of the reproductive cycles are illustrated in Fig. 8-5. The sporophyte differs from that of the liverworts by having a central column of sterile tissue, the columella, in the capsule. Growth in the capsule is continued over a long period of time.

The spore capsules contain elater cells, which are coiled in a springlike fashion for spore dispersal. The spores themselves form a tetrad of cells.

The gametophyte tissues are not so well differentiated as those of the liverworts; however, like them, they develop antheridia and archegonia on the thallus, but without stalks. This thallus often contains colonies of the alga *Nostoc* sp. growing inside cavities on its surface.

Classification. Only one order and one family belong to this class. The hornworts are a small group of less than 500 species that live on soil and rocks and, in the tropics, on the trunks of trees. Of course, they are found always in moist places.

DIVISION 17. Bryophyta—mosses

The mosses represent the most complex development of the two groups. They have a greater degree of tissue differentiation than do the others. The thallus develops false stems and leaves, with a thickened portion resembling a midrib to give it support. They all, of course, lack vascular tissue. The rhizoids are septate, branched, and more extensive than those of the other division. Their rhizoids serve as anchoring roots, and the thallus is stemlike and erect. Small leaves branch from the stems. Both the stem and the leaves contain chlorophyll and carry on photosynthesis. Because there is no vascular tissue and therefore no efficient supporting tissue, the size of mosses is limited. Exchange of food and raw materials is the result of diffusion through the cells.

Instead of single cell specialization for reproduction and other processes, groups of cells in the form of tissues carry on these functions. However, bryophytes lack true organs, which decidedly limits their development in other habitats. These plants are restricted to moist regions.

The sex organs are multicellular, with the outer layer of sterile cells, in direct contrast to the algae studied. This means that a true archegonium is developed in place of the oogonium, a distinct advance over the previous divisions.

A conspicuous protonema is formed by the germination of spores, and this persists throughout the life of the plant. Their sporangia lack elater cells, and spore dispersal is effected entirely by air currents. However, the spores are smaller and lighter than those of the other division and float away easily from the capsule. In fact, spores are often seen leaving the capsule like smoke so that a patch of moss looks as if it were on fire. A greater portion of the sporophyte is devoted to the manufacture of food so that, although it is still attached to the gametophyte, it is not so completely dependent as are sporophytes of the liverworts and hornworts.

This division is much more abundant than the Hepatophyta, with approximately 14,000 species widely distributed over the surface of the earth and sometimes in large enough patches to form a conspicuous feature of the vegetation of some areas. Although they are dependent on moist conditions for sexual reproduction, they can be found in relatively arid areas. Arctic regions also abound in mosses.

The reproductive cycle is described in Appendix B (life cycle 18) and is illustrated by *Polytrichum* sp., a common woodland group. This cycle is typical of all mosses, the only variation being that of structure.

Occurrence. All species are terrestrial; they are found on the soil, rocks, and decaying wood or bark of trees. Some bryophytes are of commercial importance, but none are disease producing, and none are edible (reindeer moss is not a moss but a lichen). However, *Sphagnum*, a genus of mosses that forms in bogs, is commercially valuable as peat moss. It is gathered and packed into bales for use by gardeners as a mulch. A unique property that prevents it from growing mold adds to its usefulness. Eventually bogs left undisturbed for many years will form peat, which in turn after many thousands of years, will form coal.

Classification. The division is divided into three groups. One, Sphagnopsida, is exemplified by the genus *Sphagnum* just mentioned, which is characterized by a single gametophyte produced from the protonema

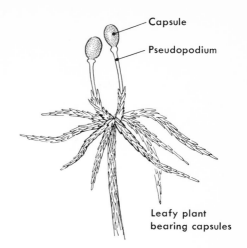

Fig. 8-6. Bog moss, *Sphagnum* sp., showing the sporophyte.

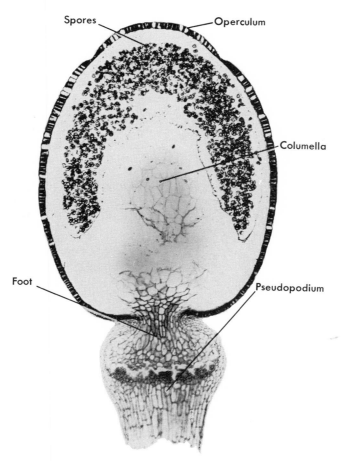

Fig. 8-7. Cross section of sporophyte capsule of *Sphagnum* sp. (Courtesy George H. Conant, Triarch Products, Ripon, Wis.)

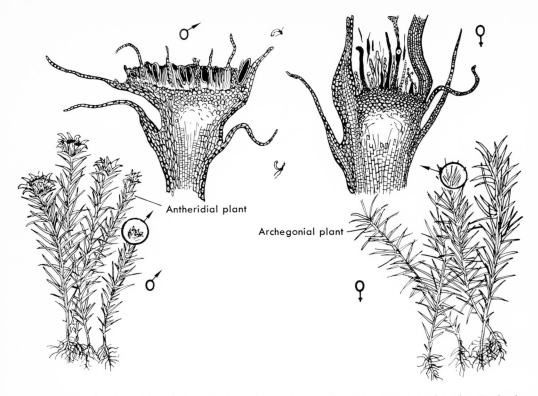

Fig. 8-8. Drawings of male and female gametophyte plants of moss; also enlargements to show longitudinal sections of antheridial and archegonial heads.

(Fig. 8-6). The sporophyte capsule (Fig. 8-7) is borne on a stalk, the pseudopodium, which is developed by the gametophyte. There is only one order, one family, and the single genus, *Sphagnum*, with over 300 species. They all grow in bogs in acid water.

The granite mosses, Andreaeopsida, are a small group with a single order, one family, and two genera. These mosses grow only on rocks in cold arctic climates or in the alpine zone of high mountains.

The true mosses belong to the Mnionopsida. The protonemata produce several plants with leaves having midribs. The pseudopodium is absent. *Polytrichum* is a typical genus. The vegetative plants, the gametophytes, are often sexually dimorphic. The male plant is devoid of parasitic sporophytes. The antheridial discs are crown shaped. The female plants (Fig. 8-8) with bristly archegonia are the hosts for the parasitic sporophytes (Fig. 8-8), which appear to be the fruiting bodies of the plant (life cycle 17).

The group is divided into about 14 orders and 80 families, which are widely distributed in all regions of the world. As inhabitants of moist places, they seem to grow best in shaded areas. The familiar rule that moss grows on the north side of a tree (at least in the northern hemisphere) is a result of this habit. However, these tree mosses should not be confused with the algae that also grow on trees. Actually, the avoidance of intense light is not attributable to any physiological aversion to light, but rather it prevents desiccation. In general, these plants prefer acid soil.

Interesting ecological relationships have been found in the mosses. Moist but bare soil is soon covered by a feltlike mat of protonemata that develop into gametophytes. A new area may be rapidly colonized by a succession of species; the early species are replaced by others as the habitat becomes established. The effect of this is important under natural conditions because of the water-holding qualities of the mosses. In tropical areas where the soil cover is thin, this plays a major role in the prevention of complete erosion and the exposure of bare rock. Even bare rock may be colonized by mosses, but this is a much slower process.

QUESTIONS AND PROBLEMS

1. What features are lacking by the divisions considered in this chapter to make them completely independent of water for reproduction?

2. How are mosses able to colonize; that is, what feature makes them available to occupy a new habitat?
3. Why are mosses rarely a dominant feature of the biota?
4. What does true alternation of generations mean?

DISCUSSION FOR LEARNING

1. Mosses, liverworts, and hornworts are referred to as a sideline development, as are the animal amphibia. Discuss why this is so in terms of a more probable origin of vascular plants.

ADDITIONAL READING

Conrad, N. S. 1956. How to know the mosses and liverworts. William C. Brown Co., Publishers, Dubuque, Iowa. (A very useful guide for the identification of the common species of bryophytes.)

Fig. 9-1. *Dryopteris austriaca,* a representative fern. (From Russell, N.H. 1958. An introduction to the plant kingdom. The C. V. Mosby Co., St. Louis.)

CHAPTER 9

Ferns and other primitive vascular plants

■ As we have seen, several divisions of plants produce spores instead of seeds, but those discussed so far lack vascular tissue. The remaining plants are differentiated into roots, stems, and leaves. Because of the development of true organs and the conduction of water from the soil, they are truly terrestrial plants. We have included in this chapter four divisions, which probably represent the only extant forms of much larger groups. In fact, more fossil than extant (now living) species are known.

Table 9-1 showing the features and classification of the vascular plants summarizes our knowledge of these groups. The plant body of the surviving species, unlike those of past ages, is an obscure part of our landscape, as can be seen in Fig. 9-1, a widespread fern, one of the living representatives of this group of primitive plants.

DIVISION 18. Psilophyta—whisk ferns

The sporophytes of these plants are rootless but have a rhizoidlike underground stem and an aerial portion forming the vegetative plant. This portion is branched, sometimes profusely in fossil species, and either it is without leaves or the leaves are small, veinless, scalelike structures. In fossil species the sporangia are borne on the ends of branches or as lateral appendages, the latter condition obtaining in the extant species.

Most of the species are extinct, and about 25 genera are known as fossils, which first appeared during the Middle Silurian period but rapidly disappeared after the Devonian period. As far as the fossil record shows, these were the first vascular plants. Only 3 rather rare species have survived and are placed in two genera, of which *Psilotum* sp. (Fig. 9-2) is the more common. Their present-day distribution is confined to tropical and subtropical and southern regions. In the United States one species can be found along the coast from Florida through South Carolina, where it is confined to sandy areas. Outside the United States the other species are epiphytes of tropical trees.

DIVISION 19. Lycophyta—lycopods and allies

The sporophyte of these plants is differentiated into true stems, leaves, and roots. The leaves, with a single, central, unbranched vein, are usually arranged spirally on the stem, but some may be opposite or in whorls. The sporangia of these and all remaining tracheophytes are located at the base of a special leaf, the sporophyll. In most cases the sporophylls are arranged in special groups, forming a strobilus.

Most species are known as fossils. They were once the dominant plants, even reaching treelike proportions. Their dead bodies contributed to our modern coal deposits. Today, no member of the group is a very familiar part of our flora.

Three distinctive orders are placed in the division.

Order Lycopodiales. The sporophyte is

Table 9-1. Classification of the vascular plants

	Division Psilophyta	**Division Lycophyta**	**Division Arthrophyta**	**Division Pterophyta**	**Divisions Coniferophyta, Cycadophyta, Gingkophyta, and Gnetophyta**	**Division Anthophyta**
Common name	Whisk ferns	Lycopods, club mosses, and ground pine	Horsetails	Ferns	Conifers	Flowering plants
Number of living species	3	780	25	8000	675	231,413
Leaves	Scalelike	Minute	Whorled	Foliaceous	Needlelike or foliaceous	Usually foliaceous
Strobilus	None	Large, apical, or lacking	Large, apical	Usually none	Cones	Flowers
Distinctive features	Rootless	Leaf with single, central, unbranched vein	Jointed stems	Spores usually in sori	Cones (usually)	Flowers
Life cycle	Sporophyte dominant; gametophyte subterranean	Sporophyte dominant; gametophyte subterranean or remaining in strobilus	Sporophyte dominant; gametophyte small green plant	Sporophyte dominant; gametophyte above ground, small	Sporophyte dominant; gametophyte parasitic	Sporophyte dominant; gametophyte parasitic
Fossil record	Abundant in Silurian and Devonian periods; now nearly extinct	Abundant during Carboniferous and Permian periods; now only a few species	Abundant during Carboniferous and Permian periods; now nearly extinct	First appeared in Carboniferous period; abundant since then, some more abundant in Cenozoic period	First appeared in Permian period; dominant in Mesozoic period; now diminishing	First known in Jurassic period and rapidly became dominant plants

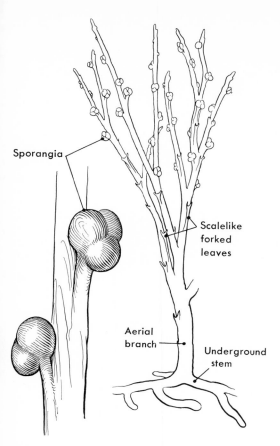

Fig. 9-2. *Psilotum* sp., an example of a rare living fossil.

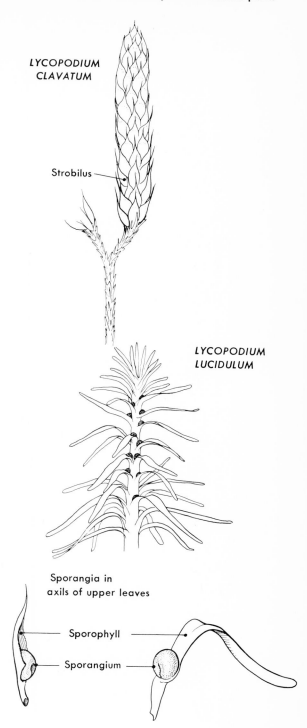

Fig. 9-3. Illustrations of two different species of *Lycopodium*, commonly called club mosses or ground pines.

herbaceous, with short stems that often run along the ground. The leaves are small and evergreen and give the plant the appearance of a small pine—hence its common name ground pine. The sporophylls are borne in strobili, and in several species are on the end of stalks resembling clubs, which has given rise to the common name "club moss." The gametophytes are either subterranean, or nearly so, and usually saprophytic or they may live symbiotically with a fungus. They are seldom seen.

Members of this order differ from those of the other two orders in having only one kind of spore instead of two. The spores are shed and may be dormant for many years, which eventually assures a new crop of plants even if unfavorable conditions temporarily exterminate the entire local population.

The order contains only two extant genera. One, *Lycopodium* (Fig. 9-3), is worldwide in distribution, including arctic regions. However, most species are tropical or subtropical. In the temperate regions it is fairly common in mountainous woodland habitats. The other genus is restricted to the Australian region.

Order Selaginellales. This order comprises the most abundant group in the class and is often confused with mosses. The sporophyte has leaves that are usually borne in four rows, two rows of small ones and two rows of large ones. At the apices of some branches are poorly developed strobili. Two types of spores are produced in the sporangia of these plants. The upper portion of the strobilus produces **microspores,** whereas the lower portion produces **megaspores.** Further details of the life cycle are described in Appendix B (life cycle 19). The separate male and

Fig. 9-4. The high rainfall in the Olympic National Park results in a dense temperate-climate rain forest. The forest is beautifully luminous with only a soft, green light filtering to the ground. The branches of the giant trees, such as these Douglas firs, are covered with *Selaginella oregona,* locally termed "clubmoss."

female gametophytes foreshadow the condition found in the seed plants.

There is only one genus with 700 extant species, worldwide in distribution. Most species are tropical, and many of them grow in damp forests (Fig. 9-4); however, several are well adapted for life in extremely arid areas. The resurrection plant is an example of one of these. This plant gets its name from its ability to remain dormant, rolled into a loose ball and entirely free of attachment to the soil. When it rains or the plant is placed on moist soil, it opens very soon and turns green. Vigorous growth takes place, and spores are produced. As soon as the water supply is used up, it returns to its dormant condition.

Order Isoetales. Small, herbaceous, rushlike sporophytes, the quillworts, form the remaining extant order of this class. The plant (Fig. 9-5) in some ways resembles garlic because of the arrangement of the sporophylls. The root system is scanty, with di-

chotomous branching. The stem is short and thick, resembling a corm. All the leaves are sporophylls, the outer ones subtending the megasporangia and the inner ones, the microsporangia. Both kinds are located at the base of the leaves; therefore the plants lack strobili. In the gametophytes the sexes are separate. The male plant produces multiciliated sperm.

The 65 species grow mostly in wet places, completely or partly submerged in ponds and lakes; a few species are found in dry areas. They are found in the temperate regions of the world.

DIVISION 20. Arthrophyta—horsetails

The modern horsetails are the last of a great group of plants with treelike proportions that once inhabited the earth. Some of them grew to heights of 60 to 90 feet. The sporophyte is differentiated into roots, stems, and small scalelike leaves. The leaves

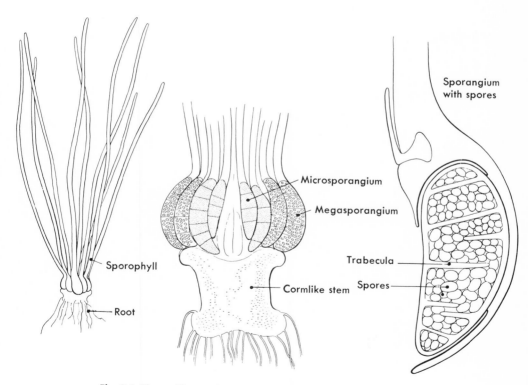

Fig. 9-5. The quillwort (*Isoetes* sp.) develops from a cormlike stem.

are borne in transverse whorls upon the stem. The unique stems are jointed, with distinct nodes, are hollow, and are longitudinally ribbed or furrowed. The cell walls of the epidermis contain silicon crystals, giving them a rough appearance. The **sporangia** are borne on special stems called sporangiophores, which are grouped into a compact **strobilus.** These branches are short lived and colorless. Other branches are green with longer aerial branches that give the vegetative part of the plant the appearance of asparagus. These may appear to be separate plants. However, they are connected by underground stems, the rhizomes.

The gametophyte develops separately as a small, green thallus. The life cycle is described in Appendix B (life cycle 20).

The horsetails are common in swamps, shallow ponds, meadows, damp shaded woods, and dry sandy embankments such as those found along railroad tracks. They grow from about 1 to 3 feet high. Only about 25

species have survived to modern times; thus they are sometimes called "living fossils."

The presence of silicon crystals in the stem has earned them the name "scouring rushes" because they were used by early settlers to scour pots and pans. The abrasive nature of the silicon is still used today by campers for the same purpose.

The fertile branches appear first in the spring with a strobilus. Spores have four spiral bands (**elaters**) that are coiled when moist but straighten out upon drying, aiding in catapulting the spores. The sporangiophores wilt and die after **sporulation,** but the vegetative branches remain throughout the growing season.

A spore germinates and develops into a small, green, somewhat branched **prothallium,** the gametophyte. This produces antheridia at the upper end and archegonia at the base of the short branches. The gametangia are produced at different times of the year, thus preventing self-fertilization.

Fig. 9-6. A large tree fern in central Fiji. Note the new frond uncoiling in the center of the frond cluster at the top of the tree.

The sperm fertilize one of the eggs during wet weather, and a new sporophyte develops.

The division is divided into three orders, two of which are extinct. The other order, Equisetales, contains two families, one of which is extinct. All living species belong to the family Equisetaceae, the genus *Equisetum*, and inhabit all regions of the world except the Australian region.

DIVISION 21. Pterophyta—ferns

The division Pterophyta is characterized by large, usually compound leaves that have a branching vein system. The spores are in sporangia that are clustered into sori usually located on the underside of the vegetative leaves. They are typical, in our modern flora, of cool woodland glades or, in tropical regions, they may reach treelike proportions (Fig. 9-6). They produce widely disseminated spores. In past geological ages the cryptogamic vascular plants were dominant, but now these have been replaced by the seed plants.

These cryptogamic vascular plants represent a separate division of extant plants, but they are the only remains of at least five ancient groups. The ancient origin of this group, known as far back as the Silurian period, makes it difficult to be sure whether they developed directly from the Psilophyta or whether they had common ancestry with the bryophytes. The former view is more generally accepted at the present time, but the lack of fossil evidence to bridge the gap between the two groups prevents any positive statements on the matter.

Ferns and their allies are generally distributed over all areas of the earth. The more primitive classes have more local distribution and now are of little importance.

A few species of ferns have some medicinal properties, but these are no longer used except locally by isolated groups. Several species are cultivated as ornamentals.

The development of true vascular tissue (Fig. 9-7) in the pteridophytes, as we have seen, made possible not only their greater size but also their firmer establishment of a terrestrial life. During the early geological periods when the ferns and related plants were dominant, conditions on the earth were more or less uniform, with an abundance of water on the land surface. One of the chief advantages of a vascular system is that the dead xylem gives rigidity to the plant as a skeleton does to an animal, and thus enables plants to grow to a larger size. This advantage aided them in their competition for space by exposing to the sun a greater area of leaf surface per individual. However, as time passed the land became dryer and the weather more severe. This resulted in a greater need for a well-developed root system to obtain water; otherwise, the plants would be forced back to swamps and shore areas. Only through this vascular system are they able to maintain their large size on land. The mosses and liverworts could not meet the competition on a drying earth.

Despite the addition of the vascular system, the pteridophytes are still dependent upon a fairly moist habitat because their sperm must swim to the egg for fertilization. Therefore the sexual part of their life cycle is restricted to the same habitat and condition as those of mosses. Thus in the life cycle there is a need for moist conditions during the gametophyte stage and dry conditions for spore dispersal during the fruiting stage of the sporophyte generation.

The ferns have the highest development of alternate generations in which the two plants, the mature sporophyte and the gametophyte, are separate. Ferns are characterized by their well-developed sporophyte stage and prothallium. The sporophyte is differentiated into a rhizome, roots, and large leaves (Fig. 9-7), called **fronds**. A well-developed **vascular system** is present and consists of a **vascular bundle** with **endodermis, pericycle, xylem,** and **phloem.** Unlike other plants, the fronds of ferns exhibit **circinate vernation,** also called "fiddleheads" (Fig. 9-8); that is, they develop by unrolling from the base to the apex. The sporangia of these plants are borne in clusters (sori) on the

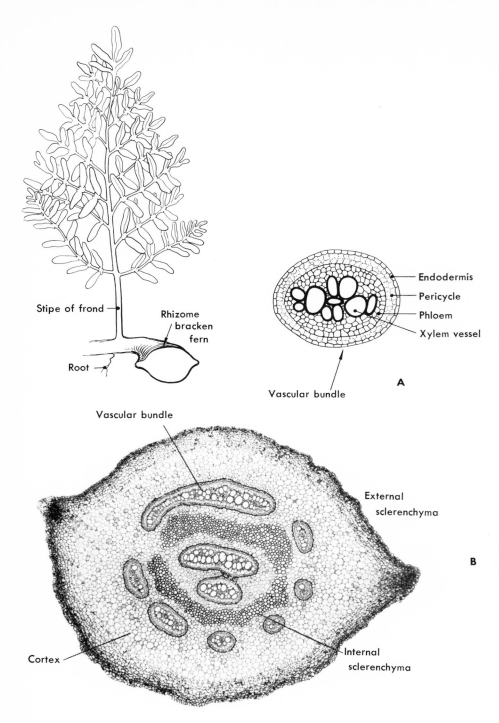

Fig. 9-7. Fern leaf and stem. **A,** Cross section of vascular bundle. **B,** Photograph of cross section of a rhizome (×30). (**B** courtesy George H. Conant, Triarch Products, Ripon, Wis.)

underside of the leaves (Fig. 9-9), on the leaf margin, or, rarely, in separate heads or in a **sporocarp**, a special leaf containing only the sori. Fronds bearing sporangia are called sporophylls (Fig. 9-9).

The prothallium is platelike, ribbonlike, or heart shaped and eventually becomes independent of the sporophyte. These prothallia are seldom observed but may be cultured from mature spores. When they can be observed, either or both kinds of sex organs are identifiable.

The ferns are the largest group of pteridophytes, having over 8000 described species. The majority of the genera are known from tropical America and the Indo-Malayan–South Pacific area.

Classification. The two classes comprising the division Pterophyta are very distinctive. The first, the class Eusporangiopsida, develops massive sporangia partially or entirely embedded on the ventral surface of the leaf, with an indefinitely large number of spores surrounded by a many-layered sporangia wall. The other class, Leptosporangiopsida have simple layered walls and produce a definite number of spores, usually multiples of two, frequently 48 to 64.

Fig. 9-8. Young fern fronds unrolling as they grow in the spring, showing circinate vernation.

Fig. 9-9. Frond of a broad leaf fern in a rain forest in Ecuador. Note the rows of sori.

Fig. 9-10. Giant tropical ferns with fronds over 10 feet high in a rain forest in Ecuador.

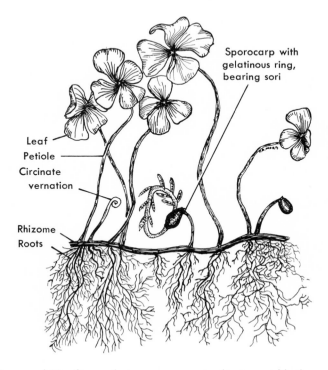

Fig. 9-11. Leaves of *Marsilea* sp. (heterosporous water fern) resemble those of clover.

Only about 110 species occur in temperate North America, representing four orders and 12 families. The plants are widespread in their ecological distribution, but the majority of the species prefer damp shady places such as deep woods. A few species are cultivated in shady gardens. They also grow on rocks, cliffs, open fields, and especially land recently cut over, and a few species are aquatic. In tropical rain forests some species grow to a height of 50 feet, with fronds 12 to 15 feet long (Fig. 9-10). Many others grow as epiphytes in these forests and cover the trunks and branches of trees.

The ferns have a long fossil history, first appearing in the Devonian period. Of the 19 or more known families, only five are extinct. Four families are unknown as fossils, and the remaining 10 families have both fossil and modern representatives. Most of the families appear to have developed during the Mesozoic Era so that they are not a principal element in the formation of coal, as is sometimes believed. The group is probably as abundant today as in past ages. Their spore dissemination has enabled them to become widely distributed, although they probably do not occupy as much area now as they did when habitat conditions were more uniform.

Many botanists classify the ferns as closely related to the gymnosperms and angiosperms. The reason for this is that leaf gaps are present in the stele just above the departure of the branch veins. This morphological feature is shared with the conifers and flowering plants and is a fundamental indication of their close relationship to each other. Ferns are classified into family, genus, and species according to their frond shape and the position, shape, and arrangement of sori on the fronds.

One primitive family, the grape fern family, has a few solitary leaves. Part of the frond is nonfoliaceous and bears the sporangia, which are not grouped into sori. The blade of the frond may be simple, lobed, or variously compound.

The osmunda ferns, including the common cinnamon fern, have some fronds that are nonfoliaceous and are interrupted in the middle by the special pinnae bearing the sori.

The curly grass or climbing ferns may not be recognized as ferns because of their vine-like appearance and small, grasslike leaves or leaves resembling those of flowering plants. The sporangia are clustered into fertile fronds that resemble strobili, a feature of the several water fern families such as Marsileaceae (Fig. 9-11). *Marsilea* sp. are one of the few fern groups that are heterosporous; that is, they produce two kinds of spore, one of which develops into the male gametophyte, and the other into the female gametophyte. The remaining several families have the more typical fern fronds.

The life cycle of ferns is given in detail in Appendix B (life cycle 21, *Polytrichium* sp.).

QUESTIONS AND PROBLEMS

1. What is meant by a "living fossil"?
2. List the new developments in the group of plants treated in this chapter.
3. What are the most likely ancestors of the vascular plants?
4. Referring back to the paleontological discussion of this group or primitive vascular plants, explain why they are no longer dominant.
5. List the new developments found in the ferns.
6. What features lead to the subjection of these plants by the seed plants, that is, what did the fern fail to develop?

DISCUSSION FOR LEARNING

1. Considering the answers to the above questions and the facts given in the paleontological chapter, discuss the reasons why the horsetails and lycopods were once a dominant group of plants.
2. Since ferns are still a widely distributed group of hardy plants, discuss the possibility of this group's again becoming dominant on the earth's surface.

ADDITIONAL READING

NOTE: *These plant groups are usually treated with the manuals of the flowering plants of states or larger geographical regions. Many are listed at the end of Chapter 11.*

Cobb, B. 1956. A field guide to the ferns. Houghton Mifflin Co., Boston, Mass. (A useful identification tool for the northeastern and central United States.)

CHAPTER 10

Primitive seed plants

■ The seed-bearing plants, sometimes referred to as **phanerogams,** plants with visible gametes or reproductive organs (as opposed to the cryptogams, plants with hidden reproductive organs), or **spermatophytes,** meaning seed plants, are the peak of development of the plant kingdom. Spores provided the best means of dispersal during the ages of more or less uniform climate, but, as we have already demonstrated, seasonal changes brought about a different mode of life, that of producing seeds, a phenomenon shared by the last five divisions of plants. Each of these divisions has an ancient origin, and it is not certain whether they developed the seed-producing structures before they diversified, or separately after they branched off from their fernlike ancestors. It is certain, however, that the seed plants developed while the cryptogams were dominant and that they remained obscure, poorly adapted plants for many millions of years. One such plant, the ginkgo (Fig. 10-4) has survived to the present day and represents still another of our "living fossil" plants.

The seed plants are the most familiar and dominant forms of plant life on the earth. Everywhere one looks on land, except in the most barren deserts or arctic wastes, these plants are present.

They are characterized by dominant sporophytes (with the gametophytes reduced to small parasitic organisms enclosed in sporophyllic tissue), by the pollen (the microgametophyte), and by a small egg with simple accessory tissue (the megagametophyte). Typical seed plants have flowers or cones, both **staminate** and **pistillate.** However, not all seed plants are flowering plants or conifers, and, generally speaking, only the more primitive are woody. In many respects the life cycle of *Selaginella* sp. (life cycle 19) foreshadows the seed habit. In seed plants a seed is produced as the result of the union of a sperm nucleus with the egg nucleus in a very complicated manner, as explained in the section on pollination in Chapter 16.

Seed plants have two principal means of seed development. The seeds of the primitive groups are exposed on the axis of a scale and are considered to have naked seeds, to which the term gymnosperm is applied. These seeds are produced in a strobilus, or in a cone (the conifers) (Fig. 10-1). The other group of seed plants, the flowering plants, have their seeds, as we have seen in previous discussions, enclosed by a covering produced by the sporophylls.

Classification. Four of the five divisions of seed plants are shown in Table 10-1. The

Fig. 10-1. The New Caledonian pine, *Araucaria columnaris,* is native to New Caledonia and New Hebrides, but is planted in other tropical and subtropical regions. It belongs to a family of conifers with about 12 species native to Australia, South Pacific Islands, and southern South America. It grows to great heights and ranks in size with the redwoods of California, although it does not have as large a trunk.

Table 10-1. Comparison of extant divisions of nonflowering seed plants

Division	Cycadophyta	Ginkgophyta	Coniferophyta	Gnetophyta
Habit	Palmlike	Tree with broad leaves	Trees or shrubs with needles	Prostrate forms with broad leaves or needles
Distribution	Tropical and subtropical only	Temperate regions under cultivation; native only in central Asia	Worldwide	Some desert regions of Old and New Worlds
No. of species	100±	1	520±	75±
Distinctive features	Sperm cell ciliate, motile; sporangia in strobili; leaves pinnately compound	Sperm cells ciliate; sporangia solitary or few, not in strobili; leaves simple	Sperm cells non-motile; staminate strobili simple; stem with resin canals	Sperm cells non-motile; staminate strobili compound; stems without resin canals

Fig. 10-2. Sago palm, *Cycas revoluta*, from Java.

flowering plants, division Anthophyta, discussed in the next chapter, completes the list.

DIVISION 22. Cycadophyta—cycads

The cycads, or sago palms (Fig. 10-2), are a tropical or subtropical group of about 100 species. They are palmlike in appearance, with woody, thick, tuberous stems that are mostly subterranean. Some of then grow to heights of 50 feet, but they are very slow growing. Specimens are known to be more than 1000 years old. Some produce very large cones (Fig. 10-3), which may weigh nearly 100 pounds, and others have individual ovules the size of large eggs. These plants are of interest to the botanist because of their retention of the primitive method of fertilization of the egg by motile sperm after they are carried to the female by windblown pollen. They are an ancient group known as fossils from the upper Pennsylvanian period.

Other than their cultivation as ornamentals, the species have little economic value. Some of the species produce edible seeds marketed only locally. One species, *Zamia floridana*, grows wild in the sandy pine barrens of Florida. It is dependent on a scarab beetle to start seed germination by feeding on the outer covering after burying it in loose sand.

DIVISION 23. Ginkgophyta—ginkgos

Only one extant species of this otherwise fossil group is known. This species has survived because it was kept in cultivation by Chinese monks. The group was known from fossils before it was discovered to have a living representative, the maidenhair tree, *Ginkgo biloba* (Fig. 10-4). It was introduced into Europe in 1730 and into the United States in 1784. Since then it has been widely cultivated as a shade tree because it can tolerate smoke and a limited water supply.

Fig. 10-3. Sago palm, *Cycas revoluta*, pistillate head.

Fig. 10-4. *Ginkgo biloba*, the maidenhair tree, a "living fossil."

However, because the female ginkgo seed is foul smelling, only male trees are propagated, and these from cuttings, so that the undesirable trees may be avoided.

The tree grows to a height of 100 feet, with broad, fan-shaped, bilobed leaves quite unlike the more familiar representatives of the class (pine, etc.). The leaves are arranged alternately on the branches, which gives the tree a distinctive symmetry that can be identified at some distance. Fertilization by a small motile sperm attests to its ancient origin. Other members of this order were abundant during the Mesozoic period.

DIVISION 24. Coniferophyta—conifers

The conifers, or true cone-bearing plants, are widely distributed, forming a dominant group, generally cold adapted, that is, adapted for life in cool regions. They live in vast forests of single species in the northern hemisphere. At higher altitudes in the mountains, pines, then spruce, take over and become dominant in place of the junipers.

Characteristics. The sporophyte is the dominant vegetative plant, producing two types of spores: the microspore, which develops into a pollen grain, and the mega-

Fig. 10-5. Alligator juniper, *Juniperus deppeana,* in Texas.

Fig. 10-6. A stand of young spruce in the White Mountains of Arizona.

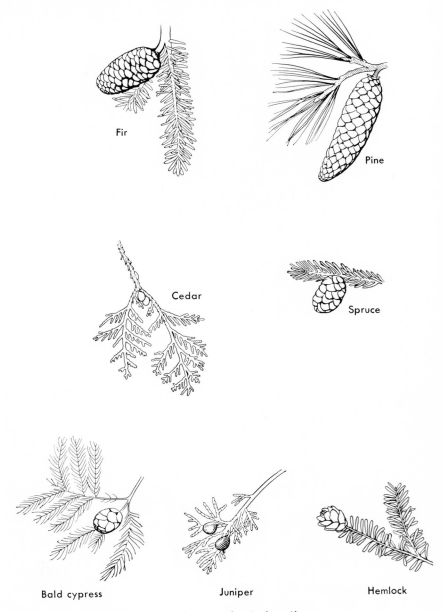

Fir

Pine

Cedar

Spruce

Bald cypress

Juniper

Hemlock

Fig. 10-7. Leaf structure and cones of typical conifers.

Fig. 10-8. Cypress and other conifers along the banks of a river in Louisiana.

Fig. 10-9. Young pistillate cones of cypress, Florida.

Fig. 10-10. Unusual gymnosperm *Welwitschia mirabilis* found in a mist desert in western Africa. Photograph of a museum habitat group. (Courtesy Field Museum of Natural History, Chicago, Ill.; from Russell, N. H. 1958. An introduction to the plant kingdom. The C. V. Mosby Co., St. Louis.)

spore, which develops into the female gametophyte. True seeds are produced. The vegetative plant has true roots, stems, and leaves. Most members of the class are woody plants, with an active cambium developing both secondary xylem (or wood) and phloem. All species are perennial. Most of these plants are evergreen; that is, they have green leaves throughout the year, with no particular time for leaf dropping. However, some are deciduous. Fertilization takes place by the transmission of airborne pollen in the same manner as spore dispersal in the cryptogams, thus eliminating the necessity of water for fertilization, as in all other groups of plants. Like both the higher vertebrates, with their membrane-protected eggs, and the terrestrial arthropods, these plants are at last free of water as a necessity for reproduction (life cycle 22 *Pinus* sp.).

This division, the largest of the nonflower-ing seed plants, includes the familiar pines, juniper (Fig. 10-5), spruce (Fig. 10-6), hemlock, cedar, yew, fir, redwood, and cypress, each with berries or distinctive cones (Fig. 10-7); all have representatives used extensively as lumber or grown for ornamentals. With few exceptions the leaves are needle-like and are borne on branchlets or in bundles. The 520 or so species are trees or woody shrubs that usually produce cones. A few, such as the juniper, produce a berrylike fruit. The sperm are nonmotile and are dispersed in pollen.

Seven families are known; of these, four are native to the United States, and the others are grown as cultivated ornamentals.

Perhaps the group is best known to the public as Christmas trees, for which several species are used. On the other hand, the giant redwood tree ranks as the largest living member of the plant kingdom. These trees

Fig. 10-11. The ephedra plant, common in the white sand desert of New Mexico and elsewhere in the western United States, is a conifer.

tower to a height of 360 feet, more than half the height of the Washington Monument, and some of them living today were seedlings when the Babylonian kings were ruling 4000 years ago. The oldest living tree is a bristle-cone pine, *Pinus aristata*, in California, that is over 4900 years old! Of equal but less spectacular interest are the knees of bald cypress trees growing with their roots submerged in water (Fig. 10-8 and 10-9).

Among the most valuable of the lumber trees are the western Douglas fir, the ponderosa pine, and the western white pine. In the east the southern white pine is used for wood pulp for paper and plastics, as well as for a source of turpentine and resins.

Only a few of the conifers are deciduous and are, therefore, not evergreen, such as the larch or tamarack and the bald cypress.

DIVISION 25. Gnetophyta—gnetes

This final division before the flowering plants includes a heterogeneous assemblage of about 75 species of plants that may well be representatives of more than one order. They are very different from the rest of the class. However, they all have naked seeds like the rest.

Included in this group is one of the most unusual plants in the world, *Welwitschia mirabilis* (Fig. 10-10) of the extremely dry deserts of southwest Africa. It consists of a huge tuberous root with a crown just above the surface. It produces two strap-shaped opposite leaves, the only ones produced during the hundred years or more of its life. These leaves continue to grow at the base and are beaten and frayed by the wind at the apex. A few members of this division also grow in the United States in the deserts of the southwest, for example, Mormon tea and joint fir. These are members of the genus *Ephedra* (Fig. 10-11), which produce cones and yield the medicinal alkaloid ephedrine formerly used to make nose drops. The synthetic product Neo-Synephrine has replaced the natural product. The plant remains an interesting member of a desert community where few persons would expect to see a gymnosperm. It may be seen at White Sands, New Mexico, as one of the few plants growing on the torrid, ever-shifting, pure gypsum mounds.

New features developed by the conifers and their relatives. The conifers introduced several features new to the kingdom Plantae. Undoubtedly this accounts for the great

155

Fig. 10-12. Typical pine cones (strobili). **A,** Staminate. **B,** Carpellate. (U.S. Forest Service.)

abundance of flowering plants that later developed by elaboration and refinement of these attributes. The production of two kinds of cones (Fig. 10-12) is a further development of the production of two types of spores, as seen in *Selaginella* sp. The permanent retention of the megaspore and the megagametophyte, with the megasporangium, and the development of a covering integument about each megasporangium permit seed development. The protection of the gametophyte by the sporophyte allows the production of fewer spores. The development of the new sporophyte within the old not only decreases the hazards of a double life cycle but also, by the addition of nutrient material and partial growth of the embryo, greatly increases the chances of survival once it has left the protection of the parent. Further simplification of the microgametophyte as it becomes a pollen grain allows it to function in place of the original spore. The process of pollination by the pollen grain and the formation of the pollen tube are permitted. These are necessary adjuncts to the new revised life cycle. The result is complete parasitism of both gametophytes upon the sporophyte and the establishment of the seed habit.

QUESTIONS AND PROBLEMS

1. Define a "seed" plant.
2. What factors might have caused the extinction of the ginkgo tree if it had been left alone, and yet today, it is considered a hardy city tree?

DISCUSSION FOR LEARNING

1. Discuss the possibilities of discovering still more "living fossils," and tell which groups might have living remains.
2. Carefully consider the statements just made in the "new features" section, and discuss.

ADDITIONAL READING

Brockman, C. F. 1968. Trees of North America. Golden Press, Inc., New York. (A useful identification manual to the major native and introduced species north of Mexico.)

Ketchum, R. M. 1970. The secret life of the forest. American Heritage Press Publishing Co., Inc., New York. (A general account of trees, especially conifers of lumber value.)

CHAPTER 11

Flowering plants

■ The flowering plants are distinguished from the conifers fundamentally on the basis of seed formation. The presence of a complete flower is not an essential feature of the division. The seeds of the conifers, as described in Chapter 10, are naked, that is, not enclosed within a fruiting structure of ovarian origin because the ovules are borne directly on the sporophyll as a part of the sporangium. The ovules of the flowering plants are located within an ovary, which also takes part in fruit formation. The seeds, therefore, are never exposed during their formation.

The flowers, seeds, and general habitus of these plants, along with their many vegetative features are used for the classification of the orders, families, and species.

The development of a flower instead of a cone is the most radical change from the conifer to the flowering plants (Appendix B, life cycle 23, lily). The advantages of this method of reproduction are primarily gained by the accessory structures, the flower parts, used to facilitate pollination, and by the development of fruit parts to aid in seed dissemination. Additional protection is afforded the ovule by this kind of structure.

The economic importance of the group is of course obvious. The esthetic value of flowers and other ornamentals (Fig. 11-1) is so much a part of our lives that it is taken for granted. The mind of man seems to be capable of understanding and of synthesis only through the organization of sensory data sorted to kinds, from which generalizations can be made. Most people take pleasure in this kind of exercise, and without entering further into more philosphical and psychological aspects, we can state that any kind of organization will broaden both the. pleasure and appreciation of any study. Therefore, since we know that these flowering plants are a part of our life, the following classification should help toward a fuller comprehension of this vast group of organisms.

We have not attempted a complete classification of the flowering plants, but rather, a classification of those plants most familiar to us, including all those used as examples in this text. Appendix A includes a list of all orders, families, and species referred to in the following discussion. Chapter 21 will refer to this same list in reference to useful or harmful plants. Even though no characteristics are given, by association of familiar plants, one will be able to note similarities and differences. We believe that a knowledge of the names of the flowering plants so often seen, used, and even eaten, offers recreation and relaxation to those who may find this of interest. By knowing the family grouping of common plants, one can further recognize plants new to them. With the aid of some of the

Fig. 11-1. Tasmanian blue gum, *Eucalyptus globulus* (Myrtaceae), Tasmania, sometimes imported as an ornamental tree in California.

many manuals now available, further identification of other plants can be made.

DIVISION 26. Anthophyta— flowering plants

The division Anthophyta (sometimes called Spermatophyta) contains one class, the Angiospermae, divided into two subclasses, the Dicotyledoneae and the Monocotyledoneae. Details of the structure of the roots, stems, leaves, and flowers of these two subclasses are included in Chapters 14 and 16. The bean plant is often used as an example of the dicots, and the lily plant, though it has certain exceptional features, is representative of the monocots.

Several major classification systems have been proposed over the past 200 years for the flowering plants, but a review of these would be beyond the scope of this text. No two books and no two monographs on these plants follow exactly the same system. There have been major revisions in the arrangement of plants in a classification system during the past few years. As taxonomists accumulate and assimilate more data, new evidence brings about changes in classification; therefore more changes can be expected. Modern classification merely reflects, by a linear listing, what is believed to be the phylogenetic or evolutionary development of plants. All such lists are limited by having two dimensions instead of three. Most of all, the student should not think of any list of extant plants as a list of ancestors and descendants: no modern plant taxon is the ancestor of another taxon.

One of the most definitive recent classifications and one that is readily available to the student is that outlined by Stebbins (1974), which makes use of simple characteristics of the flower parts and gives a complete list and summary of the flowering plant orders and families of the world. His arrangement of families is followed in this text.

Stebbins proposes that the magnolias are among the most primitive of extant species. He carefully analyzed the evolution of flowering plants, using the magnolias as a starting point and then listing the superorders and orders, with a total of 349 families accounting for 231,413 species in the division.

Certain features of the flowering plants are believed to be more primitive than others. Some of these trends are as follows:

1. Flower parts spirally arranged are primitive; those with cyclic or whorled parts are more advanced.

2. Flowers with many parts, the number of which may vary with the individual, are more primitive than those flowers with a few parts of a definite number, for example, five petals and five stamens.

3. Flowers with separate parts, for example, the individual petals free at the base, are more primitive than those with fused parts, for example, the petals fused to form a **calyx** cup.

4. Fusion of one type of flower part with another, for example, stamens fused to petals, is more advanced than those with free stamens.

5. **Hypogynous** flowers are more primitive than **epigynous** ones.

6. Plants with solitary flowers are more primitive than those with numerous flowers grouped into an inflorescence.

7. **Monoecious** plants are more primitive than **dioecious** ones.

8. Complete flowers are more primitive than incomplete ones.

9. Insect-pollinated flowers seem to be more primitive than wind-pollinated ones, although this is probably peculiar to certain groups and is not a generalization.

10. Woody plants in a given group are more primitive than the herbaceous members of that group; primitive plants tend to be perennials; biennials and annuals are more advanced.

These and a few other similar characteristics, along with a wide knowledge of the flowering plants, have been used to develop a classification. This classification is supplemented by detailed studies of all aspects of the species, such as their distribution, ecology, genetics, and morphology, all of which help to produce a sound and natural classification.

Class Angiospermae
Subclass Dicotyledoneae

The subclass Dicotyledoneae is the larger of the two subclasses of the class Angiospermae and until recently was believed to be the more advanced. Now, however, because of the application of the foregoing phylogenetic trends, it can be seen that the monocots are the more advanced.

The subclass is divided into superorders. Each of these groups is divided into several orders, some of which are further divided into suborders. Some orders contain only one family, but most have from two to as many as 20 families, with a world total of 349 extant families. Of these, about 200 occur in the United States and Canada, either as native plants, or as introduced plants that have become established under natural conditions. Many more are known as fossils. Some of the families are further divided into subfamilies and they in turn, into tribes. Every genus and species of plant is accounted for by being assigned to one of these families.

The following discussion includes plants that are familiar to all of us because either they are common members of our environment, are grown as ornamentals, or have some product of economic importance.

There is hardly a family that does not have some member well known either to the field biologist or to the gardener, but all could not be included here. Students interested should refer to the references listed at the end of this chapter.

The magnolia (Fig. 11-2) has the characteristics of a primitive angiosperm flower because the parts are arranged in spirals. The stamens (for flower parts see Chapter 16) are numerous. This beautiful tree and the tulip tree belong to the family Magnoliaceae. They are among the relatively few trees in the United States with large, showy flowers, in contrast to the many tropical trees with flowers that color the rain forest landscape. This family and several others, including the nutmeg family (Myristicaceae) are assigned to the order **Magnoliales,** which is a relatively small order in number of species.

Closely related to the magnolias is the order **Laurales,** the laurels, including the avocado with its edible fruit; the bay, the leaves of which are used for flavoring (but the fresh leaves contain cyanide!); and sassafras. The roots of the latter were used by the Indians to make an aromatic tea. Other members of the family provide us with cinnamon and camphor.

Fig. 11-2. Magnolia, *Magnolia soulangeana* (Magnoliaceae), Florida.

Fig. 11-3. Water pot vine, or pitcher plant, *Nepenthes pervillei* (Nepenthaceae), Seychelles Islands.

Pepper has long been a spice of great commercial importance. The search for this and other spices led ancient merchants on voyages of exploration. Our black pepper used for seasoning is a plant assigned to the family Piperaceae, of the order **Piperales.**

Near the top of the low tropical mountains of islands from Malagasy to New Caledonia live some short vinelike shrubs that produce pitchers somewhat of the appearance of our New World pitcher plants. These waterpot plants belong to the family Nepenthaceae (Fig. 11-3) of the order **Aristolochiales.** The pots fill with water in which live insects of various kinds, including mosquitos.

Water lilies (Fig. 11-4) are grown in ponds for decoration. They are members of a small family, Nymphaeaceae, order **Nymphaeales.** This order is related to the **Ranunculales,** believed by many botanists to be the most primitive of the dicots. The largest family, the crowfoot or buttercup family, Ranuncula-

Fig. 11-4. Water lily, *Nymphaea odorata* (Nymphaeaceae), eastern United States. (From Russell, N. H. 1958. An introduction to the plant kingdom. The C. V. Mosby Co., St. Louis.)

Fig. 11-5. Marsh marigold, *Caltha palustris* (Ranunculaceae), New York. (From Russell, N. H. 1958. An introduction to the plant kingdom. The C. V. Mosby Co., St. Louis.)

Fig. 11-6. Mayapple, *Podophyllum peltatum* (Berberidaceae), New York. (Note also the "fiddleheads," fronds of young ferns.)

Fig. 11-7. Bloodroot, *Sanguinaria canadensis* (Papaveraceae), New York.

ceae, has many common species, including the marsh marigold (Fig. 11-5), or cowslip, frequently eaten as greens. Other common species are the peony and hellebore. The latter has long been used as an insecticide. As a natural, biodegradable product, it is used by those concerned about our environment. Another family in the order is the barberry family, Berberidaceae. This family is infamous because the wild bush is one of the hosts of wheat rust. However, this family also includes the familiar Japanese barberry, an ornamental that provides a thorny fence around many dooryards, and the May apple (Fig. 11-6), a common, eastern, spring wild flower. The poppy family, Papaveraceae, order **Papaverales,** is one of the best known of the five families in this order because of the species that yields the drugs opium and morphine. Plants of this family are grown widely as ornamentals also. Among the many wild species are bloodroot (Fig. 11-7), a common, eastern, spring wild flower, and the prickly poppy (Fig. 11-8) of overgrazed, western arid lands. Bleeding heart, a delicate woods flower, belongs to the family.

Members of the order **Sarraceniales** belong to one family only and are of interest because of their remarkable insect-capturing structures. Species of the pitcher-plant family (Fig. 11-9) are native to the bogs of the United States. The leaves of these species are rolled into a fused tube partly filled with water. A nectarlike substance, which attracts insects, is secreted. They are trapped and die inside. Either bacterial action or enzymes cause their decay and digestion, after which they are absorbed by the plant to supplement the nitrogen-poor soil in which it grows.

Although trees are not a taxonomic group, several orders and families contain many trees and shrubs. These include the witch hazel and plane tree (sycamore) (order **Hamamelidales,** families Hamamelidaceae and Platanaceae), corkwood (**Leitneriales,** family Leitneriaceae), and the order **Fagales.** This order contains some of our most important forest trees as well as ornamentals. For example, the birches, alders, hazel and hornbeam belong to the birch family (Betulaceae). The beech family (Fagaceae) contains the oaks, beech, and the once important chest-

Fig. 11-8. Prickly poppy, *Argemone platyceras* (Papaveraceae), Arizona.

nut tree, which had nearly been made extinct in the 1930s by chestnut blight, a fungus disease.

A variety of plants of great food and ornamental value belong to the order **Caryophyllales.** Among the families assigned to this group is the Cactaceae (Figs. 11-10 to 11-12). Although these plants are typical of the desert and arid regions, the family is by no means confined to arid regions. The ornamental four-o'clocks (Nyctaginaceae) are portulacas (Portulacaceae) are widely planted around homes. The vegetables, beets, chard, and spinach (Chenopodiaceae) are, of course, well-known plants.

A large group of tropical plants (order **Theales**) contains the well-known tea plant (Theaceae) and a close relative, the ornamental camellia. One of the most remarkable trees is the baobab of Africa (Fig. 11-13). This

Fig. 11-9. Green pitcher plant, *Sarracenia oreophila* (Sarraceniaceae), Florida.

Fig. 11-10. Prickly pear cactus, *Opuntia chlorotica* (Cactaceae), Arizona.

Fig. 11-11. Organ-pipe cactus, *Lemaireocerus thurberi* (Cactaceae), Sonora, Mexico.

Fig. 11-12. Saguaro, *Carnegiea giganteus* (Cactaceae), Arizona.

Fig. 11-13. Baobab, *Adansonia digitata* (Bombacaceae), tropical Africa (5000 years old).

tree (order **Malvales,** family Bombacaceae) is among the oldest known trees, living as long as 5000 years. They are seriously damaged or destroyed by elephants. Closely related is the family Malvaceae, with the common hollyhock, and old-fashioned flower that adorns the backyards of many homes and is still a familiar sight, at least in rural areas. An examination of the "cheeses" or "wheels" formed by the neat ring of tender seeds in the developing seed pod is a good mark of identification of this group, particularly the mallow family, to which the hollyhock, hibiscus, and cotton belong. A near relative, the linden family, includes a large number of tropical trees, one of which is native to North America, the important lumber tree, the basswood or linden tree. Four additional families are grouped here also.

As the name implies, many species of the order **Urticales** are urticating; that is, they secrete substances that irritate the skin of mammals. However, the majestic elm, so symbolic of our peaceful village streets and now so threatened by the fungus causing Dutch elm disease that it may become extinct, belongs here also. In the same order are the mulberries, which furnish food for the silkworm moths as well as a delicious fruit, and the drugs hashish and marijuana, obtained from Indian hemp. The seeds are often included in canary bird food, and the stems also yield the hemp fiber of commerce. The tropical banyan (Fig. 11-14) belongs to this family too. The nettle family contains a variety of these weedy plants. Only the three families listed here are included in the order.

A strange combination of plant families comprise the order **Violales.** The violet (Fig. 11-15), family Violaceae, is familiar to everyone. But the ocotillo (Fig. 11-16) and the boogum tree of Baja California are bizarre desert plants of the family Fouquieriaceae. The former is distributed throughout the

167

Fig. 11-14. Banyan, *Ficus bengalensis* (Moraceae), American Samoa.

Fig. 11-15. Violet, *Viola odorata* (Violaceae), New York.

Fig. 11-16. Ocotillo, *Fouquieria splendens* (Fouquieriaceae), Arizona.

Fig. 11-17. Papaya, *Carica papaya* (Caricaceae), Tahiti.

Sonoran desert region of the southwestern United States and northern Mexico, and the latter, one of the world's most curious plants, is confined to a few square miles of Baja California, where it makes a weird forest of what appears to be giant carrots covered with spines and stuck into the ground upside down and twisted into grotesque shapes.

A tasty tropical fruit, the papaya (family Caricaceae, Fig. 11-17) is in odd contrast to the preceding family. A near relative, the begonia family (Begoniaceae) contains several ornamentals. The order is completed by still another odd, but important group, the gourd family (Cucurbitaceae), a family of vines, including squash, pumpkins, cucumbers, and the many kinds of gourds.

All common members of the order **Salicales** are trees that bear their flowers in catkins, a pendulous string of flowers, the most familiar of which are the pussy willow, willow, poplar (Fig. 11-18), aspen, and cottonwood.

A great many vegetables belong to the mustard family, the only New World family of the order **Capparales.** Among them are the radish, cabbage, turnip, kale, cauliflower, Brussels sprouts, horseradish, broccoli, rutabaga, mustard, and kohlrabi. Many other species of the family are well-known weeds, some of which are mentioned in the Bible.

The heath family is the principal family of the several included in the order **Ericales.** It includes a great variety of plants, many of which are typical of the arctic and subarctic. Fruits such as huckleberries, cranberries, blueberries, and others are members of this family. Wintergreen, used as a source of wintergreen oil, is likewise a member, along with the remarkable Indian pipe (Fig. 11-19), a flowering plant saprophytic on decaying vegetation such as a log or stump. Indian pipe is a plant that lacks chlorophyll and may be mistaken for a fungus unless it is examined closely.

The rose family (Rosaceae, order **Rosales**) includes the beautiful cultivated roses as typ-

Fig. 11-18. Quaking aspen, *Populus tremuloides* (Salicaceae), Arizona.

Fig. 11-19. Indian pipe, *Monotropa uniflora* (Pyrolaceae), New York. (From Russell, N. H. 1958. An introduction to the plant kingdom. The C. V. Mosby Co., St. Louis.)

ical members. However, equally as typical are the following fruits: blackberry, raspberry, prune, plum, cherry, apricot, peach, nectarine, loganberry, and many others. Venus's-flytrap (Droseraceae) contains species that are capable of rapid closure of traplike leaves. An insect attracted to these leaves by a clear, sticky fluid secreted by gland cells springs the trap after contact with the surface of the leaf. This causes a mechanical stimulation that sets in action special cells, so that the leaves close by a rapid change in the turgor pressure of these cells. The fate awaiting the trapped insect is the same as when one is trapped in the pitcher plant.

A variety of wild flowers belong to the large saxifrage family, which includes gooseberries, currants, and similar plants.

The large family Leguminosae, the pea

Fig. 11-20. Red clover, *Trifolium pratense* (Leguminosae), Florida. (USDA photograph.)

171

Fig. 11-21. Flowering dogwood, *Cornus florida* (Cornaceae), Virginia. Each "flower" is actually a cluster of small florets surrounded by four petaloid bracts. (From Russell, N. H. 1958. An introduction to the plant kingdom. The C. V. Mosby Co., St. Louis.)

family (order **Fabales**), contains not only the various peas, beans, and clovers (Fig. 11-20), but also a variety of trees and shrubs, many of which have spines or thorns. One of the most interesting is the sensitive plant, which has the ability to fold its leaves and drop limply on the stem when it is touched. Plants of this kind are common in the tropics where they seem to go to sleep at night because they fold their leaves together and look very much denuded. Sometimes these plants form a major portion of a lawn in the tropics. Imagine the frustration of the operator of a lawn mower who sees a brown swath behind him because these plants have folded their leaves at the first touch of the mower, only to open them again after the machine is safely past.

The order **Myrtales** contains a variety of families, most members of which are shrubs or trees. The herbaceous members belong primarily to the myrtle family, Myrtaceae, which includes the common myrtle, and the large gum tree, genus *Eucalyptus* (Fig. 11-1), of the Australian region. The evening primrose belongs to a group of plants forming

the separate family Onagraceae. A pest of waterways, the water chestnut, belongs here (Trapaceae), as well as the tropical fruit pomegranate (Punicaceae). Finally, the interesting mangroves of low coastal swamps (family Rhizophoraceae) completes the list of familiar plants belonging to this order.

A small group, the order **Cornales**, mostly trees and shrubs, includes the dogwood (Cornaceae, Fig. 11-21), so named because it was once widely used as a wood for the "dogs" (grates) of a fireplace. In addition, the sour-gum family (Nyssaceae) is placed in this order.

It is doubtful that many students are unaware of the tradition in connection with the members of one of the principal families of the order **Santales**, the mistletoe family (Loranthaceae), the members of which are parasitic on the branches of trees (Fig. 11-22). Except for sandalwood most of the other families of the order are obscure plants.

Only one North American family (Euphorbiaceae) is assigned to the order **Euphorbiales**. The spurge family is one of the largest

Fig. 11-22. Mistletoe (host is juniper), *Phoradendron juniperinum* (Loranthaceae), Texas.

Fig. 11-23. Candelilla, *Euphorbia antisyphilitica* (Euphorbiaceae), a plant growing in Big Bend National Park, Texas, produces a wax that has been much used for the production of candles and shoe and car polish.

families of plants. Among the many species are such familiar plants as the poinsettia, the castor bean (which yields the dreaded and, incidently, very poisonous castor oil), the cassava (which yields tapioca), and the rubber tree of commerce. Many members of this family are plants that superficially resemble desert cacti and are often sold in drugstores and five-and-ten-cent stores as cactus. The candelilla is a leafless desert shrub (Fig. 11-23) that has been harvested for its high-quality wax used in polishing automobiles. The so-called Mexican jumping bean is a member of the family. Seeds of several species are infected by the larvae of a species of moth, and the jumping effect is produced by their movements. Persons who appreciate authenticity are shocked to find that even the jumping bean has been synthesized, since mechanical jumping beans are now sold in novelty shops in the United States.

The small order Rhamnales has as its most important family, the grape family, Vitaceae. Closely related to the grape are the Boston ivy and Virginia creeper, both of which cov-

ered the walls of our academic institutions until the progressive engineers and economists decided that this touch of tradition was too costly to maintain. So this, like so many things of nature, is now nearly a thing of the past.

Most members of the order **Sapindales** are trees or shrubs. Among the families belonging to this order, the citrus trees and the mahogany tree are the most valuable. Spices, including the myrrh mentioned in the Bible, are obtained also from members of this order. The cashew family contains such poisonous plants as sumac. Maples (Fig. 11-24), horse chestnut, and a variety of other plants are members of this order.

The walnut family, Juglandaceae, including the hickory and pecan, is in a separate order, **Juglandales.** This is followed by the order **Geraniales.** Among the families included can be found the flax plant, which yields by fermentation of the soft parts the fiber flax from which linen is manufactured. Other members include the familiar touch-me-not of eastern United States woodlands,

173

Fig. 11-24. Flowers of silver maple, *Acer saccharinum* (Aceraceae), New York.

Fig. 11-25. Poison hemlock, *Conium maculatum* (Umbelliferae), New York. (From Russell, N. H. 1958. An introduction to the plant kingdom. The C. V. Mosby Co., St. Louis.)

174

the geranium, and the coca plant, yielding cocaine.

The carrot family, Umbelliferae, belongs to the order **Umbellales.** Familar species such as caraway, celery seed, dill, anise, and others, and many vegetables are members of this family. A number of common weeds such as Queen Anne's lace, or wild carrot, and the famous poison hemlock (Fig. 11-25) are assigned here.

The order **Gentianales** contains the gentian family, with many showy wild flowers, and the logania family with the genus *Strychnos*, which yields strychnine and curare, the latter a very poisonous substance used in the tropics to tip arrows and spears. It is used also in the physiology laboratory to demonstrate the physiological effect of such alkaloids on the muscles of frogs. Included here also is the dogbane family, which is a large family with few familiar species other than the periwinkle of gardens and lawns. The milkweed (Fig. 11-26) is included here, along with the beautiful, tropical frangipani (Fig. 11-27). The olive family belongs to this order. It may be somewhat of a surprise to learn that the familiar lilac and privet are near relatives of the olive (Fig. 11-28) and of the ash tree, which yields valuable wood for ax handles and similar wood products.

A host of familiar, closely related plants belong to the order **Polemoniales,** which con-

Fig. 11-26. Milkweed, *Asclepias syriaca* (Apocynaceae), New York. The winged seeds of the common milkweed emerging from the podlike fruits (follicles) in autumn. (From Russell, N. H. 1958. An introduction to the plant kingdom, The C. V. Mosby Co., St. Louis.)

Fig. 11-27. Frangipani, *Plumeria rubia* (Apocynaceae), Jamaica.

Fig. 11-28. Small fruited olive, *Olea africana* (Oleaceae), Kenya.

Fig. 11-29. Thorn apple, or jimsonweed, *Datura wrightii* (Solanaceae), Arizona.

tains several families, many of which are listed here. The morning glory, phlox, and a variety of garden flowers are included. The important nightshade family (Solanaceae), which provides us with a variety of plants such as the potato, tobacco, eggplant, and belladonna, from which atropine for dilation of the eyes for examination is obtained, the tomato, and the green or bell pepper. Several very poisonous plants, such as the thorn apple (Fig. 11-29) and nightshade, are related species. The borage family (Boraginaceae) contains the hound's-tongue, and the forget-me-not (genus *Myosotis*, also the name of the lake on the famous Huyck Preserve and Biological Station at Rensselaerville, New York), and others.

Several orders, including the largest of the dicots (**Asterales**) complete our review of this subclass. The order **Lamiales,** with verbena and mints, has many common species. The **Plantaginales** contains a variety of weeds, and the **Scrophulariales** includes the snapdragons. The bellflower family (Campanula-

Fig. 11-30. Dandelion, *Taraxacum officinale* (Compositae), Maryland.

ceae) contains a variety of small flowers, none of which are especially common except for the beautiful Canterbury bells. This and several other exotic families comprise the order **Campanulales.** The order **Rubiales** has only one large family, the madder family (Rubiaceae) with the very important plants Cinchona, the source of quinine formerly used in the treatment of malaria, and the coffee plant. Two common families, Caprifoliaceae, the honeysuckle family, and Dipsacaceae, the teasel family, belong to the order **Dipsacales.**

The largest of all plant families, the aster or sunflower family with about 20,000 species of rather showy flowers, many of them under cultivation, is the only member of the order **Asterales.** These plants may be readily recognized by their compound flower heads surrounded by a single outer row of petals, the ray. They are chiefly herbs. Among the many familiar examples are the aster, sunflower, ragweed, marigold, thistle, zinnia, chicory, lettuce, artichoke, and dandelion (Fig. 11-30).

Subclass Monocotyledoneae

The peak of plant development is found in the monocots if the current theories of the phylogeny of plants are correct. Most of the species of this group are herbaceous. A few, such as the palm and bamboo are woody. Their wood is much different from dicot wood and is not usable as lumber except for special ornate construction or, in the case of bamboo, for lightly constructed buildings in tropical and subtropical areas. In general the seed embryo produces a single cotyledon instead of two, as in the dicots. The leaves most frequently are lanceolate structures with parallel veins, although in the palms they may be compound structures.

The subclass is divided into many orders and families. From the standpoint of the economy of mankind the grasses far exceed in value all other plants. Grasses are one of the primary sources of food. Bread, as is obvious to everyone, comes from flour, which is made from the seed of a few species of grasses. Meat comes from livestock that feed entirely on plants, mainly many species of grasses.

The arrowhead plants (Alismataceae) of ponds and marshes are among the relatively few common species of the order **Alis-**

matales. The species are confined to aquatic or marshy areas.

Among the tropical species of the order **Najadales** is a plant important for its role in the transmission of disease. The naiad or *Najas* (Najadaceae) are plants that make large floating mats of vegetation in rivers and lakes in tropical America, especially in the Panama Canal. These plants provide hundreds of square miles of breeding ground for mosquitoes, especially for certain species of *Anopheles* mosquitoes, which act as vectors of the disease malaria. Unfortunately, it has been impossible to eliminate these plants, so that malaria remains a threat to the Panama Canal

Fig. 11-31. Blue dayflower, *Commelina dianthifolia* (Commelinaceae), Arizona.

area. Also included in the order is a group of water plants called pondweeds (Potamoge-tonaceae), common in freshwater ponds and streams.

The spiderworts are a group of attractive annuals (Commelinaceae) assigned to a separate order **Commelinales.** The orchidlike appearance of the blue dayflower (Fig. 11-31) is a good example of the group.

The order **Poales** has the single, but most important family Gramineae, the main source of most of the world's food supply. The grasses (Fig. 11-32) probably constitute the best known family of plants because of their great importance. Bamboo (Fig. 11-33) also belongs to the family. Many botanists are specialists in this group alone, as systematists, ecologists, and managers of range lands.

Several orders contain familar and inter-

Fig. 11-32. Blue grama, *Bouteloua gracilis* (Gramineae), Arizona.

Fig. 11-33. Bamboo, *Bambusa vulgaris* (Gramineae), Florida. (USDA photograph.)

esting species. For example, the order **Cyperales,** with the family Cyperaceae, the sedges, contain the plant papyrus, our original source of paper, and, of course, the origin of the name. The order **Typhales,** with the cattail family (Typhaceae) is a familiar plant of marsh lands. The pineapple family, Bromeliaceae, order **Bromeliales,** includes not only the pineapple, but many **epiphytes** so typical of tropical rain forests and the Spanish moss of parts of the southeastern United States.

The order **Zingiberales** is remarkable for the size and beauty of the plants it contains. The African bird-of-paradise flower (Fig. 11-34) is a representative of the family Strelitziaceae. The tasty banana (Fig. 11-35) and its relative, the traveler's-tree (traveler's palm) (Fig. 11-36) are examples of the Musaceae.

The order **Arecales** contains one family, the palm family, composed of hundreds of species, all of which are tropical or sub-

tropical (Fig. 11-37). Dates and coconut (Fig. 11-38) are among the many product these important woody plants produce.

A relative of the palms, the pandanus (Fig 11-39) is a prominent feature of tropica South Sea islands. They belong to the orde **Pandanales,** family Pandanaceae.

Two interesting families included in th order **Arales** occur in the United States. Th jack-in-the-pulpit (Fig. 11-40), a common spring woods flower, and the skunk cabbag belongs to the arum family (Araceae). Th duckweed family (Lemnaceae) would hardl be suspected of being even a flowering plant The common species of *Lemna* are sma floating plants consisting of a leaflike thallu and a few roots. Most reproduction is vegeta tive, and the simple little reproductiv pouches that comprise their flowers are se dom seen.

The lily family (Liliaceae, order **Liliale**

Text continued on p. 18

Fig. 11-34. Bird-of-paradise flower, *Strelitzia nicolai* (Strelitziaceae), South Africa.

Fig. 11-35. Banana, *Musa paradisiaca* var. *sapienta* (Musaceae), Ecuador.

Fig. 11-36. Traveler's palm, or traveller's-tree, *Ravenala madagascariensis* (Musaceae), Ecuador. Cuplike leaf bases hold water, which travelers are said to drink. It is a close relative of the banana, and not a palm.

Fig. 11-37. Oil palm, *Elaeis guineensis* (Palmae), tropical western Africa.

Fig. 11-38. Coconut, *Cocos nucifera* (Palmae), Tahiti.

Fig. 11-39. Screw pine, *Pandanus pacificus* (Pandanaceae), Fiji.

Fig. 11-40. Jack-in-the-pulpit, *Arisaema triphyllum* (Araceae), eastern North America.

Fig. 11-41. Australia grass trees, *Xanthorrhoea preisii* (Liliaceae), Australia.

Fig. 11-42. Century plant, *Agave americana* (Agavaceae), Arizona.

Fig. 11-43. Joshua tree, *Yucca brevifolia* (Agavaceae), California.

Fig. 11-44. Yellow lady's-slipper, *Cypripedium calceolus* (Orchidaceae), New York.

mainly tropical origin, although the lady's slipper (Fig. 11-44) lives in swampy areas in northeastern North America. Orchids are widely cultivated in greenhouses throughout the world. Many horticultural varieties of great beauty have been developed.

ADDITIONAL READING

Abrams, L. 1923-1940. An illustrated flora of the Pacific States, Washington, Oregon, and California. Stanford University Press, Palo Alto, Calif.

Baerg, H. J. 1955. How to know the western trees. William C. Brown Co., Publishers, Dubuque, Iowa.

Fassett, N. 1957. A manual of aquatic plants. University of Wisconsin Press, Madison, Wis.

Fernald, M. L. 1950. Gray's manual of botany. 8th ed. American Book Co., Boston.

Gleason, H. A., and A. Cronquist. 1963. Manual of vascular plants of northeastern United States and adjacent Canada. D. Van Nostrand Co., Inc., Princeton, N. J.

Gleason, H. A. 1963. The new Britton and Brown illustrated flora of the northeastern United States and adjacent Canada. Hafner Publishing Co., Inc., New York. 3 vol.

Kearney, T. H., and R. H. Peebles. 1960. Arizona flora. University of California Press, Berkeley, Calif.

Lawrence, G. H. M. 1951. Taxonomy of vascular plants. The Macmillan Co., New York.

Lawrence, G. H. M. 1955. An introduction to plant taxonomy. The Macmillan Co., New York.

Lloyd, F. E. 1942. The carnivorous plants. Chronica Botanica Co., Waltham, Mass.

Muenscher, W. C. 1944. Aquatic plants of the United States. Cornell University Press, Ithaca, N. Y.

Petrides, G. A. 1958. A field guide to trees and shrubs. Houghton Mifflin Co., Boston.

Porter, C. L. 1967. Taxonomy of flowering plants. 2nd ed. W. H. Freeman & Co., Publishers, San Francisco. (A textbook of plant classification.)

Small, J. K. 1933. Manual of the southeastern flora. Hafner Publishing Co., Inc., New York.

Stebbins, G. L. 1974. Flowering plants. Belknap Press, Harvard University Press, Cambridge, Mass. (The appendix to this volume lists all families of flowering plants, notes their distribution, and the number of genera and species in each.)

contains many beautiful flowers, most of which are readily identifiable on sight as members of this family. Even when the flowers are not nearby, reminders of the family are present in the odoriferous seasonings, onion, garlic, and leek. The strange grass tree (Fig. 11-41) of Australia is a lily. The century plant (Fig. 11-42), and the Joshua tree (Fig. 11-43) are all members of the family Agavaceae,

Most people agree that the members of the family Orchidaceae (order **Orchidales**), the last group of flowering plants, are the most beautiful of all plants. This is a large family of

CHAPTER 12

Energy transformations, respiration, and fermentation

■ Every living cell must obtain energy to survive, grow, and multiply. Without a constant supply of energy the cell would not be able to carry on such vital functions as growth, reproduction, movement, membrane transport, gas exchange, and biosynthesis. The study of energy changes in chemical systems is in the realm of chemical **thermodynamics.** The principles of thermodynamics were developed mostly from work done on the steam engine during the late nineteenth century. However, with increasing awareness of the chemical and physical nature of life, many scientists are studying the thermodynamics of living systems. When studying thermodynamics, the scientist considers the universe and selects a portion of it for consideration. The portion selected for study is called a **system,** the remaining portion is referred to as the surrounding . Systems are divided into two categories, closed and open. A closed system is one in which only energy can enter or leave, whereas an open system is one in which both matter and energy can enter or leave. Although living processes are studied in a closed system in the laboratory, in nature they most likely occur in open systems.

Energy, known in such forms as mechanical, electrical, and heat, can best be defined as the capacity to do work. It was observed that when chemical reactions occurred in systems, they could be divided into two types. **Exergonic reactions** give off heat to the surroundings, whereas **endergonic reactions** absorb heat from the surroundings. From this basic observation it was realized that heat in motion is a form of energy and it could do work when it moved from a hotter to a colder place. Heat is like water—it must move downhill in order to do work. For example, hot water does not boil when it is placed on a cold stove. From these concepts the first law of thermodynamics was postulated. It states that energy is neither created nor destroyed in a process but merely changes form. If we assume that the universe is a closed system and heat is a form of energy, then the heat in the universe is continually being diluted and the system is moving toward a state of maximum heat dilution. This state would represent a maximum loss of the energy available to do useful work. Energy available to do useful work is defined as **free energy.** Since, according to the first law, energy is neither created nor destroyed, free energy then is converted to another form. This form cannot

---

Fig. 12-1. The sun is the ultimate source of the earth's energy. Through evaporation by sunlight, energy had been transferred by raising seawater to form rain. Water runoff released the energy needed to cut this gorge of the Rio Grande river between the United States and Mexico. Plant life captures the energy of the sun in the process of photosynthesis. The amount of energy involved is infinitesimal compared to canyon-cutting energy. Plants store the captured energy in organic molecules. The energy stored in the organic molecules is then recovered by the processes of fermentation and respiration found in plant cells.

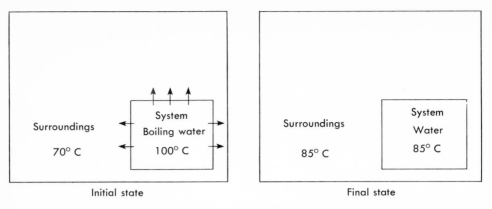

Fig. 12-2. Diagram illustrating what happens when a vessel of hot boiling water is placed in a container that is at room temperature. In its initial state the vessel contains heat, which is a form of energy and it can do useful work (free energy). In time the heat will flow from the hotter vessel to the colder surroundings; thus the flow causes the system to cool and the surroundings to warm. In the final state equilibrium has been reached so that the temperature of the surroundings is equal to that of the vessel. Heat has been diluted and the entropy has been increased by an increase in the temperature of the surroundings. Entropy cannot do useful work.

do useful work and is called **entropy.** The second law of thermodynamics states that the entropy of a system will always increase (Fig. 12-2). It makes one wonder whether or not living systems obey the laws of thermodynamics. These systems as we know them represent the ultimate in organization, whereas the concept of entropy would continually increase the degree of disorganization. Living systems do in fact obey the laws of thermodynamics by maintaining a constant supply of free energy in the form of fuel molecules. This supply of free energy continually offsets the trend toward disorganization. When the free-energy supply is cut off, the system becomes disorganized and death ensues. Living systems have also developed the capacity to couple exergonic and endergonic reactions; thus the maximum amount of free energy is conserved.

Cellular **metabolism** represents the integrated network of chemical reactions that enable the cell to extract free energy from fuel molecules. The extraction process is a controlled release and utilization of stored energy occurring through the mechanisms of **fermentation** and **respiration.** Metabolism is divided into catabolism, which is the degrading of molecules, and **anabolism,** which is the

synthesis of molecules. The energy cycle of the plant cell begins with solar energy, which is converted to chemical-bond energy during photosynthesis. The major energy-storage product formed in the plant cell appears to be sucrose. During cellular respiration, sucrose is oxidized to CO_2 and H_2O. The stored energy of sucrose is converted to a special form before it is used for cellular purposes. This special form of energy is the high-energy phosphate bond found in **adenosine triphosphate (ATP).** ATP is the special free-energy carrier of the cell formed by oxidative phosphorylation during respiration. ATP is synthesized from ADP (adenosine diphosphate) and inorganic phosphorus (P_i).

$$\text{Energy} + \text{ADP} + P_i \rightarrow \text{ATP}$$

When the energy of ATP is released, ADP and inorganic phosphate are regenerated. In addition to the production of ATP, the oxidation of fuel molecules produces NADPH, which provides the necessary reducing power for biosynthesis. When ATP and NADPH are depleted of their energy, **ADP (adenosine diphosphate)** and NADP are regenerated and used in the resynthesis of the high-energy compounds (Fig. 12-3).

The extraction of energy in cells can be

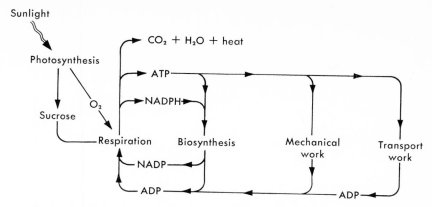

Fig. 12-3. Diagram of energy cycle of plant cell. Solar energy is converted into chemical-bond energy represented in sucrose. The energy stored in sucrose is transformed into ATP and NADPH by the process of respiration. The high-energy compounds ATP and NADPH are used to drive the various energy-requiring functions of the cell.

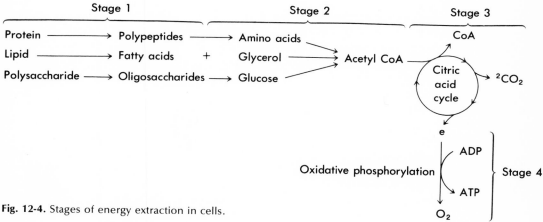

Fig. 12-4. Stages of energy extraction in cells.

separated into four stages (Fig. 12-4). The first stage is characterized by the conversion of large polymers such as polysaccharides, proteins, and fats into small molecules through a series of intermediate-sized ones. The second stage is the conversion of the small-molecule monosaccharides, amino acids, glycerol, and fatty acids into a 2-carbon fragment called acetate, which will combine with **coenzyme A** to form **acetyl coenzyme A.** The third stage is the **Krebs cycle,** which represents the pathway to complete oxidation of acetate and the passing of electrons and protons to primary acceptor molecules (NAD

and NADP). The fourth stage is oxidative phosphorylation, which couples the energy released during oxidation to the synthesis of high-energy phosphate bonds.

The generation of metabolic energy in the cell is accomplished by the complete oxidation of fuel molecules to CO_2 and H_2O. The energy generated in the process is captured by a reaction coupled to the oxidation of these fuel molecules. All heterotrophs extract their energy from a series of oxidation-reduction reactions, in which one substance acts as an electron donor and one as an electron acceptor. The electron donor is oxidized

191

when it donates an electron, and the electron acceptor is reduced when it receives an electron.

The primary fuel molecules in plant tissue are the carbohydrates synthesized as a result of photosynthesis. It appears that the major carbohydrate storage product in plant cells is sucrose. This is most probably synthesized in the cytoplasm and not in the chloroplast because of the permeability barrier of the chloroplast membrane. Biochemists are currently very intrigued with the interdependence of the chloroplast and mitochondrion in the overall energy economy of the cell. The chloroplast through the process of photosynthesis fixes carbon into organic molecules that may be transported to the mitochondrion where they are oxidized and the energy is released in the form of ATP.

The primary fuel molecule of the cell is glucose, which is formed by the hydrolysis of sucrose. The complete oxidation of glucose occurs within two metabolic pathways. The first pathway is glycolysis, a series of reactions that convert glucose into pyruvic acid. This pathway is the only mechanism that anaerobic organisms have to extract energy from organic compounds. Organic molecules act both as electron donors and electron acceptors in this pathway, and molecular oxygen is not involved. The second pathway is the Krebs cycle, which is utilized by all aerobic organisms as the final metabolic pathway for the oxidation of all fuel molecules such as amino acids, fats, and the products of glycolysis (pyruvic acid).

FERMENTATION

Glycolysis is the anaerobic oxidation of 1 mole of glucose into 2 moles of pyruvic acid. Glycolysis is often referred to as fermentation. Fermentation is the process whereby organisms extract energy from carbohydrates or other fuels in the absence of molecular oxygen. It is called **anaerobic respiration.** Alcoholic and lactic acid fermentations are accomplished by organisms using the glycolytic pathway. Once these organisms have formed pyruvic acid, they further convert it to ethyl alcohol or lactic acid. These compounds build up in the cell and are excreted into the surrounding medium. Yeast and fungi are capable of many kinds of fermentations that are of economic importance. Aerobic organisms usually continue to degrade glucose to pyruvic acid, however the pyruvic acid is then oxidized in the Krebs cycle at the expense of molecular oxygen as an electron acceptor. During glycolysis glucose is degraded in stepwise fashion into 2 moles of pyruvate, and the energy extracted is utilized to form ATP. The overall equation for glycolysis can be written as follows:

$$\text{Glucose} \xrightarrow{\text{glycolysis}} 2 \text{ Pyruvate}$$

Glycolysis can be partitioned into three stages. Stage 1 is the conversion of glucose into fructose-1,6-diphosphate.

This stage involves three steps, each of which is catalyzed by a different enzyme. The first step uses 1 mole of ATP to phosphorylate glucose at the sixth carbon. The

Glucose $\rightleftharpoons$ Glucose-6-phosphate $\rightleftharpoons$ Fructose-6-phosphate $\rightleftharpoons$ Fructose-1,6-diphosphate

(ATP → ADP)

Fructose-1,6-diphosphate $\longrightarrow$ 2(3-Phosphoglycerate)

2 ATP / 2 NADH

second step changes glucose-6-phosphate to fructose-6-phosphate, and the third step uses another mole of ATP to form fructose-1,6-diphosphate. This stage utilizes 2 moles of ATP to form fructose-1,6-diphosphate from glucose. The second stage of glycolysis, which consists of four steps, converts 1 molecule of fructose-1,6-diphosphate into 2 molecules of 3-phosphoglycerate, with the production of 2 molecules of NADH and 2 molecules of ATP.

In the last stage of glycolysis, 3-phosphoglycerate is converted to pyruvic acid in three sequential reactions with the simultaneous generation of 2 molecules of ATP.

$$\text{2(3-Phosphoglycerate)} \xrightarrow{\text{2 ATP}} \text{2 (Pyruvic acid)}$$

The overall reaction can be summarized as follows: One molecule of glucose is converted into 2 molecules of pyruvic acid (Fig. 12-5).

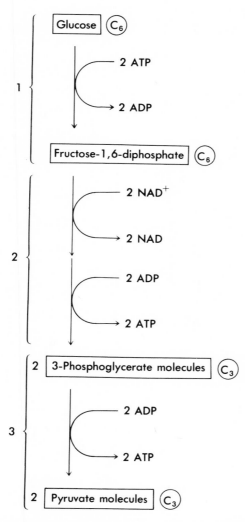

Fig. 12-5. Diagram of mechanism of glycolysis depicting one molecule of glucose being converted into two molecules of pyruvic acid with the production of two molecules of NADH and two net molecules of ATP.

Two molecules of ATP are used in the initial phosphorylations of the sugar. Four molecules of ATP are produced as a result of the oxidation; thus there is a net gain of 2 moles of ATP. Two molecules of NAD^+ are reduced to NADH, and 20 kilocalories are lost as heat.

For glycolysis to continue, it is necessary that NAD^+ be regenerated from the NADH formed during the process. In aerobic organisms this is simply accomplished by NADH passing its electron onto molecular oxygen, thus regenerating NAD^+. However, under anaerobic conditions, NAD^+ is regenerated by the reduction of pyruvate to either lactic acid or ethyl alcohol (fermentation) (Fig. 12-6).

KREBS CYCLE

The Krebs, citric acid, or TCA (tricarboxylic acid) cycle, is the final oxidative path-

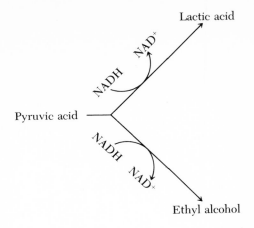

way for fuel molecules. It operates only in the presence of oxygen and is the major producer of cellular ATP. In addition to providing the cell with ATP, it also supplies it with some intermediates for biosynthesis. In eukaryotic cells the enzymes that cause the re-

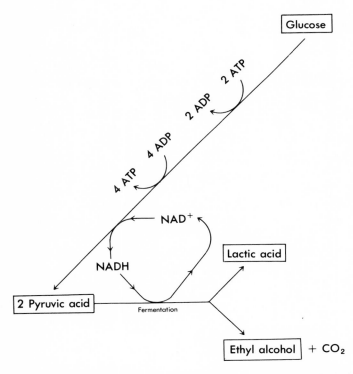

Fig. 12-6. Diagram depicting fermentation of glucose into either ethyl alcohol or lactic acid with the regeneration of NAD^+ from NADH.

194

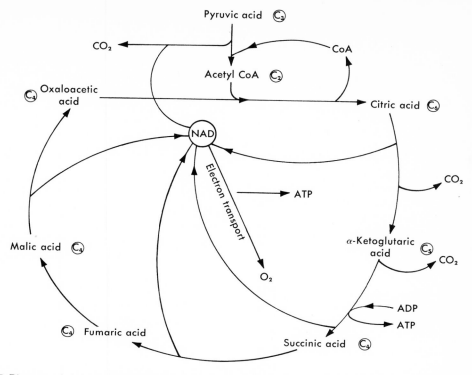

Fig. 12-7. Diagram of citric acid cycle depicting the passage of electrons to NAD and the subsequent formation of ATP.

actions of the cycle are located on the inner surface of the mitochondrial membrane, and they are arranged in a very specific pattern. The entire cycle occurs inside the mitochondria as compared with glycolysis, which is found in the cytoplasm. Glycolysis is linked to the Krebs cycle by the oxidative decarboxylation of pyruvate. In this reaction pyruvate reacts with coenzyme A (CoA—SH) to form acetyl CoA plus CO_2.

The reaction is essentially irreversible and necessary for the entrance of carbohydrate into the Krebs cycle. It is catalyzed by a complex enzyme system consisting of three

enzymes and five coenzymes. This reaction is not a part of the Krebs cycle. The first reaction in the cycle is the condensation of a 2-carbon unit (**acetyl coenzyme A**) with a 4-carbon unit (oxaloacetic acid) to form the 6-carbon unit citric acid. Through a series of successive oxidations, ATP and CO_2 are formed and oxaloacetate is regenerated (Fig. 12-7).

OXIDATIVE PHOSPHORYLATION

The fourth and final stage of energy extraction from fuel molecules is oxidative phosphorylation. In this process ATP is formed

$$CH_3-\overset{\overset{\displaystyle O}{\|}}{C}-COOH + CoA-SH \xrightarrow{\quad NAD^+ \quad NADH + H^+ \quad} CH_3-\overset{\overset{\displaystyle O}{\|}}{C}-S-CoA + CO_2$$

Pyruvate　　*Coenzyme A*　　　　　　　　　*Acetyl CoA*

Table 12-1. Electron-transport molecules

Abbreviation	Chemical name
NAD^+	Nicotinamide adenine dinucleotide
$NADP^+$	Nicotinamide adenine dinucleotide phosphate
FAD	Flavin adenine dinucleotide
FMN	Flavin mononucleotide
CoQ	Coenzyme Q (ubiquinones)
Cyto. b	Cytochrome b
Cyto. c	Cytochrome c
Cyto. a	Cytochrome a
Cyto. a_3	Cytochrome a_3 (cytochrome oxidase)

when high-energy electrons are transferred through a series of carrier molecules to molecular oxygen. In eukaryotes the system is located on the inner mitochondrial membrane and is often referred to as the respiratory assembly. The carrier molecules (Table 12-1) include the nicotinamides, flavins, quinones, and cytochromes. Some of them are coenzymes. Iron is a very important constituent of some of these molecules, it is the atom directly involved in the electron-transport process.

$$Fe^{+++} + e^- \rightarrow Fe^{++}$$

Oxidized iron   *Reduced iron*

In the cytochromes iron is found in the form of heme similar to that found in hemoglobin, although in some of the other molecules it is bound to protein and classified as nonheme iron. The electron carriers are arranged on the inner mitochondrial membrane in a specific sequence, based on their oxidation-reduction potential (**redox potential**). Redox potential is a chemical concept that enables us to measure the capacity of a substance to lose electrons. The more negative the redox potential, the greater the capacity of a substance to lose electrons or remain in the oxidized state. This substance when it loses electrons will then pass them to

a substance with less of a potential to lose electrons, and eventually to molecular oxygen. For example, in the reduction of pyruvate to lactate, NADH is oxidized or loses its electrons to pyruvate. NADH has a greater tendency to lose electrons than does pyruvate so that pyruvate will be reduced and NADH will be oxidized.

Pyruvate   +   NADH   →   Lactate   + NAD^+
oxidized   reduced   reduced   oxidized

The electron-transport carriers are arranged according to their potential to lose electrons (Fig. 12-8), NADH having the greatest potential and cytochrome a + a_3 the lowest. The initial acceptor molecule from the oxidized organic acid in the Krebs cycle is NAD^+ or in one instance FAD^+. The electron carriers between NADH and molecular oxygen consist of flavins, quinones, and cytochromes in that order.

$$NADH \rightarrow \left\{ \begin{array}{l} Flavins \\ Flavoproteins \end{array} \right\} \rightarrow CoQ \rightarrow Cyto.\ c_b$$
$$\rightarrow Cyto.\ c_1 \rightarrow Cyto.\ c \rightarrow \left\{ \begin{array}{l} Cyto.\ a \\ Cyto.\ a_3 \end{array} \right\} \rightarrow O_2$$

Cytochromes b, c_1, and c contain heme iron as the active electron transporter. The cytochrome a-a_3 complex contains a different form of heme iron and also copper. This complex is called cytochrome oxidase and is capable of passing the electron to molecular oxygen. Two hydrogen atoms from the cell then combine with the oxygen to form water.

The sequential oxidation of a series of electron carriers causes the electron to lose energy in a stepwise fashion. The oxidation process is very tightly coupled to the synthesis of ATP, one process will not occur without the other. ATP is generated at three specific sites in the electron transport chain (Fig. 12-9). The first site is the oxidation of NADH by the flavins; the second site is the oxidation of cytochrome b by cytochrome c; and the last site is the oxidation of the cytochrome a-a_3 complex by molecular oxygen. Thus for every electron passed from NADH to molecular oxygen, 3 moles of ATP are formed and for

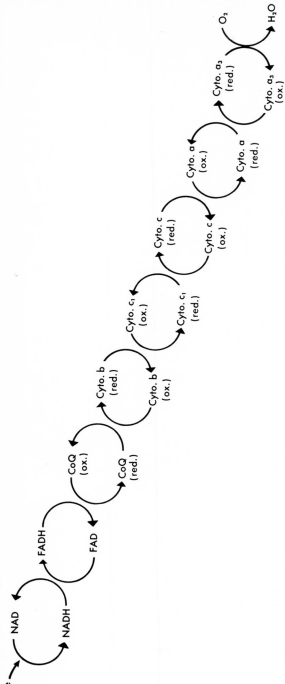

Fig. 12-8. Diagram of electron transport from a substrate to molecular oxygen. The electron-transport carriers are arranged according to their potential to lose electrons (redox potential). NAD has the highest potential and cytochrome a_3 the lowest.

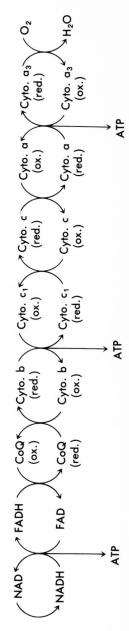

Fig. 12-9. Diagram of electron-transport chain depicting the sites of coupled ATP formation.

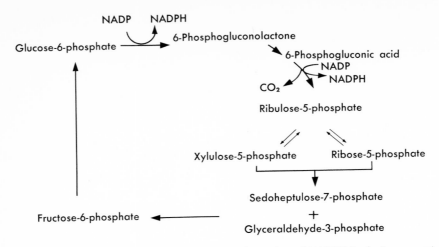

Fig. 12-10. Pentose-shunt pathway showing the formation of NADPH (reducing power).

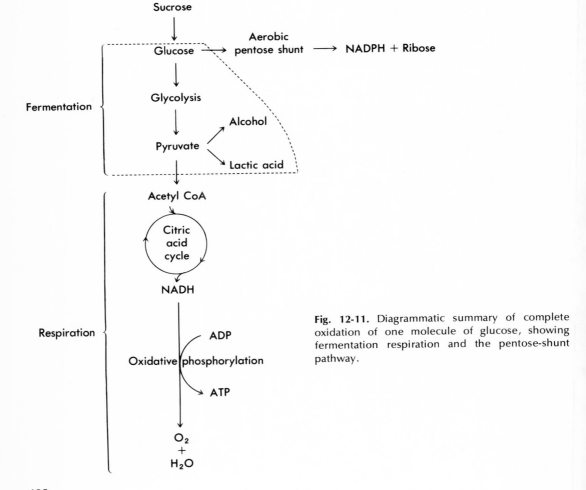

Fig. 12-11. Diagrammatic summary of complete oxidation of one molecule of glucose, showing fermentation respiration and the pentose-shunt pathway.

every electron passed from FAD to molecular oxygen, 2 moles of ATP are formed. For the complete oxidation of 1 mole of glucose, 36 moles of high-energy phosphate are formed. The exact mechanism to account for the coupling of oxidation to the synthesis of ATP is currently the subject of much biochemical research. Although three possible mechanisms have been proposed, to date the exact nature of coupling remains obscure. The reactions of glycolysis and the Krebs cycle generate high-energy phosphate bonds by the complete oxidation of glucose coupled to electron transport, as indicated in the initial discussion of this chapter, NADPH is also produced as a source of reducing power in order to drive biosynthesis. In addition, NADPH is different from NAD by a single phosphate group; however, there is a functional difference. NADH always functions in electron transport, whereas NADPH is always used as reducing power and does not participate in electron transport. NADPH is generated by the aerobic oxidation of glucose through the **pentose-shunt pathway** (Fig. 12-10). Within the first three reactions of this pathway 2 moles of NADPH are synthesized per mole of glucose-6-phosphate converted. In addition the pathway also functions by making available ribose for cellular synthetic processes. The pathway is found in the cytoplasm and is not associated with the mitochondria (Fig. 12-11).

QUESTIONS AND PROBLEMS

1. How is the energy cycle related to metabolism in a plant cell?
2. Distinguish between fermentation and respiration.
3. Describe the four stages of cellular energy extraction in a plant cell.
4. Define glycolysis. How is it related to fermentation?
5. What is the relationship of oxidative phosphorylation to the citric acid cycle?
6. List the electron-transport molecules, as they are found in the mitochondria.
7. Where are the sites of phosphorylation in the electron-transport chain?
8. How is NADPH generated and how does it differ from NADH?

DISCUSSION FOR LEARNING

1. Discuss the relationship of living systems to free energy and entropy.
2. Discuss the relationship of photosynthesis, glycolysis, and the citric acid cycle.

ADDITIONAL READING

Bronk, J. 1973. Chemical biology. The Macmillan Co., New York. (Basic biochemistry text with metabolic pathways.)

Lehninger, A. L. 1965. Bioenergetics. W. A. Benjamin, Inc., New York. (Biology-teaching monograph series, discusses the molecular basis of biological energy transformations.)

Lowy, A. G., and P. Siekevitz. 1969. Cell structure and function. Holt, Rinehart & Winston, Inc., New York. (Cell biology text with a good discussion of life and the second law of thermodynamics.)

Stryer, L. 1975. Biochemistry. W. A. Freeman & Co., San Francisco. (Biochemistry text with good discussion and illustrations of oxidative phosphorylation.)

CHAPTER 13

Photosynthesis

■ Certain organisms, particularly a few bacteria, the algae, and most of the higher plants, are capable of transforming energy from light into energy stored in chemical bonds. This process, known as photosynthesis, or the synthesis of organic food material using light as the energy source, was known to biologists by the end of the eighteenth century. Without a doubt it is the most essential natural phenomenon we know because almost all existing organisms depend on the products of photosynthesis for a source of food. With the search for new energy sources, this may well serve as a model for the conversion of solar energy directly into heat, light, and electricity. Although much is understood about the process, by no means is the entire process known.

Simply stated, certain green plant cells use two very low energy compounds, carbon dioxide (CO_2) and water (H_2O), in the presence of chlorophyll and with energy furnished by sunlight to manufacture the sugar glucose. It, like respiration, is an oxidation-reduction process. Glucose, a carbohydrate, has a much higher energy level than either of the compounds used in its synthesis. Energy from sunlight is trapped to bind together carbon atoms to form glucose, and thus energy is stored. During the process molecular oxygen is given off as a by-product. Photosynthesis actually takes place in two steps,

first the light reaction, followed next by the dark reactions. Naturally, these reactions are continuous. Once synthesized, glucose is used in the manufacture of other food materials such as protein, lipids, and nucleic acids. The chemical reaction is written as follows:

$$\text{Carbon dioxide} + \text{Water} \xrightarrow[\text{chlorophyll}]{\text{light}} \text{Glucose} + \text{Oxygen}$$

or with chemical notations:

$$6CO_2 + 6H_2O \rightarrow C_6H_{12}O_6 + 6O_2$$

SITE OF PHOTOSYNTHESIS

The location of photosynthesis depends primarily on the kind of organism. The site in bacteria, for example, is either in a series of infolded membranes or in a cigar-shaped organelle. The algae, with the exception of the blue-green algae, on the other hand, contain plastids, special organelles containing chlorophyll of one type or another. The higher green plants and some of the higher algae house the photosynthetic apparatus in cells in highly organized structures called chloroplasts (see Chapter 4). The chloroplast (Fig. 13-2) is bounded by a single outer membrane. The inner membrane is continuous and forms a series of complex infoldings arranged in the form of many flattened discs called **thylakoid discs.** These discs are stacked transversely across the chloroplast.

Fig. 13-1. Wherever there is sunlight, water, and carbon dioxide, green plants are able to manufacture their own food, even under the extremely arid conditions of a desert region such as this scene of century plants in southern New Mexico. Here water and possibly carbon dioxide are limiting factors. The plants are modified for water conservation and water storage to achieve maximum and rapid growth even when conditions are unfavorable for photosynthesis. (Courtesy U.S. Forest Service.)

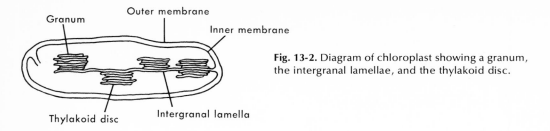

Fig. 13-2. Diagram of chloroplast showing a granum, the intergranal lamellae, and the thylakoid disc.

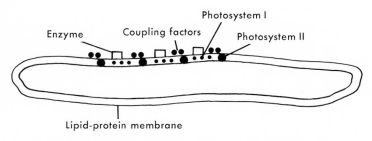

Fig. 13-3. Diagram of ultrastructure of a thylakoid disc.

The stacks are called **grana** and contain all of the photosynthetic pigments and the enzymes necessary to catalyze the light-dependent reactions. The ultrastructure of a thylakoid (Fig. 13-3) shows a fairly simple arrangement of the membranes. The positioning of specific structures within the thylakoid appears to be necessary for the energy-coupling process to take place. This is reminiscent of mitochondrial structure and their relationship to electron transport and oxidative phosphorylation. Photosynthesis may take place in chloroplasts experimentally isolated from living cells, showing the independence of these organelles. Some protozoans have been known to carry on the process after ingestion of plant chloroplasts.

From evidence of the rate of photosynthesis under different conditions, it is concluded that photosynthesis is not a single photochemical reaction. It took many long and complex but interesting experiments to unravel the intricate biochemical pathways of the photosynthetic process. Early investigators such as Jan van Helmont (1604) proved that plants get some of their nourishment from the air and that the gaseous exchange in plants is affected both by light and by darkness; thus he recognized two distinct reactions, the light reaction and the dark reaction. At least one dark reaction is necessary. It is established that, in the light, carbon dioxide is absorbed and oxygen is given off (it is also partially reabsorbed for plant metabolism). In the dark, carbon dioxide is given off and oxygen is absorbed (for metabolism). Although early experiments were conducted with use of the leaves of land plants, recently much of the work has been done with a one-celled green plant, *Chlorella vulgaris*. One reason is that this plant is most easily kept under controlled conditions. By making use of the isotope ^{14}C, or radioactive carbon, Melvin Calvin (1956) was able to work out the sequence of reactions involved in the carbon cycle of photosynthesis. Still other workers pieced together other parts of the process. In other words, some part of the process of photosynthesis is not affected by light. It is believed, then, that a dark or thermochemical reaction follows the light or photochemical reaction and that these are consecutive reactions as is known from temperature coefficients and from the use of flashing light.

So that we might better understand the

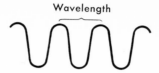

Fig. 13-4. Diagram of a wave, illustrating wavelength.

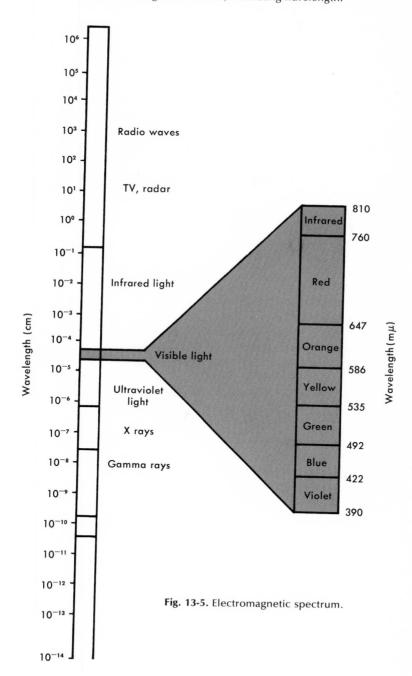

Fig. 13-5. Electromagnetic spectrum.

flow of energy from sunlight to the chemical bond and to the driving of biosynthetic reactions, we will dissect the photosynthetic process into its components, the light reaction and the dark reaction.

THE NATURE OF LIGHT

Prior to a discussion of the photosynthetic light reactions, we need to consider the nature of light. Light possesses the dual characteristics of a particle and of waves. Initially Isaac Newton (1660) postulated that luminous objects were emitting a series of discrete particles, which composed light. However, there were many properties of light that could not be explained on the basis of this theory. By the midnineteenth century, Maxwell (1850) and others concluded that light also has the characteristics of a wave. Wave motion can best be described as the transfer of energy from one point to another without the physical transfer of materials between the two points. When the medium used for the energy transfer is either air or water, the resultant wave is a mechanical one. Good examples of this are the shock wave of an explosion and the waves in water that cause a boat to oscillate. When the disturbance is electrical and magnetic instead of mechanical, electromagnetic waves result. Light has the characteristics of an electromagnetic wave. The wavelength of a wave is measured from crest to crest (Fig. 13-4). The shorter the wavelength, the more energy it contains.

On the average the wavelength of light is about 2000 waves per millimeter. The position of light in the electromagnetic spectrum (Fig. 13-5) is between 10^{-4} and 10^{-5} cm. The wavelength that affects plant growth is in the range of 300 to 800 millimicrons (or nanometers). When light strikes the surface of a conductor metal, it causes the ejection of an electron from the metal (photoelectric effect). This observation was not consistent with the wave theory of light. When an electromagnetic wave strikes an atom, some of its energy is imparted to an electron. The energy from the wave is much greater than the energy holding the electron to the nucleus of the atom. As a result of this collision, the electron can be ejected from the atom. To explain the photoelectric effect, one must employ a particulate theory of light. During absorption or emission, light behaves as a particle. It is absorbed or emitted in discrete packets called **quanta.** A single packet of electromagnetic radiation is called a **photon.**

PIGMENT SYSTEMS

Generally there are two distinct photoreaction centers in the leaf surface capable of absorbing photons and raising the energy level of an electron. Each photocenter is found in the chloroplast, and it consists of several pigments, the most active of which is chlorophyll. One of the most characteristic properties of chlorophyll is that it shows a brilliant fluorescent band in the red part of the spectrum when illuminated. After illumination, chlorophyll will emit light in the dark for as long as 2 minutes, a characteristic called **bioluminescence.** All green plant cells grown in the dark appear white, or etiolated, but contain protochlorophyll, which is rapidly reduced to chlorophyll when the plant is placed in the light. At least eight types of chlorophyll can be distinguished. Green plants contain chlorophyll a and b, which absorb the wavelengths of red and blue light. Chlorophylls c, d, and e are found only in algae also containing chlorophyll a. Several other pigments, found in the photosynthetic bacteria, are classified as bacteriochlorophylls. The molecular structure of chlorophyll (Fig. 13-6) is closely related to that of the blood pigment heme and the cytochromes. Heme and the cytochromes have iron instead of magnesium and do not contain the phytyl side chain of chlorophyll. All chlorophylls absorb light in the visible spectrum because of the presence of many double bonds in their molecular structure. The absorption spectrum of chlorophyll from spinach leaves (Fig. 13-7) indicates that light is absorbed in the blue (400 nm) and red (640 to 650 nm) region of the visible spectrum. Also associated with the photoreaction centers and capable of absorbing light are two classes of accessory pigments, carotenoids and phycobilins. The carotenoids are orange-yellow pigments with a complex molecular structure. They are differentiated into two main groups, the carotenes and xanthophylls. The carotenoid most often found in plants is β-carotene, which has received more attention as vitamin A than as a photosynthetic component. Although the chlorophylls alone absorb sufficient light for photosynthesis, it has been shown that the carotenoids can absorb light, an ability that in turn may also be used in photosynthesis. Although the absorption of light by carotenoids is not too important in the higher plants, it is of considerable importance in certain algae and bacteria. It is not known whether they transfer the radiant energy directly to the carbon dioxide reduction process or whether it is transferred to the chlorophyll and then used in photosynthesis. Nevertheless, it is evident that light absorption by pigments other than chlorophyll is not necessary for photosynthesis in the higher plants. Tissues that contain carotenoids and lack chlorophyll do not photosynthesize. The phycobilins are red or blue pigments found only in algae and function in the same manner as do the carotenoids.

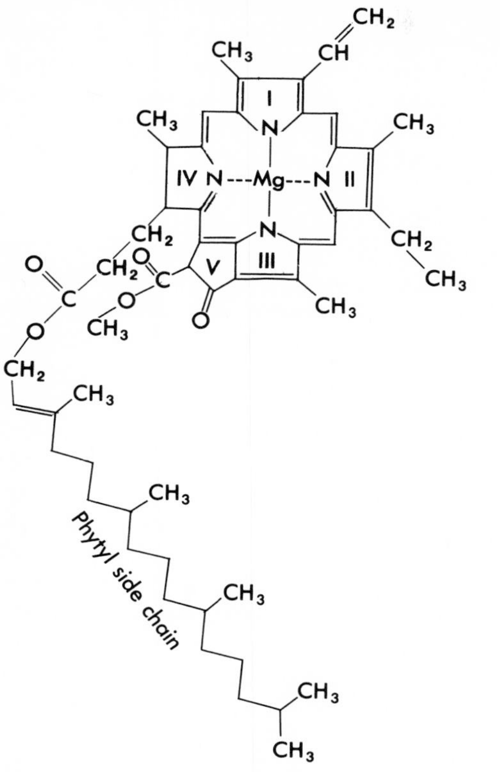

Fig. 13-6. Structural formula of chlorophyll A molecule.

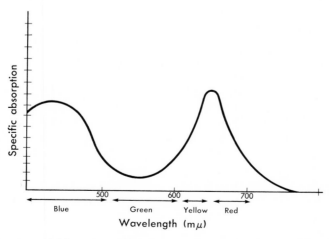

Fig. 13-7. Absorption spectrum of chlorophyll showing absorbance in the blue and red regions of the spectrum.

LIGHT REACTIONS

The light reactions are those that convert the energy of the photon into chemical bond energy of adenosine triphosphate (ATP) and reduced nicotinamide adenine dinucleotide phosphate (NADPH). The ATP formed during the process remains in the chloroplast and is used in the dark reaction. The NADPH is used as reducing power in the dark reaction, and it can also be transported out of the chloroplast and used in the cytoplasm. The light reaction is also involved in the generation of oxygen. Hill (1937) demonstrated that chloroplasts in the presence of water, light, and a hydrogen acceptor would cause the evolution of oxygen. Prior to that time it was assumed that the oxygen was derived from carbon dioxide. Overall, two photocenters interact to reduce $NADP^+$, using water as an electron and hydrogen source. During the electron transport involved, ATP is generated. Photosystem I is characterized by an active chlorophyll molecule with a maximum absorbance at 700 nm.

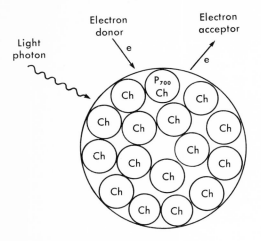

Fig. 13-8. Diagram of a photosynthetic unit at the photoreaction center. A photon of light strikes the center causing the center to accept an electron from an electron donor and to donate one to an electron acceptor. *Ch,* Chlorophyll.

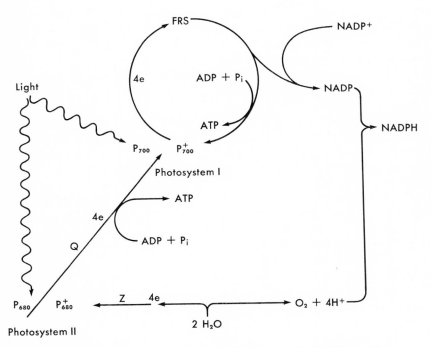

Fig. 13-9. Schematic of electron flow in the photosynthetic light reactions.

This molecule (P_{700}) is chemically identical to the other molecules in the photosynthetic unit. Its activity and uniqueness is attributable to its position, because it is situated in proximity to its electron donor and electron acceptor. It accepts electrons transmitted to it by the nearby chlorophylls when the unit has been photoexcited. It can then pass this electron to the electron acceptor nearby. To restore electroneutrality, a donor molecule passes an electron to P_{700} (Fig. 13-8).

Photosystem II has not been characterized as well as photosystem I. However it appears to act in the same manner as photosystem I, with the active chlorophyll having a maximum absorbance peak at 680 nm (P_{680}). Oxygen evolution is associated with photosystem II. Let us now follow the flow of electrons in the light reactions (Fig. 13-9). A photon of light strikes chlorophyll P_{680} in photosystem II, exciting an electron and causing its transfer to an acceptor molecule called Q. P_{680} regains the lost electron from an electron donor Z. After 4 electrons have been transferred, Z reestablishes electroneutrality by reclaiming 4 electrons from 2 molecules of water. In this process 1 molecule of oxygen and 4 protons are formed. The excited electron moves from Q through an electron-transport system to a molecule of plastocyanin, which is the electron donor for P_{700} in photosystem I. During the transport of this electron, ATP is generated. At photosystem I a photon excites P_{700}, causing the ejection of an electron to an acceptor molecule of ferrodoxin-reducing substance (FRS). At this juncture the electron can take one of two paths. Through electron transport molecules it can be cycled back to P_{700}^+, with ATP being generated in the process. It is then referred to as a cyclic electron. The second path is the passage to $NADP^+$, with its reduction in NADP. NADP secures a proton to form NADPH and acts as reducing power. In summary, electrons from water are used to reduce $NADP^+$. This is accomplished by two photoreactions in series. When the water molecule gives up the electrons, it becomes oxidized and forms oxygen and protons.

When an electron is cycled from P_{700} back to P_{700}^+ to generate ATP, this is called cyclic photophosphorylation.

THE DARK REACTION

Early experiments demonstrated that chloroplasts illuminated in the absence of carbon dioxide would fix (that is, incorporate into an organic compound) carbon dioxide when the chloroplasts were placed in the dark. The carbon dioxide fixation in the dark was of short duration but significant. It became apparent that there was a dark reaction and a light reaction. The fixation of carbon dioxide into carbohydrate is the dark reaction. There are two distinct mechanisms responsible for the initial fixation process in green plants. The first is the 3-carbon pathway, which was elucidated mainly by Melvin Calvin and his co-workers. They initially observed that when algae were exposed to carbon dioxide that contained radioactive carbon 14, carbon dioxide was rapidly fixed. After 5 seconds, 80% of the radioactive carbon was found in the molecule 3-phosphoglyceric acid.

$$^{14}COOH$$
$$|$$
$$HCOH$$
$$|$$
$$CH_2OPO_3H_2$$

3-Phosphoglyceric acid

Initially investigators believed that carbon dioxide was fixed by a 2-carbon compound. However further research proved that ribulose-1,5-diphosphate is the compound that reacts initially with CO_2. The reaction is catalyzed by the enzyme ribulose-1,5-diphosphate carboxylase, which is bound to the surface of the thylakoid.

$$CH_2OPO_3H_2$$
$$|$$
$$C=O$$
$$|$$
$$H-COH$$
$$|$$
$$H-COH$$
$$|$$
$$CH_2OPO_3H_2$$

Ribulose-1,5-diphosphoric acid

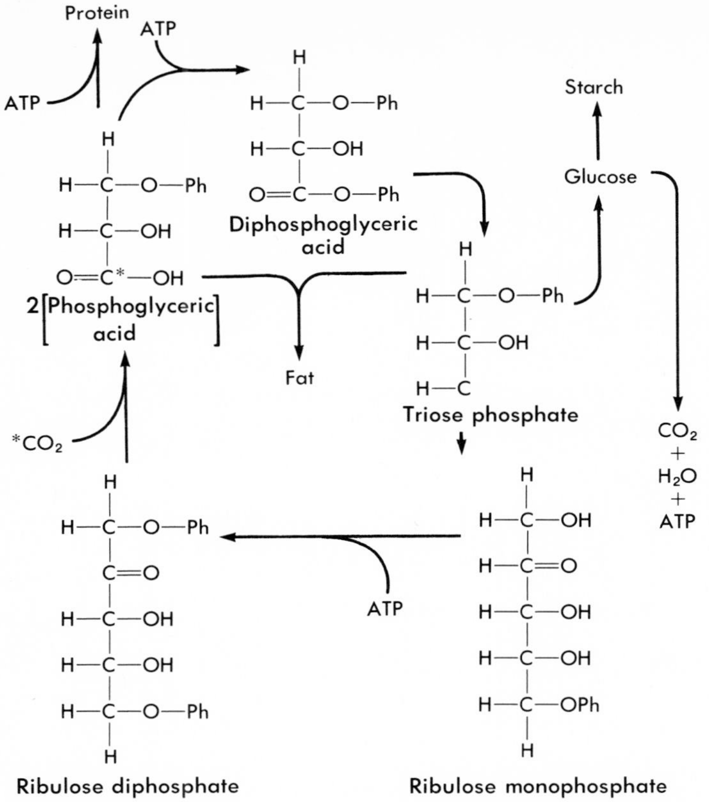

Fig. 13-10. Diagram of abbreviated Benson-Calvin cycle.

The identification of 3-phosphoglyceric acid as an early product of carbon dioxide fixation led to the assumption that this compound was a precursor of glucose. This assumption was essentially correct, and it led to the formulation of the Benson-Calvin cycle (Fig. 13-9), which is the pathway of carbon assimilation in green plant cells. This cycle provides a mechanism whereby the plant can accumulate glucose and regenerate ribulose diphosphate for further carbon dioxide fixation. Two very important points are illustrated in the Benson-Calvin Pathway (Fig. 13-10). First, 1 mole of ribulose diphosphate fixes (combines with) 1 mole of carbon dioxide to produce 2 moles of 3-phosphoglyceric acid, which is eventually used in the synthesis of glucose. Second, energy in the form of ATP is used to drive the reaction. As we have

seen, this energy is generated by the photosynthetic light reactions.

HATCH-SLACK PATHWAY

Over 100 genera and at least 10 families of plants (including corn, sugarcane, sorghum, crabgrass and Bermuda grass) grow well in an hot arid environment and have an increased potential for efficient water utilization. This adaptation is due to the development of a carbon-fixation pathway that enables them to increase the efficiency of photosynthesis under more or less adverse conditions. This is known as the **Hatch-Slack pathway** described below (Fig. 13-11). There was much experimental evidence to indicate that the Benson-Calvin mode of carbon dioxide fixation was not the only one operating in photosynthetic systems. In some pho-

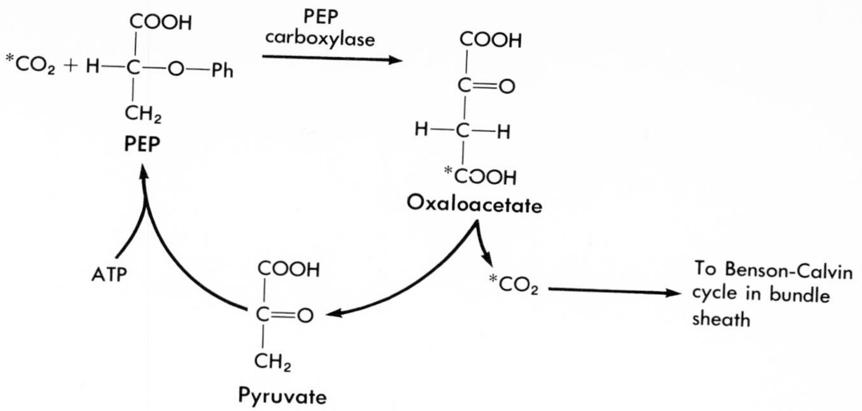

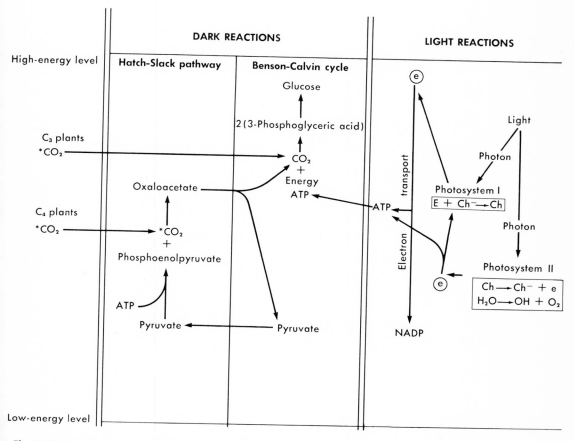

Fig. 13-11. Diagram of Hatch-Slack pathway found in C_4 photosynthetic dark reactions.

Fig. 13-12. Summary of photosynthesis showing the dark and light reactions, the relationship of the Hatch-Slack and Benson-Calvin cycles, the energy levels, and the mechanism employed by C_3 and C_4 plants. e, Electron; *Ch*, chlorophyll.

tosynthetic bacteria, the first detectable ^{14}C-labeled compound is alanine. When some higher plants, particularly tropical grasses, were exposed to radioactive carbon dioxide the first compounds labeled were the 4-carbon acids, oxaloacetate, malate, and aspartate. In addition to this unique fixation process, such plants also show two forms of chloroplasts. In the bundle sheath cells are large open chloroplasts with membraneous lamellae and few grana. In the surrounding mesophyll the chloroplasts are tightly packed, are much smaller, and exhibit an abundance of grana. The mesophyllic chloroplasts take up $^{14}CO_2$ at a very rapid rate, but do not store starch. In contrast, the chloroplasts in the bundle sheath store starch and fix carbon dioxide at a much slower rate. It was also observed that the enzyme phosphoenolpyruvate (PEP) carboxylase was bound loosely to the surface of the mesophyllic chloroplasts. This enzyme was almost totally absent from the bundle-sheath chloroplasts. This was an extremely significant discovery because this enzyme catalyzes the initial fixation of carbon dioxide into PEP, forming oxaloacetic acid. Then depending on the species of plant being observed, oxaloacetate may be converted to other 4-carbon acids such as malate and aspartate. The 4-carbon acid is then transported to the bundle sheath where it is hydrolyzed to pyruvate with the simultaneous release of carbon dioxide. The carbon dioxide is then fixed into 3-phosphoglycerate by the first reaction of the Calvin-Benson cycle. Plants containing the Hatch-Slack pathway are referred to as C_4 plants. Three of the main distinguishing characteristics of the C_4 plants are as follows: (1) The unique distribution and kind of chloroplast found in the mesophyll and bundle sheath cells. (2) The high concentrations of certain enzymes such as PEP carboxylase in the leaves. (3) The ability to fix carbon dioxide at extremely low concentrations. The elucidation of the Hatch-Slack pathway has far-reaching agricultural implications. Through plant-breeding experiments and genetic manipulation it will be possible to produce agriculturally useful plants that grow in hot arid environments (Fig. 13-12).

FACTORS INFLUENCING THE RATE OF PHOTOSYNTHESIS

The rate of photosynthesis is dependent on light, concentration of carbon dioxide, temperature, water, and other factors in the environment, such as the organic and inorganic materials available. It is evident that one or a combination of these factors may decrease or increase the rate of photosynthesis and that any one factor is probably influenced to some extent by the others. In addition, the general physiological activities of the cell will likewise have some bearing on the photosynthetic rate. The photosynthetic rate is most often determined by measuring either the oxygen evolution or the carbon dioxide uptake.

Three aspects of light are important for photosynthesis: intensity, quality, and duration. Intensity depends on the number of quanta falling on a given chloroplast per second, regardless of quality. This intensity affects the life cycle of some plants. Quality refers to the color bands making up light. A tungsten bulb produces light rich in red but low in blue, whereas a fluorescent bulb is richer in blue than in red.

The amount of light (intensity) upon which a cell can maintain itself with the sugar produced in a day is called the critical light intensity. The rate of photosynthesis will be increased if the light intensity is increased up to its critical or maximum intensity. It has been shown that maximum carbon dioxide fixation (photochemical reaction) occurs at light intensities varying from 200 to 10,000 footcandles.

When a suspension of the alga *Chlorella* is exposed to flashing light, more photosynthesis occurs per unit dose of light than when a suspension is exposed to continuous light. This indicates that light is used more efficiently when it is followed by a dark period. The lower the temperature, the longer will be the dark period required for the completion of the thermochemical reaction. Therefore, if the light intensity is sufficient to bring about the maximum photochemical reaction, increasing the temperature will increase the rate of photosynthesis. However, if the light intensity is low, thereby limiting the photochemical reaction, the temperature may be varied from 15° to 32° C. without greatly altering the rate of photosynthesis.

The rate of photosynthesis in cells exposed to full sunlight for 10-minute periods varies with temperature and carbon dioxide concentration. When carbon dioxide concentration is increased, the rate of photosynthesis will increase in proportion to the increase in temperature.

The water available to a cell also influences the rate of photosynthesis. Since water is one of the raw materials essential for photosynthesis, the rate of carbohydrate synthesis will be reduced according to the decrease in available water. Water is also necessary for the numerous other chemical processes in a plant cell, and any decrease will affect its metabolism, in addition to affecting those processes connected with photosynthesis.

Extremes of temperature will diminish, if not entirely inhibit, photosynthesis. However, temperature is interrelated with the concentration of carbon dioxide.

Although water, temperature, carbon dioxide concentration, and light are the most important factors influencing photosynthesis, there are also various other environmental and structural factors, which are discussed in Chapters 15 and 20.

It is clear, then, that photosynthesis by green plant cells, in addition to being the only ultimate source of food on earth, is also the only source of animal energy. Since wood, peat, coal, and oil have been our principal sources of heat, light, and industrial power, photosynthesis may be considered responsible for fulfilling many of the energy requirements of modern civilization except, of course, those now being supplied by water power and, more recently, by nuclear reactions.

QUESTIONS AND PROBLEMS

1. What wavelengths of light are most beneficial for photosynthesis?
2. What is meant by photosystems I and II?
3. What is meant by cyclic photophosphorylation?
4. Describe the importance of water in the photosynthetic light reaction.
5. What is the source of oxygen produced during photosynthesis?
6. What is meant by carbon dioxide fixation?
7. Would it be possible to fix carbon dioxide in the absence of the light reactions? Explain.
8. Describe the Benson-Calvin pathway.
9. What are the distinct advantages of the Hatch-Slack pathway?
10. Describe the relationship of the Benson-Calvin pathway to the Hatch-Slack pathway.

DISCUSSION FOR LEARNING

1. Today we are faced with the problems of energy conservation and world hunger. Discuss how photosynthesis can be used to investigate possible solutions to these problems.

ADDITIONAL READING

Beiser, A., and K. Krauskopf. 1969. Introduction to physics and chemistry. The McGraw-Hill Book Co., Inc., New York.

Bjorkman, O., and J. Berry. 1973. High efficiency photosynthesis. Sci. Am. **229**:80.

Devlin, R. M. 1969. Plant physiology. D. Van Nostrand Co., Princeton, N.J.

Govindjee, G., and R. Govindjee. 1974. The absorption of light in photosynthesis. Sci. Am. 231(6):68.

Lehninger, A. 1973. Biochemistry. Worth Book Publishers, New York.

Levitt, J. 1974. Introduction to plant physiology. The C. V. Mosby Co., St. Louis.

Mahler, H. R., and E. H. Cordes. 1971. Biological chemistry. Harper & Row, Publishers, New York.

Shortley, G., and D. Williams. 1959. Principles of college physics. Prentice Hall, Inc., Englewood Cliffs, N.J.

White, A., and P. Handler. 1973. Principles of biochemistry. The McGraw-Hill Book Co., Inc., New York.

CHAPTER 14

Roots, stems, and leaves of flowering plants

■ All living organisms in the world around us, and we ourselves, are the result of adaptation. This chapter treats of the results of this process as they affect the vascular plants. The particular adaptations of the lower plants are discussed in those chapters dealing with those plants.

Programmed growth results in the relatively uniform habitus of the species. At the same time, the great variation possible in the arrangement of the DNA molecule and in its genetic processes provides for the vast array of plant life of the present and in the past. Certain features, however, are common to many kinds of plants, and these features may be grouped together to demonstrate the phenomenon of adaptation. Also, many of these growth structures are useful in classification schemes. In fact, it is only after a recognition of the adaptive structures and their distinction from the nonadaptive structures that we are able to construct a classification based on evolutionary results. On the other hand, it is mostly because of these interspecific adaptive features that we are able to recognize the species of plants. All the main growth structures—roots, stems, leaves, and flowers—and the resulting fruit and seeds are involved. These are typified by a tree (Fig. 14-1) and by the diagram in Fig. 14-2. Form and

adaptation are not restricted to the external shape of a plant. Cells show many different shapes according to their function and position (Fig. 14-3).

Seeds develop into a plant with the growth habits of its species. Some plants grow from seed and produce flowers, fruit, and new seed in one growing season. These plants are termed **annuals** and are characteristic of herbaceous plants. At the end of their growth period they die down. The same cycle of seed to seed is repeated the next year. Other plants produce seeds that germinate and produce a rosette of leaves and a taproot the first growing season. The second year they produce a flower stalk and set seed, using the food reserves in the tap root. This type of plant growth is called **biennial.** Many of our common vegetables, carrots, beets, and parsnips, for example, are biennials. They are harvested, of course, before the second year of flower and seed production. The final type of growth habit is termed **perennial.** Plants of this type are usually woody, and the plant body persists for several or many years, producing flowers and seeds only after a few years of growth. After the initial growth, the shrub, bush, vine, or tree continues to grow year after year, but more slowly. Once they have reached their maximum height these

Fig. 14-1. A tree, such as this Brisbane box tree in Queensland, represents the maximum of functional development of simple cells into roots, stems, and leaves.

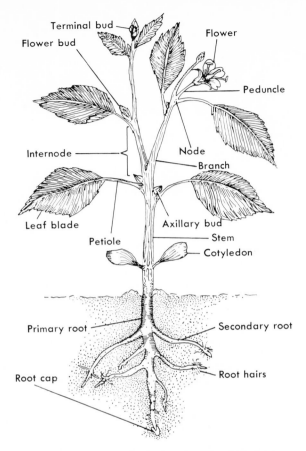

Fig. 14-2. Organs and systems of a flowering plant.

plants continue to replace limbs and twigs that may die back through disease or insect damage or that are broken during storms. Both evergreen and deciduous trees replace their leaves as they are lost.

ORGANIZATION OF THE PLANT BODY

Whether a plant is a single cell or is made of billions of cells, such as complex vegetation, the functions of obtaining raw materials, internal transportation, elimination and secretion, synthesis, liberation of energy, support and protection, locomotion, and, finally, self-perpetuation are necessary to sustain life. A one-celled organism must carry out all these complex processes within the confines of its cell membrane, whereas in a

many-celled organism certain cells become specifically adapted to carry on one or more processes exclusively; in other words, in a many-celled organism the cells become specialized. Thereby, through a division of labor among the cells, it appears that the vital functions are performed more efficiently in a multicellular plant than in a single-celled plant that perhaps may be functionally more versatile.

Whereas there may be varying degrees of specialization among cells, there is very little specialization in the embryonic tissues of plants, the cells of which differentiate as the requirement arises. All in all, there appears to be great versatility as well as great economy in the resources of a multicellular organ-

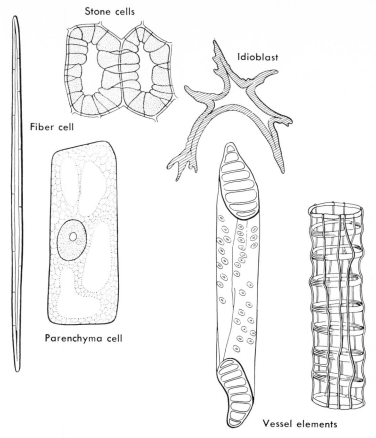

Stone cells

Idioblast

Fiber cell

Parenchyma cell

Vessel elements

Fig. 14-3. Cells of plants show many different shapes according to their function and position. Those above show some of the many adaptations at the cellular level.

ism. (Similar cells specialized to perform a particular function are grouped together to form **tissues.** The cooperative specialization of a number of tissues leads to the formation of an **organ.** Finally, the close cooperation of several organs develops into the **organ system.** Several of these organ systems compose the organism in higher plants.) The structure and function of the cells constituting the tissues, the organs, and the organ systems in large measure determine the organism. However, the difference between organisms is more than that, since it lies in structural organization rather than in the sum of each organism's parts. For example, a mass of carbon atoms may form a diamond, graphite, or soot. The secret of each lies in its structural

organization. In much the same way this holds true for the variety of plant life found on the earth today.

THE SEED AND ITS GERMINATION

The embryo of dicot plants is differentiated into five main parts: the **hypocotyl,** which is the lower primitive stem; the **epicotyl,** or upper stem; the **radicle,** or primitive root; two seed leaves, the **cotyledons,** which are attached to the upper portion of the hypocotyl; and the **plumule,** which is a bud located between the cotyledons (Fig. 14-4). In the dicot, such as the garden bean, all the stored food is in the tissues of the embryo, particularly the cotyledons. Therefore many dicot cotyledons appear thickened and have lost

215

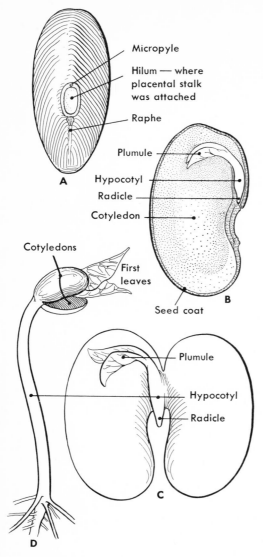

Fig. 14-4. Bean seed (dicot), its parts and development.

water, the embryo and seed coat swell to rupture the seed coat. No matter how the seed is oriented in the soil, the root always grows downward into the soil (positive **geotropism**) and rapidly develops root hairs. The initial or **primary root** readily produces branch or **secondary roots,** thus anchoring the young plant firmly in the ground. In the bean the cotyledons emerge from the seed, turn green, and serve as the first leaves of the seedling plant (Fig. 14-5). Eventually the cotyledons wither and drop off when their reserve supply of food is depleted. The more or less permanent leaves of the plant develop on the shoot located above the cotyledons. The young plant is now capable of sustaining itself through food manufactured by photosynthesis.

The seeds of all plants have embryos, but not all are without an **endosperm.** In a grain of corn the endosperm persists and is not absorbed until germination begins. The endosperm of the corn seed is a prominent tissue, rich in accumulated food in the form of oil, starch, or protein. The embryo of the corn is embedded in the endosperm (Fig. 14-6), which developed from the polar nuclei in the embryo sac (life cycle 23). The milk of the coconut consists of a pint or more of fluid containing numerous nuclei and is the free nuclear stage in the development of its endosperm. As the seed matures, these nuclei migrate toward the outer side, cell walls form around them, and the mature endosperm or coconut meat develops in much the same way as that of dicots, except that only one cotyledon (Fig. 14-6) is formed, rather than the two described for the bean.

The primary function of this one cotyledon is the digestion and translocation of the reserve food in the endosperm to the growing parts of the seedling. During germination of the corn seed (Fig. 14-7) the **radicle** grows downward and the plumule grows upward to produce the stem and leaves of the new plant. The corn grain, with its endosperm and cotyledon, both storage tissues, remains in the soil. Protecting the plumule from injury as it grows upward through the soil is a

their leaflike characteristics. The plumule of dicot embryos is large and readily develops a leaf-bearing shoot upon germination.

The processes occurring from the time the embryo resumes its growth until the seedling is established are called, collectively, **germination.** The first noticeable evidence of the germination of a seed is the bursting of the seed coat and the protrusion of the **root** tip (Fig. 14-5). As a result of the absorption of

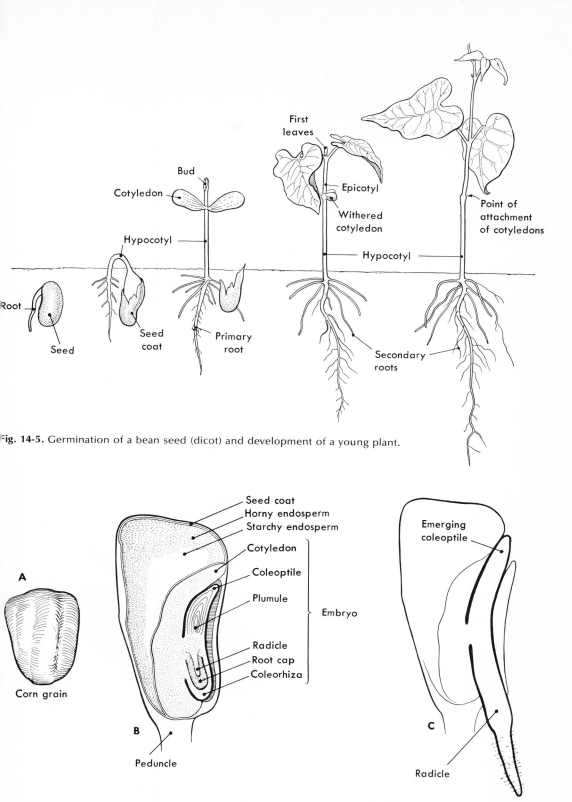

Fig. 14-5. Germination of a bean seed (dicot) and development of a young plant.

Labels (Fig. 14-5): Bud, Cotyledon, Hypocotyl, Root, Seed, Seed coat, Primary root, First leaves, Epicotyl, Withered cotyledon, Hypocotyl, Secondary roots, Point of attachment of cotyledons

Labels (Fig. 14-6): A — Corn grain; B — Seed coat, Horny endosperm, Starchy endosperm, Cotyledon, Coleoptile, Plumule, Embryo, Radicle, Root cap, Coleorhiza, Peduncle; C — Emerging coleoptile, Radicle

Fig. 14-6. Grain of corn (monocot seed) showing parts and early development of the embryo.

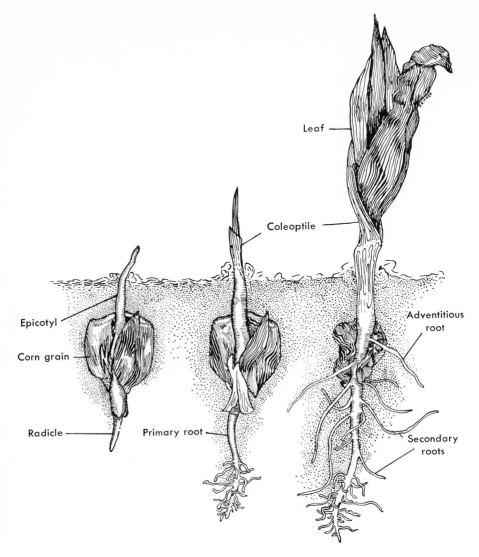

Leaf

Coleoptile

Epicotyl

Corn grain

Adventitious
root

Radicle

Primary root

Secondary
roots

Fig. 14-7. Germination of a corn seed (monocot) and development of a young plant.

pointed sheath called the **coleoptile.** This is regarded as the first leaf of the seedling, but one that does not expand to become a leaf characterististic of corn. The radicle does not persist in monocots, although it develops into the primary root and even produces branches or secondary roots. It is supplemented by a stronger root system that is adventitious in origin. The adventitious roots develop from the lower nodes of the stem just under the soil, and the plumule, the first node, and the coleoptile are pushed from the seed principally by elongation of the first internode of the seedling's stem. Elongation in an embryo is caused by the translocation of auxin, a growth hormone, from the tip of the coleoptile downward to the stem (Fig. 14-8). Auxin inhibits the growth of roots and leaves from the first node while stimulating the elongation of the internode. Once the coleoptile leaves the soil and enters the light, the production of auxin is reduced, and leaves and roots will develop. This makes certain that no matter how deeply a seed is planted, neither adventitious roots nor leaves develop until the coleoptile reaches the surface.

The radicle regularly becomes the primary root, and the epicotyl becomes the primary shoot in both dicots and monocots. However, the functions of hypocotyl and cotyledons are not the same in all seeds. In some, such as the peach (Fig. 14-9), the hypocotyl remains short, and the cotyledons never emerge from the seed. The epicotyl will appear above the ground only by its own growth. This type of seed exhibits hypogeal germination, in contrast to epigeal germination, in which the hypocotyl elongates, bringing the epicotyl and the cotyledons with it.

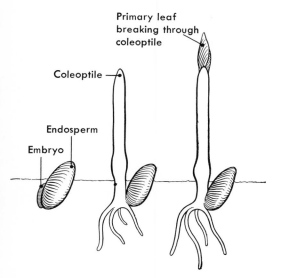

Fig. 14-8. Elongation of the coleoptile of a seed and the formation of the first leaves.

PARTS OF A SEED—THE POTENTIAL PLANT

Plumule—leaf bud
Epicotyl—upper stem
Cotyledons—Food storage for embryo tissue
Hypocotyl—lower stem
Radicle—root

The requirements for germination of all seeds are adequate amounts of moisture and oxygen, as well as a suitable temperature. There are some seeds, however, that require light for germination, but light inhibits the development of most seeds. The water content of most seeds is very low, and the water

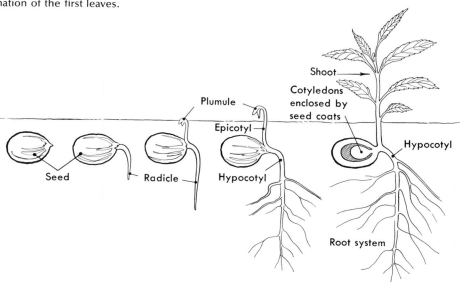

Fig. 14-9. Development of a peach seed (shell removed) into a young tree.

requirements for the physiological processes necessary for germination are high. Germinating seeds respire rapidly; therefore an adequate supply of oxygen is required. Until the time the seed coat is ruptured, respiration is almost entirely anaerobic. Aerobic respiration is more prominent as soon as the embryo ruptures the seed coat. The optimum temperature for the germination of seeds varies. For all, there is a maximum and a minimum, above or below which germination will not take place. The effect of light upon germination also varies since some seeds require light, whereas others are retarded by it.

Some seeds will not germinate when initially produced, even under favorable environmental conditions. Seeds that require a certain period to elapse before germination begins are said to be dormant. This may be attributable to several factors. The following are the most common: (1) seed coat is not permeable to water or oxygen or both, (2) seed coat is so tough that it resists rupture, (3) embryo is not mature, or (4) chemicals inhibit germination. The seed coat may have to decompose partially before oxygen or water can enter the seed to induce growth. This is true of very hard, stony seeds, such as the coconut. Seeds with hard, stony coats generally remain viable for longer times than do those with softer coverings. Most of our common crop plants produce seeds that remain viable one 1 year or at most a few years. However, others, such as the Oriental lotus, have been reported viable for what appears to have been 800 to 900 years.

DIFFERENTIATION IN HIGHER PLANTS

The term **differentiation** is used to describe any kind of change in a cell, either morphological or physiological or both, that occurs after the cell has ceased elongating. Originally, it was used to describe the morphological changes occurring in cells as they developed to maturity. Therefore, today, differentiation is used to designate the formation of cell substances as well as the structures produced by the cells.

It is appropriate to begin the discussion of differentiation with those cells capable of indefinite division of mitosis. Those tissues made up of cells undergoing division are called **meristem.** Two types are recognized: (1) meristematic tissues at the tip of roots and stems and (2) tissues in which cell division occurs laterally. However, any living cell may become meristematic if injury or some other abnormal condition arises.

The terminal or apical meristem of roots and stems is commonly called a growing point. The cells that make up the meristem are immature and have walls composed of a few thin layers of cellulose. There are no thickened secondary walls. The nucleus is large and surrounded by dense cytoplasm, in which may be found small and scattered vacuoles. The cells are isodiametric in shape. In other words, the cells have approximately the same length, width, and depth. Few, if any, intercellular spaces can be found in meristems. The most striking characteristic of the meristem is its capacity for repeated cell divisions.

The terminal meristem of stems possesses the potentialities for producing both vegetative and reproductive organs. The formation of leaves or flowers depends on either heredity or environmental factors, such as photoperiodism and temperature or both. It is also known that important growth regulators are also produced in the terminal meristematic tissue (see Chapter 17). The terminal meristem in both stems and roots is alike and is responsible for the increase in length of these structures. However, in roots the meristem produces cells that form the protective root cap, whereas in stems the growing point is protected primarily by the bud scales and leaf primordia.

The meristem is embryonic tissue that by cell division and enlargement will differentiate into various groups of tissues. A **simple tissue** is one in which the cells are essentially alike in structure and function. On the other hand, there are **complex tissues,** which consist of two or more different kinds of cells that perform a common function.

Simple permanent tissues

There are five kinds of simple permanent tissues in plants: **parenchyma, collenchyma, sclerenchyma, cork,** and **epidermis.**

The **epidermis** is a tissue covering the external surface of leaves, young roots, and stems. Primarily a protective tissue, it consists of a single layer of cells in stems compactly arranged and uniform in thickness. Epidermal cells are without chloroplasts, except in the cells that surround the openings (stomas) through which an exchange of gases takes place in leaves. Ordinarily the outer surface of epidermal cells is covered with a layer of a waxlike substance known as **cutin.** This waxlike secretion of the epidermis prevents excessive loss of water through evaporation.

A common tissue found throughout a plant is the **parenchyma,** cells that are large, thin walled, and relatively undifferentiated in shape and contain conspicuous intercellular spaces. If they do have chloroplasts, they are classified as **chlorenchyma.** Parenchyma cells are living and often serve to store foods.

Collenchyma (Fig. 14-10) is a supporting tissue frequently found in the outer layers of stems and in the petioles of leaves. Like parenchyma, the collenchyma cells are living cells, but elongated, and the cell walls are thickened with cellulose, particularly at the corners. Collenchyma cells are found in the stems of herbaceous plants and in the younger stems of woody plants.

Like collenchyma, **sclerenchyma** cells (Fig. 14-10) are supporting tissues that give

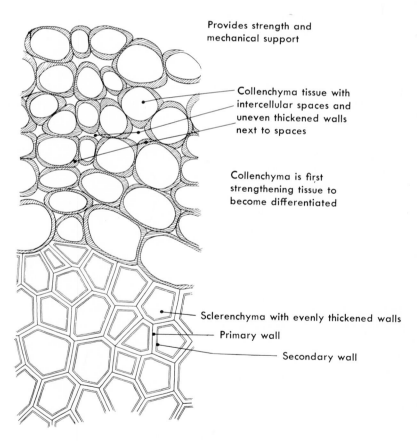

Provides strength and mechanical support

Collenchyma tissue with intercellular spaces and uneven thickened walls next to spaces

Collenchyma is first strengthening tissue to become differentiated

Sclerenchyma with evenly thickened walls

Primary wall

Secondary wall

Fig. 14-10. Two of the fundamental supporting tissues in plants are collenchyma and sclerenchyma shown here. Note the cell walls of each.

rigidity, toughness, and hardness to the plant, but unlike collenchyma, sclerenchyma cells are dead when they are mature because the entire wall is heavily and uniformly thickened. The wall of the mature cell is composed largely of cellulose and often, in addition, a hard substance called **lignin.** Thus sclerenchyma cells are said to be dead, lignified cells. Some sclerenchyma cells are long, slender, and pointed at both ends. These are called fibers and have great tensile strength and elasticity. Plants with very long fibers are useful in manufacturing cordage, linen, and cloth. Another type of sclerenchyma cell is the **stone cell.** They are similar to fibers, but much shorter, and sometimes form isolated groups of cells in the outer parts of stems, the hard shells of nuts, and the stones in cherries and peaches. They also account for the gritty texture of a pear.

Finally, there is the **cork tissue,** which consists of cells that, when mature, are dead because of the fatty substance in their walls (suberin). As a protective layer, cork eventually replaces the epidermis in older stems and roots.

Complex permanent tissues

Xylem and phloem are **complex permanent tissues.** Both are conductive in function, forming the greater part of the vascular system of plants. These tissues are described in detail for each major plant part. Their structure and arrangement differs from region to region and with the kind of plant.

ROOTS

The root of a flowering plant always begins in the seed. It starts its development from the hypocotyl of the embryo within the seed (Fig. 14-9). Once a seed germinates and absorbs water, the radicle grows out of the seed, producing the first or primary root of the newly developing plant. Soon the primary root produces branch or secondary roots, which in turn give rise to branches of their own. Roots ordinarily grow downward since they are positively geotropic (geotropism), even though the secondary roots

may at first grow out horizontally. By turning a potted plant over on its side, one can demonstrate that the roots will turn downward. Occasionally roots will develop from leaves or stems (Fig. 14-30). Roots that develop from parts other than the primary root or one of its branches are called **adventitious roots,** such as the prop roots of corn. Other examples are roots that develop from bulbs or rhizomes (underground stems). Nevertheless, all root systems, with the exception of aerial roots, find their way into the ground to form a complex root system.

If we examine any root, it will invariably be cylindrical, although tapered toward its free end. This narrowed end of the root is called the **root tip** and is that area of a root

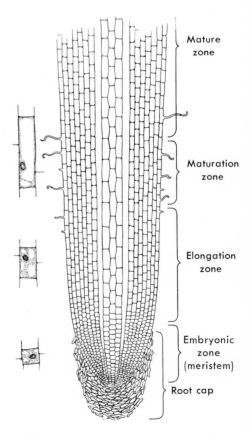

Fig. 14-11. Root tip and root cap with the major growing regions indicated.

(primary, secondary, or adventitious) in which unlimited growth and elongation take place (Fig. 14-11). Growth in diameter occurs only in the older parts of a root. Each root tip is covered by a cap of epidermal cells, the function of which is to protect the more delicate dividing cells that are found immediately behind it. Thus the **root cap** fits over the end of the root tip like the short finger of a glove. A region of actively dividing cells undergoing mitosis may be found just back of the root cap. This area of meristematic cells is called the **embryonic region,** an area in which the number of cells is being increased. It is possible to see cells in all stages of mitosis in this embryonic region when the root tip is properly prepared, sectioned, and stained to show chromosomes. Since these cells are continually undergoing mitosis, they remain the undifferentiated **apical meristem** of the root. The cells in the embryonic region are, therefore, the origin of those cells that contribute to the permanent tissues of the root, as well as of the root cap cells, which are constantly being sloughed off as the root tip pushes its way through the soil. Because the cells are rapidly dividing, they are thin walled, are mostly cubical in shape, and have a large nucleus within a dense cytoplasm. This region of the root tip appears dense in stained preparations and, in the living condition, has a yellowish color.

The cells of the embryonic region cease dividing as they become farther removed from the tip and gradually merge into a **region of elongation.** In this part of the root tip the cells, typically columnar in shape, mature and increase in length. Because of the absorption of water, the cells develop large vacuoles so that the cytoplasm appears less dense than in the mitotic cells of the embryonic region. However, the cells in the region of elongation have slightly thicker walls.

A more extensive area of the root tip is found behind the region of elongation. It is an area in which the cells are becoming differentiated and can easily be detected because of the appearance of numerous **root hairs** (Fig. 14-11). This area of the root tip is called the **region of maturation** or the **root hair region.** It is in this region of the root tip that the cells become specialized in structure and function, thus becoming differentiated into the permanent tissues of the root. The root hairs first appear as small swellings on individual epidermal cells. These swellings increase in length with the age of the cell, forming slender outgrowths of the epidermal cells of the root tip. Root hairs are easily demonstrated on grass seedlings grown on moist paper. They appear as a cottony mass just behind the region of elongation, and under the microscope they can be seen as tubular extensions of a single cell. Root hairs have thin walls lined with a thin layer of cytoplasm that surrounds a large central vacuole (Fig. 14-12). Usually a nucleus can be seen lying in the cytoplasm near the tip of the root hair. Root hairs are short lived, lasting only a few days or weeks. Therefore they are not present on the older parts of a root. As the root tip grows, new root hairs arise be-

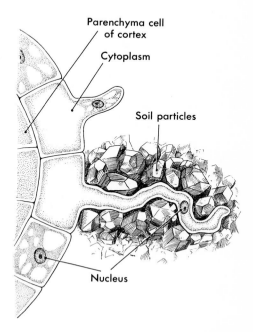

Fig. 14-12. Root hair cells showing growth of root hairs.

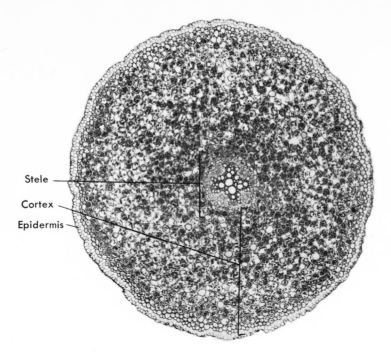

Fig. 14-13. Cross section of an older root of a buttercup showing the three distinct regions of a root. (×40.) (Courtesy George H. Conant, Triarch Products, Ripon, Wis.)

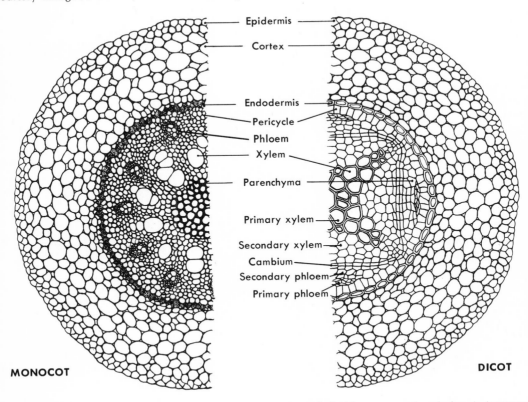

Fig. 14-14. Comparison of monocot and dicot roots showing stele inside cortex and endodermis (cross section).

224

hind the region of elongation, thus assuring a supply on every growing root.

The root hair region gradually merges into that part of the root in which cell differentiation is complete and in which the permanent tissues of the mature root have formed. This area of the root reveals three distinct regions (1) the epidermis, which is a superficial layer of cells completely surrounding the root, (2) the **cortex,** which lies inside the epidermis and occupies an extensive region of many layers of cells, and (3) the **stele** or conducting portion, which occupies the center of the root (Fig. 14-13).

The epidermis is a single layer of cells, the walls of which are thin and composed almost entirely of cellulose. There is no cuticle on the outer surface of the epidermis of a root, as appears in leaves and stems.

The cortex is composed of many layers of large parenchyma cells (living and thin walled), between which are intercellular spaces (Fig. 14-14). The cortical cells transport water and mineral nutrients in a young root from the epidermis to the stele. In an older root the cortex serves primarily to store food and water. The innermost layer of the cortex consists of a layer of smaller cells, the walls of which become thickened with suberin and form the **endodermis** (Fig. 14-14). The endodermis may be a ring of thick-walled cells around the stele or may appear as a broken ring in which only some of the endodermal cells have thick walls. The thinner-walled endodermal cells provide a pathway for the radial transportation of water and other nutrients. Although its function is not precisely known, the endodermis probably prevents loss of water from the stele.

Two kinds of conducting tissue, xylem and phloem, are found in the stele. The xylem appears as a star-shaped group of thick-walled cells in the center of the dicot root, whereas in a monocot there are numerous

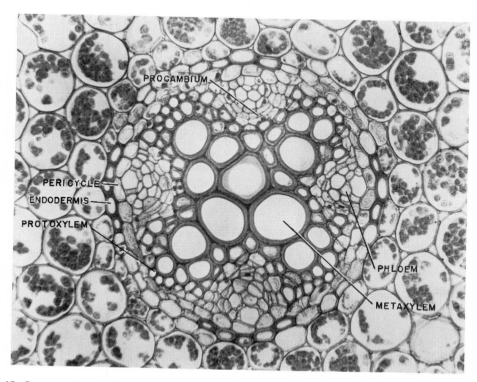

Fig. 14-15. Cross section of the stele of an older root of a buttercup. (×260.) (Courtesy George H. Conant, Triarch Products, Ripon, Wis.)

225

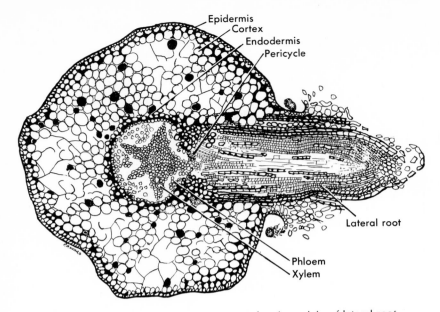

Epidermis
Cortex
Endodermis
Pericycle

Lateral root

Phloem
Xylem

Fig. 14-16. Cross section of primary root showing origin of lateral root.

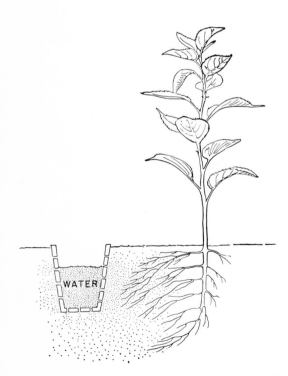

WATER

Fig. 14-17. Roots in the test shown here grow toward water, a demonstration of their hydrotropism.

xylem ridges (Fig. 14-14). The larger cells in the center are called **metaxylem,** whereas the smaller cells between the large ones are **protoxylem** (Fig. 14-15). Between the star-shaped arms of the xylem in dicots are small group of phloem cells. A layer of parenchyma cells, the **pericycle,** lies between the phloem and the endodermis to form the outermost border of the stele. The pericycle is sometimes several layers thick. As the cells of the pericycle divide, they push their way outward, dissolving and crushing the cells of the cortex to form **lateral roots** (Fig. 14-16). Lateral roots always arise from the pericycle of the stele, unlike branch stems, which arise from the outermost layers of the stem. Lateral or secondary roots arise from the points of the star-shaped xylem and therefore produce as many vertical rows of lateral roots as there are points to the xylem. Dicot roots commonly have three, four, or five protoxylem points, but there can be more. Most roots have a solid mass of conducting tissue that consists of a center of xylem and a peripheral area of phloem but no pith, as is found in some stems (Fig. 14-17). In roots this type of stele is called a **protostele.**

Most fibrous root systems will have only the **primary permanent tissues** just described. These are the tissues that have arisen from the embryonic cells of the tip that are undergoing elongation and differentiation. However, in older roots and tap-root systems a special layer of meristematic cells develops from the pericycle, the **cambium.** A dicot root develops a cambium; a monocot rarely forms one. The cambium forms **secondary permanent tissues** in older roots, and, as a result of their activity, the root is able to increase in diameter. The cambium arises between the initial xylem and phloem (Fig. 14-14). It is a layer of thin-walled cells that retain the capacity for division. Initially, the cambium consists of a broken band of cells, between the xylem and phloem, that eventually becomes continuous around the entire protoxylem to form a cylinder. It will form new xylem and phloem (secondary, in contrast to the original primary xylem). The secondary xylem forms toward the center of the root and the secondary phloem toward the outside. As the layers of xylem increase, the cambium is pushed farther toward the periphery. The xylem layer becomes much thicker than the phloem layer, and both are traversed by thin radial layers of parenchymal cells called **vascular rays.** The thick xylem layers differentiated into seasonal growths recognizable as **annual rings.** However, they are not defined as clearly in woody roots as in woody stems. In some roots, for example, old trees, another cambium develops in the pericycle to form cork. A thick layer of cork causes the cells of the cortex and epidermis to die and slough off, leaving the corky layer as a protective covering on the outside of the root. It is this cork layer on thick and old roots of trees that makes them so difficult to cut with an ax.

ROOT ADAPTATION

Roots are adapted to function as anchors for plants and in some cases as food-storage reservoirs (the biennials mentioned above). All the roots produced by the vascular plant spread through the soil not only to anchor the

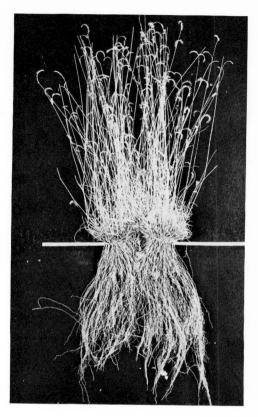

Fig. 14-18. Fibrous root system of grass. (USDA photograph.)

plant, but also to absorb water (Fig. 14-17), much the same as the stem grows toward light. They have developed two major forms of root systems. A **fibrous root** system (Figs. 14-18 and 14-19) is composed of many slender roots, most of which are nearly equal in size. In grasses the primary root is lost in the maze of adventitious roots that have developed from the stem, whereas in other plants the secondary roots have grown so large as to be indistinguishable from the primary root. Therefore the mass of roots, whatever its origin, constitutes the fibrous root system. In the other type of root system, called a **taproot,** the primary root is enlarged and the secondary roots are few and slender (Fig. 14-19). Common examples of taproots have been mentioned previously. The type of root system and size vary with different species of plants and also with environmental

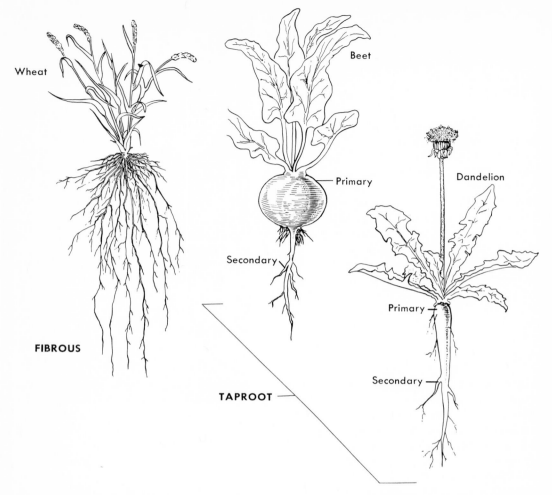

Fig. 14-19. Root systems of two types. The one on the *left* is fibrous (see also Fig. 14-18). Taproots, shown in the *middle* and on the *right,* are used for food storage by biennials during their first year of growth.

physical factors such as moisture and soil composition.

Root formation is the result of environmental stimulation. It is well known by the gardener that "rootings" can be made from slips or cuttings from the stems or even leaves of many kinds of plants (Fig. 14-20). Hormone secretion is induced by placing the vegetative part in moist soil or sand, and this same process often occurs in nature when growth conditions permit it. The gardener may speed up the process by adding com-

mercially prepared rooting hormone to the water.

STEMS

The stem or **shoot** of a flowering plant has its origin in that part of the embryo within the seed known as the epicotyl, which is a continuation of the hypocotyl (Fig. 14-9). The epicotyl is a cylindrical mass of meristematic tissue that is narrower at its apex or stem tip. Frequently, as in the bean embryo, a pair of tiny leaves can be seen at the tip of the young

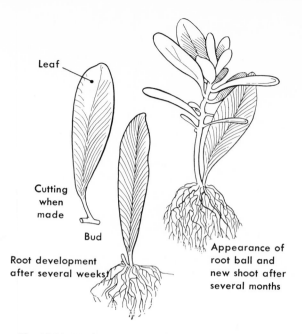

Leaf

Cutting
when
made

Bud

Root development
after several weeks

Appearance of
root ball and
new shoot after
several months

Fig. 14-20. Method of rooting rhododendron leaves.

stem (Fig. 14-2). When the seed germinates, the root emerges first so that it can absorb the water and nutrients necessary for further development before the food stored in the seed is entirely consumed. The young stem emerges from the seed to grow upward (Fig. 14-2) because most stems react negatively to gravity.

The growth of stems, like that of roots, is unlimited. However, the length of a stem, when growth is taking place, is greater than that of a root. Furthermore, the arrangement of the permanent tissues is different. The regions in the growing point of a stem are very much like those of a root, except that a stem tip has no protective cap of cells comparable to the root cap; instead, the stem is covered by the leaf primordium. At the tip of the growing stem is a peak of actively dividing cells called the apical meristem (Fig. 14-21), the cells of which are thin walled and many in active mitosis. Immediately below this are lateral outgrowths or leaf primordia that also consist of undifferentiated embryonic cells. The youngest and smallest leaf primordia are

nearest the apex of the stem; the oldest and largest are near the base of the stem tip. Behind the apical meristem is the **zone of elongation** where certain cells are undergoing some modification but where most are increasing in size. This region may extend for several inches behind a terminal bud. It merges into the **zone of differentiation,** a region in which differentiation occurs more rapidly in leaves and internodes of the stems than at the nodes. It is in this region that the primary permanent tissues are being formed (Figs. 14-21, 14-22, and 14-24). Three kinds of primary meristematic tissues can be initially recognized as developing from the undifferentiated cells or **promeristem** at the tip of the stem: (1) protoderm, a layer of surface cells, (2) procambium, strands of cells that retain the capacity for division, and (3) **ground meristem** (Fig. 14-24), cells that surround the procambial strands. The protoderm is a single layer of cells on the outer surface of the developing stem, which will form the future epidermis. Procambial strands become xylem and phloem, but a single layer of this

229

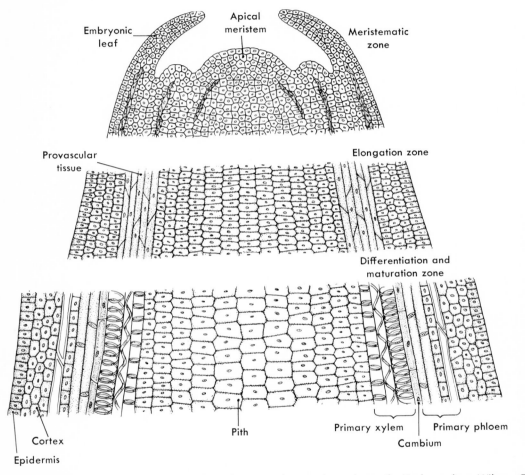

Embryonic leaf

Apical meristem

Meristematic zone

Provascular tissue

Elongation zone

Differentiation and maturation zone

Cortex

Epidermis

Pith

Primary xylem

Primary phloem

Cambium

Fig. 14-21. Longitudinal section through the tip of stem to show regions of growth. (Redrawn from Wilson, C. L., and W. E. Loomis. 1957. Botany. Rev. ed. The Dryden Press, New York.)

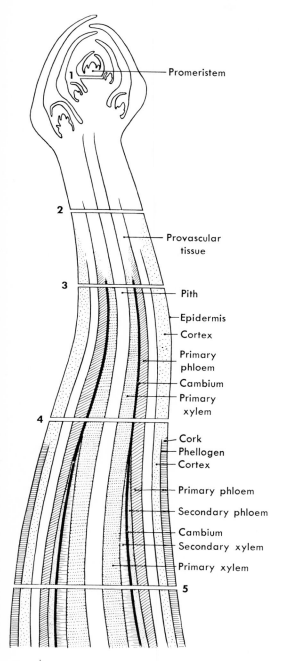

Promeristem

Provascular tissue

Pith

Epidermis

Cortex

Primary phloem

Cambium

Primary xylem

Cork

Phellogen

Cortex

Primary phloem

Secondary phloem

Cambium

Secondary xylem

Primary xylem

Fig. 14-22. Longitudinal section of growing stem. (Numbers refer to sections shown in Fig. 14-24.)

tissue may retain the ability to produce new xylem and phloem. This layer is called the **vascular cambium** (Fig. 14-23, *A*). It functions in stems as it does in roots—to produce secondary xylem and phloem—thereby being responsible for the increase in the diameter of the stem. The ground meristem develops into a number of layers of cells located between the epidermis and the phloem. These layers of cells are called the cortex. The ground meristem is also responsible for the formation of a group of cells in the center of the stem that becomes completely surrounded by xylem. This central group of cells is known as **pith** and may be connected with the cortex by strands of cells betweeen the vascular tissue (xylem and phloem). The strands are called **pith rays** or **medullary rays.**

The vascular cambium consists of a single layer of cells, which produce, by division, new cells toward the center of a root or stem called the xylem (Fig. 14-23, *B*) and other cells toward the outer surface, phloem (Fig. 14-23, *C*). The xylem is a water-conducting tissue, and the phloem a food-conducting tissue. Since the cambium consists of one layer of cells, it is obvious that one daughter cell of the division remains meristematic. The other daughter cell will develop into either a xylem or a phloem cell. Why a cell will develop one time into xylem and the next into phloem is a matter of speculation. It has been suggested that when vascular cambium is acid, xylem is produced; when it is not, phloem is produced. Regardless of the reason, during the active growing season more xylem cells are produced than phloem cells, thereby accounting for the greater amount of xylem than of phloem in gymnosperm and dicot roots and stems (Fig. 14-24). The xylem cells have walls heavily impregnated with lignin. Thus they are capable of withstanding crushing. The softer-walled cells of the phloem are easily crushed and destroyed from year to year.

Any lateral growth of monocot roots and stems is largely attributable to the enlargement of cells, since the vascular cambium is

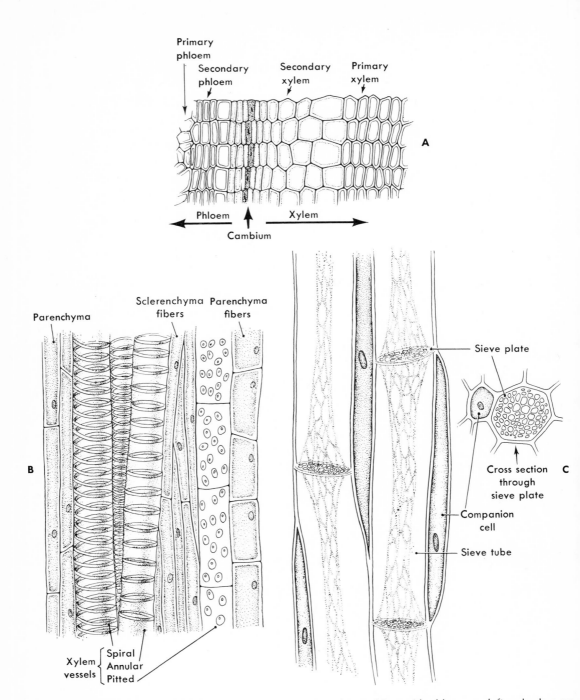

Fig. 14-23. Details of the cells of the vascular system. **A,** Vascular cambium with phloem on *left* and xylem on *right,* both specialized cells produced by the growing cambium. **B,** Xylem vessels in detail. **C,** Phloem cells in detail. Note the remains of the cytoplasm in the sieve tubes of the phloem.

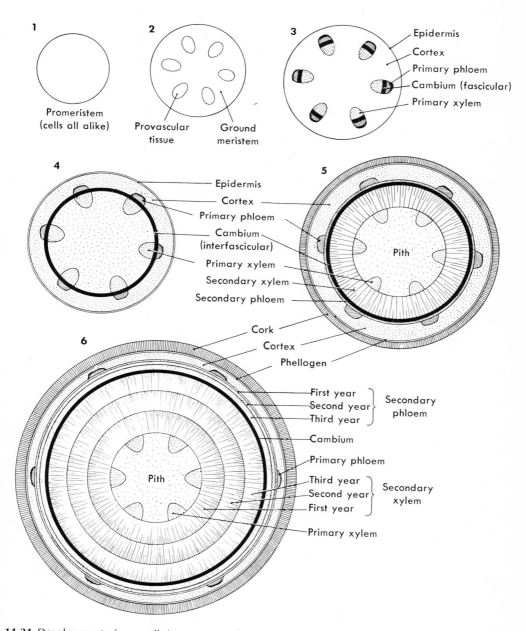

1
Promeristem
(cells all alike)

2
Provascular
tissue
Ground
meristem

3
Epidermis
Cortex
Primary phloem
Cambium (fascicular)
Primary xylem

4
Epidermis
Cortex
Primary phloem
Cambium
(interfascicular)
Primary xylem
Secondary xylem
Secondary phloem

5
Pith

Cork
Cortex
Phellogen

First year
Second year
Third year
} Secondary
phloem

Cambium

Primary phloem

Third year
Second year
First year
} Secondary
xylem

Primary xylem

6
Pith

Fig. 14-24. Development of new cells in young woody stem. Various cross sections of the stem are numbered to correspond to the regions in the longitudinal section. (See Fig. 14-22.)

either poorly developed or not present at all. In perennial plants, in which a vascular cambium does exist, the cambium increases in diameter as the root or stem enlarges. This is accomplished by the division of the cambium cells in a radial plane and the tangential enlargement of the daughter cells. Wherever vascular cambium exists, it is derived from the terminal meristem and remains meristematic to form xylem and phloem.

Another meristematic layer of cells is the **cork cambium,** which develops between the phloem and the outer surface of the plant organ. It usually originates in the cortex layer just below the surface of a stem or root. Cork cambium cuts off cells on the outer surface, which then develop into cork cells. The walls of cork cells become impregnated with **suberin,** a fatty substance, which makes the cell impermeable to food and water. Therefore mature cork cells are dead and form a major part of the bark of older stems and roots of trees and shrubs. The death of the cork layer also results in the death of all tissues outside the cork layer. Subsequent growth in diameter of the stem or root will cause the dead tissues to crack and peel off, resulting in the deep furrows so characteristic of the bark of some trees.

Xylem consists of several kinds of cells, most of which are elongated dead cells, with walls composed largely of lignin. Lignin is deposited on the inner surface of the cell wall, replacing most of the cellulose. The cells become hard and strong, but this does not interfere with the passage of water upward from the roots to the stems and leaves. The simplest of these xylem cells is the **tracheid** (Fig. 14-25). On the basis of the manner in which lignin becomes deposited, the secondary walls may appear as spirals, rings, parallel bars, irregular networks, or pits. Pitted tracheids are the most common. For example, in gymnosperms the wood is made up almost entirely of pitted tracheids. In some plants, such as angiosperms, long **vessels** (Figs. 14-26 and 14-27) develop from a longitudinal row of cells, the end walls of which break down to form conducting tubes. Simi-

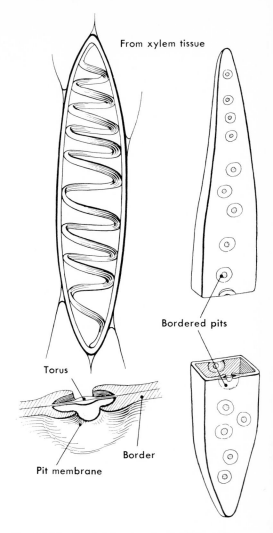

Fig. 14-25. Tracheids showing details. Note cell walls, pits, accessory structures, and absence of cytoplasm.

larly, they may be classified as spiral, annular, or pitted. Fibers are like tracheids but have greater length, more pointed ends, and few pits. Both fibers and vessels are absent in gymnosperms, but the fibers do give the hard quality to the wood of angiosperms. Parenchyma cells may also be found in xylem.

As a part of the conducting tissue in plants and associated with the xylem are those tissues called phloem. Phloem conducts food

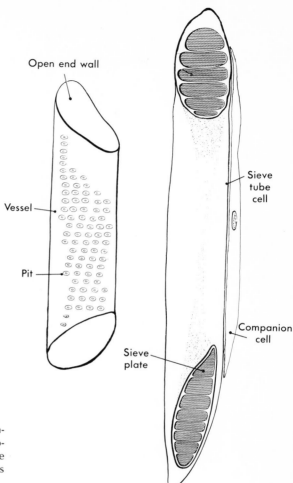

Fig. 14-26. Conducting vessels, principal conducting structure in the xylem of most angiosperms. **A,** Open vessel. **B,** Sieve tube. Note strand of "slime" in the sieve tube, the remains of cell contents.

from the leaves to the stems and roots. The principal cells in phloem are **sieve tubes**—elongated cells, the end and side walls of which may be perforated with sieve plates. They have thin walls of cellulose, cytoplasm, but no nucleus. Because they have no nucleus, other living cells, **companion cells** (Figs. 14-26 and 14-27), are usually associated with them. These are narrow cells that have a nucleus and cytoplasm. Companion cells are not present in gymnosperms. In addition to sieve tubes, some fibers and parenchyma cells may be present.

The epidermis is usually one layer of cells in thickness. The stem epidermis may last throughout the life of the plant, or it may be replaced by other protective tissues, such as cork. The epidermis consists of relatively unspecialized cells with living protoplasm and plastids but rarely any chlorophyll. The outer epidermal cell wall is usually thicker than the inner one. The most characteristic feature of the epidermis is the presence of **cutin,** a waxlike substance that forms as a separate layer on the outer wall of the cell, the cuticle. The cuticle appears to be highly impenetrable and thereby protects the stem against excessive loss of water.

The cortex of stems is largely composed of parenchyma cells, which occasionally may

235

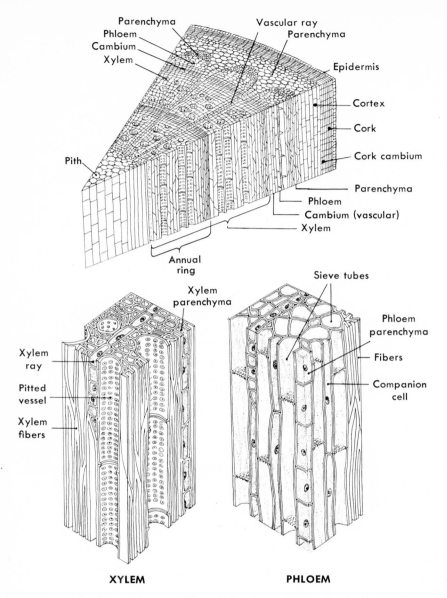

Fig. 14-27. Details of a woody stem (*Tilia* sp.) showing arrangement of cell types (see text).

have chloroplasts. Intercellular spaces are prominent. The outer layers frequently contain collenchyma cells in groups or in a more or less continuous ring. The delimitation of the cortex from the vascular region is an important morphological criterion in identifying both stems and roots. In roots the innermost layer of cells of the cortex is called an en-

dodermis, which only rarely is found in stems. The stems of gymnosperms and angiosperms lack a morphologically differentiated endodermis, but lower vascular plants may have one. In some young stems the innermost layer of the cortex contains large amounts of starch and becomes recognized as **a starch sheath.** The innermost group of cells

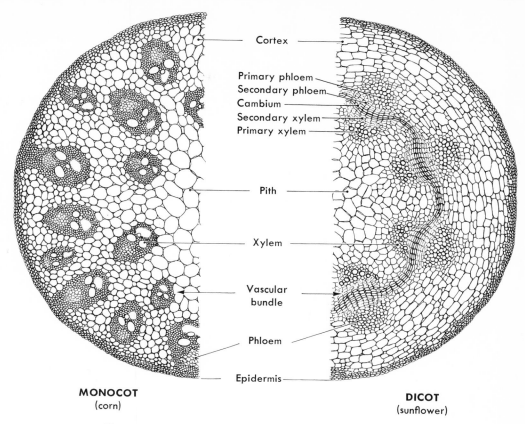

Cortex

Primary phloem
Secondary phloem
Cambium
Secondary xylem
Primary xylem

Pith

Xylem

Vascular
bundle

Phloem

Epidermis

MONOCOT
(corn)

DICOT
(sunflower)

Fig. 14-28. Cross sections of monocot and dicot (typical) stems.

in stems is the pith, which is composed of parenchyma cells. In many stems the central pith is destroyed during growth. In some stems this destruction is limited to internodes, but the nodes retain their pith as a diaphragm (for example, corn and bamboo stems). The cells of the pith have the prominent intercellular spaces (Fig. 14-28) characteristic of parenchyma.

The vascular tissue, composed of xylem and phloem, commonly appears as a cylinder between the cortex on the outside and the pith in the center of the stem. In young dicots the vascular tissue is separated by interfascicular regions into smaller or larger units commonly referred to as vascular bundles. The xylem and phloem ordinarily occur in a collateral arrangement, with the phloem located outside the xylem.

The vascular system of the stem is always

continuous with that of the leaf. The vascular connection of the leaf and stem at the node may be one or more strands of vascular tissue that lead from the stem to the leaf. This close association of the stem with the leaves makes this part of the plant more complex than the root. However, like the root, the stem consists of three tissue systems: the epidermal, the ground meristem, and the vascular. The variation of these primary tissues in stems of different species or in the larger groups of plants is based chiefly on differences in the relative distribution of these fundamental tissues. The stems of gymnosperms and woody dicots have a vascular system that appears as a cylinder, with cortex on the outside and pith in the center. In monocots and herbaceous dicots the vascular tissue does not appear as a cylinder. It appears as a single ring or as more than one ring (Fig. 14-28). If

237

it is a single ring, the bundles are widely scattered (Fig. 14-28).

Secondary growth increases the amount of vascular tissue in a stem. However, it contributes only to the thickness of the axis, which is characteristic of gymnosperms and woody dicots. On the other hand, most monocots and herbaceous dicots ordinarily have no secondary growth, except that herbaceous dicots have limited lateral growth and a cambium. If secondary growth does occur in monocots, it is of a special type. The term secondary growth includes the formation not only of secondary vascular tissues but also of the **periderm,** which is a protective tissue that replaces the epidermis of many woody plants.

In a cross section of a young woody stem, the vascular bundles can be seen arranged in groups around the stem. The **vascular cambium** will eventually form a complete cylinder. It arises in part from the procambium within the vascular bundles, where it is known as the **fascicular cambium,** and in part from the parenchyma between the bundles, where it is called **interfascicular cambium.** Therefore, in the most common type of secondary growth the vascular cambium, while becoming a complete cylinder, will produce continuous cylinders of secondary xylem and secondary phloem to ultimately increase the diameter of the stem.

The development of secondary vascular tissues between the primary xylem and the primary phloem exerts a great amount of stress on the internal tissues of the stem, particularly on those located outside the cambium. As the cambium produces secondary vascular tissues, the primary phloem becomes isolated from the primary xylem. The primary xylem and the pith are enveloped by the secondary xylem so that in an older stem the small amount of remaining pith may eventually lose its identity.

Slowly the primary phloem is pushed outward and becomes nonconducting because its thin-walled elements are thoroughly crushed. The cortex may persist for many years, increasing in circumference through cell enlargement and division. The increase in circumference may also be aided by the growth of the parenchyma ray cells that, in some stems, can be seen expanding or fanning out into the cortex. The epidermis may persist; its cells enlarge and divide as the diameter of the stem increases.

In some stems a protective tissue of secondary origin called periderm may develop. This eventually replaces the epidermis in stems (and roots) where secondary growth takes place. The best examples of periderm development can be seen in gymnosperms and woody dicots. However, it does occur in the oldest parts of the stem (and roots) of herbaceous dicots, as well as in some monocots.

Periderm must be distinguished from the bark of a stem. **Bark** is used most appropriately to designate all those tissues outside the vascular cambium (for example, phloem, cortex, periderm, and epidermis when present). The development of the periderm in the cortex isolates its outer cells so that they become nonliving. Consequently, it is convenient to differentiate between the outer, nonliving bark and the inner, living bark of a stem.

The first periderm commonly appears during the first year's growth of a stem (or root). In some plants only one periderm develops, whereas in others several may be found. The first periderm usually originates in the outermost layer of the cortex just under the epidermis. Subsequent layers of periderm arise in successively deeper layers of the cortex beneath the first periderm. Periderm consists of a layer of meristematic cells called the cork cambium or **phellogen,** the cells of which produce the cork or phellem, a protective tissue of dead and suberized cells. They are formed by the phellogen and the **phelloderm,** which is a living tissue of parenchyma cells formed inwardly by the meristem. The presence of cork outside the cortex accounts for the death of any cells outside it.

The vascular cambium of stems is inactive during the winter months in temperate climates. In the spring it becomes active again to produce new xylem and phloem. This activity is determined by the seasonal activity of the vascular cambium and has been found to be closely correlated with the growth of buds in the spring. This relationship was easy to comprehend when it was learned that the auxin produced by buds was transferred toward the base of the stem, thereby inducing its growth.

Seasonal growth of xylem is more readily demonstrated than that of phloem. A larger number of xylem cells (wood) is produced in late summer than in spring. Cells are smaller and more numerous in summer wood, whereas in spring wood the cells are fewer and larger. Hence it is easy to differentiate one year's growth from another, since the differences in xylem growth appear as **annual rings** in a cross section of a stem (Fig. 14-24). The oldest part of the stem is nearer the center and the youngest nearer the outside. The width of the annual ring will vary from year to year, giving some indication of growth or the environmental conditions existing at the time of growth. This fact has been used to determine changes in climate during the years before weather records were kept. Even the climate during prehistoric times can be determined in this way by examining the annual rings of petrified wood.

Since stems differ in their primary and secondary structure, it becomes convenient to distinguish between types of stems. Therefore it is customary to identify them as gymnosperms, woody and herbaceous dicots, monocots, and vines.

The stem of the pine is a good example of a gymnosperm stem. Initially the vascular tissue is in discrete bundles, which later will form a continuous cylinder of secondary xylem and secondary phloem. The rings of annual growth are easily recognized in an older stem. Some pith remains in the center, the cortex and xylem contain resin ducts, and the first periderm arises beneath the epidermis and is not replaced by deeper ones for many years.

In all woody dicots the secondary vascular tissues form a continuous cylinder. Rings of annual growth are present through which many wide rays of parenchyma (vascular rays)

can be found, as well as narrow rays. The secondary phloem appears as alternate bands of fibers and bands of sieve tubes, companion cells, and parenchyma. The periderm originates just beneath the epidermis and persists for years, and the cortex remains intact. The pith is composed of parenchyma cells in which cavities appear, and the pith cells next to the xylem act as storage tissue.

A herbaceous dicot (Fig. 14-28) resembles a young woody stem in which the vascular tissue occurs as a ring of separate vascular bundles. However, in some stems a continuous and narrow cylinder may form, which in turn may exhibit some secondary growth. The buttercup genus *Ranunculus* has little, if any, secondary growth, and there is no periderm. The pith, in which a large central cavity may be present, makes up the major part of the diameter of the mature stem.

A typical monocot has widely spaced vascular bundles that are not restricted to a single ring (Fig. 14-28) but are scattered throughout. Other monocots, such as *Avena sativa* (oats), have bundles in only two rings. In both these monocots there is a peripheral ring of sclerenchyma cells near the epidermis, with which some vascular bundles are associated. Stomas appear in the epidermis; the pith may be broken down to form large or small spaces, and all the vascular bundles are enclosed in sheaths of sclerenchyma. Although monocots lack secondary growth, they may develop thick stems through primary growth, as in the palms.

STEM MODIFICATIONS

The stem system of a flowering plant may be a distinctive mark of the plant; for instance, a pine tree, a cornstalk, and an asparagus are obviously different. The primary function of a stem is to support the leaves and distribute them in space so that they receive adequate light. In addition to the conduction of materials to and from the leaves, it often stores food. Sugar cane and corn stalks are two of the best examples of this.

Some stems are underground, but most typically they start at the soil level. Those stems that grow above the soil are called **aerial stems** (Fig. 14-29) to distinguish them from those that grow underground as rhizomes (Fig. 14-30), as in the iris. Therefore an aerial stem, characteristic of most plants, is erect, but some may be more flexible in their growth habits and form climbing stems or **vines** (known as **lianas** in tropical regions), and still others are prostrate and run along the surface of the ground (Fig. 14-31). The stems of plants may vary greatly in form, size, structure, and even length of their life as we see in the **herbaceous stem** in contrast to the **woody stem.** The growth form distinction between **shrubs** and **trees** is dif-

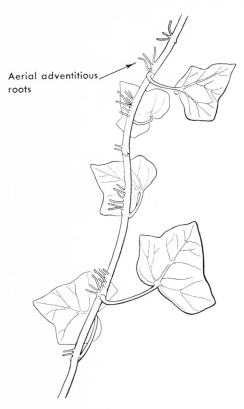

Aerial adventitious roots

Fig. 14-29. Adventitious roots on English ivy stem.

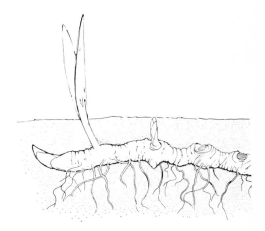

Fig. 14-30. Underground stem of iris.

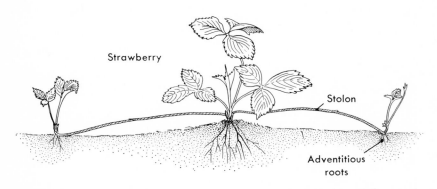

Strawberry

Stolon

Adventitious roots

Fig. 14-31. Vegetative reproduction runners or stolons. Horizontal stem of the strawberry.

ficult to make. Both are woody stem, and, in general, shrubs have many stems, all approximately the same size. All the main stems of most shrubs branch profusely at about the ground level. However, plants may grow as shrubs in one locality and as trees in another. Unlike a root, the stem is made up of distinct areas to which leaves or buds are attached. These areas are called nodes, and the area between two successive nodes is called an internode (Fig. 14-32). The number of leaves or buds found at each node, depending on the season of the year, will help to identify the stem. If only one leaf or bud appears at a node, the stem is described as having **alternate** leaves or buds (Fig. 14-33). When two leaves or buds are located at each node on opposite sides of the stem, it is said to have **opposite** leaves or buds (Fig. 14-33). The stem is designated as having leaves or buds in **whorls** if three or more are located at each node (Fig. 14-33). Elms and apples have al-

ternate arrangements of leaves, maples and dogwood have opposite arrangements, and catalpas are whorled. Whorled leaves are the least common arrangement.

When leaves are arranged alternately, they occur in spiral succession. In other words, each successive leaf arises slightly to one side of the one below it so that they form a spiral around the stem. When leaves are arranged on the stem in whorls or opposites, each group is not directly above the one below it. Each successive pair of opposite leaves is arranged at a right angle to the previous pair, whereas those in whorls are one third the distance around the stem from those below. This pattern of leaf attachment to the stem orients all the leaves in such a way that the maximum amount of light reaches each leaf. With such an arrange-

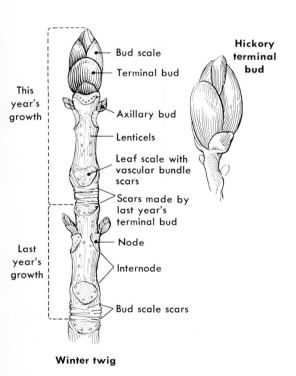

Fig. 14-32. *Left,* Nodes and internodes of a stem. *Right,* Opening of a terminal bud.

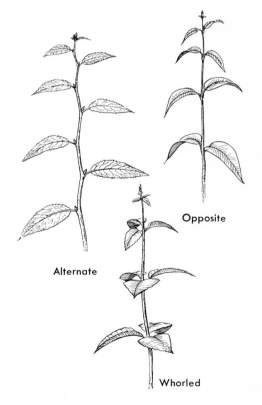

Fig. 14-33. Various leaf arrangements on stems.

ment, the greatest amount of leaf surface is exposed to light to form the **leaf mosaic.**

The surface of an active, growing stem reveals a number of structures. Most conspicuous are the leaves and buds. The leaves may be broad and flat, narrow and elongate, or needlelike and are attached at the nodes. A **bud** is usually found in the angle of the leaf attachment to the stem. This angle between the leaf and the stem is called the **leaf axil,** and the buds located there are called **axillary** or lateral buds (Fig. 14-32). A **terminal bud** may be located at the very apex or tip of the stem. The distinction between lateral and terminal buds is primarily one of position, since they are usually alike in structure and function. Although buds may be located in the axils of leaves, occasionally they appear on other parts of the stem and are then called **adventitious buds.** Structurally and functionally, adventitious buds are like other buds but sometimes appear as the result of an injury, for example, when a branch is cut.

A bud is an embryonic or young shoot that has the capacity to produce leaves. It is essentially a cone-shaped mass of meristematic cells that produces leaf primordia laterally as the bud grows. Therefore a bud possesses nodes, internodes, and leaf primordia. It is a miniature stem on which all the leaves of the following season are already formed. Most buds have overlapping **bud scales** that cover them and protect them from drying, injury, and cold. Buds, like leaves, can be classified as alternate, opposite, or whorled, depending on the number located at each node. They may also be classified as to their activity and the kinds of structures they produce. Buds that grow are called active buds to distinguish them from the occasional ones that are inactive or dormant. A dormant bud may remain embedded in the bark, only to develop years after it has formed. Some buds produce only flowers at particular points on the stem. These are called flower buds to distinguish them from buds that produce leaves and branches, the leaf or branch buds.

When the leaves drop off in the fall of the

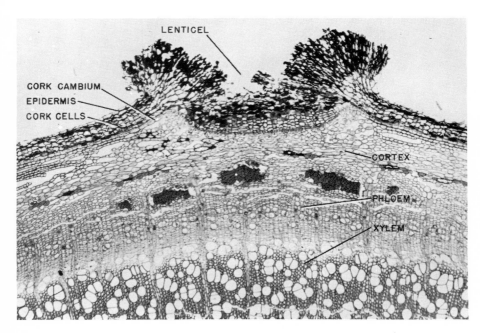

Fig. 14-34. Cross section through lenticel of stem. (×35.) (Courtesy George H. Conant, Triarch Products, Ripon, Wis.)

242

year, **leaf scars** will remain (Fig. 14-32). In those plants that lose their leaves each year, buds are quiescent during the winter season and become active again in the spring. New buds appear in the spring in the axils of the leaves in preparation for each following year. By proper pruning of the stem, a bushier growth of the plant can be induced. The removal of buds causes the remaining ones to form sturdier branches or larger flowers. Also by pinching off lateral buds, one can induce a plant to grow tall and erect.

Besides the leaves and buds present on a stem, there may be leaf scars, **bud scars,** and often blisterlike lines or dots called **lenticels** (Fig. 14-34). These broken areas in the bark or epidermis of the stem are for the entrance of air necessary to the living tissues beneath.

It is, of course, the stems that serve as food in the case of asparagus, and lumber is made from the stem of a woody plant. Stems sometimes undergo a particular modification and appear as **thorns, prickles,** or **tendrils** (Figs. 14-35 and 14-36). Some stems occur as

stolons, producing at intervals erect stems and adventitious roots as a modification of the hortizontal or prostrate stems mentioned previously. If they are long, slender, nearly leafless stolons, they are known as runners (Fig. 14-31). The rhizomes of some plants, such as the potato, may be thick and fleshy or partially thickened and are called **tubers** (Fig. 14-37). Some plants are propagated by a short, stout, erect underground stem in which food is stored. Such a stem, usually broader than it is long, is known as a **corm** (Fig. 14-38). At the end of a growing season the plant dies, but the corm sends out adventitious roots, and the terminal bud develops into the new stem, which grows and eventually produces a flower. The few dry leaves surrounding a corm are the remains of leaf bases, which do not store food. However, one of the principal storage organs is a **bulb** (Fig. 14-39), in which layers of thickened leaves (bulb scales) surround the short, erect stem. The tulip and onion are examples of bulbs.

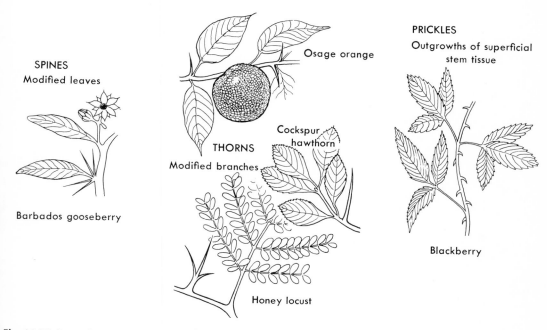

Fig. 14-35. Examples showing the different stem adaptations. Thorns and prickles are modified stems or stem parts; spines are modified leaves.

243

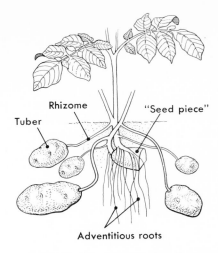

Fig. 14-37. Tuber, rhizome, and adventitious roots of potato.

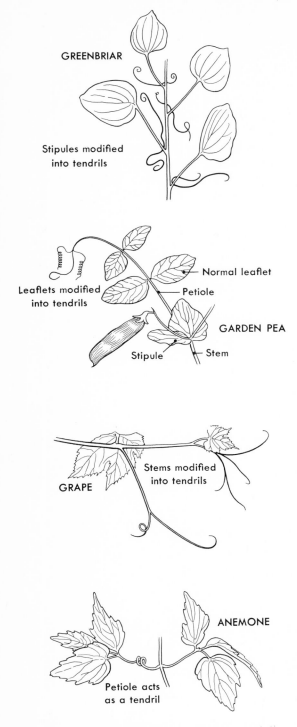

Fig. 14-36. Stem modifications that form tendrils.

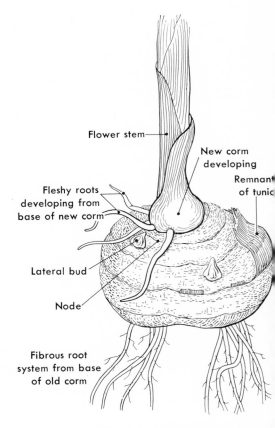

Fig. 14-38. Corm of a galdiolus.

LEAVES

The leaf develops during localized meristematic activity on the side of the apex of a stem. As a leaf progresses in its development, it grows outward from the leaf base, revealing the final shape and form characteristic of the species.

Leaf tissues and structures. A continuous outermost layer of nonliving material covering the epidermal cells is secreted. This is the **cuticle.** The blade of a leaf is protected by an upper and a lower layer of cells called the epidermis. Between these two layers of protective cells is the **mesophyll.** It consists of several layers of cells containing chloroplasts. The mesophyll is transversed by numerous vascular bundles (veins) of different sizes. The monocot leaf vein has a greater amount of mechanical tissue (sclerenchyma fibers) and xylem fibers than does the vein of the dicot leaf (Fig. 14-40).

The epidermis may develop a variety of appendages called **trichomes** (Fig. 14-41). These may include secretory (glandular) and nonsecretory hair, scales, and papillae. Trichomes are useful as morphological structures in classification.

Between the two layers of the epidermis are a number of loosely arranged cells that

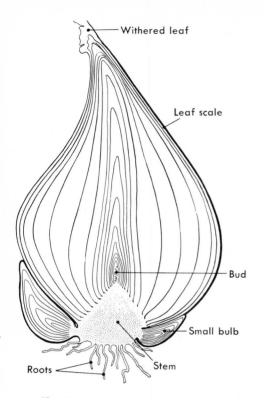

Fig. 14-39. Onion bulb and stem.

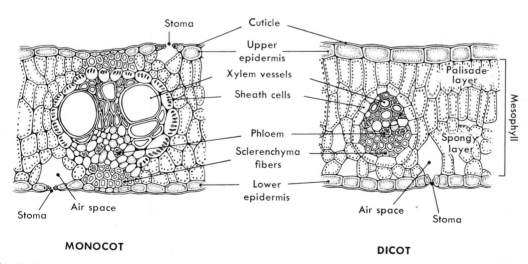

Fig. 14-40. Cross sections of typical leaves, monocot and dicot, showing differences between the conducting (vein) tissue and the mechanical tissue surrounding them.

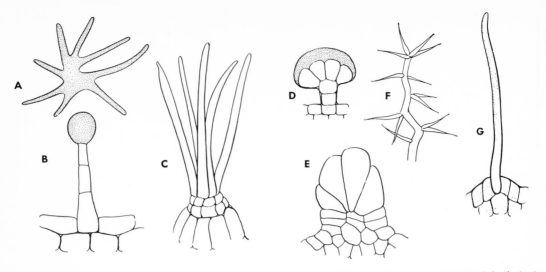

Fig. 14-41. Epidermal appendages and trichomes of leaves. **A**, Stellate hairs in *Alyssum* sp. **B**, Glandular hairs in *Pelargonium* sp. **C**, Floccose hairs in *Malva* sp. **D**, Short glandular hairs in *Lavandula* sp. **E**, Glandular hairs in *Solanum* sp. **F**, Urticating hairs in *Verbascum* sp. **G**, Stinging hair in *Cistus* sp.

comprise the mesophyll (Fig. 14-40) of the leaf. This layer varies in thickness with the thickness of the leaf. The cells of the mesophyll are primarily parenchyma cells containing chloroplasts (chlorenchyma). The mesophyll can be differentiated into two easily recognizable layers; one is called the **palisade layer** and the other the **spongy layer.** The palisade layer ordinarily will consist of one or two rows of elongate cells just beneath the epidermis. If they are present just beneath the upper epidermis and not the lower epidermis, they cause the leaf to appear darker green on its upper surface and lighter green on its lower surface. If a palisade layer is present on both surfaces of the leaf, it will then be the same color on both upper and lower surfaces. A leaf, the blade of which is broad and structurally different on its upper and lower surfaces, is called a dorsiventral leaf. If it is structurally the same on both sides, it is known as an isobilateral leaf. The cells of the palisade layer are closely set together and contain many chloroplasts.

The spongy layer of the mesophyll is composed of irregularly shaped parenchyma cells containing fewer chloroplasts than those

of the palisade layer. There are many intercellular air spaces in the spongy layer. Therefore, in a dorsiventral leaf the lower epidermis is easier to strip off than the upper. Near each stoma is an extra large air space. The cells of the spongy layer are not arranged in rows, and it appears porous. Mesophyll may vary in different species of plants; for example, monocots (Fig. 14-40) have a more compact mesophyll layer than do dicots.

Since the primary function of leaves is photosynthesis, it appears that the mesophyll is well adapted for this function. The densest tissue is near the upper surface in a dorsiventral leaf and contains the greatest concentration of chloroplasts. This is the part that receives most of the light. The air spaces are largest near stomas that connect to the atmosphere in order that carbon dioxide can easily enter the intercellular spaces and be rapidly distributed throughout the leaf. The compact palisade layers somewhat retard the loss of water by evaporation because only a small area of their surface is exposed to air.

Many interconnected and branching veins are present in the leaves of flowering plants. A typical vein (Fig. 14-40) is composed of

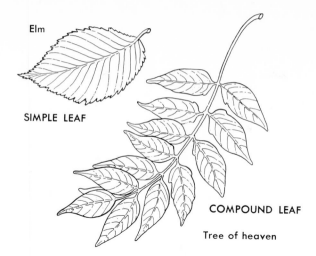

Elm

SIMPLE LEAF

COMPOUND LEAF

Tree of heaven

Fig. 14-42. Simple and compound leaves.

xylem and phloem surrounded by a layer of sheath cells. The xylem is closer to the upper surface and the phloem is closer to the lower surface of the leaf. The smaller veins consist of only primary vascular tissues, whereas the larger veins of dicots may show secondary growth. The amount of xylem and phloem decreases with the size of the vein; for example, the smallest veins may have only a single tracheid as their xylem tissue, and a modified parenchyma cell will represent the phloem.

The largest vein in a net-veined leaf is usually located in the center and is called the **midrib.** Large veins such as the midrib are embedded in a mass of compact parenchyma cells, rather than in a sheath. To strengthen the vein, collenchyma tissue may be present, as well as sclerenchyma tissue. It is these that make the vein stand out on the leaf.

ADAPTATION OF LEAVES

Although the principal function of leaves is photosynthesis, they may have other functions, such as storage of food and water, reproduction, root formation, climbing, insect-catching, protection, or flower formation. Therefore some leaves are not photosynthetic at all but are highly specialized in other ways.

Many of our economic plants are useful because they store water, salts, food materials, and vitamins in their leaves, for example, spinach, onion, and cabbage.

The blade of the leaf of flowering plants is usually flat and expanded in form. It arises from the stem at a node and ordinarily has a bud located in the **axil.** Leaves are arranged on stems in the same way that buds are—alternate when one leaf occurs at a node, opposite when two leaves arise at a node, and whorled when there are three or more leaves at a node (Fig. 14-33).

Leaves are appendages of the stem that carry on photosynthesis; they usually have a distinct stalk or **petiole** and a thin, flat, expanded portion called a **blade.** When the blade is attached directly to the stem and the petiole is absent, the leaf is said to be sessile. In some leaves there may be a pair of small leafy structures at the base of the petiole called **stipules.**

The form and structure of the leaf blade may vary in different plants or even on the same plant. A number of different terms have been coined to describe the kinds of leaves. These terms, though useful in recognizing and distinguishing different species, are very numerous. Therefore only the most common ones will be mentioned. If the blade of the leaf is all in one piece, the leaf is called sim-

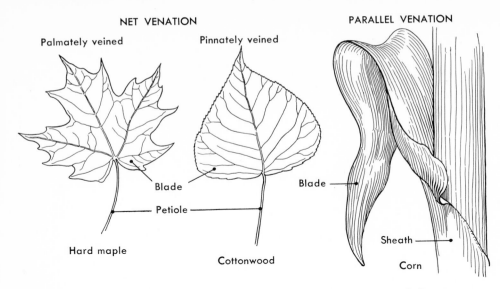

Fig. 14-43. Examples of leaf venation (net and parallel venation) in simple leaves.

ple, and if it is composed of several leaflets, the leaf is compound (Fig. 14-42). The part of the **compound leaf** to which leaflets are attached is known as the **rachis,** and each leaflet is attached by a petiole.

The simple leaves of angiosperms have a characteristic vascular system made up of vascular bundles called **veins.** The arrangement of veins in a blade may vary from species to species. Venation alone generally will distinguish the two classes of flowering plants. Most dicot leaves have net venation, whereas monocots, such as corn, will exhibit parallel venation (Fig. 14-43). A net-veined leaf may have a single main vein with numerous side branches, as does the cottonwood leaf. Vein arrangement can be more specifically identified, for example, that of the cottonwood or poplar leaf as **pinnate** venation. Another type of net venation is **palmate,** typical of maple leaves, in which a number of prominent veins radiate from the petiole. Between the larger veins in the maple leaf is an obvious network of smaller ones. A compound leaf can also exhibit either pinnate or palmate venation. Compound leaves in which the leaflets are attached on opposite sides of the rachis, as in walnuts,

hickories, ashes, and roses, are termed **pinnately compound** leaves. Those in which the leaflets are attached at the same point to the petiole, as in the horse chestnut and buckeye, are palmately compound (Fig. 14-44). Pinnately compound leaves of the honey locust are **pinnately bicompound,** since the primary divisions are themselves compound. Leaves in which three leaflets are attached to the petiole are **trifoliolate.** They may be pinnately compound, as in poison ivy, or palmately compound, as in clover.

Many terms derived from plane geometry and other sources are used to describe the forms of leaves. Some that are self-explanatory are linear, oblong, elliptical, and orbicular. If it is heart shaped, the leaf is called cordate; if kidney shaped, reniform. An ovate leaf is shaped somewhat like an oval, with one end narrower than the other.

The edges of leaves may vary anywhere from a smooth or entire margin to a deeply lobed one. Those having sharp, toothlike edges are called serrate, and deep indentations of the margin make the leaf lobed.

Leaves usually have a limited life-span, since most grow only one season; however, others persist for several years. Woody plants

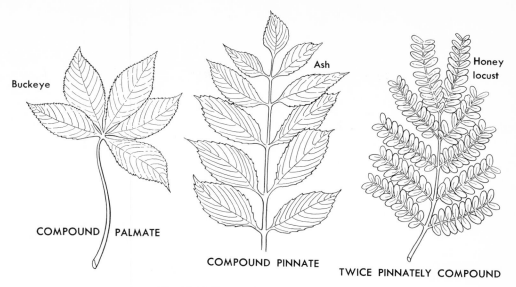

Fig. 14-44. Types of compound leaves.

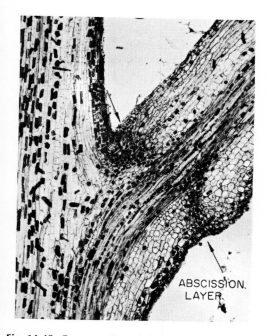

Fig. 14-45. Cross section of abscission layer in petiole.

that have aerial stems that survive year after year are classified as either deciduous or evergreen, according to the duration of the leaves. If the leaves formed during one season remain on the plant until the leaves of the following year are produced, the plant is called an evergreen. Nearly all cone-bearing trees are evergreen, as well as many angiosperm trees in moist tropical climates. Only a few angiosperms in temperate regions are evergreen—the holly, live oak, and sagebrush.

Trees and shrubs that lose their leaves at the end of each growing season, leaving the branches bare during the winter months, are called **deciduous.** In a temperate climate leaves form in the spring and are shed in the fall. The fall of leaves in tropical or warm areas coincides with the wet and dry seasons of the year, rather than with warm and cold temperatures. Leaves fall from most angiosperms (and gymnosperms) because of the development of a specialized transverse region at the base of the leaf called the **abscission zone** (Fig. 14-45). This region generally contains a large number of parenchyma cells and a few thick-walled cells, making it the weakest part of the petiole, or of the blade if the leaf is sessile. This zone sooner or later becomes differentiated into two layers. One is a protective layer, several cells thick on the side nearer the stem. The other, a separation layer, is the layer in which changes occur that eventually result in the fall of the leaf. The cells of this layer swell and become gelatinous, and the cells may or may not divide. Regardless of what occurs, the weight of the leaf and the force of the wind or rain will break the leaf away from the stem along the separation layer. Strictly speaking, the term "abscission layer" is used to designate the separation layer after it has begun to undergo the changes described.

Although these changes are taking place in the separation layer, other changes are occurring in the protective layer. The cells of the protective layer become suberized, making them waterproof. A layer of lignin permeates their cell walls, as well as the intercellular spaces, and the healing process is complete. Eventually, the periderm of the stem becomes continuous with the periderm formed over the abscission zone.

Some leaves, such as those of the rhododendron can be induced to produce roots, whereas others may have some of their leaves modified into slender, twining tendrils, also mentioned earlier, for example, the pea or climbing plants such as the grape.

There are several families of plants with leaves that have become modified so that they can trap and digest insects. These plants, such as Venus's-flytrap and the pitcher plant, are commonly found in marshy or boggy habitats, where the supply of available nitrogen is limited.

The leaves of Venus's-flytrap (*Dionaea muscipula*) are hinged along the midrib. Along the leaf are a number of trigger hairs that, when touched, cause the leaf to snap shut so that the fringed leaf margins interlock (Fig. 14-46). The closure of the leaf appears to be the result of a combined enzyme-and-turgor pressure mechanism that is initiated by chemical and physical stimulation from the presence of the insect. The pitcher plant (*Sarracenia purpurea*) has vaselike leaves

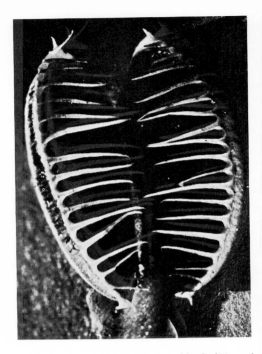

Fig. 14-46. Close-up of partly closed leaf of Venus's-flytrap. (Courtesy Carolina Biological Supply Co., Burlington, N.C.)

lined with stiff, downward pointing hairs so that insects seeking the water at the bottom of the leaves are trapped. The water contains digestive enzymes capable of destroying the insect and of making available the nitrogen necessary to the plant.

Small modified leaves called bud scales cover the winter buds of most deciduous trees. Bud scales do not have chlorophyll and usually have more lignified cells than do ordinary leaves. Thus they are capable of protecting the developing leaves and stem they surround. When the bud opens up in the spring, the scales fall off and leave a ring of bud scale scars on the twig. Other plants, such as cacti, have leaves modified into slender, sharp spines that protect the plant from grazing animals.

We are most familiar with those leaves that add beauty to our homes or gardens. Common flowering plants, such as pansies and asters, are attractive only because the leaves have become modified into attractive shapes and colors. However, even uniformly green leaves can change their color in the fall of the year, especially those of deciduous trees in a temperate zone. The leaves of many woody plants take on yellow and red colors before they fall, particularly if the days are warm and the nights cool. In the chloroplasts of green leaves are two yellow pigments, the carotenes and the xanthophylls, in addition to chlorophyll. They cannot be seen in a green leaf because the more intense green pigments mask them. However, if the green pigments disappear, as they do in the leaves of many plants in the fall, the yellow pigments become visible. It is then that the leaves of birches, poplars, and elms appear golden over the landscape. Other pigments, such as **anthocyanins,** are dissolved in the cell sap and are responsible for the red coloring of leaves. In most leaves anthocyanins are not present during the summer but are formed in the autumn. The same kinds of trees can be different colors in different localities. In one area the leaves may appear crimson; in another area they may not. Soils low in nitrogen, bright skies, drought, and low temperatures favor the formation of anthocyanins so that under these conditions the leaves appear bright red in the autumn. The development of anthocyanins is therefore influenced by environmental conditions that favor a high concentration of sugar.

QUESTIONS AND PROBLEMS

1. What is the function of a seed? What are the factors needed for germination?
2. What happens to the cotyledons of a bean seed? How does this differ from that of a corn seed?
3. Why are the shoots of grasses (like corn) protected when the young stem emerges from the ground and those of beans are not?
4. What is the function of root hairs? Of secondary roots? Of prop roots?
5. How can a seed planted bottom side up grow properly? Does this happen in nature?

6. What are the principle functions of the stem? Of the leaves?
7. Describe the stem in cross section of an herbaceous plant and a woody plant.
8. Describe growth in a woody plant.
9. What protects a plant?

DISCUSSION FOR LEARNING

1. What are some of the many factors that determine the shape and size of plants? Discuss both the individual variation and the features that differentiate the species.
2. Discuss some of the advantages to a plant of the following growth habits: tall tree; vine; broad shrub.
3. In this particular chapter we have dealt mainly with the morphology of the higher vascular (seed) plants and some of the simpler adaptation. Since these plants all have similar parts, we believe them to be homologous parts. Discuss the structure of some of the common plants about the home, for example, food plants, in terms of modifications of these parts from the standpoint of homology (not special adaptations, which will be discussed in a later chapter). For example, compare the parts of a carrot with those of a cabbage and point out the same parts in each.

ADDITIONAL READING

Bierhorst, D. W. 1971. Morphology of vascular plants. The Macmillan Co., New York. (A textbook, well illustrated with photographs of structure, with the central theme of the evolution of these structures. It includes the anatomy of many extinct kinds of plants.)

Bold, H. C. 1973. Morphology of plants, 3rd ed. Harper & Row, Publishers, Inc., New York. (This text includes all the groups of plants from the algae to the flowering plants and discusses each division of plants in phylogenetic sequence.)

Foster, A. S., and E. M. Gifford, Jr. 1974. Comparative morphology of vascular plants, 2nd ed. W. H. Freeman & Co., San Francisco, Calif. (A shorter, concise treatment of the anatomy of vascular plants.)

Lott, J. N. 1976. A scanning electron microscope study of green plants. The C. V. Mosby Co., St. Louis. (A set of excellent photographs showing details of plant anatomy.)

Fig. 15-1. Water flow through vascular plants is a constant need for orderly metabolism of these plants. The root system for absorption must exceed that of the leaf system for transpiration. The stem functions to transport water and solutes from the roots to the leaves and from the leaves throughout the plant. The proper balance of these systems is critical for plant growth. Many adaptations have evolved to maintain this balance. The mesquite tree from southern Arizona as shown here is one of the most successful plants because it is able to maintain its water balance through an extensive root system that seeks deep desert waters and by the small leaves that keep transpiration to a minimum while permitting maximum growth in the bright desert sun.

CHAPTER 15

Absorption, translocation, and transpiration

■ Plants and animals need food for growth and energy. Green plant nutrition is different from animal nutrition because photosynthesizing plants combine atmospheric carbon dioxide and soil minerals to produce organic compounds, whereas animals depend on plants for their organic food. Little evidence is available to show that plants absorb organic molecules. As many as 16 chemical elements are essential for plant nutrition (Table 15-1). Water and minerals are absorbed and circulated throughout vascular plants, and water is released from the leaf surface. Essential minerals, their function in the plant cell, the absorption of minerals and water, their translocation, and the transpiration of water from the surface of leaves are the topics of this chapter.

CHEMICAL REQUIREMENTS OF GROWTH

The various elements necessary for the normal growth of plants were determined by growing plants in nutrient solutions either containing or lacking certain mineral salts dissolved in water. The growing of plants in nutrient solutions without soil is known as **hydroponics**. The role of each of the known essential elements is given in Table 15-1.

Although some of these essential elements are necessary only in minute amounts (**micronutrients**), any deficiency or absence of them will have a distinct effect on the appearance of the plant (Table 15-2). The major elements (**macronutrients**) are discussed first.

MACRONUTRIENTS

Carbon is obtained from the atmosphere in the form of carbon dioxide. Oxygen is obtained from atmospheric carbon dioxide and soil water. Green plants obtain essential mineral elements dissolved in water from the soil (the chemical composition of soil is discussed in Chapter 20). Although salts are occasionally absorbed through the leaves, the roots of vascular plants are the organs responsible for mineral absorption. Aquatic plants absorb salts through their entire surface: therefore, the roots are of minor importance in mineral absorption. Bryophytes growing in moist woodland humus absorb minerals both through their leaves and through their scanty root system. Mineral elements enter the root cell by a process called absorption. To be absorbed by the root, the minerals must be in aqueous solution. The mineral salts are dissolved in soil water and move by diffusion to the root hair, the site of absorption. Water is absorbed by the roots and is translocated through the plant to the leaf surface where it evaporates, a process called **transpiration**.

ESSENTIAL MINERALS

All of the essential mineral elements have a critical level, amounts above which there is no increase in growth or yield. However when these essential elements are below this critical level in the soil, increase in growth and yield can be produced by increasing the concentration of the elements. The critical

Table 15-1. Role of the essential elements that form compounds used in plant growth

Macronutrients	Role	Micronutrients	Role
Carbon	Entire cell and parts	Iron	Enzyme (cytochrome, formation of chlorophyll)
Oxygen	Entire cell and parts		
Hydrogen	Entire cell and parts	Boron	Enzymes, synthesis of glutamine
Nitrogen	Protoplasm (especially proteins)	Copper	Activation of enzymes, synthesis of ascorbic acid
Phosphorus	Protoplasm, enzymes, fats, and nucleoproteins	Manganese	Formation of chlorophyll, photosynthesis, activation of enzymes
Sulfur	Protoplasm, synthesis of chlorophyll, and proteins		
Calcium	Constituent of cell wall, permeability	Molybdenum	Change nitrates to nitrites, synthesis of tannin
Potassium	Enzyme, citric acid cycle, respiration, and photosynthesis	Zinc	Formation of auxin, enzymes
		Chlorine	Unknown
Magnesium	Formation of chlorophyll, enzymes, and photosynthesis		

Table 15-2. Effects of the lack or deficiency of mineral compounds containing the essential elements needed for proper plant growth

Deficiency	Characteristics
Nitrogen	Pale green; older leaves become yellow and dry; stems short and slender
Magnesium	Chlorotic leaves; edges of leaves curled
Phosphorus	Dark green; older leaves become yellow and dry; stems short and slender
Potassium	Chlorotic leaves; tip of leaf dies; dead, dry spots in leaf between veins as well as on edges
Zinc	Large, dead, dry spots over entire leaf; leaves become thick
Copper	Wilting of young leaves; tip of stem droops
Calcium	Terminal buds die
Boron	Terminal bud light green; stalk withered near apex
Manganese	Leaf becomes spotted by dead, dry areas
Iron	Chlorosis of young leaves; stems short and slender
Sulfur	Light green young leaves

level of elements associated with maximum growth must be determined for each species of plant. Furthermore the concentration requirements can vary with the season, and anything above the critical level is termed luxury consumption. Lack or deficiency in the available amounts of nitrogen, phosphorus, and potassium causes reduction in the growth of the plant. Because the symptoms for a deficiency of more than one element may be the same, it becomes difficult to diagnose exactly which element is lacking. In a general way, how the lack or deficiency of each of the better-understood elements taken from the soil by plants affects their growth is listed in Table 15-2.

Table 15-2 shows that several mineral deficiencies cause similar abnormal growths and dwarfing of the plant. The yellowing of leaves, which may occur in a variety of patterns, is called chlorosis. It appears to be a lack of chlorophyll in various parts of the leaf and causes them to have a variegated appearance. A deficiency of minerals such as boron and calcium affect the meristem, resulting in distorted growth of the tip of stems and buds.

Carbon, hydrogen, oxygen, and nitrogen are the most abundant elements in plant tis-

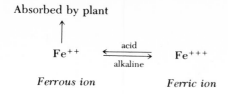

Absorbed by plant

$$Fe^{++} \underset{\text{alkaline}}{\overset{\text{acid}}{\rightleftarrows}} Fe^{+++}$$

Ferrous ion   *Ferric ion*

Fig. 15-5. Diagram depicting reduction of iron in the soil under acid conditions and oxidation of iron in alkaline conditions. The plant root cell absorbs iron in the ferrous state.

Iron is found in the soil in both organic and inorganic forms. Inorganic iron in the ferrous form (Fe^{++}) is absorbed readily by plants. Ferric iron (Fe^{+++}) is also absorbed but to a lesser degree. The availability of ferrous ion for plants is determined by soil pH. In acid soil the ferric ion is converted to a ferrous ion, which is readily absorbed. When the soil is alkaline, the reverse reaction occurs and less inorganic iron in ferrous form is available to the plant (Fig. 15-5). Iron is a constituent of several crucial molecules in the cell. Cytochromes contain iron, which functions in electron-transport processes. Also the synthesis of chlorophyll is dependent on iron. Ferrodoxin, an important electron acceptor in photosynthesis, contains iron. Hemoglobin, an iron-containing compound, is found in the root nodules of nitrogen-fixing organisms. Iron deficiency is characterized by extensive chlorosis, in which the veins of the leaf remain green the longest. Young leaves are affected first. There is also an abrupt halt in cell division when plants are iron deficient.

MICRONUTRIENTS (TRACE ELEMENTS)

Boron is apparently absorbed by the plant as the borate ion (BO_2^-). The soil concentration of this element is very low and its biological function is not clear. Research indicates that boron may be involved in sucrose transport and the synthesis of gibberellic acid.

Copper is available to plants in the inorganic form only. Organic copper cannot be utilized by plants. Many important plant enzymes, such as cytochrome oxidase, contain copper, which appears to function as an electron acceptor. Copper deficiency causes the leaves of citrus fruits to become twisted and turn dark green. The whole plant may appear wilted and degeneration or necrosis occurs at the tip of young leaves.

Manganese is available to plants only in the inorganic form, the divalent cation (Mn^{++}) being most readily absorbed. Respiration is dependent on manganese as an enzyme activator. Manganese is also essential for nitrate reduction because it is an activator for nitrate reductase. When a manganese deficiency occurs, the chloroplast is the first part of the plant affected. Visible signs of a deficiency are chlorosis and leaf degeneration.

Molybdenum is absorbed as the divalent cation (Mo^{++}). It is essential to the nitrate reductase reaction where it combines with the enzyme protein and functions in electron transport. A molybdenum deficiency causes chlorosis and decreases flower formation.

Zinc is available to plants only as an inorganic salt. It is absorbed as the zinc ion (Zn^{++}). It functions primarily as an enzyme activator. Enzymes involved in the synthesis of tryptophan, an auxin precursor, are activated by zinc. It also activates carbonic anhydrase, an enzyme that forms carbon dioxide and water from carbonic acid. Zinc deficiency causes "little leaf disease," which is characterized by stunted growth of leaves and stems.

The elements of sodium and chlorine are so ubiquitous in nature that it has been difficult to establish a specific function for each. However sodium appears to be essential for plants showing the C_4 photosynthetic pathway.

ABSORPTION

Plants absorb or take up water and mineral elements from the soil. Absorption is the process that involves the passage of either water or minerals, or both, from the soil into the plant cytoplasm. Various forces cause the movement of an ion from the soil into the cell cytoplasm. If the concentration of potassium ions is 10 times greater on the outside of the cell than on the inside and if we assume that the cell membrane is freely permeable to potassium, the ion will move from an area of its higher concentration to an area of its lower concentration. This is an example of movement down a concentration gradient and an example of passive absorption. For an ion to move up or against the concentration gradient, energy is required. Movement against a gradient is called **active absorption,** or active transport. Plant cell absorption is classified as either active or passive (Fig. 15-6).

When plant cells are initially placed in an ion solution, there is a rapid uptake of the ions. The rapid uptake is attributable to the movement of ions by diffusion into the free

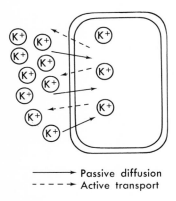

Passive diffusion
Active transport

Fig. 15-6. Diagram depicting active and passive absorption. Active absorption, *dotted lines,* is the movement of an ion against a concentration gradient. Passive absorption, *solid lines,* is the movement of an ion down a concentration gradient.

space of the cell. The barrier to further movement or the delineation of the free space from the nonfree space is the plasma membrane (Fig. 15-7). The plasma membrane regulates the passage of ions into the cytoplasm. Some ions may freely diffuse across the membrane (passive), whereas other ions are restricted and must be transported across (active transport).

Passive absorption is attributable to purely physical forces, and the cell does not expend energy in the process. On the contrary, active transport depends on cell expenditure of metabolic energy.

Passive absorption can be accomplished by one of three processes. The first process is ion exchange. For example, the solution surrounding the cell contains potassium chloride (KCl), which ionizes into the potassium ion (K^+) and the chloride ion (Cl^-). The plasma membrane of the cell has hydrogen ions (H^+) attached to it. During ion exchange the hydrogen ions will detach from the membrane and go into solution. The potassium ions (K^+) replace the hydrogen ions on the membrane. The cell will concentrate potassium ions against a gradient without expending metabolic energy (Fig. 15-8). The second process of passive ion absorption is called **Donnan equilibrium.** Let us assume that a negatively charged ion (B^-) is inside the cell and it cannot pass to the outside through the plasma membrane. The plasma membrane is not permeable to the negatively charged ion (anion) B^-. As a result of this, the charge on the inside of the cell will be negative and pos-

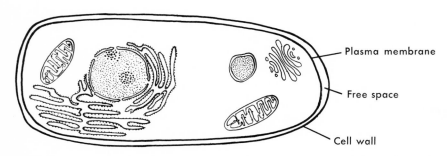

Fig. 15-7. Diagram of plant cell depicting the free space between the plasma membrane and the cell wall.

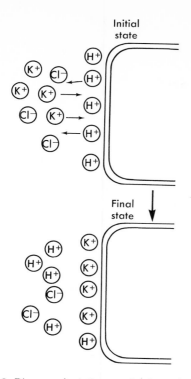

Fig. 15-8. Diagram depicting mechanism of ion exchange at the cell membrane. In the initial state hydrogen ions are attached to the cell membrane while potassium and chloride ions are in the surrounding medium. An exchange occurs during which hydrogen ions leave the membrane and are replaced by the potassium ions from the medium.

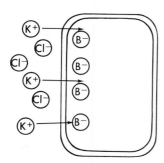

Fig. 15-9. Diagram depicting phenomenon of Donnan equilibrium at the cell membrane. Inside of the cell, large molecules depicted as B have a negative charge and cannot pass through the cell membrane. Outside of the cell the medium contains potassium ions K^+ and chloride ions Cl^-, both of which can pass through the membrane. As a result the potassium ions are drawn through the membrane by the strong negative charge inside the cell.

itively charged ions from outside the plasma membrane will move through the membrane and into the cytoplasm. The cell will concentrate the positive ions (cations) against a concentration gradient and will not expend metabolic energy (Fig. 15-9).

A third possibility that has been proposed as a mechanism for passive absorption is the mass flow of water. Water is passed through the plant by being absorbed through the roots and evaporated at the leaf surface. Substances dissolved in water would be carried into the plant cell as a result of the movement of water. The passage of water through a plant is a passive process, therefore substances would be accumulated against a gradient without the expenditure of metabolic energy.

Active absorption depends on the expenditure of metabolic energy in order to accumulate substances against a gradient. The accumulation of mineral ions by plant cells cannot be accounted for on the basis of passive transport alone. Active transport is definitely involved. Active transport mechanisms are inhibited by low temperatures, decreased glucose levels, and metabolic inhibitors. Research has proved that mineral ion accumulation by plants was considerably decreased when any of the active-transport inhibitors were applied. We therefore assume that mineral ions are accumulated by active-transport mechanisms.

The basic mechanism or mechanisms of active transport have not been elucidated. Research however indicates that plasma membranes contain proteins that will specifically combine with cations and anions. For example, a sulfate-binding protein has been isolated in bacteria. It has been postulated that the binding proteins are carrier molecules, which will combine with the cation or anion at the outer surface of the membrane and release it at the inner surface or into the cytoplasm. The energy for the process is supplied by the hydrolysis of ATP, catalyzed by a membrane-associated ATPase. Membrane-associated ATPase has been isolated and characterized from a variety of organisms.

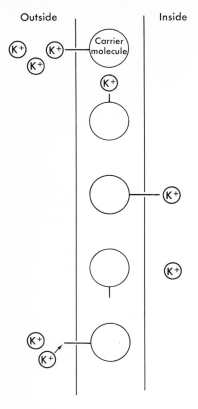

Fig. 15-10. Diagram depicting the transport of potassium across the cell membrane by a carrier molecule. The carrier molecule combines with potassium on the outside of the cell membrane and releases it on the inside.

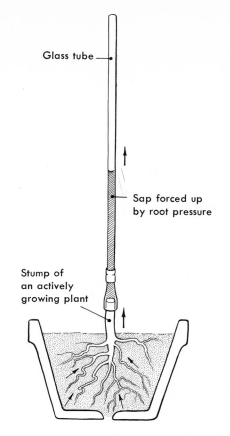

Fig. 15-11. Test to show the effect of root pressure on a column of fluid (sap).

Oat roots (*Avena sativa* var. *goodfieldi*) contain a membrane ATPase that is activated by potassium ions and may be involved in potassium transport in this organism. There is also a considerable amount of evidence for the active transport of carbohydrates in bacteria. A model for the carrier hypothesis of active transport is illustrated in Fig. 15-10.

Most water is absorbed at the root-hair zone of the root. Young secondary roots have a diffusion-pressure deficit of water greater than that in the soil solution. Therefore there will be a net movement of water molecules from the water solution into the vacuoles of the cells of the functional secondary roots. The movement of water molecules is from a region of lesser diffusion-pressure deficit to a region of greater diffusion-pressure deficit and not from a region of greater **osmotic pressure** to one of lesser osmotic pressure. The pressure resulting from the net movement of water molecules from the soil to the vacuoles will bring about an increase in diffusion pressure of the water in the root cell vacuoles that will be greater than that of the adjacent cells. As a result, a series of diffusion-pressure gradients will occur across a root and into xylem vessels. This is the absorption of water through a root, and although it is most rapid in the region of young secondary roots, it may occur to a lesser extent in some other parts of the root.

The absorption of water is possible in roots only when the diffusion-pressure deficit of the root is greater than that of the soil water. However as water is removed from the soil water solutions, its diffusion-pressure deficit will increase unless it is reduced by rainfall or irrigation. This increase could retard the uptake of water by the root system uless there is an increase in the diffusion-pressure deficit of the root cells to offset that of the soil water. Absorption is responsible for the oozing of fluids from the cut surface of a plant or from specialized structures in leaves called **hydathodes** (meaning 'water way'). Each hydathode will have one or more pores that resemble stomas, but unlike stomas they are permanently open.

The movement of water into roots produces what is called **root pressure** (Fig. 15-11) and may cause the formation of droplets of water on leaves, a phenomenon known as **guttation.** Both root pressure and guttation are a result of active absorption when transpiration (loss of water by evaporation) is at a minimum and water is being absorbed by the roots in greater amounts than is being lost at the leaf surface. Root pressure can be inhibited by metabolic inhibitors; thus an active-transport process is indicated. It is not an important force in the ascent of sap, although it may be used to refill xylem during dormancy periods of the plant. The importance of roots cannot be minimized, since they are the plant's chief absorbing structures. During transplanting, their detachment can be responsible for the unsuccessful replanting of a growing plant.

When roots are grown by suspension, experimentally, in moist air, they are uniformly arranged by age and location, but those growing in soil become distorted as they push their way around and between soil particles. It seems that the root hairs actually become firmly united to soil particles that are surrounded by films of water. This contact of the root hair with the soil particle enables the root to absorb the maximum amount of moisture and minerals dissolved in the water. In addition to the secondary roots, elongated conducting cells can be seen developing in the center of the root.

The absorption of water is attributable to passive transport mechanisms. Active transport influences water absorption indirectly by causing changes in the physical forces that cause passive transport. Water absorption is limited by soil conditions such as temperature, aeration, pH, osmotic pressure, and available water. Evaporation at the leaf surface will also affect water absorption.

Minerals and water are both absorbed by the leaf surface. Aquatic plants obtain the majority of their nutrients in this way. Foliar fertilization with mineral salts has been successful, but it does not substitute for soil fertilization. Water is absorbed both as a vapor and as a liquid. Mineral salts are absorbed as ions. The pathway of ion entry is apparently both the cuticle and stomas. The cuticle is much more active in this process than are the stomas. The absorption of water and minerals by plant organs other than roots is minimal.

TRANSLOCATION

The growing plant requires an extensive system of circulation in order to transport nutrients and water. In unicellular algae, transportation is by diffusion and does not present a problem because the photosynthetic site is adjacent to the absorbing site for water and minerals. However in multicellular plants the photosynthetic site in the leaves is separated from the water and mineral supply. The plant must have a circulatory system to carry the organic molecules synthesized during photosynthesis to the remainder of the plant and also carry water and minerals from the soil to the upper parts of the plant, particularly to the leaves. One of the functions of the stem is the conduction of materials upward to the leaves, downward to the roots, and transversly to stem tissues. The **translocation** of materials in stems is not completely understood, since the mechanisms behind the rise of sap in a stem and the path it follows are very complex. Xylem and phloem are the tissues of the circulatory sys-

261

tem. The morphology of these tissues is described in Chapter 14.

The term **sap,** as used by a botanist, is usually applied to the water and minerals that move upward in the younger annual rings of woody plants, particularly the xylem. The fact that xylem is involved in the upward movement of substances in the plant has been known for over 100 years. Researchers gradually unraveled the anatomy of the xylem and revealed that the conduits are composed of dead cells without cytoplasm, thus making it possible for the translocation of large bulk solutions. The most convincing evidence for the function of xylem is from a chemical analysis of the xylem sap. Analysis of xylem sap indicates that it contains 0.1% to 0.4% solid material. One third of the solid material is ash. The essential mineral elements can always be detected, in addition to sugars, amino acids, keto acids, alkaloids, and occasionally coumarins. The observations that xylem sap contains many mineral ions and that cut stems take up certain dyes in the xylem led researchers to investigate the possible role of xylem in the transport of inorganic minerals. Prior to the use of radioactive isotopes for analysis the most convincing evidence in favor of this concept was the demonstration that lithium and cesium were transported upward in the plant from the roots. The removal of a small piece of xylem completely inhibited translocation while the removal of a piece of phloem did not interfere with the upward movement of these cations. The radioactive isotope that best demonstrates the translocation of xylem is potassium (^{42}K). Small plants such as cotton, geranium, and willow were used in this experiment. They were planted in a sand culture containing ^{42}K. Then by delicate surgery on a 9-inch portion of the stem, the xylem was physically separated from the phloem by wax paper. The plants were allowed to absorb the radioactive isotope ^{42}K for 5 hours. After this time period various sections of the stem were removed and analyzed for the isotope content. Isotope concentrations were most abundant in the xylem where it was separated from the phloem by the wax paper. However where it was not separated there appeared to be some lateral transport of potassium from xylem to phloem. The small amounts of potassium found in the phloem indicated that this tissue does not have the capacity to transport potassium (Fig. 15-12).

Plant physiologists have demonstrated in numerous experiments that water and ions

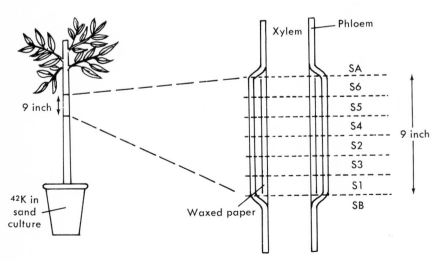

Fig. 15-12. Method of detecting the translocation of ^{42}K in a rooted cutting of willow *(Salix).* Lateral transport of isotope from xylem to phloem is prevented by a cylinder of waxed paper. (Modified from Stout, P. R., and D. R. Hoagland. 1939. Am. J. Bot. **26:**320-324.)

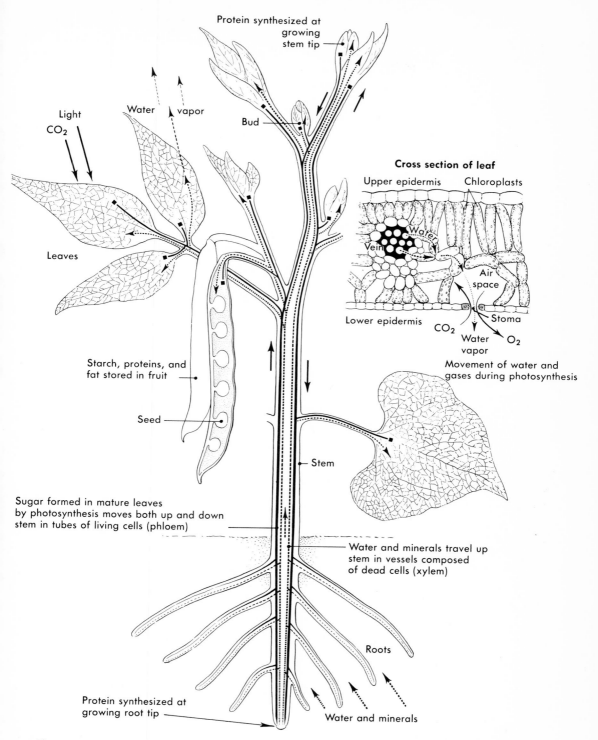

Cross section of leaf

Upper epidermis   Chloroplasts

Vein

Water

Air space

Lower epidermis CO_2 Water vapor O_2 Stoma

Movement of water and gases during photosynthesis

Protein synthesized at growing stem tip

Light CO_2

Water vapor

Bud

Leaves

Starch, proteins, and fat stored in fruit

Seed

Stem

Sugar formed in mature leaves by photosynthesis moves both up and down stem in tubes of living cells (phloem)

Water and minerals travel up stem in vessels composed of dead cells (xylem)

Roots

Protein synthesized at growing root tip

Water and minerals

Fig. 15-13. Path of circulation of gases and liquids in plants. *Upper right,* Details of interior of a leaf.

move upward in plants, largely in the xylem. However to explain the cause of this ascent is not easy. It is partly attributable to transpiration. It has been suggested that capillarity is involved in the rise of sap, as well as are the rhythmic expansions and contractions of the living parenchyma cells associated with xylem. A more satisfactory explanation to account for the rise of liquids in a stem is based on the idea that liquids are pulled up rather than pushed up from below. The pulling action, caused by evaporation of water from leaf cells, and the cohesive action of water appear to be able to account for the rise of liquids to great heights in trees. This mechanism is known as the cohesion-tension theory.

Water evaporates from the tiny capillaries of the walls of the leaf parenchyma cells. As a result the leaf parenchyma cells draw water from neighboring cells so that an osmotic gradient reaches across the leaf to the xylem, where water is drawn up the xylem ducts. The xylem tissue of the leaf is directly connected with the xylem tissue of the stem. The stem xylem tissue is directly connected to the xylem tissue of the roots, which receives water from root hair absorption of soil water. Thus we have a continuous system from root hair to leaf surface (Fig. 15-13).

COHESION-TENSION THEORY

The theory most accepted by botanists for the forces controlling the ascent of sap is termed the cohesion-tension theory. The theory, simply stated, says that sap is pulled up the xylem of the plant by the forces or negative pressure created by the evaporation of water at the leaf surface. When a long capillary tube is filled with water, it is possible for one to pull a continuous stream of water through the tube by evaporating water at the end or by creating a negative pressure at the end, like sucking through a straw. Water moves in a continuous stream through the capillary tube as the result of two forces, cohesion of water molecules for each other and the adhesion of water to the sides of the capillary tube. Water molecules have a large molecular attraction for one another so that

when they are pulled, they tend to remain together, thus maintaining a continuous stream. The second force of adhesion for the sides of the capillary tube creates a tension between the wall of the tube and the stream of water. If the tube were flexible, you would expect it to contract. Researchers have demonstrated a decrease in the diameter of trees during peak hours of transpiration. Evidence such as this lends support to the cohesion-tension theory. One of the most important objections to the cohesion-tension theory is that the presence of air bubbles will interrupt the continuous stream of water. It seems very likely that trees get air bubbles in their xylem. Researchers have shown, using saw cuts and ^{32}P, that the stream in translocation moves laterally, avoiding the xylem with air bubbles (Fig. 15-14).

TRANSLOCATION IN PHLOEM

The food manufactured by the leaves is transported through the living sieve tubes and, to a lesser degree, through the companion cells of the phloem. Therefore translocation of food is chiefly downward through the phloem in summer. The vascular rays (Fig. 14-27) transport dissolved materials and water transversely across a stem by diffusion and by the protoplasmic streaming of the ray cells. Transverse conduction is extremely important because it makes possible the movement of foods from the sieve tubes into the dividing cells of the cambium. Carbon dioxide moves outward through the lenticels and oxygen moves inward, both movements thereby aiding respiration of cambium cells and other living cells in both xylem and phloem. Many early researchers demonstrated that the phloem was the primary conduit for the translocation of carbohydrates manufactured during photosynthesis. When bark was removed from a tree, the tissue above the removed area swelled. However swelling occurred only when the tree had leaves. Sap from the phloem is very rich in carbohydrates, and increased rates of photosynthesis were associated with increased carbohydrate content of phloem sap. How-

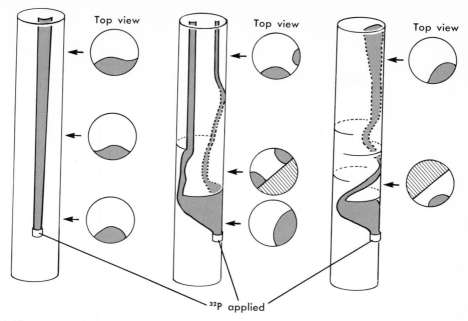

Fig. 15-14. Movement of ^{32}P around saw cuts in the xylem of pine trees. **A,** No saw cuts. **B,** Two overlapping saw cuts. **C,** Four overlapping saw cuts. (After Postlethwart, S. N., and B. Rodgers. 1958. Am. J. Bot. **45:**753-757.)

ever the problem of lateral transport plagued early researchers. In lateral transport, carbohydrates are transported from phloem to xylem. As a result, carbohydrates were also found in xylem and because of this fact it was difficult to establish the phloem as the transporter of carbohydrates. With the use of radioactive isotopes and aphids this problem was resolved and the sieve tubes of the phloem was definitely established as the main conduit carrying carbohydrates to the other parts of the plant. The mechanism of the movement of materials through the phloem has not been clearly defined. However recent research using radioactive isotopes, viruses, aphids, and growth hormones have aided in a better understanding of the process.

MASS-FLOW HYPOTHESIS

The mass-flow hypothesis was first postulated in 1860 and later reformulated in 1930. Currently it is accepted by many botanists. During photosynthesis the carbohydrate concentration in the leaf cell increases, while the carbohydrate concentration in an actively growing cell, such as a root cell, decreases. As a result a gradient of hydrostatic pressure is created. Water moves into the leaf cell and out of the root cell. The water moving into the leaf cell drives a solution with dissolved carbohydrate from the leaf cell through the phloem to the root cell until the concentration of carbohydrate in each cell is equal. Carbohydrate then will flow from an area of higher concentration, the leaf, to an area of lower concentration, the root cell. Water moves in the opposite direction, from the root cell to the leaf cell. An example of this is illustrated in Fig. 15-15. The leaf cell, A, is connected by the phloem, C, to the root cell, B. Both cells are freely permeable to water but not permeable to carbohydrate. If the leaf cell and root cell are immersed in a common water bath, and the concentration of carbohydrate in the leaf increases, water will flow into the leaf from the bath. As a result water with dissolved carbohydrate will move from the leaf through the phloem to the root. Further tests show that when growth hor-

265

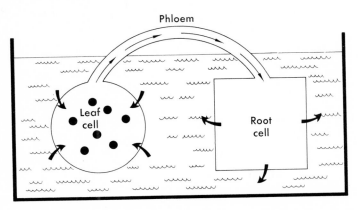

Fig. 15-15. Diagram depicting mass-flow hypothesis. The two vessels are connected by a tube and submerged in distilled water. The large molecules in the circular vessel cannot pass through the wall of the vessels; however water can pass the vessel walls easily. Water will move into the spherical vessel and through the tube to the retangular vessel. In the process the large molecules will be carried from the spherical vessel to the rectangular one until equilibrium is reached.

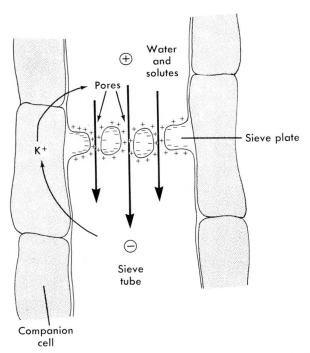

Fig. 15-16. Electro-osmotic flow of water and assimilation through the pores of phloem sieve plates. (From Spanner, D. C. 1958. J. Exp. Botany. **9:**332-342.)

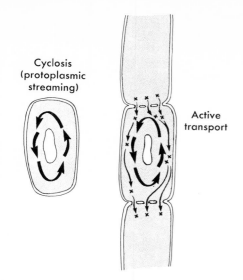

Cyclosis
(protoplasmic
streaming)

Active
transport

Fig. 15-17. Diagram depicting the movement of substances in plant cells caused by protoplasmic streaming.

mones and viruses are applied to leaves, they are translocated through the phloem only during photosynthesis. This observation is interpreted as support for the mass-flow hypothesis because during photosynthesis the concentration of carbohydrates increase in the leaf. The application of sugar solutions to the leaves caused the translocation of both viruses and growth substances in the dark. Recent research however indicates that an initial step in sugar translocation involves an active-transport mechanism using ATP from respiration as an energy source.

A modification of the mass-flow hypothesis is the electro-osmosis and mass-flow concept. This concept postulates that the phloem companion cells actively transport potassium ions through the cells and around the sieve plates, creating an electro-osmotic force that pushes water and solutes through the pores in the sieve plates (Fig. 15-17).

The second major theory for the mechanism of phloem translocation is the protoplasmic streaming hypothesis. The protoplasm of a living cell is constantly streaming in circles within the plasma membrane. As the protoplasm streams, it carries particles and sol-

utes from one end of the cell to the other end. The theory states that protoplasmic streaming in the cytoplasm of the sieve tube element carries solutes from the sieve plate area at one end of the cell to the sieve plate area at the other end. The solutes then pass from one cell to the adjacent one by diffusion through the sieve plates (Fig. 15-17).

In summary, then, at present two major theories predominate of phloem translocation: (1) mass flow and electro-osmosis and (2) protoplasmic streaming. Neither theory alone accounts for the phenomenon of phloem translocation. Experimental evidence supports and refutes both theories. It definitely appears that phloem translocation is dependent on metabolic energy. Phloem cells contain high levels of ATP and rapidly convert inorganic phosphorus into high-energy phosphate. Also limiting the nutrient phosphate halts phloem translocation. Translocation in the phloem is attributable to a combination of active- and passive-transport mechanisms. The nature of the individual forces and their interaction with each other and with cellular structures remains to be elucidated.

TRANSPIRATION

The primary function of leaves is photosynthesis. At the same time, they provide the site for transpiration. This effect makes a forested area cooler in the summer than an open area with little vegetation. The epidermis of a leaf protects it from excessive loss of water, mechanical injury, and parasitic infection. Epidermal cells are usually covered with a definite cuticle that aids in preventing evaporation of water. Ordinarily, the cuticle is thicker on the upper surface of a leaf than on the lower surface. Plants growing in the sun have a thicker cuticle than do those growing in the shade. Some leaves, such as mullein, may have a dense covering of hair over the epidermal surface, which also tends to retard evaporation.

The cells of the epidermis are irregularly shaped, have no intercellular spaces or chloroplasts, and include **guard cells.** Scattered

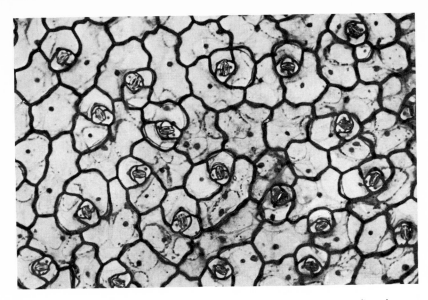

Fig. 15-18. Surface view of lower epidermis of leaf to show the guard cells surrounding the stomas. (Courtesy General Biological Supply House Inc., Chicago, Ill.)

among the epidermal cells of most leaves are numerous openings called stomas (Fig. 15-18). Each stoma is surrounded by two specialized cells known as guard cells (Fig. 15-19), which regulate the size of the opening. Frequently the cells adjoining the guard cells are ordinary epidermal cells, but in some leaves they may be differentiated into supporting cells. Although the term "stoma" is sometimes used to apply to the entire stomatal apparatus, it is more correctly restricted to the opening.

The guard cells are usually kidney shaped, and their concave surfaces form the intercellular space or stoma. When guard cells are turgid (Fig. 15-18), they are more bowed apart, causing the opening to enlarge; when they lose their turgidity (Fig. 15-18), they become more closely approximated throughout their length, and the opening closes. The mechanism of the opening and closing of the stomas is discussed more fully later. The structure of the guard cells and the presence or absence of supporting cells, as well as their arrangement, are two of many criteria used by the taxonomist to aid him in classifying plants.

Usually stomas are more abundant on the lower surface of a leaf than on its upper surface. However, in some leaves they are equally numerous on both surfaces. In other plants the upper surface may lack stomas entirely, whereas floating leaves such as those of water lilies have stomas only on the upper surfaces. In most dicot leaves the stomas are scattered uniformly over the leaf surface, whereas in monocots they are arranged longitudinally along the veins. The number of stomas per unit area varies in different species from as low as 10 to as high as 1300 per square millimeter. The larger the stomas, the fewer they are in number and vice versa. The stoma, relatively small, varies from 7 to 40 μ in length and from 3 to 12 μ in width when completely open. It is an advantage for a leaf to have numerous small stomas rather than a few large ones, since diffusion of gases through many small openings is much greater than through a few large ones.

Passive translocation of water occurs through the root cells as a result of forces in the shoot system of the plant—the leaves. As the water evaporates from the cells between the upper and lower epidermis of leaves, or

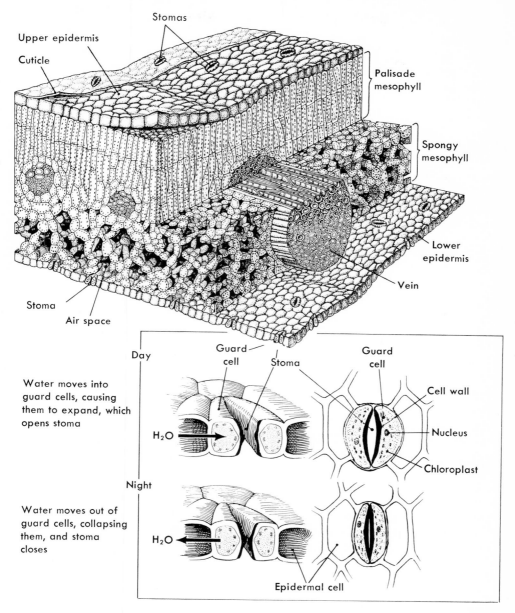

Stomas

Upper epidermis

Cuticle

Palisade mesophyll

Spongy mesophyll

Lower epidermis

Vein

Stoma

Air space

Day

Guard cell

Stoma

Guard cell

Water moves into guard cells, causing them to expand, which opens stoma

Cell wall

H₂O

Nucleus

Night

Chloroplast

Water moves out of guard cells, collapsing them, and stoma closes

H₂O

Epidermal cell

Fig. 15-19. Details of the cells and tissues of a leaf. *Top,* Cross section of a leaf showing structural details. *Bottom,* Enlargement of the stomas showing functional details.

as water is used in their growth processes, or both, an increasing amount of water is withdrawn from xylem tubes. This increased movement of water (Fig. 15-13) increases the diffusion- pressure deficit in the cells, which in turn reduces the deficit in the leaves, stems, and root cells. The water is more or less pulled through the plant by transpiration. In other words, it is the result of evaporation of water from leaves. Almost 95% of all water taken in by a plant is lost through transpiration. The remaining 5% is used in cell growth, photosynthesis, or other life processes.

In transpiration the loss of water vapor is through the stomas of leaves and young stems. When the vapor pressure is less in the atmosphere than it is in the spaces of the leaves, water molecules diffuse into the atmosphere. This loss of water in the leaf spaces causes water to diffuse from the mesophyll cells, which in turn causes diffusion from the xylem tubes of the veins of the leaf that extend down into the roots. The cohesion force of the narrow column of water extending from the leaf to the root is so great that there is actually a pulling force on the column. This force originating in the leaves from the transpiration is responsible for the passive translocation of water through the plant and for its absorption from the soil.

The rate of transpiration is affected by many factors. The energy of the sun will affect the internal temperature of the leaves, and it is not unusual for a leaf to be 10 degrees warmer than the air around it. This increased temperature as a result of radiation from the sun will cause greater vaporization of the water in the leaf. Therefore transpiration will be greater on sunny days than on cloudy days. Circulation of air also affects transpiration. When air movement increases up to approximately 15 miles per hour, the rate of transpiration increases. When the circulation of air is above that rate, transpiration will be reduced because the temperature of the leaf is reduced. Another factor affecting the rate of transpiration is the amount of

moisture in the soil. Decreasing the amount of water in the soil will decrease transpiration.

In addition to all the factors mentioned that affect transpiration, the principal one is the opening and closing of the stomas and their number and position on the leaf. Stomas are found on the upper surface of a leaf only, on the lower surface only, or on both surfaces, depending on the kind of plant and its habitat. Most trees have leaves in which the stomas are found only in the lower epidermis and in large numbers. For example, more than 70,000 stomas per square centimeter can be found on the lower epidermis of an oak leaf.

Each stoma, or opening, is surrounded by two guard cells. Although the epidermal cells are normally without chloroplasts, they are present in the guard cells. Several explanations are given for the manner in which stomas are opened or closed. The walls of the guard cells are thicker toward the center of the stoma. Turgor pressure, as a result of osmosis, may increase and cause the guard cells to become more turgid, thus enlarging the stomatal opening. When fully distended, each guard cell is C shaped. When the guard cells lose turgidity, their centers tend to meet, and the size of the opening is reduced. Pressure exerted by the surrounding epidermal cells may also aid in pushing the relaxed guard cells closer together.

A considerable amount of evidence indicates that growth regulators control stomatal movement. Gibberellic acid and cytokinin cause opening, whereas auxin and abscissic acid cause a closing. The mechanism of stomatal movement is not fully understood. However as a result of recent research, a theoretical model may be proposed. Water moves into the guard cells, causing them to expand, which opens the stoma. When water moves out, the guard cells collapse and the stoma closes. The movement of water is the direct force causing stomatal movement. The osmotic contents of the cell determine whether water will flow in or out. In the light stomas open; therefore water moves into the

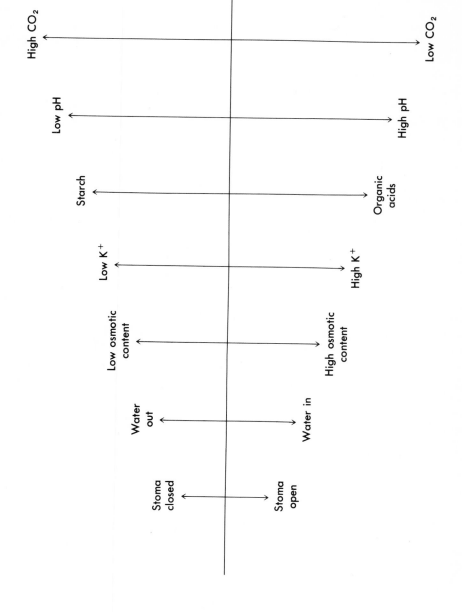

Dark

Light

Fig. 15-20. Model depicting the cellular events associated with stomatal movement in the light and dark.

guard cells and the osmotic content of the guard cells has increased. The light stimulating photosynthesis causes a decrease in the carbon dioxide concentration (which is acid), thus raising the pH. The raise in pH stimulates the conversion of starch (osmotically inactive) to organic acids (osmotically active). Simultaneously light stimulates the active transport of potassium ions into the cell. The transport of potassium is dependent on the metabolic energy of the cellular ATP. Potassium ions combine with the organic acids; thus potassium organic salts are formed and the pH continues to rise. As night descends, photosynthesis ceases, carbon dioxide concentration increases, pH decreases, potassium transport is no longer activated, potassium ions leave the cell, organic acids are converted into starch, osmotic contents of the guard cells drop, water moves out of the guard cell, and the stoma closes (Fig. 15-20). This is a general model that does not explain the action of growth hormones or the opening of stomas in the dark. Further research will help establish the role of growth hormones, carbohydrate metabolism and the photoperiod.

The longer and wider the stomas are open, the greater will be the rate of transpiration. The number of stomas makes the surface of a leaf resemble a sievelike membrane. Evaporation from such a surface is many times greater than that from a perfectly flat surface. Evaporation of water from the leaf surface is less during the night than during the daylight hours for the reasons already given. At night since absorption exceeds transpiration, the turgidity of the plant is replaced. When this does not occur, the plant becomes permanently wilted. Excessive soil water or salts can be responsible for preventing the uptake of water to balance the loss of water by transpiration. Why this occurs is not fully understood. One reason may be that a decrease in permeability of a root cell to water could be caused by a lack of oxygen and, hence, reduced respiration. High temperature causes stomatal closure by increasing respiration, thus increasing the carbon dioxide concentration. Temperatures above 25° C cause closure under all circumstances. Transpiration is also affected by environmental factors such as the relative humidity of the atmosphere, movement of air, light intensity, and soil conditions. Morphological factors that affect transpiration include the position and number of stomas, length of roots, water-storage tissue, and the effective transpirational surface.

QUESTIONS AND PROBLEMS

1. List some difficulties that you would expect to encounter when studying translocation experimentally.
2. Distinguish active transport from Donnan equilibrium.
3. Indicate how the evaporation of water from the surface of the mesophyll cells in the leaf is somehow involved in the mechanism of water transport.
4. Design an experiment that would demon-

strate the necessity of iron (Fe^{++} or Fe^{+++}) as a plant-growth substance.

5. Plants have a high requirement for nitrogen, and consequently soil without nitrogen will not support plant growth. Describe the absorption and assimilation of nitrogen by plant cells.

6. Proteins are indispensible for the survival of a plant cell. List the nutrients that supply the basic atoms necessary for protein formation.

7. Describe the major theories for the mechanism of phloem translocation.

8. Where would you expect to find the greatest concentration of carbohydrate (xylem or phloem) when the plant is in bright sun? Why?

9. What is the relationship of osmosis to the proposed mechanism of stomatal movement?

10. Describe an experiment that demonstrates the upward translocation of potassium.

ADDITIONAL READINGS

Bidwell, R. G. S. 1974. Plant physiology. Macmillan Publishing Co., Inc., New York. (An in-depth plant physiology textbook.)

Nobel, P. S. 1974. Biophysical plant physiology. W. H. Freeman & Co., San Francisco. (An excellent quantitative treatment of the movement of water and ions in plants.)

Sutcliffe, J. F., and Baker, D. A. 1974. Plants and mineral salts. Studies in biology, no. 48. Edward Arnold, Publishers, London. (An elementary discussion of the supply, requirement, and absorption minerals by plants.)

Richardson, M. 1968. Translocation in plants. Studies in biology, no. 10. Edward Arnold, Publishers, London. (A good discussion of the experimental evidence supporting the function of xylem and phloem tissues.)

Fig. 16-1. Nectaries on flower petals attract insects that often act as pollinators. (USDA photograph.)

CHAPTER 16

Reproduction by flowers and their fruits and seeds

■ Life itself is defined by its unique ability to reproduce, and this is often symbolized by a bee or a flower (Fig. 16-1). A philosopher might refer to a reproductive series as the chain of life, a maze of organisms linked together since the first replication of nucleic acid. Profound studies have been made to determine how life originated. These interesting and scholarly deductions probe deeply into biochemical reactions. It is apparent that the process was slow to develop, but it also seems obvious that the real beginning of life was at the moment of replication, through the synthesis and arrangement of nucleic acids to preserve an orderly structure capable of growth and further replication. It is in this sense that we define **reproduction.** The very nature of the biota, as we shall see, is based on reproductive structures and processes, and the ordinary connotation of reproduction is signified by the union of gametes, as will be seen throughout this chapter.

All plants from the simplest to the most complex have structures for reproduction and for getting food. Plants that have a complex organization are able to live in situations uninhabitable by plants lacking these complexities. Every organism, as an individual, is equipped with some means of reproduction, and that organism is occupied with growth, preservation of self, and reproduction of its kind. The most distinctive characteristics of any plant are its structures for, and means of, reproduction. This chapter presents the basic principles and details of reproduction of the flowering plants.

Selected life cycles of representative plants of each major division of plants are given in Appendix 2. Reference to these and to each division as described in Chapters 5 to 11 will supply further details of plant reproduction.

BASIC PRINCIPLES

The reproduction of all plants is very similar in some respects. Means are usually provided for a **gamete** union to form a zygote. This is necessary to mix the genetic material and thus provide for individuality (variation). The gametes usually have the haploid chromosome number. The union of two gametes forms a zygote, which is the diploid chromosome number. All the complexities of plants simply involve refinements of structure for facilitation of the union of gametes or the dispersal of spores and seeds.

Reproduction of most plants is basically an alternation of generations, that is, an asexual plant body alternating with a sexual plant in the succession of generations. Relatively few of the simple plants have sexual generation, and most of the higher plants (seed plants) have so reduced the sexual reproductive plant body that it is negligible. It is the asexual plant that becomes dominant in the higher plants; the gametophyte is so greatly reduced that it is, for all practical purposes, a part of the sporophyte.

Most organisms are either single-celled, **unicellular,** or divided into several to billions of cells, **multicellular.** Considerable difference in terminology is necessary in considering these two types of organization. A third condition exists in certain fungi and slime molds that have two to many millions of nuclei without cytoplasmic walls separating the body into cellular units. Finally, the cells of bacteria and related organisms (that is, the prokaryotic organisms) lack an organized nucleus. Each of these types of organisms maintains its reproductive processes in a different manner, utilizing quite different structures.

Despite this great diversity, a general plan is seen that is, no doubt, inherent in the universally present control system provided by the DNA molecule. This uniformity is seen as follows:

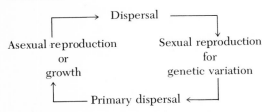

Each of these stages is considered in detail in the following discussion.

ASEXUAL REPRODUCTION

Asexual reproduction is not a single process or even a homologous process in plants. In unicellular plants, cell division or mitosis is a form of asexual reproduction. It is asexual because it does not involve a mixing of genes; that is, the organism is duplicated by the mitotic process, but each of the offspring is genetically identical with the original cell. This same process in multicellular plants is called growth. When the cells adhere to each other after mitosis, a multicellular organism is formed. This growth process is not considered to be a form of reproduction because no new organism is produced.

True asexual generation is evident by the production of spores, sporulation. In the lower plants this is the primary dispersal stage. The sporophyte produces large quantities of cells that may be specially fit to resist adverse conditions and that are at the same time capable of wide dispersal. In the lower aquatic forms these cells are provided with cilia or flagella that permit them to swim for

some distance and thus populate new areas if conditions are suitable for growth. Swimming spores do not have a protective cover. They may have either the haploid or diploid chromosome number. These forms continue to produce new plants without sexual reproduction. The same plant also produces the gametes, and sexual reproduction takes place. In the higher but nonflowering plants these spores are produced by a sporophyte that has the diploid chromosome number. The spore itself is formed after meiosis has taken place. Such spores are light and airborne. They are provided with a protective covering resistant to adverse growing conditions. These spores, instead of producing new sporophytes, produce the gametophyte. True alternation of generations is then present.

There are other types of asexual reproduction. Some of the bryophytes are provided with special organs, the cupules, which develop **gemmae.** A small clump of cells breaks away from the parent plant and grows into a new gametophyte. This type of vegetative reproduction is well organized. In lower plants simple fragmentation of the **vegetative body** will result in the formation of new individuals. In flowering plants the vegetative parts of many plants will reproduce new individuals by means of **runners,** which may be either stems or roots, or even by means of rooting of leaves or parts of leaves. A clump of plants developed in this manner, each with identical genetic composition, is termed a **clone.** The airplane plant, a common succulent house plant, normally reproduces with the development of small plants along the margin of the leaves. The plantlets are budded off, fall to the soil, root, and grow into a new plant.

SEXUAL REPRODUCTION

The key to **sexual** reproduction is meiosis. The formation of gametes must be preceded by the reduction of the chromosome number. The sporophyte is usually diploid in chromosome number. If the formation of gametes were not preceded by the separation of the homologs (chromosome pairs), the fusion of the gametes would cause a doubling of the chromosome number at each fertilization. Obviously this process could not continue indefinitely. In many lower plants meiosis takes place immediately after fertilization. In these plants the cells of the **gametophyte** are now prepared for fusion without the need for reduction in the chromosome number. The true sporophyte is absent in such plants. Because of this, it is assumed that the haploid condition of the chromosomes is primitive, and that diploid plants are caused by a delay in the time of meiosis, which is the result of the development of a spore-producing plant, the sporophyte.

In many of the algae the production of gametes occurs only toward the end of the growing season and cannot be regarded as the normal means of increasing numbers. With more complex development of body structure in the higher algae and the rest of the plant groups, gamete production plays a more prominent role in reproduction. As may be expected, special structures are developed in these plants for gamete production. The gamete chamber is characteristic of particular groups of plants. The gamete chamber of the algae and related plants is always unicellular, whereas the gametangia of higher plants are multicellular. The simple algae, and also many fungi, produce isogametes with no morphological distinction between the two gametes that unite. Certainly, however, in all cases there are as yet undetectable physiological differences. The gametes of the higher algae and most of the rest of the plant kingdom show distinct morphological differences. The female gamete or egg cell is larger than the male gamete or sperm cell. The **sperm** is always motile, but the **egg** remains in situ until fertilized.

In the algae the cell that produces the egg is referred to as an oogonium; in the higher plants the gamete chamber producing the egg is called an archegonium and is many celled. The male gamete-producing chamber, whether unicellular or multicellular, is referred to as an antheridium. Many sperm cells are produced by the antheridium, but only a single egg cell develops in the oogonium or in the archegonium. Collectively these structures are referred to as gametangia.

SPOROPHYTES

The production of spores as a form of asexual reproduction is the function of the sporophyte. In the higher forms this plant is also the vegetative or food-producing plant. In some it remains as a parasite on the gametophyte. Gradually the sporophyte evolved to become independent until in the seed plant its role as the vegetative plant is dominant, with the gametophyte remaining as a parasite on the sporophyte.

GAMETOPHYTES

The role of the gametophyte is much different from that of the sporophyte. The essential purpose of the gametophyte is the production of gametes, haploid cells of opposite types, that is, genetically different. Upon fusion they produce a new individual with a gene complement unlike that of the parents. The gametophytes of algae are vegetative, in which any cell is capable of producing the gametes. In others, the gamete-producing plants are distinct from the spore-producing plants.

Various types of gametophytes are known. The bryophytes have vegetative gametophytes to which the sporophyte remains attached. However, in some of the bryophytes the sporophyte becomes partially independent. The lower vascular plants have separate, independent gametophytes with relatively limited vegetative functions. This is typical of the ferns and horsetails, but not of all the others. The club mosses, for example, have partially parasitic gametophytes on the sporophytes. This is a forerunner of the seed habit of the conifers and the flowering plants, the remaining groups in which the gametophyte is completely dependent on the sporophyte. Spore production in such plants is extremely limited, since the spores are retained in the plant and mature into a tiny gametophyte. Thus the two generations of the plant life cycle merge into one as the ultimate development of plant reproduction, the seed habit. This development best fits the conditions presently on earth and accounts for the success of the conifers and the flowering plants, which have become the dominant element of our modern floras.

EVOLUTION OF THE PLANT BODY

Considerable attention has been given to the reproductive habits of plants because an understanding of these processes is necessary in order to understand the role of plant life in the balance of nature. Efficient reproductive processes are necessary for the survival of the plant species. In addition to the evolution of the reproductive organs, there is a parallel evolution of the vegetative body.

The organization of the single-celled vegetative plant is the least complex because each of the cells carries on all life processes, independent of each other. Plants organized in this way vary greatly in size but within definite limits. Although some are just visible to the unaided eye, most are microscopic in size. These plants are able to populate certain habitats in great abundance, but they are strictly limited regarding the types of habitat in which they can thrive. They are all restricted to an aquatic niche, either a body of water, such as the ocean, the most uniform of ecological habitats, freshwater, or a boggy area that rarely dries out. Therefore, as a rule, a generalized body organization demands a generalized ecological condition. Lack of specialization prevents most single-celled vegetative plants from inhabiting locations that are subject to great variation. However, the land forms have developed one feature that permits them to emerge from the aquatic habitats. This adaptation is the ability to form spores that can resist extremes of temperature and fluctuations in moisture. The introduction of sporulation in the life cycle is probably the first specialization of the plant body. The existence of this in bacteria is probably indicative of the specialized rather than generalized nature of these plants.

The next development in the specialization of the plant body is not clear, but the formation of filaments is the most logical one. A **filament** is easily formed by successive cell divisions without the breaking away of the newly formed cells. If a certain cell grows and divides rapidly and the cells remain attached to each other, the plant body becomes multicellular and filamentous. Frequently a sheath is secreted around this string of cells, affording common protection. If some of the cells in this multicellular plant become specialized for spore production, or for gamete production, a division of labor within the plant body results. The remaining cells in the body contribute food to the reproductive cells, and the filament becomes an individual. Additional specialization among the cells may be the formation of a basal holdfast cell that holds the plant to the substratum.

The difference between filamentous organization of the plant body and colony formation is slight. A true **colony,** more simply organized than a filament, is a collection of unicellular organisms clumped together and perhaps providing some mutual protection. There may be an exchange of food material, with the individuals of the colony sharing in the food-getting process. However, this is not evident in photosynthetic organisms. Colony formation occurs but rarely in the plant kingdom, although it is complex in the animal kingdom, in which it is very difficult to distinguish between a colony and an organism because of the complex interaction among the individuals. The only way to test this is to determine whether the individual cells are able to survive and reproduce when broken apart from the main mass. Even this is a poor criterion at times because the process cannot always be distinguished from regeneration.

The change from a filament to a **thallus** organism, that is, a plant body made of several parallel filaments, is accomplished by the lateral division of cells as well as by the apical division, to form a sheet one or more cells thick. Once this plant body is formed, specialization within the new individual becomes complex. This type of plant body is the most common.

The plant body does not always develop from cells in the usual sense. The third type, mentioned previously as occurring in some algae, slime molds, and fungi, often show a multinucleated organism composed of only one or only a few very large cells. In these forms only the nucleus divides; the cytoplasm remains as a common material throughout the large cell. The individual nucleus controls the cytoplasm in its immediate vicinity, but the cytoplasm carries on the common functions of the plant body. Multinucleated cells are called **syncytia** and may be filamentous, as are the hyphae of bread mold, or they may be in the form of a sheet, the plasmodia, as in the slime molds, in which individual cells unite to form the plant body.

Cell specialization soon leads to tissue formation in the higher algae, the bryophytes, and the fungi. Several tissue then becomes associated to form the major parts of the higher plants. These plants have functional tissues that serve much the same purpose as the integumentary, reproductive, excretory, regulatory, skeletal, and circulatory systems of the higher animals. Higher plants never have even a simple receptor system, but some of the lower forms have the same structures for reception as those found in the Protozoa, and some of the higher plants react to touch and changes in light intensity. Plants always lack digestive, muscular, and nervous systems. Therefore, compared to animals, plants are more simply organized. The movement of fluids in the vascular plants, although somewhat analogous to the circulatory system of animals, is poorly developed. Even in trees it is still restricted to a thin cylinder of living tissue. The ability of most plants to manufacture their own food seems to have eliminated the necessity for developing the complex body systems found in animals and yet, at the same time, plants are well adapted for life on this earth. Plants can easily survive without animals, but animals cannot live without plants.

FLOWERS

The flowering plants produce certain stems with "leaves" modified for the production of seeds. These modified leaves comprise the flower. Roots, stems, and leaves are strictly vegetative organs concerned with the nutrition and growth of the plant body. The **flower,** however, is a group of modified leaves specialized for reproduction and is responsible, either directly or indirectly, for bringing into existence new plants to perpetuate the species.

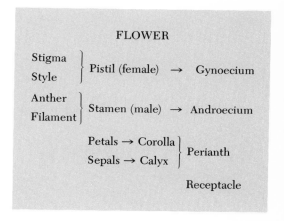

FLOWER

Stigma ⎫
Style ⎬ Pistil (female) → Gynoecium

Anther ⎫
Filament ⎬ Stamen (male) → Androecium

Petals → Corolla ⎫
Sepals → Calyx ⎬ Perianth

Receptacle

The flower develops sex cells and accessory structures that indirectly contribute to the process of reproduction (life cycle 23).

A **complete flower** (Fig. 16-2) has four groups of parts arranged symmetrically and in whorls at the end of an enlarged stem called the **receptacle.** The flower stalk is a stem that supports the floral parts in a position favorable for eventual pollination. When the stem bears a single flower or is the main stalk of a cluster of flowers, it is known as a **peduncle.** In clusters the individual flower stalks are called **pedicels.** The flower develops as a branch from a bud in the axil of an ordinary leaf or in a much reduced leaf, a bract. The presence of four circles of parts makes up a complete flower, as illustrated in Fig. 16-2. The outermost circle is composed of individual leaves, which in most cases are

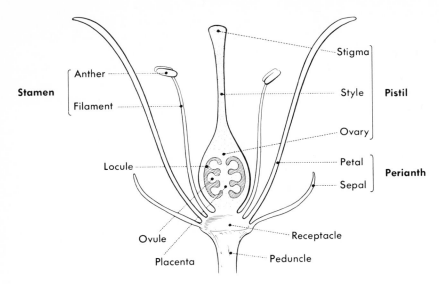

Fig. 16-2. Parts of a typical flower.

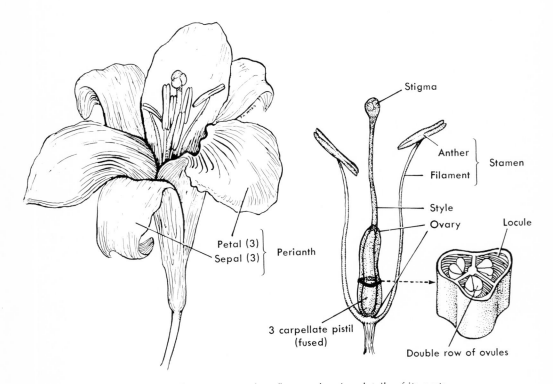

Fig. 16-3. Lily flower, a complete flower, showing details of its parts.

green. These are the **sepals** and collectively are called the calyx. Their function is to protect the inner and more delicate parts of the flower as they develop from the bud. In some plants sepals may be colored like the petals. Inside the calyx are a group of more delicately textured and often brightly colored **petals.** In their attachment to the receptacle, petals usually alternate with sepals. Collectively, the petals are known as the **corolla.** The petals have attractive odors and colors and secrete a sugary solution, **nectar.** The glandular hairs at the base of the petal, called **nectaries,** produce this substance. Petal color ranges through the entire spectrum. Green is an uncommon floral color, but white, yellow, pink, red, and blue are common colors, and even black flowers are not unknown. The various pigments that give color to petals may be dissolved in the cell sap, such as the

reds and blues of the anthocyanins, or they may by contained in special plastids in the cells, such as the yellow and orange colors (carotenes). When the plastids are colored other than with chlorophyll, they are known as chromoplasts. A petal appears white because the light is reflected from many air-filled spaces in loose cellular tissue. The **perianth** (Fig. 16-3) of the flower includes both the calyx and corolla.

The third and next innermost whorl of leaves in a flower consists of the stamens. All the stamens of the flower, collectively, are known as the **androecium.** A **stamen** is composed of a stalk or filament that supports at its terminal end an organ known as an **anther,** containing two **pollen sacs.** Within the pollen sacs are a large number of **pollen grains,** which are liberated when mature. Pollen grains vary in size and shape for each species

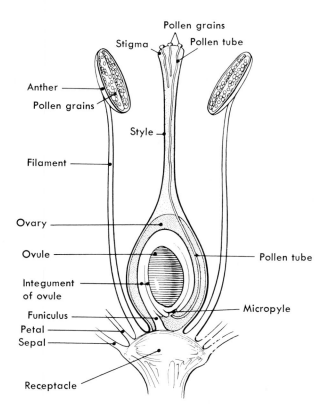

Fig. 16-4. Cross section of a fully developed flower showing the growth of the pollen tube in preparation for fertilization.

of plant. They are readily identified as belonging to a particular kind of plant by a person who is trained in pollen morphology, or **palynology.**

In the center of the flower is one to several **pistils** in which **ovules** or potential seeds may be found. Collectively, the pistils are termed the **gynoecium.** Each unit of the pistil is a **carpel,** surrounding ovules located in a cavity. The base of the pistil is enlarged to form an **ovary** from which extends a slender filament, the **style.** The ovary is a hollow structure in which there are spaces called **locules.** One or more ovules, which will develop into seeds, are located in each locule. At the terminal end of the style is located an organ, the **stigma,** which is either sticky, roughened, or otherwise modified to catch and hold pollen.

For fertilization it is necessary for the pollen tube to grow down the style and reach the **micropyle** (Fig. 16-4). Further details of this process are discussed in life cycle 23.

FLOWER ADAPTATION

The greatest variation and adaptation of the flowering plants is seen in the flowers, which are, as we have seen, really modified leaves. The arrangement of the floral parts is used extensively in classification. Some flowers, for example, the lily, are radially symmetrical in the arrangement of sepals, petals, stamens, and pistils. Others are asymmetrical, and others may lack some parts (**incomplete flowers**). If both stamens and pistils are present in the same flower, it is a **perfect** flower. If a flower lacks either stamens or pistil, it is called **imperfect.**

Many variations may appear in the number, shape, size, color, texture, and relative position of the floral parts. Monocot and dicot

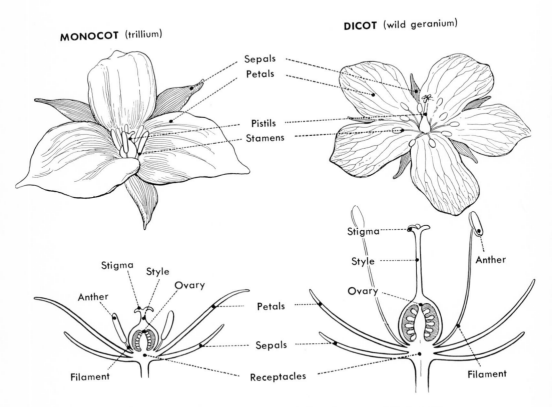

Fig. 16-5. Comparison of monocot and dicot flowers and their parts.

flowers show certain differences (Fig. 16-5). For example, monocot floral parts commonly occur in threes or multiples of three, whereas dicot floral parts are usually in fours or fives or their multiples. These variations are morphological characteristics that are important in distinguishing one species from another. More often than not, there is only one pistil consisting of several carpels (Fig. 16-6). The stamens are frequently the most numerous of floral parts and, in general, are separate from one another. When flower parts fuse, they are said to be **coalescent.** However, even in a bilaterally symmetrical flower such as the sweet pea (Fig. 16-7), the parts may be partially fused. The sepals are generally the same in number as the petals, but a flower is **gamosepalous** when the sepals are fused. The petals are usually fewer in number than are the stamens, and they rarely exceed ten. In some flowers the petals are fused to form a continuous corolla, a characteristic termed **gamopetalous.** In others they may be fused to form a keel (Fig. 16-7) or a similar modification. Some of the floral parts of the sweet pea are different; that is, the parts are not symmetrical. This makes the pea flower **irregular,** a characteristic of most legumes. Sometimes the sepals, petals, or even the stamens will fuse to one another. When two or more series of parts are united, they are referred to as **adnate.** The lack of sepals makes the flower **asepalous;** the lack of petals makes the flower **apetalous;** and the flower is said to be **naked** if both are missing. A flower that has stamens but no pistil is said to be **staminate;** when it has a pistil but no stamens, it is pistillate. If both staminate and pistillate flowers are distinct from one an-

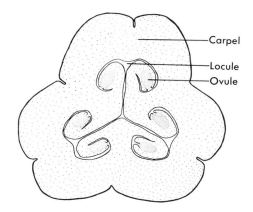

Fig. 16-6. Cross section of an ovary showing the relationship of the parts. This ovary has three locules.

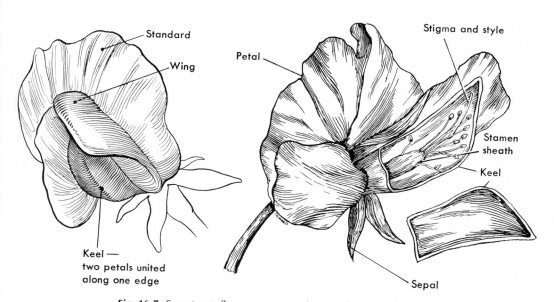

Fig. 16-7. Sweet pea flower, an example of an irregular flower.

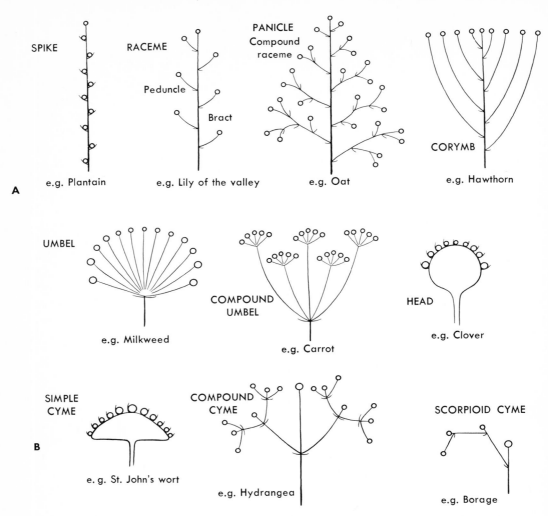

Fig. 16-8. Types of inflorescences. **A,** Indeterminate (racemose). **B,** Determinate (cymose).

other but on the same plant, as in corn, the plant is called monoecious. If the flowers are on separate plants, the plant is called dioecious.

If the flowers arise directly from the base of the plant or are produced singly in the axil of a leaf, they are said to be solitary. When flowers are grouped together into clusters, they are called an **inflorescence.** An example of an open inflorescence is the dandelion. The shape and arrangement of inflorescences vary considerably. The inflorescence is **indeterminate** if the first flowers to open are at the

base or the outside of the cluster (Fig. 16-8, A). If the first flowers to open in the cluster are at the tip or the center of the inflorescence, it is called **determinate.** Typical examples of both determinate and indeterminate inflorescences may be found in Fig. 16-8.

Other ways to classify flowers are by the variations in the relationship of the ovary to other flower parts and by the type of placentation. In some plants all the floral parts arise below the pistil. If the calyx, corolla, and stamens originate under the ovary of the pistil, the flower is hypogynous and the ovary is

284

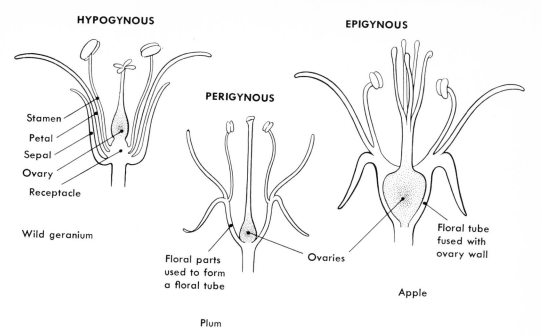

HYPOGYNOUS

PERIGYNOUS

EPIGYNOUS

Stamen
Petal
Sepal
Ovary
Receptacle

Wild geranium

Floral parts
used to form
a floral tube

Ovaries

Floral tube
fused with
ovary wall

Apple

Plum

Fig. 16-9. Types of ovaries (diagrammatic).

superior; if the floral parts arise above the ovary, which is embedded in the receptacle, the flower is epigynous and the ovary is inferior; or if the floral parts are fused at their base and arranged around and above a "free" ovary, the flower is **perigynous** (Fig. 16-9). The ovule is attached to the **placenta** of the ovary wall by a short stalk known as a **funiculus** (Fig. 16-10). If the placenta is on the ovary wall, the placentation is **parietal;** if it is on a central axis, the placentation is **axial;** and if it is borne on a stalk that extends from the base of the ovary into the cavity, the placentation is **central** (Fig. 16-10).

Pistils vary in different plants. It is called a **simple pistil** when it consists of only one carpel such as in the bean. When a pistil represents the fusion of two or more carpels, as in the lily, it is called **compound** (Fig. 16-11). A compound pistil may have more than one style or stigma (Fig. 16-11).

The number of carpels and the arrangement of seeds vary according to the species of flower. The carpels may coalesce to form a single chamber as found in bean and pea

pods, or they may remain separate as in the apple.

All these variations of flower parts are used to classify as well as to identify plant orders, families, and species.

Pollination. Except on rare occasions, ovules will not develop into seeds unless pollen is transferred from the anther of the stamen to the stigma of the pistil. This transfer of pollen is called **pollination.** It cannot occur until after pollen is produced in the pollen sacs of the anther.

The first step in the accomplishment of reproduction in seed plants is the transfer of pollen from the anther to the stigma. If the pollen from the anther to the stigma of the same flower, the process is called self-pollination. If it is transferred from the anther of one flower to the stigma of a flower of another plant, the process is **cross-pollination.** When **self-pollination** occurs, the anther usually lies so close to the pistil that the pollen is transferred directly and even before the flower opens. In the majority of plants the transfer requires some external

285

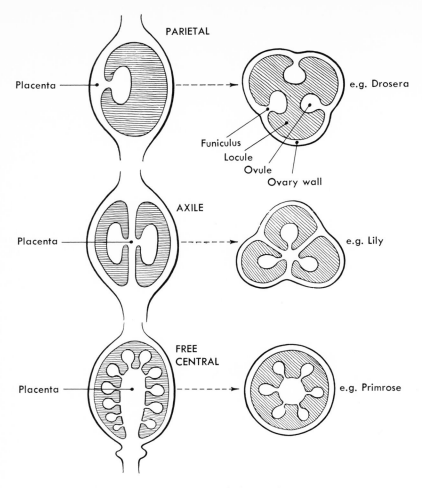

Fig. 16-10. Types of placentation.

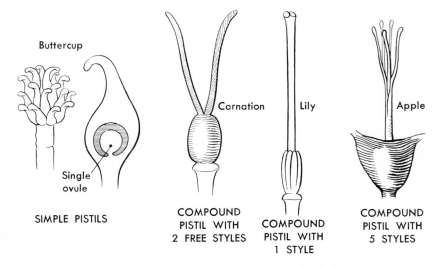

Fig. 16-11. Different kinds and arrangements of pistils.

Fig. 16-12. Insect pollination (pollen carriers) in salvia. The bee travels from flower to flower carrying pollen from one to another. The anthers of young flowers ripen and release pollen and then dry. The older flowers have a mature style that receives the pollen.

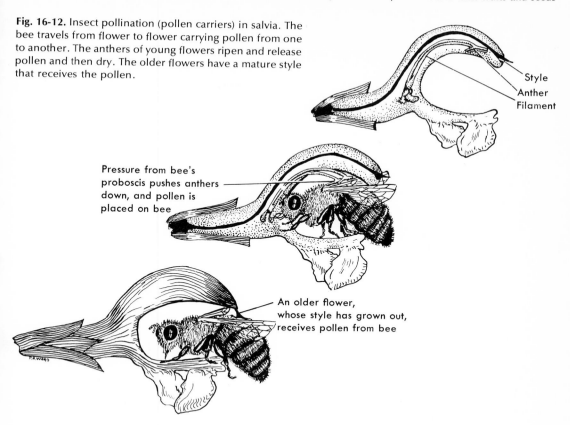

Style
Anther
Filament

Pressure from bee's proboscis pushes anthers down, and pollen is placed on bee

An older flower, whose style has grown out, receives pollen from bee

agent. As a result, pollen is frequently carried from one flower to another.

Wind and insects are the two most important pollinating agents. Wind-pollinated flowers are inconspicuous, are exposed prominently on the plant, and have poorly developed perianths. Their pollen is light and dry and is produced in tremendous quantities. In the spring the yellow dust (pollen) from wind-pollinated plants can be seen covering automobiles and forming a yellow blanket over ponds. It can be carried hundreds of miles by the wind and is one of the principal causes of hay fever and asthma.

The flowers of insect-pollinated plants are usually conspicuous and brilliantly colored and possess characteristic odors or nectaries (Fig. 16-11), which attract insects to them. The pollen is usually sticky and adheres readily to the body of an insect. In this way it is carried from flower to flower and deposited on the stigma (Fig. 16-12). Bees are more important than any other insect in pollination, although flies, butterflies, moths, and even beetles and bats account for some pollen transfer. Few flowers are pollinated by water, compared with those pollinated by insects and wind.

The subject of pollination and pollination ecology is much too involved for a further discussion in the space allowed.

FRUITS

Although we commonly think of apples, oranges, grapes, and the like as **fruits,** to the botanist a fruit is a fully developed and ripened ovary. Fruits are so varied that it is difficult to fit them all into even a most detailed system of classification. Fig. 16-13 illustrates one kind of fruit. Fruits are important not only to plants for seed dispersal and survival, but as a source of food for other organisms.

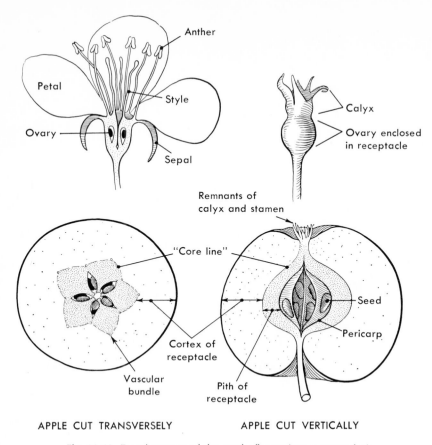

Fig. 16-13. Development of the apple flower into a pome fruit.

The development of a fruit from an ovary is a complex process that involves many physiological activities. It can be shown that auxins and growth hormones play an important part in a fruit's development. Various sugars, soluble proteins, and other foods accumulate in the ovary walls during its maturation. The increase in sugars accounts for the sweetness of some fruits. In others, fats and oils accumulate in large amounts (olives, for example). Pulpy fruits accumulate large amounts of water. The growth of the ovary is stimulated by auxins that are either present in the pollen or produced by it in the ovary tissue. Other changes, such as pigmentation, may accompany the physiological activities involved in developing fruits. The chlorophyll in apples (Fig. 16-13) disappears when the

fruit ripens and is replaced by other pigments. Anthocyanins accumulate in the ripening of fruits such as apples or plums to give them a characteristic red or blue hue.

The growth of the vegetative parts of a plant is frequently inhibited by the development of fruits. This could be attributable partly to the translocation and accumulation of foods from other parts of the plant, partly to hormones, to the fruits, or to some other physiological process. It was discovered that ethylene gas would hasten ripening, particularly of citrus fruit, as well as development of flowers and leaves. Apparently this artificial ripening process is caused by acceleration of otherwise normal physiological processes.

The ovary wall, or **pericarp,** usually thickens when it develops into a fruit and

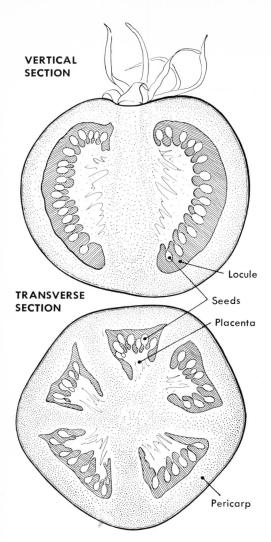

VERTICAL SECTION

TRANSVERSE SECTION

Locule

Seeds

Placenta

Pericarp

Fig. 16-14. Example of a berry (the fruit of the tomato plant).

becomes differentiated into three distinct layers known as the **exocarp, mesocarp,** and **endocarp.** The outermost covering (exocarp) consists of one or more layers of parenchyma cells. The middle layer (mesocarp) may vary in thickness from one to several layers. The innermost layer (endocarp) varies in structure, texture, and thickness in the fruits of different species. In some, parts other than the ovary contribute to the formation of a fruit.

Kinds of fruits

Fruit is the usual result of the growth and development of a fertilized flower; therefore it is to be expected that great variation will be found in these structures, paralleling those found in the flowers. When parts other than the ovary adhere to or enclose it, the fruit is called an **accessory fruit.** A familiar example is the strawberry, in which individual fruits or achenes are borne upon a sweet and fleshy receptacle (Fig. 16-17). A strawberry is an accessory fruit that is also designated an **aggregate fruit.** The apple is another type of accessory fruit, but it is called a **pome** (Fig. 16-13). The ovaries of the apple become surrounded by the enlarged receptacle and the tissues of the floral tube. The true fruit of the apple is the **core,** which is not eaten. Therefore, in accessory fruits such as the strawberry, apple, pear, and quince, the edible portions are not mature ovaries, but instead are the stem and floral tubes in which the ovaries are embedded.

Fruits are usually classified as **simple, aggregate,** or **multiple.** A simple fruit is the result of the maturation of the ovary of a single pistil, whether it is simple or compound. The ovary of the pistil may be superior or inferior; thus the fruit may be composed of the ovary alone or of the ovary and any other part of the flower. Simple fruits may also be dry or fleshy. A **fleshy fruit** is one in which all or most of the pericarp is soft and fleshy. An example of a simple fleshy fruit is the apple, a pome fruit. A fleshy fruit in which the entire pericarp is soft is called a **berry.** Good examples of berries are grapes, bananas, and tomatoes (Fig. 16-14). Watermelon, cucumber, and cantaloupe are berries having a hard outer covering, and citrus fruits are berries with leathery coverings. Another kind of fleshy fruit is called a **drupe.** Peaches, cherries, and plums are drupes (Fig. 16-15) in which exocarp is thin, the mesocarp thick and fleshy, and the endocarp hard and stony. Drupes are fruits that have a stone or pit in the center and only one seed.

A second kind of simple fruit is called a **dry fruit.** In a dry fruit the pericarp becomes ei-

289

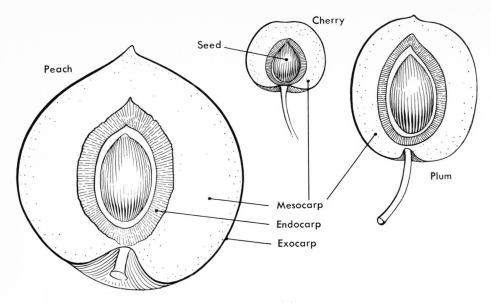

Fig. 16-15. Examples of drupes.

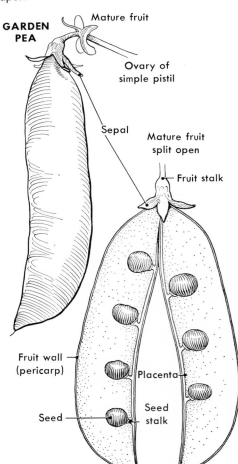

Fig. 16-16. Pea pods with peas (the fruit), an example of a legume.

ther leathery, papery, or woody. Dry fruits are further differentiated by the manner in which they split open to expel the seeds. If they split in a definite pattern, they are **dehiscent fruits.** If they do not split open along a definite seam they are **indehiscent fruits.** Legumes (pea and bean) are good examples of dehiscent fruits (Fig. 16-16). The **samara** seeds of the maples and the acorns of the oak are examples of indehiscent fruits.

An aggregate fruit is a cluster of several to many ripened ovaries produced by a single flower (Fig. 16-17) and borne on a single receptacle. Each individual ripe ovary may be a drupe, for example, blackberries, or an **achene,** in which a single seed is inseparable

from the ovary wall except where it is attached to the pericarp, as in the dandelion.

Finally, multiple fruits are a cluster of several ripened ovaries produced by many flowers on the same inflorescence. The mulberry, fig, and pineapple (Fig. 16-18) are multiple fruits. Although many fruits are popularly regarded as vegetables, few, if any, vegetables are considered fruits. However, there is one vegetable, rhubarb, that is used as a fruit. The edible portion of that plant is the petiole of the leaf and hence does not meet the botanist's definition of a fruit. The date and the grapefruit are among the many economically important fruits used by man.

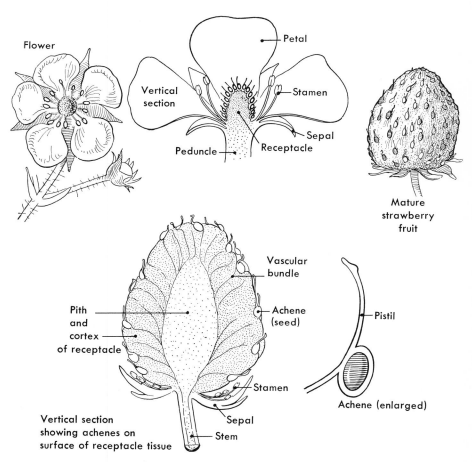

Fig. 16-17. Development of an aggregate fruit (strawberry) from the receptacle of the flower.

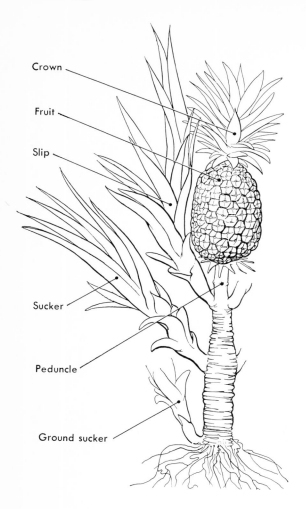

Crown

Fruit

Slip

Sucker

Peduncle

Ground sucker

Fig. 16-18. Pineapple plant showing parts, including the fruit.

SEEDS

A seed is a ripened ovule. Seeds vary in size from those that are barely visible to the naked eye (tobacco) to those that are several inches in diameter (coconuts). The **seed coat** and the **embryo** are the two essential parts of seeds. The seed coat develops from the integuments of the ovule, and the embryo from the fertilized egg. Sometimes the embryo is embedded in the endosperm, which furnishes food for it. In most mature seeds the **nucellus** is no longer recognizable unless it has developed as food-storage tissue. It is evident that a seed represents three generations—the old sporophyte (nucellus), the fe-

male gametophyte (**embryo sac**), and a new sporophyte (embryo).

Kinds of seeds

Seeds vary in size, shape, and structure. In fact, many plants may be recognized by their shed seeds alone.

The structure and texture of the seed coat vary. In some it is thin and fragile; in others it is hard and stony or even fleshy. The attachment of the funiculus to the placenta is recognized externally as a scar or **hilum.** Leading away from the hilum may be a short ridge called the **raphe,** and on the opposite side of the hilum may be the micropyle opening.

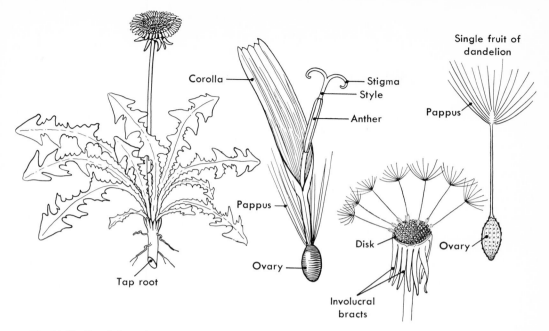

Fig. 16-19. Dandelion plant, a common weed, and a good example of a plant producing achenes.

Although the new generation of the flowering plant begins with the embryo in the seed, it is a limited beginning. The growth of the embryo is usually arrested until the seed becomes detached from the parent plant and is dispersed. It is after the seeds are disseminated that germination takes place. In the meantime the seeds may have survived extremes of heat, cold, and drying.

Many plants develop novel and valuable structural modifications for the dispersal of seeds over great distances. This usually accounts for the many different kinds of seeds. It becomes difficult to separate the methods of dissemination of seeds from those of fruits, since the same or similar modifications are noted in both. Furthermore, it is often difficult to distinguish the seeds from the fruit in some plants, since they are often both disseminated as a unit. Seeds are dispersed by agents such as wind, water, birds, insects, and mammals, including man.

Seeds that are small and light in weight are usually dispersed by the wind. Orchid seeds are so small that more than a million may be found in a single capsule. Thus they are easily airborne. Others (Fig. 16-20) may have plumes, modified floral parts that aid in their dispersal by the wind. Winged seeds, such as those of the maple (Fig. 16-19) are a common sight in the spring of the year.

The action of water can be almost as important as that of wind in disseminating seeds. Large and small seeds and fruits are transported over long distances by the washing of rain, especially during flash floods, into streams and rivers, particularly alpine rivers. Seeds that have light, spongy coats or those relatively unaffected by water are modified water-dispersed seeds. Contrary to popular belief, coconuts are not fit for transportation by ocean currents and owe their widespread distribution instead to native transportation.

Animals are responsible for the dispersal of many seeds. Birds and other animals eat the fruits and seeds of plants, which pass through the digestive tract unharmed, to be dispersed during their travels. Other seeds have

293

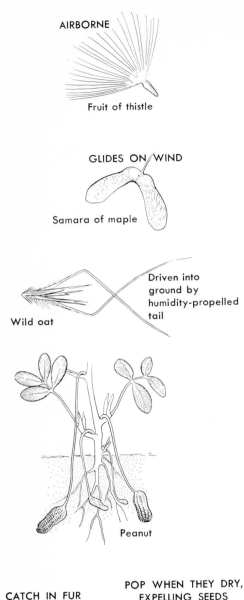

AIRBORNE

Fruit of thistle

GLIDES ON WIND

Samara of maple

Driven into ground by humidity-propelled tail

Wild oat

Peanut

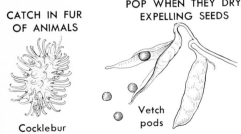

CATCH IN FUR OF ANIMALS

POP WHEN THEY DRY, EXPELLING SEEDS

Cocklebur

Vetch pods

Fig. 16-20. Examples of seed and fruit dispersal.

hooks, bristles, or spines, such as the cocklebur (Fig. 16-20), which catch in the fur of animals or the clothing of man, to be transported over a wide area. Through commerce and travel, man accounts for the dissemination of a variety of seeds. Other adaptations for seed dispersal are the exploding of the dry pod of the vetch (Fig. 16-20). Many forms of seed-catapulting take place, but the most striking is the famous squirting cucumber of Italy, *Ecballium elaterium*. The internal turgor pressure extends the outer wall of the fruit, and at a certain point the stalk end is forced out exactly like the cork of a champagne bottle. The fruit wall contracts, and the entire juicy contents, including the seeds, are squirted out. In contrast, some seeds, such as the peanut, are prevented from being dispersed. The peanut fruit contains seeds that develop on a long "stem" that extends into the ground (Fig. 16-20).

QUESTIONS AND PROBLEMS

1. Distinguish between reproduction and growth.
2. Three terms are rather close in meaning: filament, colony, and clone. How are they distinguished? What is the difference between multicellular and thallus?
3. Distinguish between plasmodium and syncytium.
4. Give the function for each of the following: root, stem, leaf, flower, fruit, and seed.
5. What type of adaptation is found in roots?
6. What type of adaptation is found in stems and leaves?
7. What type of adaptation is found in the flower?
8. What type of adaptation is found in the fruit?
9. How does each of these adaptations correlate with the effective environment?
10. A showy flower is likely to be an adaptation of what? Spines on the leaves and stem? An underground stem?

DISCUSSION FOR LEARNING

1. Alternation of generations seems to take place in many plants. Is there any reason, considering the basic principles of reproduction, why this is an advantage?
2. Sex to the human mind usually implies male-

ness and femaleness. Is this a basic feature of reproduction?

3. As we have seen, dispersal is usually closely associated with reproduction, yet it has no direct part of reproduction as such. Why, however, is dispersal considered to be a link between the two generations?

4. From the standpoint of the basic parts of the vascular plants and their anatomy that has been reviewed so far, diagram and discuss the simplest or least complex plant you can and still have a complete plant.

5. From the discussion and basic plan developed in answer to the above problem, develop a mechanical plant, that is, a system that will do all the things that your simple plant above will do, except use artificial materials instead of tissues. What is the most difficult function to design?

ADDITIONAL READING

(For additional reading see end of Chapter 14.)

Hawkes, J. G. (editor). 1966. Reproductive biology and taxonomy of vascular plants. Pergamon Press, Long Island City, N.Y.

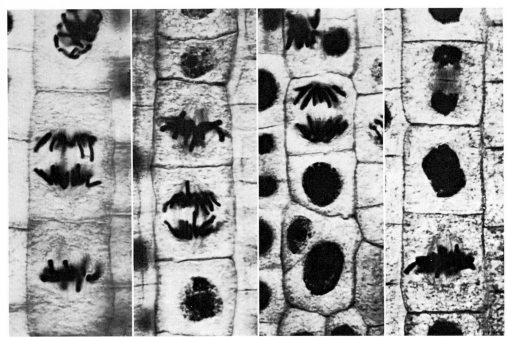

| Typical prophase and anaphase | Typical metaphase and anaphase | Late anaphase | Telophase and metaphase |

Fig. 17-1. Mitosis in onion root-tip cells in various stages. Compare with Fig. 17-2, to which the following letters refer. *Upper left,* Medium prophase, **B;** *middle left,* anaphase, **F;** *lower left,* early metaphase, **D;** *top left center,* two new daughter cells, **I;** *next down,* **D;** *next,* **F;** *lower left center,* early anaphase, **B;** *right center,* interphase, **A;** and another anaphase; *upper right,* telophase, **G;** *lower right,* later metaphase, **E.** (Courtesy Carolina Biological Supply Co., Burlington, N.C.)

CHAPTER 17

Growth, development, and hormones of flowering plants

■ Differentiation and **growth** appear to be mutually exclusive. Whereas growth is for the most part an endless process of multiplication of individual units by mitosis, differentiation is the segregation of these units from the rest, thereby making the former different and tending to prevent the cell from further multiplication. Therefore the more highly specialized the cell becomes, the less likely it is to divide. Thus growth and differentiation are the two processes by which development of a living organism is achieved. Although development is ultimately controlled at the genetic level, the gene action is not isolated. The control of gene transcription, hence development, is attributable to a number of parameters, such as temperature, photoperiod, and chemicals produced by the cell or neighboring cells.

When development is considered in terms of cells, it is found that initially simple cells of a uniform type gradually progress to more complex and varied ones. The actively dividing cells of the cambium are plastic and versatile, eventually becoming stabilized and having an unalterable structure with highly specialized functions, such as the permanent tracheids of the xylem, and the sieve tubes of the phloem. The whole course of development in a plant (or an animal) from the time of its beginning until its death is characterized both by unity and harmony in its structural unfolding and functional behavior. The completed organism is functionally no more an individual than is the sex cell, embryo, or seed that produces it. In all instances each cell behaves as a separate, living, functional entity and develops as such regardless of whether or not it is an aggregation of cells making up a whole organism.

This phenomenon is called integration, but it is neither easy to explain nor to define. To adequately comprehend such a complex phenomenon, one would require more detailed knowledge of the organization of living plants at the organ, tissue, cellular, and molecular levels than is currently available. Various biochemical factors, such as **hormones** are known to regulate this integration. But the exact reason why a cell stops dividing, or ceases growing, or even begins again to divide after it has differentiated cannot be simply stated. What determines how long a cell, tissue, organ, or organism will live? What is responsible for determining the size of one structure on a plant in relation to the others? All these are questions regarding the **morphogenesis,** or growth-form processes, of plants and are for the biologist to solve in future research. To understand growth, we must first study the process of cell division.

MITOSIS

All cells at one time or another undergo **cell division,** or **mitosis** (Fig. 17-1). They either divide regularly, as in unicellular organisms and growing tissue, or they divide only when stimulated under specific accidental, abnormal, or experimental conditions, as in X irradiation. It is probable that all cells

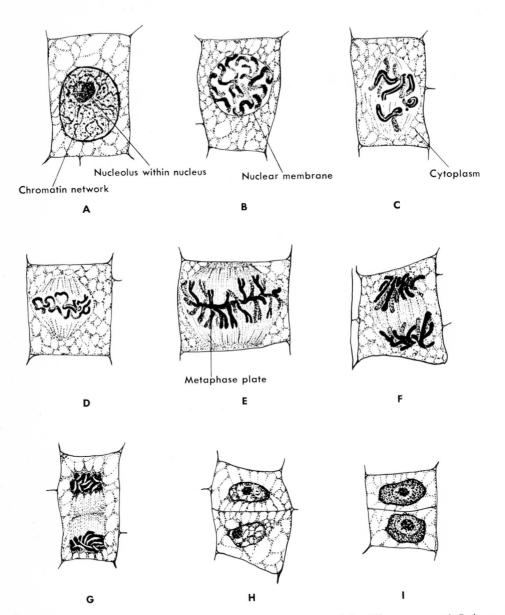

Fig. 17-2. Mitosis or cell division diagrammatically represented to show all the different stages. **A,** Early prophase (chromatin begins to coil; nucleus present). **B,** Medium prophase (chromosomes forming, becoming visible, nucleus disappears). **C,** Late prophase (nuclear membrane disappearing, spindle forming). **D,** Early metaphase (chromosomes splitting on equator of spindle). **E,** Later metaphase (chromosome showing definite longitudinal split). **F,** Anaphase (chromosomes moving toward spindle poles). **G,** Telophase (chromosomes arranged at poles; cell plate forming on spindle equator). **H,** Late telophase (nuclear membrane forming around nuclei; nucleolus appearing; chromosomes no longer visible). **I,** Two new daughter cells.

possess the potentialities for mitosis, but whether and with what frequency they do divide depends, under normal conditions, on whether they are in contact with other cells and on their growth state, that is, volume in relation to normal size. In those cells having a definite nucleus, mitosis requires duplication of the nucleus and division of the cytoplasm into two parts.

Duplication of the nucleus involves duplication of the chromosomes. The chromosomal changes and the events associated with them are grouped under the term "mitosis." However, in those cells, such as in *Oscillatoria formosa* (Monera), in which a distinct nucleus and chromosomes are not evident, cell division includes gene replication without the characteristic mitotic changes. Sometimes division of the cell may take place without these visible, characteristic mitotic events.

The immediate stimulus necessary to start cell division is probably enzymatic. Some cells divide when they have grown to double their size; other cells grow only after mitosis has occurred. Some chemicals are known to promote cell division; others inhibit the process. All that can be said is that certain environmental factors stimulate cells to divide. Acknowledging that by one means or another the process of mitosis begins, we can arbitrarily distinguish four stages that gradually merge into one another; it is difficult to determine where one stage begins and another ends. These four stages are called, in proper sequence, prophase, metaphase, anaphase, and telophase (Fig. 17-2). Prior to any visible evidence of cell division, complete biochemical replication of DNA has taken place.

Of all the activities that can be observed in a cell, mitosis and meiosis are the most complicated, and their machinery appears to be the most elaborate one occurring in the cell. During cell division a beautiful geometric design is created out of the seeming formlessness of the cell (Fig. 17-1). This design operates with extreme coordination. It begins with the resting cell (**interphase**),

resting only as far as its division is concerned, but certainly not in any other way. From chemical and genetic data it is known that chromosomes are present, although they are not visible as such.

Prophase. As the prophase begins, the previously invisible chromosomes become visible as slender threads, which thicken and become demonstrable as distinct chromosomes. Upon close examination each chromosome is actually a double filament, a single filament that has replicated during interphase. Gene replicas were formed on the template of existing genes, and the new genes have been incorporated into the new chromosomes. The duplicate chromosomes are so closely apposed that they appear as one. The disappearance of the nuclear membrane has resulted in the coalescence of nuclear and cytoplasmic contents. The presence of long, thin threads of visibly distinct chromosomes in the cell and the absence of a nuclear membrane constitute the early prophase stage of mitosis. The chromosomal threads separate longitudinally into identical halves called **chromatids**. The increasing visibility is probably caused by loss of water, thereby making the stainable parts appear more compact and dense. Equally important is that the chromosomes shorten and thicken by a process that transforms the slender chromatids into a structure coiled like a spring. The chromatids continue to shorten during prophase, and the coils decrease in number as they increase in diameter. The nucleoli, which are formed by particular chromosomes and which are initially prominent, diminish in size, and have disappeared by the end of prophase. The entire prophase takes, on the average, about 88 minutes, by far the longest stage of mitosis.

The complete disappearance of the nuclear membrane, which occurs in late prophase, coincides with the appearance of a new structure in the cytoplasm called a **spindle.** Chemically, the spindle consists of long-chain protein molecules oriented longitudinally between two **poles.** This introduces the next stage.

Metaphase. Although these phenomena mark the beginning of the second stage, metaphase, one should keep in mind that each stage is merely the continuation of a process begun in the preceding stage. Once the spindle has formed, the chromosomes move through the cytoplasm to a region midway between the two poles, called the equator, and become attached to the spindle by their **centromeres.** The centromere (or kinetochore) is a small, clear circular zone on the chromosome at the point where the arms of the chromosome are joined. It has a functional relation to the movement of chromosomes during mitosis and is constant in its position. The division of the centromeres occurs simultaneously in all chromosomes during metaphase. Daughter centromeres move apart and in turn separate the chromatids, which then begin their migration to the poles. Thus the chromatids cannot separate from each other later, or would the chromosomes have the characteristic V or rod shapes that identify them should the centromere be absent. The position of the centromere is visible in a metaphase chromosome as an area of constriction, and since the position is characteristic for each chromosome, the chromosome is divided into two arms of the same or varying lengths. Very few chromosomes have terminal centromeres. In the onion this stage lasts only 1.4 minutes, compared to the 88 minutes of the previously mentioned prophase.

Anaphase. In the mitotic cycle, metaphase is followed by anaphase. The centromeres have divided so that each chromatid will have its own centromere. Each chromatid (chromosome half) will move away from the other so that sister chromatids will eventually be found at opposite ends of the cell. Anaphase terminates when the new chromosomes are densely packed at the two poles. The process lasts, on the average, about 3 minutes.

Telophase. Essentially, the events that occur during telophase appear to be opposite to those found during prophase. The nuclear membrane forms anew, the chromosomes uncoil and become slender threads again, the nucleoli reappear, and the nucleus again takes on the appearance found in a resting cell. A new cell wall, which forms at the equator, initially appears as a cell plate within the spindle and soon reaches to the outer walls of the mother cell, roughly dividing the cytoplasm into two equal parts. The spindle disintegrates, two new cells have been formed, and mitosis is complete. This final stage lasts, on the average, about 4.6 minutes.

As a result of mitosis, two daughter cells, the chromosomes of which are exactly like the parent cells in number, in characteristic shape, and presumably in genetic makeup, have been formed in approximately 97 minutes. Cells that appear to be resting between successive mitoses are said to be in interphase.

A description of the mitotic process as a cell passes from one interphase to another by no means explains the complex aspects of cell division. The entire process is a delicately balanced and coordinated drama in which the nucleus, cytoplasm, and all parts of both come in on cue to play their individual roles. Above all, mitosis is a process of growth that involves the assimilation of materials from the outside and their transformation through breakdown and synthesis into new cell parts, all of which utilize energy. Not only are cells undergoing enlargement during mitosis, but also, more significantly, a continuous succession of genetically identical cells is ensured

The control center of the cell is the nucleus, and since the nucleus is composed mainly of chromosomes, they are the regulators both of cellular metabolism and of the structural characteristics of the cell. Since the chromosomes are composed of nucleic acids and proteins that have been intricately fabricated, the two daughter cells, which are alike, must have the same amount and type of nucleic acids as did their parent cell. The duplication of chromosomes into identical chromatids and their separation toward the poles must be exact in every detail, a mechanism provided by mitosis. Even though exact lineal heredity is established from cell to cell and from organism to organism, accidents and variations, some of which are important if evolution is to take place, can and do occur, a fact established by experiment.

Mitosis in plants, as in animals, occurs at a rate characteristic for each organism, but it is affected by temperature and is governed by such factors as nutrition. A hair cell of a plant, *Tradescantia* sp., will go through one cell division at 10° C. in 135 minutes, but at 45° C. it will take only 30 minutes. Other plants, such as bacteria, will divide every 15 to 20 minutes at 36° C., whereas some root-tip cells may take 22 hours at room temperature. Among some organisms in which the mitotic cycle takes 45 minutes, more than 17 hours elapse between mitoses, indicating that the time spent by a cell in preparation for division is an important factor, a factor about which our knowledge is limited. Growth, then in multicellular plants is the multiplication of cells, through mitosis, usually with the addition of nutrient materials and water. This causes the plant to increase in size. Many factors control the rate of growth as well as the differentiation of the tissues.

REGULATION OF GROWTH BY PLANT HORMONES

Many observations and experiments have led plant physiologists over the years to conclude that growth and development in one part of a plant is coordinated with growth and development in other parts of the plant. It was obvious then that there were regulatory mechanisms that were native to the plant and actually controlled and coordinated growth and development. Over the last 40 years it has been established that growth is controlled in plants by small quantities of organic compounds called plant hormones. These hormones are found generally in all seed plants, and it is assumed that they are synthesized by the plant's tissue. The hormones either exert their influence individually or in concert with another hormone. In general there are five substances classified as plant

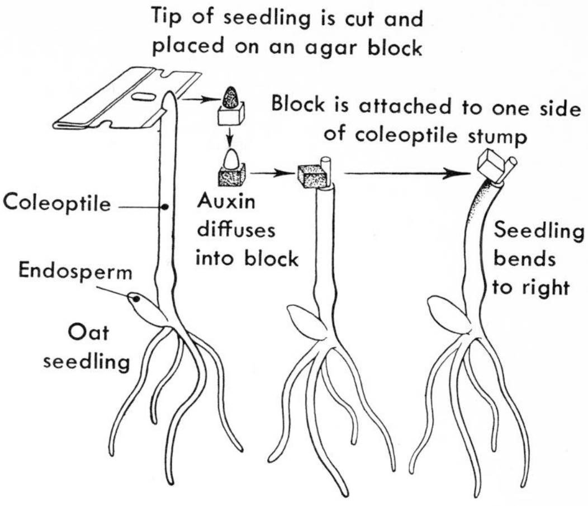

Fig. 17-3. Experiment to show the effect of auxin on plant growth.

hormones: **auxins, gibberellins, cytokinins, ethylene,** and **abscisic acid.**

Auxins

Auxins (from the Greek word *auxein,* 'to increase') were the first growth regulators discovered. It was long known that some chemical substance diffused from the coleoptile tip of the oat to the region of cell elongation, thereby causing a phototropic response. When this substance is collected and added to small agar blocks, which in turn are applied to coleoptiles in which the tip is removed, it can affect the growth of the plant (Fig. 17-3) in various ways. It is now known to be necessary for cell elongation. The principle auxin found in plants is **indoleacetic acid,** which has the following formula:

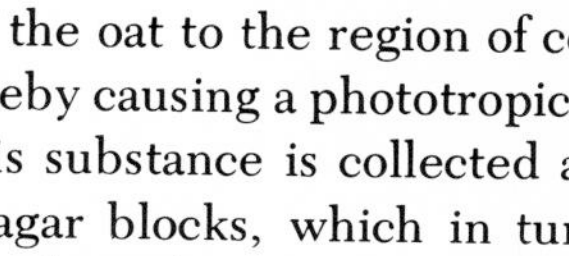

In plant tissue indoleacetic acid is found as the free acid as well as attached to glucose, aspartic acid, or protein. The physiological significance of its conjugation to other compounds is unclear. However, it is probably stored in a stable form ready for use when it is conjugated to proteins. Other compounds found in plants with auxinlike activity are the following:

Indoleacetaldehyde
Indolepyruvic acid
Indoleacetonitrile
Glucoabscisin
Indole ethanol

It is believed that the auxinlike activity of these compounds is attributable to conversion to indoleacetic acid.

Although indoleacetic acid primarily appears to cause cell elongation, it does have other influences on plant growth, some of which stimulate or inhibit growth. It is responsible for cell elongation in oat coleoptiles, stems, veins, and petioles of leaves, for cell division in adventitious roots, for parthenocarpic fruits, for callus tissues (Fig. 17-4), and for cambium growth. It influences the inhibition of growth in root cell elongation, as well as apical dominance, and suppresses the separation of the abscission layer.

Auxin both inhibits and stimulates growth according to its concentration in the tissue. It requires much greater amounts of auxin to initiate optimum growth in a stem than in a root. The same amount necessary to promote optimum growth in a stem would inhibit growth in a root. One of the tests for auxin and its effects is the curvature test, in which the coleoptile of an *Avena sativa* (oat) seedling (Fig. 17-3) is used. Initially, the seedlings are grown in the dark and are then exposed to red light for 2 to 4 hours before the coleoptile tip is removed. A block of auxin-filled agar is placed on one side of the decapitated coleoptile, and the young stem bends toward the side opposite to that on which the agar block is placed. If the tip of a coleoptile is removed, there is no response to either gravity (geotropism) or light (**phototropism**). The auxin in small amounts will stimulate cell elongation, and it is found to be produced in very small amounts at the extreme tip of stems and roots. From these tips, the auxin

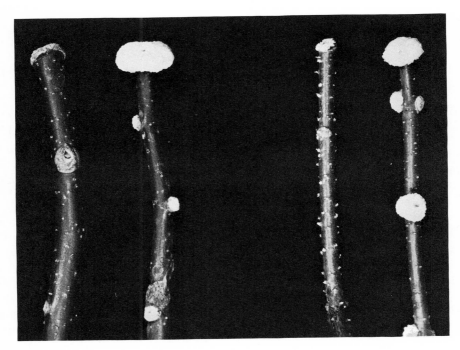

Fig. 17-4. Callus formation produced by indoleacetic acid. (Courtesy Boyce Thompson Institute, New York, N.Y.)

diffuses backward from the region of production to the region of response, that is, only in one direction, farther away from the tip. When a seedling is exposed to light on one side only, light causes a decrease in the amount of auxin on that side. Therefore cells on the darker (opposite) side will elongate more rapidly than those on the light side, and so the stem bends toward the light. The oat seedling is very sensitive to light, particularly the sheath or coleoptile surrounding the shoot of the young seedling.

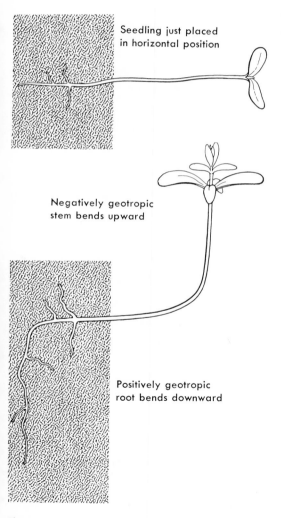

Seedling just placed in horizontal position

Negatively geotropic stem bends upward

Positively geotropic root bends downward

Fig. 17-5. Tests to show the effect of gravity on plant growth (geotropism).

The auxin is responsible for geotropic responses in plants. If a young plant is placed in a horizontal position, the stem tends to bend upward because gravity causes the auxin to accumulate on its lower side, thereby stimulating cell elongation on that side (Fig. 17-5). However, in roots the auxin inhibits cell elongation; thus the upper side grows faster than does the lower side (Fig. 17-6). In addition, auxins stimulate cambium growth, especially in those regions farthest away from the developing buds. It inhibits the growth of lateral buds, expecially those that form branches. Furthermore, the auxin stimulates the formation of roots on cuttings (cut end of stem). If the cutting is immersed in a solution of naphthaleneacetic acid for several hours and the stem is then planted in soil, many more adventitious roots will appear on the cut end of the treated stem than on an untreated one (Fig. 17-6). Another type of plant-growth substance stimulates the development of a protective layer of tissue over a cut or injury in plants. This is generally designated a wound hormone and produces wound calluses on stems and roots.

Auxins also affect the abscission of leaves and the development of fruit. If the auxin content of a leaf falls, the leaf will drop from the tree or plant. One can state then that the auxin content keeps the leaf on the plant. If a flower is not pollinated, the seed and fruit will not develop. When a seed sprouts and grows, a number of processes occur, among them the enzymatic digestion of the pectin layer of the middle lamellae and the conversion of starch into sugar. The growth and ripening of the fleshy fruit and seed is dependent on the continued production of auxin by the meristem embryo; if the seed is removed, the fruit will not ripen but it will continue to mature if it is treated with auxin. The auxin control over the abscission of leaves or fruit stems is an indirect one, usually by the control of ethylene production. Auxin also controls flowering, such as in the pineapple, because of the control over ethylene production. Also some isolated cases have been reported where auxins in-

303

Fig. 17-6. Effect of naphthaleneacetic acid. Both tomato plants have the tops removed. The plant on the right was treated with the growth substance, and the one on the left was not. Note the growth of adventitious roots on the stem of the treated plant. If this plant stem had been placed in soil after treatment, a new plant would have developed. (Courtesy Boyce Thompson Institute, New York, N.Y.)

hibit the breakdown of protein and chlorophyll in senescence of leaves.

Synthesis and mode of action of auxin. Indoleacetic acid is synthesized in the tip of the coleoptile of the root and shoot from the amino acid tryptophan. The richest sources of tryptophan are dying plant tissue and bacteria found on some nonsterile plant tissue, which is capable of forming indoleacetic acid. Continued production of auxin depends on the constant activity of the cambium and vascular differentiation. This would appear to be a self-perpetuating system whereby auxin stimulates further xylem differentiation, which in turn stimulates the production of more auxin. Once synthesized in the tip of the stem auxin moves down the stem toward the roots. It exerts its elongation effects on the stem at a point removed from its site of synthesis. The basic mechanism or primary site of auxin action is the object of much research. Experimental evidence suggests several possibilities for the primary effect of auxin. The first possible mechanism is causing an increase in the activity of a specific enzyme or enzymes in the cell's cytoplasm. Auxin could effect enzyme activity either directly or indirectly. In a direct stimulation, auxin would combine with an already formed enzyme causing a change in tertiary structure and thus enabling it to function catalytically. An indirect stimulation would be attributable to an alteration of membrane permeability. It has been demonstrated that auxin increases the activity of a specific enzyme, β-1,6-glucanase by at least 60%. When this enzyme was applied to oat coleoptile sections it caused elongation. A second enzyme activated by auxin is acid cellulase, an enzyme that hydrolyses cellulose under acid conditions. It is reasonable to expect cellulase to play a role in any cell wall changes. Acid cellulase also appears to be associated with plant cell differentiation, a role that would be consistent with the effect of auxin. It has also been suggested that auxin may activate the enzymes engaged in the synthesis of pectin because this substance is necessary for the continued synthesis of the cell wall.

A second possibility for the basic mechanism of auxin action is a direct stimulation of the genetic material. As a result of this, dormant genes would be turned on and begin functioning. Experimental evidence indicates that auxin effects an increase in the synthesis of protein and nucleic acids. However, because of the immediacy of the auxin effect, it appears that the stimulation of the genetic material is a secondary and not a primary effect. Nevertheless, we must emphasize that continued protein synthesis is necessary for the maintenance of auxin stimulation.

A third possible mechanism for auxin action is a direct effect on the cell membrane through an interaction with a specific receptor site on the membrane. Experiments with protoplasts, which are plant cells with the cell walls removed, demonstrate that auxin will cause enlargement of both vacuoles and protoplasts. This type of experiment shows that auxin does indeed have an effect on the plasma membrane. Also experiments done with specific membrane-transport inhibitors indicate that the plasma membrane contains a specific site for auxin attachment. Researchers have postulated that auxin causes a drop in the pH of the cell wall and the growth rate of the cell increases as a result. This hypothesis was based on the observations that low pH–induced and auxin-induced growth has several similarities. The only dissimilarity was the fact that low pH–induced growth did not need continued protein synthesis. Also auxin-stimulated elongating plant tissues lowered pH of the surrounding medium. In view of these observations, the

proposal was made that cell wall softening is caused by enzymes that are activated by low pH. The pH change would occur in the cell wall and it would depend on the ability of the plasma membrane to pump hydrogen ions from the cytoplasm into the cell wall. The pump would not become active until it was stimulated by auxin. It would receive its energy from the hydrolysis of ATP. Although this hypothesis has experimental support, controversy has arisen over it. Recent evidence has been presented against auxin activating a hydrogen-ion pump. Microelectrodes were not able to detect a pH change in the cell wall prior to auxin-induced cell elongation. Also the first measurements were made of the short-term auxin effects on elongation, radius, and volume of lupine hypocotyl segments. Auxin, low-pH, and carbon dioxide saturated solutions all cause an increase in the elongation rate of the cell (Fig. 17-7). However, auxin-induced and carbon dioxide–induced growth both cause an increase in the radius and volume of the cell in the first 20 minutes. However, low pH–induced growth causes an initial decrease in radius and volume (Fig. 17-8). This is evidence that the growth induced by auxin and low pH are not effecting the same function. Auxin in this case is not acting by activating a pump in the plasma membrane. It has recently been found that plant lysosomes contain enzymes that are released or activated by extremes of pH. It is possible that auxins might act on or release the enzymes attached to the lysosomal membrane, thus giving the auxin effect.

A fourth possibility is that auxin exerts its effects on cell growth through the intermediate of adenosine monophosphate (AMP). There is evidence that AMP will act with auxin to activate the genetic material.

Gibberellins

During the late nineteenth and early twentieth centuries, Japan's rice crop was destroyed by the bakanae (foolish seedling) disease. Crop loss was estimated in some cases to be as high as 40%. The disease was characterized by plants that were taller, thin-

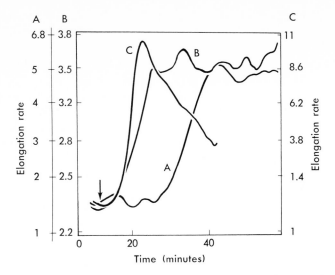

Fig. 17-7. Relative elongation rate for lupine hypocotyl segments. Twenty-five millimeter segments were changed at the downward pointing of the *arrow* from phosphate buffer pH 6.5 to the experimental treatments, which were **A,** indoleacetic acid; **B,** solution of a lower pH, that is, pH 4; **C,** carbon dioxide–saturated solution (Data from Perley, J. E., Penny, D., and Penny, P. 1975. Plant Science Letters, 4:133-136.)

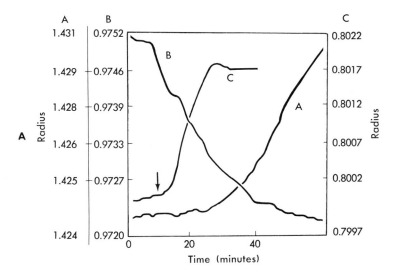

Fig. 17-8. A, Radius of the same lupine hypocotyl segments before and after the experimental treatments shown in Fig. 17-7. *A* is the radius of the segment for which the elongation rate was shown in Fig. 17-7, and similarly for *B* and C. Units are in millimeters. **B,** Volume of hypocotyl segments shown in Figs. 17-7 and 17-8, **A,** estimated from the radius and length. (Data from Perley, J. E., Penny, D., and Penny, P. 1975. Plant Science Letters 4:133-136.)

ner, and paler than normal and sometimes devoid of fruit. Scientists were obviously very interested in finding the cause of the disease and a means of prevention. As a result of this research, gibberellic acid was discovered. Bakanae disease is caused by the fungus *Gibberella fujikuroi*, and gibberellic acid was first isolated from culture filtrates of this organism.

Chemically, the gibberellins are substances related to gibberellic acid, which is generally abbreviated GA_3. Gibberellic acid is based on the gibbane skeleton, as illustrated below.

Gibberellins have been found in all plants. Their function in lower plants is uncertain. They are concentrated in the most rapidly growing parts of a plant—usually expanding leaves, germinating seeds, and the stem below the apical meristem. There seems to be variable movement of gibberellins within the plant. Some seem to move freely, whereas others seem to be restricted to a par-

ticular region. The most striking hormonal effect of gibberellins is the promotion of stem elongation. In addition, however, they affect tropic responses and stimulate cambial division, xylem differentiation, and cell enlargement in tissue culture. They are responsible for promoting fruit growth and flowering in the rosettes of biennials. However, inhibition of flowering is reported in some cases, particularly in woody plants (apple). They reverse dwarfism, release buds from dormancy, accelerate seed germination, and usually increase apical dominance. It has also been reported that in some cases they inhibit the breakdown of protein and chlorophyll in senescence. A large number of gibberellins have been isolated and characterized, and they all have the ability to cause stem elongation in the light. Some substances have been reported with gibberellin-like activity, but to date they have not been isolated and characterized and are referred to as gibberellin-like substances.

One of the most notable effects of gibberellic acid at the molecular level is the induction of hydrolytic enzymes in barley grain (Fig. 17-9). When applied endogenously, it will cause the production of proteases, nucleases, and α-amylase in the barley grain. We know most about the production of

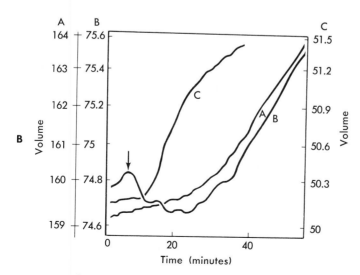

Fig. 17-8, cont'd. For legend see opposite page.

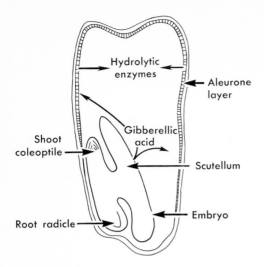

Fig. 17-9. Diagram of barley grain showing the action of gibberellic acid. The embryo releases gibberellic acid, which stimulates the aleurone cells to synthesize and release hydrolytic enzymes.

α-amylase, which catalyzes the hydrolysis of starch into sugar. A lag period of 8 to 10 hours usually occurs between the time of application of gibberellic acid and the appearance of α-amylase. During this lag there occur certain structural and chemical changes that indicate that the increased α-amylase and hydrolytic enzyme activity are secondary and not primary effects of gibberellic acid. The changes that occur during the lag period are the following: Within the first 15 minutes after application there is an increase in soluble carbohydrate and in the first 2 hours, an increase in the enzymes involved in the synthesis of phospholipid. There is also an increase in the amount of endoplasmic reticulum and the number of ribosomes on the endoplasmic reticulum between 2 and 4 hours, indicating active protein synthesis. Increased levels of cyclic AMP are evident at about 4 hours, and it has been suggested that this substance plays a definite intermediate role in the primary action of gibberellic acid.

It has been postulated that gibberellic acid acts directly on the genetic material causing gene transcription. The most convincing evidence in favor of this hypothesis is the observation that gibberellic acid stimulation or effect in barley grain is inhibited by compounds (**actinomycin D**) that inhibit the transcription of the genetic material. Although the primary site of action of gibberellic acid remains to be elucidated, it appears as if it acts directly on the genetic material (DNA).

Cytokinins

Cytokinins are substances produced by cells that promote cell division. Most of those isolated and characterized chemically have an adenine skeleton with a substitution at the sixth nitrogen on the ring. Several natural cytokinins have been isolated from plant tissue. Following is the structure for **zeatin:**

$$HN-CH_2-C=C\begin{matrix}CH_2OH\\CH_3\end{matrix}$$

Some substances such as diphenylurea promote cell division but do not have the adenine skeleton. They probably act by being broken down into an adenine derivative. It has been suggested that cytokinins are synthesized in the roots and transported to the shoot to stimulate cell division. Experimenters have demonstrated that the cell sap of the sunflower contained the greatest amounts of cytokinins during the period of active cell growth, but that the concentration fell to a tenth of the original value after maximum growth was reached (Fig. 17-10). It has also been demonstrated that cytokinin prevents chlorophyll loss and the mechanism of senescence. In addition some other effects noted on plants are induction of parthenocarpy in some fruits, stimulation of water loss by transpiration, stimulation of tuber formation in the potato, and promotion of bud formation in some plants.

It has been suggested that cytokinins are produced by the breakdown of differentiating

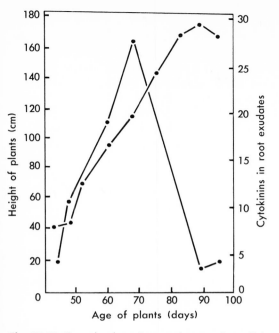

Fig. 17-10. Growth of sunflower plants and parallel changes in endogenous cytokinins in the bleeding sap obtained by cutting off the tops of the plants and collection of the root exudate for 24 hours. (Data from Sitton, D., Itai, C., and Kende, H. 1967. Planta **73:**296-300.)

xylem and phloem much like auxin. Actually we know very little about the synthesis and mode of action of cytokinins. Research is this area will help unravel many of the unanswered questions in developmental biology.

Ethylene

The structure of ethylene is as follows:

$$\begin{array}{ccc} & H & H \\ & | & | \\ H- & C= C & -H \end{array}$$

It has been known for many years that ethylene gas affects the geotropic responses of plants and is produced by ripening fruits. However, its role as a plant-growth regulator has been difficult to establish primarily because it is produced in very small quantities and it is extremely difficult to isolate and localize with a specific plant tissue. Once

produced in a cell, it will move into the intracellular space and then evaporate into the atmosphere. With the discovery of gas-liquid chromatography, a method became available for the detection of minute quantities of ethylene. Research is now underway to determine the role of this compound as a growth regulator.

Ethylene is synthesized in most plant tissues from the amino acid methionine. Its synthesis is light sensitive and, being a volatile gas, its action must depend on continued syntheiss. The effects are to inhibit auxin transport, plumular-hook opening, and apical bud expression in pea and bean shoots. It stimulates the ripening of fruits and is used for this purpose in greenhouses. Also stimulated is the release of certain enzymes such as peroxidases and α-amylase in germinating barley seeds.

Ethylene also enhances the development of the male stage of the *Annona*, a tropical fruit flower, but not the female. In the same flower, greater quantities are produced in the reproductive organs than in the petals. The growth of two perennial water plants, arrowhead *(Sagittaria pygmaea)* and pondweed *(Potamogeton distinctus)*, both bothersome weeds in Japan's rice paddies, is promoted in the dark by ethylene and carbon dioxide. Each of these plants are capable of synthesizing ethylene. The response of plant growth and development to mechanical stimulation (**thigmomorphogenesis**) may also be mediated by ethylene. Observations show that young cotton leaves produce more ethylene than do older ones.

Ethylene production in plants can be stimulated either chemically or physically. Such stimulants can be auxins, stress such as wounding, ionizing radiation, disease, and physical restriction. A good example of physical restriction would be that encountered by the plumular hook when the germinating seed is breaking through the soil.

The actual site of action of ethylene has not been determined. Although there have been reports that it stimulates the synthesis of certain macromolecules such as proteins and

309

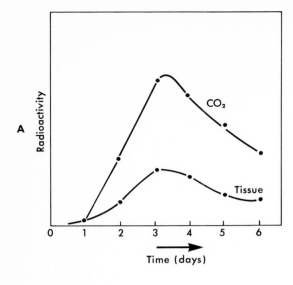

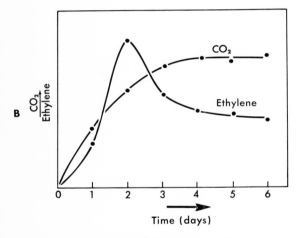

Fig. 17-11. A, Graph depicting the conversion of radioactive ethylene into carbon dioxide and tissue components in etiolated pea seedlings over a 6-day period. **B,** Production of ethylene and carbon dioxide by etiolated pea seedlings during a 6-day period. (Data from Beyer, E. 1975. Nature, vol. 255, May.)

nucleic acids, the effects on plants are so rapid that it is unlikely that a new synthesis of these compounds is provoked. Recent research with ethylene labeled with radioactive carbon, ^{14}C, has demonstrated that when ethylene is applied to etiolated pea seedlings under aseptic conditions, it is converted metabolically to carbon dioxide and other tissue

components (Fig. 17-11, *A*). If the cotyledons are detached from the pea seedlings, ethylene metabolism is reduced by 80%. During the first day the conversion of ethylene to carbon dioxide is very slow. However, the conversion becomes rapid and reaches a peak by the third day and gradually declines thereafter (Fig. 17-11, *A*). Also a period of natural ethylene production preceded the conversion by a day and carbon dioxide production gradually increased to a maximum at 3 days and never declined (Fig. 17-11, *B* and *C*). It is interesting to speculate that ethylene action and its metabolism may be linked. It has long been known that carbon dioxide is an inhibitor of ethylene action and metabolism. The fact that it accelerates senescence and does not inhibit the breakdown of protein and chlorophyll during senescence might somehow link carbon dioxide concentrations to the inhibition or control of the enzymes involved in senescence.

Inhibitors

Plant growth is frequently inhibited by a group of substances known as plant-growth inhibitors. The phenomenon of growth inhibition is clearly evident in such cases as the inhibition of vegetative growth by developing fruit. The best characterized growth inhibitor is abscisic acid. The other known inhibitors, xanthoxin and lunalaric acid, are compounds chemically related to abscisic acid. Another inhibitor not yet isolated and purified is called "senescence factor."

Abscisic acid is a plant-growth inhibitor that was isolated simultaneously by three different laboratories. One group, working with the problem of abscission in cotton plants, found a substance that increased during premature abscission of young fruits and also during abscission of mature ones. The substance was purified and called "abscisin II." A second group isolated an inhibitor associated with the control of dormancy in deciduous trees. This compound was identical to abscisin II, and it was called **dormin.** In the third study, a group isolated a compound associated with fruit abscission in lupine pods and this was identical to abscisin II. The inhibitor in all three cases was abscisic acid, the formula is given below.

Abscisic acid inhibits the growth of hypocotyls, radicles, leaf discs, leaf sections, and root sections. It inhibits chlorophyll synthesis and causes an increase in cold hardiness of *Acer negundo*. It promotes the germination of some fungal spores and also causes an increased rate of uptake of some gibberellins. In addition it has been associated with wilting in leaves. When a leaf-water deficit has been induced by any one of a variety of conditions, wilting of the leaves occurs, and this is associated with a tenfold increase in the concentration of abscisic acid. This large increase in abscisic acid causes a reduced transpiration rate and stomatal closure. The accumulated compound disappears very slowly so that stomatal closure and transpiration rate are affected for a number of days.

Xanthoxin is a natural plant-growth inhibitor, discovered during research on the biosynthesis of abscisic acid. Its molecular structure illustrated below is similar to abscisic acid.

It has been found to antagonize the effects of indoleacetic acid, gibberellins, and kinetin. It inhibits cress-seed germination and wheat coleoptile growth. It is found in the young shoots of dwarf and tall peas, dwarf beans, buttercup, and wheat. It is present in the first bean leaves, but its concentration decreases with age. The most primitive plant in which it was detected were two species of fern. The presence of xanthoxin is not associated with the orange-yellow pigment xanthophyll because some plants that are rich in xanthophyll do not contain detectable amounts of xanthoxin.

Biologically xanthoxin can be converted to abscisic acid. However, it appears that it is an inhibitor in its own right and not by conversion to abscisic acid. It has been demonstrated that 8 days' illumination of pea seedlings with intermittent red light causes a rise in the levels of xanthoxin but not of abscisic acid. Stem elongation is inhibited slightly within 1 day and altogether in 3 days. It is possible that light triggers the synthesis of xanthoxin by activating enzymes in its biosynthetic pathway.

Lunalaric acid appears to be the biological equivalent of abscisic acid in the primitive green plants called liverworts and algae. This compound also has a structure that is similar to abscisic acid.

It has been found in all the liverworts and algae but not in mosses or in higher plants, with the only exception being *Hydrangea macrophylla*. Evidence has also been presented for the presence of a substance called senescence factor, isolated from spinach leaves and associated with accelerating abscission in certain plants. This factor is very closely associated with abscisic acid but is distinct from it. It stimulates ethylene production but it has not yet been isolated chemically.

PRACTICAL USES OF PLANT-GROWTH REGULATORS

The practical uses of plant-growth regulators or their chemical equivalents is important to farmers, horticulturists, florists, and gardners. Currently a great variety of plant-growth regulators are marketed for both commercial and home use. Indoleacetic acid is combined with indolebutyric acid to stimulate the rooting of cuttings. This method is

311

used extensively in nurseries to propagate plants that do not root easily. A mixture of auxins is often used to promote food preservation in onions, potatoes, and other vegetables by prevention of sprout formation and shrinkage. Broccoli will keep longer and stay greener if it is dipped into benzyladenine right after harvest. This compound prevents chlorophyll breakdown. Ethylene is used to stimulate fruit ripening in greenhouses. Synthetic auxins are used to destroy broad-leafed weeds found in grass. In lower concentrations these auxins would actually stimulate growth, but when used on a grass that contains broad-leaved weeds, they kill the weeds but do not effect the grass because the growing tip of grasses is covered and that of the broad-leaved plants is exposed. Premature fruit drop is prevented by spraying with indoleacetic acid, which prevents the abscission of fruit. This is commonly done in apple orchards. Auxin sprays are also used to prevent the development of flowering stalks in some vegetables (bolting), especially lettuce. They are also used to delay the ripening of fruit for shipment. With an increased knowledge of growth regulators and their mode of action, in the future man may use it to his advantage in the production and shipping of fruits and vegetables.

PHOTOPERIODISM

The earth undergoes a variety of rhythmic changes as a result of its relationship with other bodies in our solar system. The earth orbits the sun while rotating around its axis and causing day-night cycles and the regular progression of seasons, the yearly cycle. These rhythmic changes cause variations in the amount of light striking the planet and in the temperature. Generally all eukaryotic organisms have developed a mechanism or mechanisms that mark the passage of time. Whatever the fundamental nature of such a biological clock (or clocks), it appears to be amazingly accurate. It supplies the organism with such vital information as the movement of the sun and the length of the day. It is evident that plant development, especially flow-

ering, is affected by the number of daylight hours. The length of the solar day is 24 hours, the illuminated portion (sunrise to sunset) of which is referred to as the **photoperiod.** The response of a plant to the relative lengths of the dark and light periods is called photoperiodism. However, photoperiodism can be modified by such factors as light intensity, quality, and quantity. In addition, the duration of the dark period appears to be of greater importance to some biological processes such as flowering than does the duration of the light period. In general, the duration and order of sequence are most important in initiating the photoperiodic response in plants. For example, the Japanese morning glory will not flower if it is exposed to a short period of light during the dark period. Other photoperiodic responses in plants are vegetative growth, internode elongation, seed germination, and leaf abscission.

PHOTOMORPHOGENESIS

As we discussed in photosynthesis (Chapter 12), chlorophyll is a photoreceptor molecule that converts solar energy into chemical-bond energy. In addition, plants can also sense light, and in many instances it will evoke a change in the pattern of development. The response of tissue morphology to light stimulation is called **photomorphogenesis.** At present it is known that light regulates every aspect of plant development and the phenomenon of photomorphogenesis is found in every plant. For a plant to respond to light, it must contain photoreceptors, like chlorophyll, capable of converting light energy into a usable physiological form. Radiant energy is passed to specific molecules, which in turn relay it to the mechanism or mechanisms regulating plant development.

Plant cells contain at least two photoreceptor substances, distinct from chlorophyll, that respond to light and in turn cause morphological changes. One substance absorbs blue light and is responsible for phototropisms by causing an alteration in hormone levels. The blue-light receptor has not yet

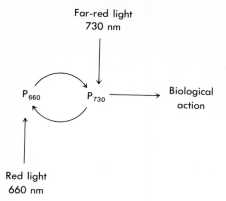

Fig. 17-12. Diagram of phytochrome system in plants. Red light strikes the P_{660} receptor causing it to change to the P_{730} form, which causes biological action. Far red light causes the P_{730} form to change back to P_{660}.

Fig. 17-13. Diagram of phytochrome activation of PAL. Phytochrome causes the conversion of the inactive form to the active one. Thus enzyme activity is increased.

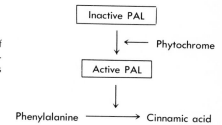

been isolated and characterized, however it appears to be a flavin or carotenoid.

The second photoreceptor system has been isolated, purified, and chemically characterized. It is a bluish protein called **phytochrome,** and it can exist in two stable forms. The forms are interconvertible by a specific wavelength of light (Fig. 17-12). One form, which is abbreviated P_{660}, absorbs red light at 660 nanometers and is converted to the other stable form P_{730}. When P_{730} absorbs far-red light at 730 nm, it is gradually converted back to P_{660}. Apparently P_{730} is the biologically active form, though this fact has never definitely been proved.

Although much progress has been made in the understanding of the mechanism of photoreception and the description of the induced morphologic changes, the pathways that link one process with the other have not been clearly elucidated. There are certain rapid effects of light that are associated with

membranes. When phytochrome perceives light, the electric charge on the membrane is known to change, and as a result it causes root-tip adhesion to the surface of negatively charged glass vessels. It appears that phytochrome acts through a second messenger and it is this messenger that causes the observed physiological changes. For example **acetylcholine** has been proposed as a second messenger; thus phytochrome 730 would cause the production of acetylcholine, which would alter the membrane properties. Whatever the mechanism, it appears likely that phytochrome regulates the movement of potassium ions across the membrane. The most rapid effects of light not associated with a change in membrane properties are an immediate increase in the cellular ATP levels and increased gibberellin activity.

Light can also control the activity of certain enzymes within the plant cell. Phenylalanine ammonia-lyase (PAL) causes the conversion

313

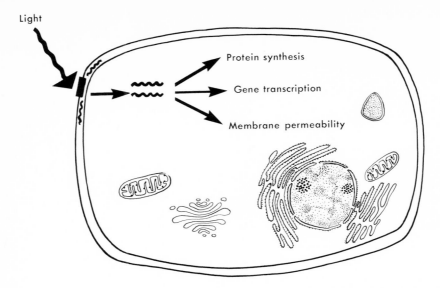

Fig. 17-14. Diagram of phytochrome acting on a second messenger molecule. The second messenger molecule is released from the plasma membrane by phytochrome action. The released messenger molecule can then act at a variety of sites.

of phenylalanine to cinnamic acid with the release of ammonia. PAL is subject to photocontrol in a wide variety of species. The photoreceptor is usually phytochrome, but occasionally it is the blue-light receptor. Light apparently causes the conversion of an inactive PAL to an active one in gherkin hypocotyls (Fig. 17-13). However, it does not stimulate a second synthesis of the enzyme in this organism, because its action is not inhibited by known protein-synthesis inhibitors. Lipoxygenase is another enzyme that is light sensitive. It catalyzes the oxidation of unsaturated fatty acids. Light appears to control the level of this enzyme within the cell rather than its synthesis. There is not sufficient evidence to pinpoint exactly the site of action in this particular system. One of the most significant control functions of phytochrome is its control over the synthesis of chlorophyll. One of the final steps in the biosynthetic pathway of chlorophyll is the conversion of protochlorophyll to chlorophyll. This step is regulated by the photoactivation of pro-

tochlorophyll, which is a photoreceptor. However, one of the early steps in the pathway is under the control of light and is mediated by phytochrome.

It has been postulated that the primary site of action of phytochrome is a cellular membrane system. The primary reaction with the membrane would cause the release of the secondary messenger, which could act at multiple sites, one of them being the site of protein synthesis (Fig. 17-14).

FLOWERING RESPONSE

Flowering plants produce special buds that have the capacity to differentiate or develop into the specialized structure of the flower (Chapter 16). This developing process is referred to as the **flowering response.** Presently there appear to be three environmental factors that control the induction of this response: photoperiodism, temperature, and **vernalization.**

According to the photoperiodic response, plants are generally placed in one of three

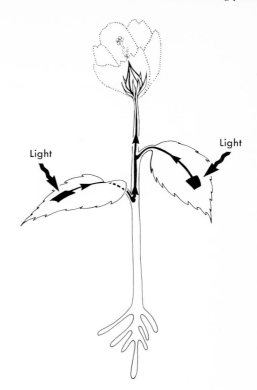

Fig. 17-15. Diagram of flowering response. After exposure for a sufficient dark period, red light is perceived by the leaves. The perception of red light by phytochrome causes the activation or synthesis of the flowering hormone, florigen. Florigen is then translocated to the site of flower formation where it promotes flower development.

categories: short-day, long-day, or neutral-day plants. Short-day plants are found mostly from the equator through the temperate zones. They flower only when they have light for less than a critical period. For example, poinsettias are short-day plants that require less than 12 hours of light in order to flower. The critical light period appears to be species-specific, such as the chrysanthemum, which has a critical day length of 11 hours. Other short-day plants are ragweed, salvia, and goldenrod. Ragweed and goldenrod pollen provoke allergies in the northeast when the days begin to shorten in the month of August. Long-day plants are those that will flower when day lengths are longer than a critical minimum. For example, spinach is a long-day plant that will flower when day length exceeds 12 hours. For this reason farmers plant spinach at a time in spring or fall when the day length is shorter and max-

imum vegetative growth is possible. Another long-day plant is lettuce. However, hybrids of lettuce have been developed that will not flower on a long-day; thus they allow vegetative growth during summer months. The third category composes the neutral-day species, which flower regardless of the day length. Examples of this are tomato, cucumber, dandelion, string bean, sunflower, and cotton.

The most effective light receptor, phytochrome, for flowering is found in recently matured leaves, and the red light is most effective in promoting flowering. Leaves are definitely the receptor for the photoperiodic response, if a short-day plant is induced to flower and then a leaf from that plant is grafted to a noninduced shortday plant, it will flower. Also if the induced leaf is grafted to a long-day plant, that plant will also flower. Apparently the induced substance is the same for both long-day and short-day plants. The **flowering factor** produced in induced leaves is referred to as the flowering hormone and given the name **florigen.** Florigen has not been isolated, but there is evidence that it is a steroid compound.

The dark period definitely has an effect on flower initiation. *Xanthium* is a short-day plant that will flower when it receives 16 hours of dark and 8 hours of light. However when it receives 8 hours of dark, 2 hours of light, and then another 6 hours of dark prior to the 8-hour light period, it does not flower. When the light period is interrupted by a 1-hour dark exposure, *Xanthium* still flowers. The length of the light period has no effect on flower initiation, however, it does effect an increase in the number of floral primordial tissue. In summary, then, the effect of light on flowering appears to involve an exposure to a dark period of sufficient length, the perception of red light by the phytochrome system in the leaves, the activation of synthesis of the flowering hormone, florigen, the transport of florigen to the site of flower formation, and the final action on either the genetic material or a pre-existing substance that would cause the differentiation to occur (Fig. 17-15).

Temperature will also control the flowering response in some plants. Tomato for example is a neutral-day plant in which flowering is controlled by temperature. Night temperatures of 15° C. and day temperatures of 25° C. will promote a maximum flowering response. There does not appear to be a specific hormone or receptor tissue involved in this response. It appears to be related to the amount of carbohydrate that has accumulated in the tissue. This response has also been observed in other plants.

The third environmental control of flowering is vernalization, which is the process of long exposure to low temperature. One must sow some plant seeds in the fall to produce flowers the following spring. It appears that the mechanism of vernalization involves the production of a hormone that has not yet been isolated, but it is at present called

vernalin. Vernalin may be either a precursor to florigen or an enzyme involved in its formation.

Gibberellins have also been implicated in the flowering response. They can be substituted for a cold requirement or a long-day requirement in flowering plants. When the long day–requiring strain of henbane was kept at short days, it had only a rosette of leaves and a single gibberellin-like substance. However application of gibberellic acid to the plants caused stem elongation and flowering. When the plant was kept on long-day conditions, the stem elongated, flowered, and increased the amount of gibberellin-like material present.

LEAF SENESCENCE AND ABSCISSION

The growth, development, and death of a leaf represents one of natures best examples of conservation. Many of the nutrients that are manufactured in the leaf such as carbohydrates, amino acids, and lipids plus essential minerals are translocated back into the permanent parts of the plant before it falls from the tree. This removal is absolutely essential for the future growth of the tree the following spring.

The most visible sign of leaf death or **senescence** is the appearance of the bright-colored autumn pigments anthocyanin and carotenoids. In some trees the pigments are always present but are only seen when the predominant green chlorophyll disappears. In others they are not synthesized until the cool autumn temperatures appear. The leaf gradually loses the ability to retain water and remain turgid. Its membranes rupture and fluid is leaked. It then becomes rigid and appears tan to dark brown in color. When it falls to the ground, soil bacteria will use the remaining minerals, especially calcium, for growth and they also hydrolyze leaf cellulose. By the time the leaf has been shed (**abscission**), it has returned to the permanent parts of the tree more than 50% of its nitrogen, potassium, and phosphorus plus small amounts of magnesium, sulfur, and manganese.

Leaf senescence is governed by three factors—environment, plant hormones, and genetics. It is an interaction of these factors that will cause a leaf to senesce. The first internal indication of senescence begins when the day length declines to less than 14 hours, usually at the beginning of August in the northern climates. The importance of the daylength is evident when it is observed that deciduous trees that shed leaves in the north do not do so in Florida where the days remain at a constant length all year. Temperature and weather do not appear to play a significant role in senescence. Temperature however, does play a role in the development of the anthocyanin pigments. When the days shorten, there is a gradual shutdown in the synthesis of chlorophyll and thus photosynthesis slackens. Gradually the decay of chlorophyll allows the bright autumn pigments of red anthocyanin and yellow carotenes to show through. Net protein synthesis declines, different proteins are synthesized, and the leaf begins to translocate nitrogen back to the permanent parts of the tree. The shortened days also affect the hormonal balance within the leaf. When the day length shortens, the leaf begins to synthesize the growth inhibitor, abscisic acid, and the concentrations of auxin, gibberellins, and cytokinins decrease. As a result the balance grad-

ually tips in favor of the growth inhibitor, abscisic acid. It has been demonstrated that indoleacetic acid will retard senescence and abscission. As more abscisic acid is synthesized in the leaf, some of it is transported to the permanent parts of the tree where it promotes winter hardiness.

Genetics is a definite factor in senescence. If the leaf does not contain the information for senescence in its genetic material, regardless of the environmental and hormonal factors, it will not senesce. For example, rhododendrons and the younger leaves of evergreens do not senesce. The genetic material provides the potential; the environmental and hormonal factors cause its expression.

DORMANCY

Dormancy is the period of the plant life cycle during which growth is either temporarily suspended or retarded to the extent that it cannot be visibly detected. It is most often observed in the propagative parts of plants such as buds and seeds. It has also been demonstrated that some organs of a plant experience dormancy independently. When one side of a tree is exposed to continuous artificial light, the leaves on the lighted side will not become dormant. However, the

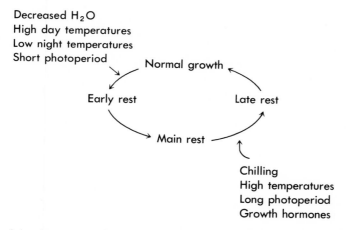

Fig. 17-16. Diagram of the dormancy cycle in plants and the stimuli that promote the transition from normal growth into early rest and the stimuli that cause the transition back to normal growth.

leaves on the unexposed side will gradually pass into dormancy.

Dormancy in trees has been arbitrarily divided into three phases (Fig. 17-16). Early rest is the first phase. When a plant in early rest is removed to a greenhouse, it will continue its normal growth pattern. It has not lost the ability to continue vegetative growth. The second phase is main rest (winter rest). A plant in main rest has lost the ability to immediately resume normal growth when it is removed to an environment with optimal growth conditions, for example, when it is brought into a greenhouse. The last phase is late rest, which is characterized by the plant's regaining the ability to resume normal growth when it is stimulated by being removed to a greenhouse. For a plant to pass from main rest to late rest, it must experience a **chilling** requirement while in main rest. The chilling requirement is a 30- to 90-day period of winter cold or temperatures of less than 6° C. Once this requirement has been met, the plant can be stimulated to resume normal growth.

Certain environmental stimuli act on plants in both the vegetative and dormant states. When the stimuli act on plants in the vegetative state, dormancy is induced. In the late rest period of dormancy, it is broken. The stimuli that cause the transition from the vegetative state to early rest are high day temperatures combined with low night temperatures, shortened photoperiod, and the availability of water. Stimuli that break dormancy are the chilling requirement, high temperatures, natural growth hormones such as indoleacetic acid, gibberellic acid, and ethylene, and long photoperiods (16 to 20 hours). It is therefore reasonable to assume that the plant must contain a number of receptors that are sensitive to the various environmental stimuli. At present the receptors have not been identified, but current research indicates that the phytochrome system is definitely involved in some aspects.

The most notable biochemical changes that mark the transition from vegetative state to early rest are a slackening of cellulose synthesis and an increase in the synthesis of lignin, starch, and abscisic acid. Anthocyanins also begin to accumulate. The most prominent morphological changes that occur are the formation of the abscission layer in leaves, the halting of cambial activity, and the translocation of minerals from the leaves. The plant will then gradually go into a period of main rest (winter rest) during which it will be exposed to winter cold.

Physiological changes that occur during the breaking of dormancy are the cessation of the synthesis of abscisic acid and certain pigments. Also the beginning of photosynthesis, respiration, transpiration, and increased concentrations of indoleacetic acid and gibberellic acid are noted. The plant will complete the cycle of dormancy and renew vegetable growth.

In summary, the basic mechanism of dormancy remains a mystery. It appears from the many reports in the literature that phytochrome and abscisic acid are definitely involved at some point or in some capacity in the total response. There are many genes involved in the dormancy response; therefore one would assume that many proteins are being synthesized. Many of the physiological and morphological changes are somewhat similar to those observed during senescence. The most popular current hypothesis is that the dormancy–normal growth cycle is controlled internally by a delicate balance of abscisic acid and gibberellic acid.

QUESTIONS AND PROBLEMS

1. Diagram and label the successive stages of mitosis in a plant cell containing two pairs of chromosomes.
2. What are the possible mechanisms for the action of auxin?
3. Describe the effect of gibberellic acid on barley grain.
4. Discuss plant-growth inhibitors.
5. Discuss and describe the effect of light on plants.
6. Describe in detail the flowering response and senescence in leaves.

ADDITIONAL READING

Hamner, K., and Hoshizaki, T. 1974. Photoperiodism and circadian rhythms: an hypothesis. Bioscience, vol. 24, no. 7. (A discussion of dawn and dusk clocks controlling the biological rhythms of organisms.)

Hill, T. 1973. Endogenous plant growth substances. Studies in biology, no. 40. Edward Arnold Publishers, London. (A discussion of the growth hormones and inhibitors.)

Levitt, J. 1974. Introduction to plant physiology. 2nd ed. The C. V. Mosby Co. St. Louis. (A basic plant physiology textbook.)

McMahon, T. 1975. The mechanical design of trees. Sci. Am. **233**:92, July. (Laws that govern the relationship between a tree's trunk length and width.)

Perry, T. 1971. Dormancy of trees in winter. Science, **171**:29. (A discussion of the variables that interact to control leaf fall and other dormancy phenomena.)

Fig. 18-1. Varieties of tomatoes produced by selection and crossbreeding. (USDA photograph.)

CHAPTER 18

Plant genetics

■ The mechanism of cell division, sexual reproduction, and the control of the development of the individual plant body as well as its variation will be considered in this chapter. The most striking characteristic of living things goes farther than mere reproduction—it is the ability to produce offspring essentially like themselves in their fundamental structures, yet, at the same time, individuals with slight differences. "Like begets like" is a common phrase. An organism will inherit the characteristics peculiar to its species. However, despite the fact that parents and offspring are unmistakably similar to each other, they will usually differ in certain respects. Fig. 18-1 shows a number of tomato berries. Although each is recognizable as a tomato, it is apparent that they are not all alike. In fact, they are an example of several varieties of berry that differ as much in taste and hardiness as they do in size and shape. Not only does one plant differ from another, but also variations can be found on the same plant. For example, on the same hibiscus plant can be found several differently shaped leaves (Fig. 18-2). This tendency of organisms to differ slightly, or to a greater degree, from their parents is called variation. Variation is the one characteristic that makes possible the adaptation of an organism, and eventually a species, to various habitats. If all offspring were exactly alike, environmental changes would cause rapid extinction instead of evolutionary change.

MEIOSIS

The sexual reproductive cycle starts with the formation of gametes through the process of meiosis and ends with the union of these gametes during fertilization to form the new individual, the zygote. In mitosis the chromosome number is maintained constant during successive nuclear divisions, and the chromosomes divided lengthwise after duplicating themselves. The heredity determiners, or **genes,** are arranged in a linear fashion on the chromosomes. Obviously, then, there is more than one gene on each chromosome. Therefore, when a chromosome duplicates itself during mitosis, the genes are also duplicated, and each new set of chromosomes will carry identical and corresponding genes. In those organisms that originate as a result of sexual reproduction, there are two sets of chromosomes, one from the female parent and one from the male parent. Each pair of chromosomes constitutes a homologous pair, and the maternal chromosome has a corresponding paternal one that is similar in structure, size, and gene makeup. The members comprising this pair are termed "homologs." Thus, any organism arising from sexual reproduction will have

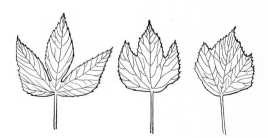

Fig. 18-2. Variations in hibiscus leaves from the same plant.

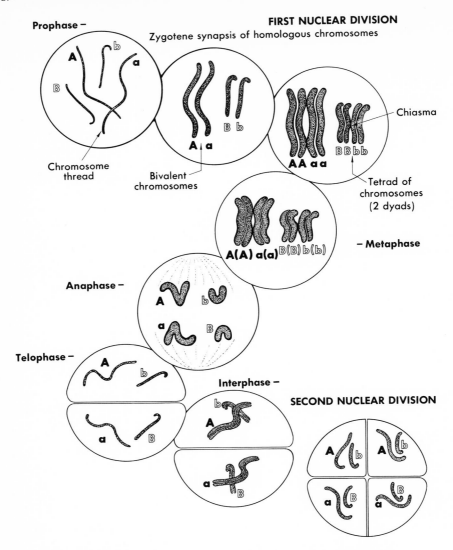

Fig. 18-3. Details of process of meiosis showing fate of chromosomes.

the **2n** or **diploid** number of chromosomes. Two sex cells unite to form the diploid zygote, and each will contribute the **n** or **haploid** number of chromosomes to that union.

The process by which the chromosome number is reduced from the diploid to the haploid is termed "meiosis." It is quite different in plants and in animals. The plant's production of sex cells is by a special haploid stage, the gametophyte. We shall consider

here only the cell division process that results in the haploid chromosome number in the sex cell. The entire reproduction process is described in Chapter 16. The body cells of a plant are produced by mitosis as we have seen (Chapter 17), but during the production of sex cells of the higher plants and sometimes spores of lower plants a series of divisions occur in which the diploid chromosome number is reduced to one half, or the haploid number. Meiosis (Fig. 18-3) occurs in two steps, one immediately after the other, that is, there are two successive nuclear divisions in which the chromosomes divide only once. During the early prophase of the first nuclear division the chromosomes appear as long, single-stranded, slender threads, the chromatids, which are made of a series of granules. This is the **leptotene** stage of prophase. Only the homologous threads come together in pairs (a phenomenon known as **synapsis**), and this is termed the **zygotene** stage of the prophase. Each **bivalent chromosome** then will duplicate itself to form a **tetrad.** The four chromatids remain united, being held together more firmly at some regions than at others. An area at which they are held closest together is called a **chiasma** and is cytological evidence that **crossing over** (the exchange of homologous regions of a chromatid) has taken place; this is known as the **pachytene** stage. Soon after this, the attraction of the chromosomes gives way to repulsion, and they tend to separate, the stage termed **diplotene.** When the tetrads become shorter, thicker, and more compact, the division between homologous chromatids is barely or no longer visible, and the **diakinesis** stage is reached, the stage during which the loops are formed between the successive chiasmas. The half-loops lie at right angles to one another. At the end of the prophase, successive stages take place that are similar to the metaphase, anaphase, and telophase stages of mitosis. As a result, the daughter nuclei receive a **dyad** of each kind of chromosome. After a short interphase, a second division occurs; the dyads divide so that each of the four daughter nuclei will have one of each chromatid type present. Thus meiosis is complete.

In summary, meiosis consists of two nuclear divisions, one of which is a reduction division. The chromosome number is halved, and four nuclei (later four cells) are produced, each having the haploid number of chromosomes. The union of the haploid sex cells accomplishes two things: (1) it restores the diploid chromosome number and (2) it provides the genetic traits of the new individual. In addition, it results in the random distribution of chromosomes in the gametes or sex cells. Thus the physical basis for the new combination of hereditary material that occurs in every generation is found in the process of meiosis.

MENDELISM

Even though the theory of heredity, frequently termed **Mendelism,** was actually published many years before 1900, it remained unused and buried in the literature. Actually our knowledge of the sex of plants grew out the work of Rudolf Jakob Camerarius (1665-1721) published in 1694. The first intentionally artificial plant hybrid was produced by Thomas Fairchild in 1717. As early as 1779, color blindess in man was reported to the Royal Society of London as being inherited. Other rather unorganized and poorly demonstrated examples of thoughts on heredity could be cited, but none can be considered as contributing to genetic theory.

Gregor Mendel (1822-1884) (Fig. 18-4), an ordained priest of the Augustinian order, lived in the monastery at Brunn, Austria. In

Fig. 18-4. Gregor Mendel (1822-1884), the first person to publish on the laws of heredity.

the gardens of this monastery he carried on his epoch-making experiments, which practically ceased with his election to position of abbot in 1868. He is best known for his experiments with the garden pea plant, *Pisum sativum*, although he experimented with several other kinds of plants as well as with bees. Mendel was fortunate that he studied characteristics that were conspicuous and easily distinguishable. He was successful because he studied single differences one at a time as well as in combination with others, because he kept accurate records of all the progeny, and because he chose a plant that was normally self-pollinated and therefore self-fertilized but that also could be cross-pollinated. Self-pollination leads to genetic purity so that each of the plant varieties he used for his experiments initially were genetically pure for the characteristics he chose.

The results of Mendel's pea experiments were published in 1866 in the *Proceedings of the Brunn Society for the Study of Natural Science* but were completely overlooked for 34 years. At the time his work was published, there was no theory of the germ plasm, no description of chromosomes, and no theory of the gene, and, furthermore, little was known about cell structure and behavior. Mendel's ignorance of what is today common knowledge makes his conclusions about heredity much more significant. His many experiments extended over a period of 8 years. The methods he used and the conclusions he reached are now well known. Mendel died in 1884, unrecognized and disappointed that he had been unable to continue his investigations. His great contribution to the field of heredity was rediscovered independently in 1900 by three botanists whose research had led to similar discoveries. Hugo De Vries (1848-1935) in Holland, Erich Tschermak von Seysenegg (1871-1962) in Austria, and Karl Correns (1864-1933) in Germany recognized the significance of Mendel's work. Many important research papers followed the disclosure of these discoveries, and what came to be known as the Mendelian principles were found in many organisms. Thus

Mendel's work led to the modern study of genetics.

MENDELIAN HEREDITY

MENDEL'S LAWS
1. Law of segregation of characters
2. Law of independent assortment of characters

Although Mendel experimented with crossing seven pairs of characteristics in peas, the one that is most familiar deals with their tall and dwarf habit. He crossed a variety having tall stems with one having dwarf stems (Fig. 18-5) and found that all the offspring of these two parents were tall. After allowing the tall offspring to undergo self-pollination and hence self-fertilization, he realized that there were two kinds of offspring from this cross. Most of the offspring resembled their original tall parents, whereas a few were dwarfs like their dwarf parent. Upon counting the numbers of each and carefully recording them, he discovered that 787 of the plants were tall and 277 were dwarf. Thus, by crossing a tall plant and a dwarf plant, he learned that the progeny of the first filial generation (F_1) resembled one of the parents and not the other. Also the F_1 plants breeding among themselves produced a second filial, or F_2, progeny that resembled both parents, occurring in a ratio of three tall plants to one dwarf plant. The trait that disappeared in the F_1 reappeared about one fourth of the time in the F_2. From the data of this and similar experiments, Mendel formulated what has become known as the first law of heredity, *the principle of segregation*. According to this law, the heredity units (genes) occur in pairs that are located at corresponding identical sites on homologous chromosomes. During the formation of gametes the pairs separate from one another. Let us symbolize the unit for tallness as T and that

Parents

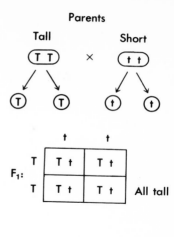

F₁: All tall

Parents

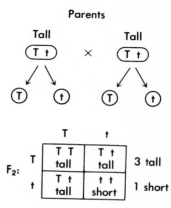

F₂: 3 tall   1 short

Fig. 18-6. Diagram of Mendel's original cross between tall and short peas showing gamete formation, genotype, and phenotype.

Parents

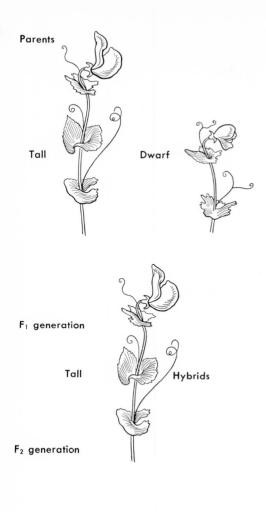

F₁ generation

F₂ generation

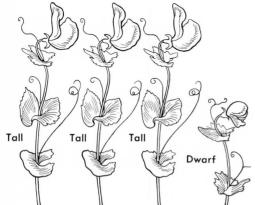

Fig. 18-5. Illustration of cross between tall and short pea plants (monohybrid cross).

for dwarfness as t. The pairs may be alike, TT or tt, in which the individual is said to be **homozygous,** or they may be different, Tt, and the individual is **heterozygous.** When the members of such pairs control alternate expressions such as Tt, they are called **alleles.** Alleles are alternate expression of the same gene. The diagram (Fig. 18-6) illustrates Mendel's initial cross showing gene constitution (**genotype**), gamete formation and visual appearance (**phenotype**). The diagram more than adequately shows all possible gamete combinations to form zygotes. It also shows that t will not be expressed in the presence of

Backcross: monohybrid to homozygous dominant parent

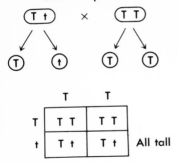

Testcross: monohybrid to homozygous recessive parent

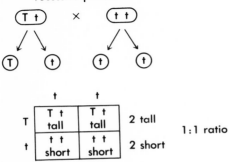

Fig. 18-7. Diagram of backcross to a homozygous dominant parent and a testcross.

T, as evidenced by the phenotype of the heterozygote Tt. This is an example of allelic interaction, whereby T completely masks the effect of or dominates over t. Geneticists say that T is the **dominant** characteristic and t is the **recessive** one. A cross between two individuals that differ because of the presence of a single allelic pair, as Mendel's cross, produce an F_1 monohybrid and the cross is referred to as a **monohybrid** cross. When a dominance relationship exists, the phenotypic ratio of a monohybrid cross is 3:1 (three dominant characters to one recessive one). If we cross the monohybrid to either parent or parental type, a different ratio will be observed (Fig. 18-7). A cross to either parent is

called a **backcross,** but a cross to the homozygous recessive parent is a **testcross.** All testcrosses are backcrosses but not all backcrosses are testcrosses.

INCOMPLETE DOMINANCE

Allelic interactions may occur where dominance is not always the rule. In plants, for example, when a red-flowered four-o'clock is crossed with a white one, one may expect that the F_1 generation would be either all red or all white. Such is not the case, since the plants are all intermediate between red and white-pink (Fig. 18-8). When the hybrid pink plants are self-pollinated to produce an F_2, they do not produce all pink offspring, but one fourth are red (like the original parent), one fourth are white (like the other parent), and one half are pink (like the hybrid F_1 generation). If this cross is diagrammed with letters for the genes, it is found that the phenotypic ratio of the F_2 generation is 1:2:1 and that it corresponds to the genotypic ratio (Fig. 18-8). Again, as in Mendel's crosses involving complete dominance, the genes do not lose their identity in the hybrid pink plant but segregate normally in a subsequent generation. This type of inheritance illustrates what can be referred to as **incomplete dominance,** or blending inheritance. Initially, it may appear to refute Mendel's first law of inheritance, but actually it strengthens it.

DIHYBRID

By experiments using other pairs of contrasting characters in peas, Mendel discovered that round seeds were dominant over wrinkled seeds (Fig. 18-9). In his experiment he found 5474 round-seeded plants and 1850 wrinkle-seeded plants in the F_1 generation. Again, it was an approximate ratio of 3:1, such as he had observed for the height of stems. Cotyledon color in peas was also inherited in the same way. Mendel's records show that he got 6022 yellow cotyledons and 2001 green ones in the F_2 generation, when he crossed homozygous yellow and green parents (Fig. 18-9). All previously mentioned

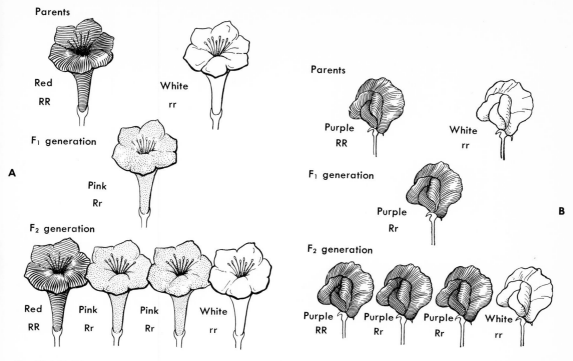

Fig. 18-8. A, Diagram showing incomplete dominance in a cross between red and white four-o'clocks compared to one, **B,** showing complete dominance between purple- and white-flowered pea plants.

crosses were monohybrid in nature, with the organisms differing in only one allelic pair. Mendel now made crosses in which the plants differed in two allelic pairs (dihybrid). From the **dihybrid** cross of a yellow-smooth and a green-wrinkled seed plant, he found that the F_1 plants had yellow-smooth seeds (dihybrids) and the F_2 generation gave the following results:

Yellow-smooth	315
Yellow-wrinkled	101
Green-smooth	108
Green-wrinkled	32

If the inheritance of each characteristic is determined separately, it is found to follow a simple monohybrid ratio of 3:1, indicating segregation of alleles.

Yellow	= 315 + 101	= 416
Green	= 108 + 32	= 140
Smooth	= 315 + 108	= 423
Wrinkled	= 101 + 32	= 133

In addition, he also observed that the F2 generation followed a 9:3:3:1 ratio.

No zygote will have more than four genes (two pairs), and no gamete will have more than one of each of the two pairs. If the four genes are allowed to assort independently, then four types of gametes are possible with equal frequency. As a result, there were 9 plants with yellow-smooth seeds, 3 with yellow-wrinkled seeds, 3 with green-smooth seeds, and 1 with green-wrinkled seeds (Fig. 18-10).

The dihybrid cross thus illustrates Mendel's second law, *the law of independent assortment,* which holds true if the genes are located on separate chromosomes, but will not be true, if they are located on the same chromosome, a phenomenon known as linkage.

The fact should be emphasized, since we are assuming that the genes are located on the chromosomes, that it is the chromosomes

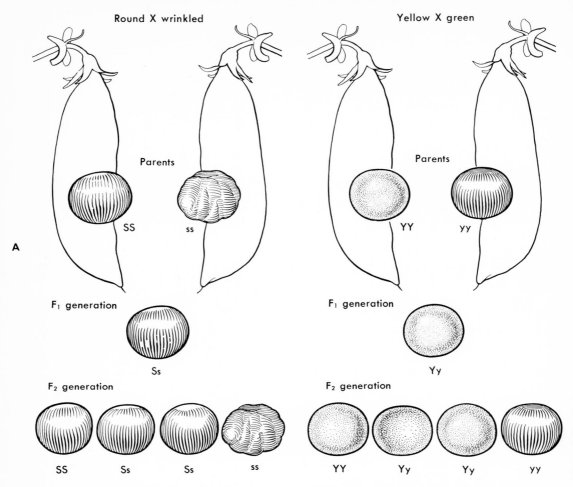

Fig. 18-9. A, Cross between round and wrinkled peas. **B,** Cross between yellow and green peas.

that are assorting at random in each generation. Each of the seven pairs of factors used by Mendel to demonstrate the segregation of unit characters and independent assortment was actually located on seven separate chromosome pairs. Therefore, when plants are crossed in which two characteristics differ and two of the alleles are completely dominant over the other two, the characteristic ratio of 9:3:3:1 will occur.

MODIFIED DIHYBRID

Should one pair of alleles be incompletely dominant (such as the red and white four-o'clocks), the 9:3:3:1 ratio will be modifed

into 3:6:3:1:2:1, giving six phenotypes instead of four. Should both be incompletely dominant over the other two, the number of phenotypes will then be increased even more.

All these ratios can be obtained by simple arithmetic. For example, in a 3:1 ratio, three fourths of the plants have yellow seeds in the F_2 generation, and so also will three fourths of the plants have smooth seeds. Three fourths of the plants with smooth seeds also have yellow seeds, the three fourths of that three fourths, or nine sixteenths, will have smooth seeds as well. Of the three fourths having yellow seeds, only one fourth of

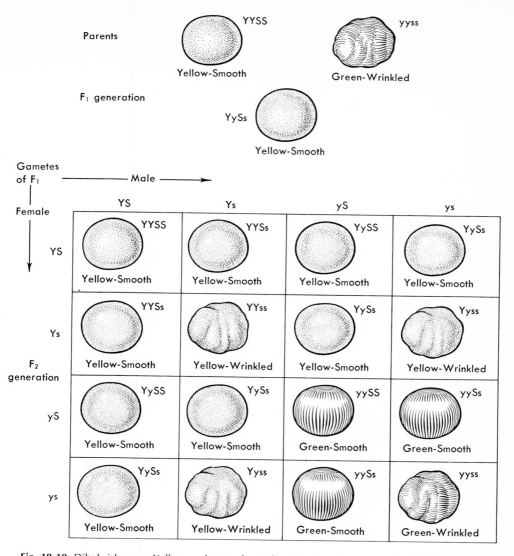

Fig. 18-10. Dihybrid cross. Yellow and smooth seeds crossed with green and wrinkled seeds.

those, or three sixteenths, will have wrinkled seeds; one fourth will have green seeds, and three fourths of those, or three sixteenths, will have smooth seeds; and of the one fourth of the plants having green seeds, only one fourth will have wrinkled seeds, in other words, one sixteenth. Thus the phenotypes can be calculated, or they can be worked out as in Fig. 18-10.

Epistasis is the effect of nonallelic genes on each other's expression. The production of

cyanide in white clover is an example of an epistatic interaction. High-cyanide strains of clover are agriculturally desirable because they generally grow much better. Cyanide in white clover is produced by the action of an enzyme on the specific substrate cyanogenic glucoside. The enzyme that causes this reaction is determined by a specific gene, H. Gene H causes the production of an active enzyme whereas, gene h, its allele, does not produce an active enzyme. So in the pres-

329

ence of cyanogenic glucoside, individuals of HH and Hh genotype cause the production of cyanide, whereas the genotype hh will not have any cyanide. The trait is inherited as a simple monohybrid giving a 3:1 ratio of high to low in F_2.

$$\text{Gene H}$$
$$\downarrow$$
$$\text{Cyanogenic glucoside} \xrightarrow{\text{Enzyme}} \text{Cyanide}$$

Another gene in clover, the L gene, determines the production of an enzyme that catalyzes the reaction that produces cyanogenic glucoside. It is also dominant to l, its allele, and behaves as a monohybrid in the same manner as gene H.

$$\text{Gene L}$$
$$\downarrow$$
$$\text{Precursor} \xrightarrow{\text{Enzyme}} \text{Cyanogenic glucoside}$$

The production of cyanide then in white clover is governed by the interaction of two genes that are not linked. To have cyanide produced, one must have a precursor converted to cyanogenic glucoside by an enzyme determined specifically by gene L and one must have cyanogenic glucoside converted to cyanide by an enzyme specifically determined by gene H. At least one gene H and one gene L must be present in an organism for the production of cyanide. When a cross is made of LLHH (high-cyanide plants) with llhh (low-cyanide plants), the F_1 is all high and the F_2 would be in a ratio of 9 high to 7 low (Fig. 18-11). This is an epistatic interaction.

Parents: High   Low
LLHH × llhh

F_1: LlHh

F_2:
L-H-	9	high
llH-	3	
L-hh	3	7 low
llhh	1	

Ratio: 9 high
7 low

Fig. 18-11. Diagram of a cross between high and low cyanide strains of white clover.

CONVERTER GENES

In 1959 a gene was discovered in corn that governed the color of the plant. Ordinarily, the gene for color retains its ability to produce the color through an unlimited number of crosses. However, a gene capable of changing the color gene from one color to another color, a **converter gene,** was discovered. The results of such a cross involving a converter gene as compared with a normal color gene are shown in Fig. 18-12.

MODIFYING GENES

In plants some genes are not responsible for a specific character in themselves but can modify the expression of other genes in a quantitative manner. These are called **modifying genes.** The phenotypic effect of such a gene is evident when it causes a dilution or concentration of a color characteristic, such as seed coat color in corn. Modifying genes can also change the expression of certain mutant characteristics, in which case they are called suppressors. In some cases the action of a suppressor gene can be explained at the molecular level. However, in many instances, the basic mechanism of modifying gene action remains obscure.

SEX DETERMINATION

A special kind of linkage in certain chromosomes occurs in both plants and animals, generally referred to as sex determination. Most higher animals are bisexual and have a special pair of sex chromosomes that determine the sex of the individual. Some plants have a similar condition, but instead of being called **bisexual,** they are referred to as dioecious. Actually, since only the gametophyte produces sex cells, or gametes, only these plants can be termed bisexual, and in many cases they will produce both kinds of gametes. The lower plants, for example, the bryophytes, have dioecious individuals. In the flowering plants where the sporophyte is dominant, the plants that show bisexuality, or dioecism, are termed either staminate or ovulate plants, that is, those that produce anthers or pistils, but not both (Chapter 16).

How surprising CONVERTER GENE changes corn plant color

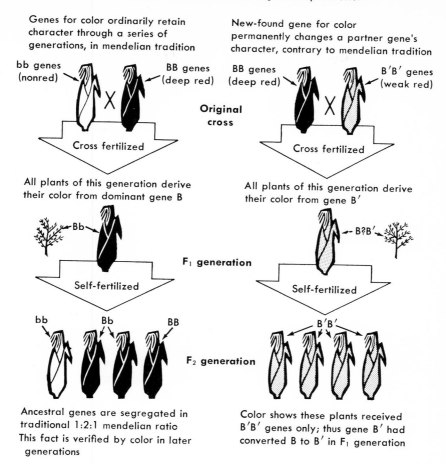

Genes for color ordinarily retain character through a series of generations, in mendelian tradition

New-found gene for color permanently changes a partner gene's character, contrary to mendelian tradition

bb genes (nonred)   BB genes (deep red)

BB genes (deep red)   B'B' genes (weak red)

Original cross

Cross fertilized

Cross fertilized

All plants of this generation derive their color from dominant gene B

All plants of this generation derive their color from gene B'

Bb

B?B'

F₁ generation

Self-fertilized

Self-fertilized

bb   Bb   BB

B'B'

F₂ generation

Ancestral genes are segregated in traditional 1:2:1 mendelian ratio
This fact is verified by color in later generations

Color shows these plants received B'B' genes only; thus gene B' had converted B to B' in F₁ generation

Fig. 18-12. Converter genes used to change the color of corn ears. By adding a gene of this kind to the breeding stock, one may control color and other traits. (USDA drawing.)

However, regardless of the situation, sex is determined by X and Y chromosomes. If the cells contain only X chromosomes, only ovules are produced. However, if the cells contain both X and Y chromosomes, the pollen that is produced is of two types, some with an X chromosome and some with a Y chromosome.

LINKAGE

The pairing of homologous chromosomes and their subsequent separation at meiosis implies that the genes are arranged in linear order and that corresponding genes, whether they are identical or not, also pair. The position they occupy on the chromosome is fixed whether they are or are not alike. This position is termed its locus. Either member of a pair of genes that occupies the same locus on two paired chromosomes is called an allele.

Any organism that exhibits the characteristic of one or more alleles (heterozygous) is a hybrid, which is the result of crossing two individuals that differ in one or more genetic characteristics, as we have seen. Natural hybrids occur wherever there is population variation, which always is the case under wild conditions. But because of the position of the

331

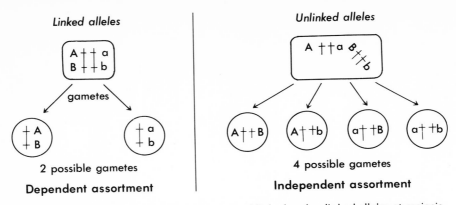

Fig. 18-13. Diagram depicting the assortment of linked and unlinked alleles at meiosis.

genes, hybridization cannot go on indefinitely because of linkage.

The Mendelian ratios discussed thus far depend on the independent assortment (random assortment) of the genes at meiosis and on their recombination at fertilization. Such a chance assortment will occur only if the genes are located on separate and different chromosomes. Plants, like animals, contain hundreds and probably thousands of genes, whereas there may be only a limited number of chromosomes (*Secale cereale*, rye, has only seven pairs). Since the number of chromosomes is limited, each chromosome must of necessity carry more than one gene. If two or more genes are located on the same chromosome, they will tend to be transmitted together to the progeny. As a result, independent assortment of these genes cannot take place. When genes are located on the same chromosome and hence tend to be inherited together, they are said to be linked (Fig. 18-13). Note use of parentheses to mean 'linked.' If two plants are crossed, one having two pairs of linked dominant genes homozygous for them (AB)/(AB) and the other being homozygous recessive (ab)/(ab), the F$_1$ generation will then be heterozygous for both (AB)/(ab). Four kinds of gametes would be produced if they were not linked, for example, AB, Ab, aB, and ab. However, if they are linked, only two kinds of gametes can be produced, AB and ab. As a result, the F$_2$ generation will have a monohybrid ratio. Three

times more plants will exhibit the two dominant characteristics than exhibit the recessive: (AB)/(AB), (AB)/(ab), (AB)/(ab), and (ab)/(ab). Therefore, instead of the expected 9:3:3:1 ratio characteristic of independent assortment in a dihybrid, the phenotypic ratio will be 3:1. A linkage group is the number of genes that tend to be inherited together. Although more than 360 different sets of alleles in corn are known, there are only 10 linkage groups, which represent the 10 pairs of chromosomes common to corn.

CROSSING-OVER

Even though complete linkage is known in some organisms (male *Drosophila*, for example), the genes originally found linked on the same chromosome do not always remain that way. There is a tendency for the genes on the same chromosome to become separated during the tetrad stage of meiosis as a result of a process of crossing-over, which is evident cytologically by the chiasma.

The process of crossing over is illustrated in Fig. 18-14. In the tetrad stage of meiosis four linked genes are on the same chromosome. If we assume that they are four dominant ones, A, B, C, and D, and that the homologous chromosome has the four recessive alleles, a, b, c, and d, then during the diplotene stage of the first prophase each has divided to form a tetrad. At diakinesis there is evidence that they may cross over so that, as a result, one chromatid will have the four

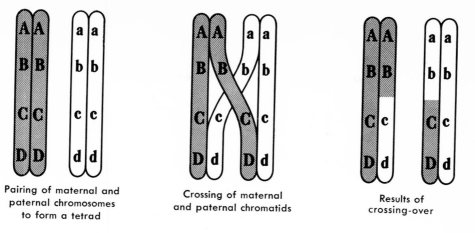

| Pairing of maternal and paternal chromosomes to form a tetrad | Crossing of maternal and paternal chromatids | Results of crossing-over |

Fig. 18-14. Diagram of crossing-over.

dominant genes, one the four recessive, one the two dominants and two recessives, and one the other two dominants and two recessives. The genetic consequence of crossing-over is the transfer of hereditary material from one chromosome to its homolog. Two of the chromosomes (in either spores or sex cells) will have the genes that were formerly linked to the other pair. Should the linked genes on one chromosome be those that are responsible for disease resistance and late seed production and the others responsible for no resistance and early seed production, it is possible, through crossing-over, to get plants that are disease resistant and early seed producers. Phenomena such as these are of interest to the plant breeder who is ever alert to the improvement of crop yields.

Although crossing-over between two or more genes may occur, it does so in only a small percentage of cases. The percentage of crossing-over between two genes is proportional to the distance between them. If genes are close together, the frequency of crossing-over is low; if they are far apart, the percentage of crossing-over is high. Thus it is possible to determine how far apart genes are from their crossover values and at the same time to define their positions on the chromosome relative to other genes.

The percentage of crossing-over between two genes is determined by testcrossing the F_1

generation. If genes B and C are linked in a homozygous individual and the parent containing them is crossed with another homozygous individual having the recessives b and c, the F_1 generation will then have the genotype (BC)/(bc). When an organism in which linkage is suspected is crossed with the homozygous recessive, the offspring having the linkage groups of the parent will occur in numbers greater than 50% (if all were to appear in equal numbers, it would be an independent assortment). The following diagram illustrates how this phenomenon occurs:

Parents:	(BC)/(BC) × (bc) (bc)	
F_1:	(BC)/(bc)	
Backcross:	(BC)/(bc) × (bc) (bc)	
F_2:	(BC)/(bc)	Noncrossover 35%
	(Bc)/(bc)	Crossover 15%
	(bC)/(bc)	Crossover 15%
	(bc)/(bc)	Noncrossover 35%

If the genes were on separate chromosomes, independent assortment would prevail, and the offspring in the F_2 generation would appear as a ratio of $1:1:1:1$. When this ratio does not occur, it is known that the genes are linked. If only noncrossovers resulted, then complete linkage would exist.

CHROMOSOME MAPPING

The frequency of recombination between linked genes is used to construct the genetic

Test cross: ABC/abc × abc/abc

		Gametes	abc
A	No crossover	{ ABC { abc	ABC/abc abc/abc
	Single crossover Region I	{ Abc { aBC	Abc/abc aBC/abc
	Single crossover Region II	{ ABc { abC	ABc/abc abC/abc
	Double crossover	{ AbC { aBc	AbC/abc aBc/abc

Parental chromosomes   **Gametes**

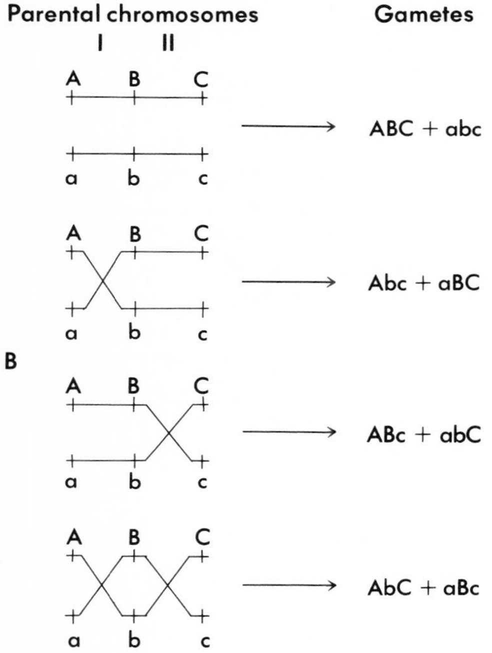

Fig. 18-15. Three-point testcross involving single and double crossovers. **A,** Possible offspring, **B,** Mechanisms of gamete formations.

map of a chromosome. The genetic map indicates the location of genes on a chromosome with respect to one another. The **map unit** is based on the frequency or recombination between two genes, 1% recombination is equal to one map unit. Genes are mapped in diploid organisms by using a technique called the three-point testcross.

The nature of the three-point testcross is illustrated by the following example (Fig. 18-15). An organism contains three very closely linked alleles, A/a, B/b, and C/c, with ABC dominant to abc. The triple heterozygote (ABC/abc) is crossed with the homozygous recessive (abc/abc). The homozygous recessive will produce only one kind of gamete, whereas the triple heterozygote will produce eight varieties.

The parental gametes are produced as a consequence of no crossing over and will represent the most frequently obtained data. A crossover in the region between A and B designated "region I," will produce gametes designated "region I crossovers." A crossover between B and C will produce region II crossover gametes. A double crossover simultaneously in regions I and II will produce double crossover gametes. The map distance between genes A and B is equal to the frequency of crossovers in that region. Crossovers in region I are equal to the sum of the single crossovers in the region plus the double crossovers. Total crossovers in region II are also computed in the same manner. The total crossovers in a region divided by the total progeny will equal the frequency of recombination for that region. One percent recombination is equal to one map unit. Once the frequency of crossing over or recombination for a region has been established, it is possible to determine whether crossovers in one region interfere with crossovers in the adjacent region or vice versa. For example, let us assume that the frequency of crossing over in region I is equal to 0.5 and for region II it was 0.4. The expected frequency of the double crossover event would be equal to the product of the individual frequencies ($0.5 \times 0.4 = 0.2$). A ratio comparing the actual frequency (that actually counted in the data) to the expected frequency of double crossovers is equal to the **coefficient of coincidence.**

Coefficient of coincidence =

$$\frac{\text{Actual-frequency double crossover}}{\text{Expected-frequency double crossover}}$$

If the coefficient of coincidence is equal to 1, no interference has occurred, less than 1 is positive interference and greater than 1 is negative interference.

HYBRID VIGOR

Frequently crosses between two supposedly different species or varieties of the same species produce offspring more vigorous than either parent (for example, a mule).

Although it is not always true in hybrids, there are some outstanding examples of **hybrid vigor,** or **heterosis.** Among those plants exhibiting heterosis is hybrid corn (Fig. 18-16), which is widely grown on farms throughout the country. There are several explanations (theoretical, however) for this increased vigor of hybrids. It may be that

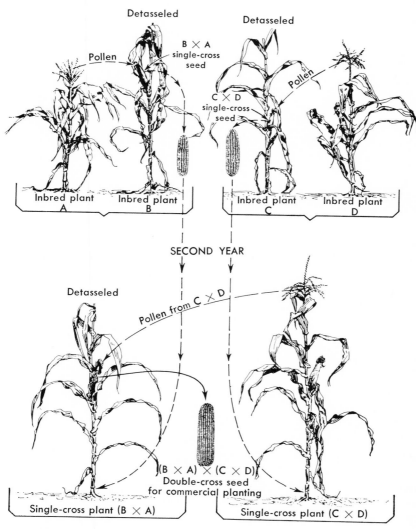

FIRST YEAR

SECOND YEAR

Fig. 18-16. Method of producing new hybrids of corn by controlling fertilization. The first year 2 plants are crossed, each with desirable characteristics, and the second year, another cross is made by use of offspring of the 2 first year's plants. Such hybrids show heterosis. (USDA photograph.)

through a chance combination of genes, the hybrid has the better genes from the two parents in which minor homozygous defects appeared. In the hybrid these defects are covered up by normal, dominant genes from each parent. Another explanation is that a heterozygous makeup is in itself an advantage, conferring greater survival rate on the bearer. A variety of genes may contribute to the plant's greater versatility in thriving under different environmental conditions.

MUTATION

Mutations are also responsible for variations in an organism. They are the result of sudden heritable changes that are not explained by either hybridization or environmental factors. Mutations may be slight and insignificant or so distinct and striking that they often seem to be new varieties or even new species.

Anything that will disturb the precise duplication of genes during meiosis will also have its effect on heredity. A mutation is any heritable change in the genetic material. Most genes have a defined mutation rate; in some this may be as high as 1 in 2000 or as low as 1 in several million gametes in plants. Anything that alters the mutation rate is referred to as a **mutagenic agent.** The mutation rate can be increased by chemicals such as mustard gas, extremes of temperature, or radiations ranging from ultraviolet light to X rays and the radiations produced by radioactive substances. Many mutations are distinctly harmful, others have little or no obvious importance in survival, and a few are obviously beneficial. Finally, most mutations are recessive; therefore they may not appear for several generations.

LETHAL GENES

One of the most harmful characteristics a green plant may have is the inability to produce chlorophyll. Seedings without chlorophyll will die, since they cannot manufacture their own food. Inability of green plants to produce chlorophyll is caused by the presence of a **lethal gene.** It is a recessive characteristic so that plants heterozygous for it are phenotypically normal. However, when they are self-pollinated, they will produce seeds in a ratio of 3 normal to 1 albino. The albinos die early from starvation.

MULTIPLE FACTORS (POLYGENES)

In all the genetic crosses illustrated, the expression of the characteristic was governed by a particular pair of genes. Many phenotypic expressions are the result of a series of genes known as **multiple genes,** in which each gene has its own effect, which is frequently cumulative. The most striking effect of multiple gene inheritance is that some of the offspring may be more extreme than either parent. If a cross is made between two varieties of wheat, one having red kernels and the other white, the F_1 offspring may be intermediate in color. The F_2 generation, however, will vary in intensity of color from red to white, but most of the offspring will be intermediate. The range may be from red, like the original red parent, to lighter red, to pink, to pale pink, and to white. The number of classes will be dependent on the number of genes involved. Should one sixteenth of the offspring be red, four sixteenths light red, six sixteenths pink, four sixteenths pale pink, and one sixteenth white, it then means that two pairs of factors are involved, as shown in the following diagram:

Parents:	$R_1R_1R_2R_2$ $\times$ $r_1r_1r_2r_2$		
	red	white	
F_1:	$R_1r_1R_2r_2$		$\Big\}$ $1/16$
	pink		
F_2:	$R_1R_1R_2R_2$	red	
	$R_1R_1R_2r_2$	light red	
	$R_1r_1R_2R_2$	light red	$\Big\}$ $4/16$
	$R_1R_1r_2r_2$	light red	
	$R_1R_1r_2R_2$	light red	
	$R_1R_1r_2r_2$	pink	
	$r_1r_1R_2R_2$	pink	
	$R_1r_1R_2r_2$	pink	
	$r_1R_1r_2R_2$	pink	$\Big\}$ $6/16$
	$R_1r_1r_2R_2$	pink	
	$r_1R_1R_2r_2$	pink	
	$r_1r_1R_2r_2$	pale pink	
	$r_1R_1r_2r_2$	pale pink	$\Big\}$ $4/16$
	$R_1r_1r_2r_2$	pale pink	
	$r_1r_1r_2R_2$	pale pink	
	$r_1r_1r_2r_2$	white	$1/16$

Each gene contributes to the ultimate color of the seed. The more red genes present, the darker will be the final color of the seed. If one plant out of 64 were as extreme as either original parent, it would indicate the action of three pairs of genes; if one plant out of 256 were extreme, four pairs would be indicated, and so on.

CHROMOSOMAL VARIATION

All the cells of a plant body have, as a rule, an identical number of chromosomes, and the genes are arranged in linear order as shown by studies of linkage groups. However, there are some exceptions. Some genes are known to mutate, and the chromosomes undergo certain rearrangements. To do this, the chromosomes must break. Sometimes this occurs without outside influence; other times chemicals or ionizing radiation causes the changes. The kinds of variation are summarized in the following outline:

I. Changes in numbers of chromosomes
  A. Changes involving whole sets of chromosomes
    1. Haploid (n) in which n is monoploid or basic number; each chromosome represented singly
    2. Polyploid in which each chromosome represented by more than two homologs (Fig. 18-17)
      Triploid $(3n)$ Pentaploid $(5n)$
      Tetraploid $(4n)$ Hexaploid $(6n)$
      a. **Autopolyploid** (multiples of diploid but of single homozygous or pure strain)
      b. **Allopolyploid** (multiples of hybrid derived from two different diploids)

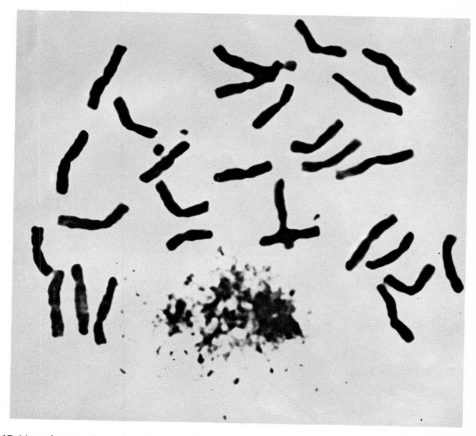

Fig. 18-17. More than two homologs for each chromosome are apparent in this polyploid oat. Polyploid crosses increased resistance to crown rust in oats. (USDA photograph.)

Fig. 18-18. Chromosome polyploidy may sometimes be recognized by flower size, as in this case. The tetraploid (4*n*) lily flower on the right is larger than the diploid (2*n*) lily flower on the left. (USDA photograph.)

Fig. 18-19. Fruit may be larger as a result of polyploidy. For example, several species of the blueberries (*Vaccinium* spp.) of eastern United States are well-known polyploid complexes in which enhanced fruit has resulted. (USDA photograph.)

B. Changes involving numbers of chromosomes within set, **heteroploidy**
   1. **Monosomic** (loss of one chromosome from set; in diploid, $2n - 1$)
   2. **Polysomic** (addition of one or more chromosomes to set)
      Trisomic $= 2n + 1$
      Tetrasomic $= 2n + 2$
      and so on
   3. **Nullisomic** (loss of both chromosomes of pair)
II. Changes involving number of genes or their arrangement and position on chromosome
   A. Changes involving number of genes
      1. Deficiency or deletion (loss of one or more genes)
      2. Duplication (addition of one or more genes so that same gene is repeated in haploid complement)
   B. Changes involving arrangement of genes
      1. Translocation (exchange of parts between nonhomologous chromosomes to form two new ones)
      2. Inversion (180-degree rotation of block of genes within chromosome)

Of all the chromosomal aberrations, **polyploidy** is the most adaptable for the improvement of economic plants. The most common polyploids are tetraploids (Fig. 18-18). Polyploidy may influence the phenotype of the individual as well as its genetic and evolutionary potentialities. One of the most common effects of polyploidy is increase in cell size and frequently in size and vigor of the plant. However, the direct effects of increasing the chromosome numbers are not always predictable. Some annual plants may become perennials, the polyploids may be less fertile than diploids, and other physiological factors may be changed. Occasionally it is possible to make fertile plants from sterile hybrids through polyploidy. Blueberries are a well-known polyploid complex (Fig. 18-19).

Polyploidy has been studied extensively by plant geneticists, who have learned that it originates through the suppression of the meiotic process. In other words, the gametes have the unreduced number of chromosomes. Polyploidy may also arise through failure of cell division but not of chromosome division. In nature these phenomena are accidental, being favored by extremes of temperature, particularly low temperatures.

Fig. 18-20. Colchicine, used by plant geneticist to induce hybridization after crossing a tobacco plant having 48 chromosomes with one having 24 chromosomes. (USDA photograph.)

Polyploids are more frequently found at high altitudes than at low ones. Sometimes scientists use a drug called **colchicine** to produce polyploidy in plants. Colchicine is an alkaloid derived from the autumn crocus (*Colchicum autumnale*) and is commonly used by physicians to alleviate gout. It is a deadly poison except when used in minute amounts. It can be applied to plants as a paste (Fig. 18-20) or in some other form to inhibit the formation of spindles in mitotic cells, thus inducing polyploidy.

CYTOPLASMIC INHERITANCE

In plants some characteristics are inherited through the cytoplasm instead of through the nucleus. Since an egg contains much more cytoplasm than a sperm, **cytoplasmic inheritance** is usually maternally dominated. Inheritance through the cytoplasm is recognized by its maternal effect, particularly when plastids are involved. There is a yellow-leaved variety of primula that contains less chlorophyll than the normal variety. The seeds of this yellow-leaved form transmit the characteristic but the pollen does not. There are many instances supporting the interpretation that cytoplasmic inheritance is involved in maternal cytoplasm.

It is now firmly established that the basis for cytoplasmic inheritance is the DNA found in the chloroplasts and mitochondria. Al-

though it has been established that both of these organelles possess the requirements for genetic autonomy, their structure and function depend to a large extent on nuclear genes. In barley, 86 nuclear genes that control the evelopment of the chloroplast have been found. In yeast, nuclear genes that control the development and function of the mitochondria have been found. Nuclear genes control at least seven components of the photosynthetic system. It has been shown that the enzyme in tobacco, involved in carbon dioxide fixation (ribulose-1,5-diphosphate carboxylase) is partially determined by a cytoplasmic gene presumably in the chloroplast and partially by a nuclear gene (Fig. 18-21). It also appears that the synthesis of the chloroplast membrane is dependent on both the chloroplast and mitochondrial genes.

Many investigators are currently studying the genetic systems of chloroplast and mitichondria. These organelles contain some of the genetic information necessary for the cell but it does not appear as if they are autonomous.

CHEMICAL NATURE OF HEREDITY

We have seen in the previous section that genes, the units of heredity, are carried on chromosomes and behave in predictable fashion in controlled crosses (segregation of alleles and independent assortment). As il-

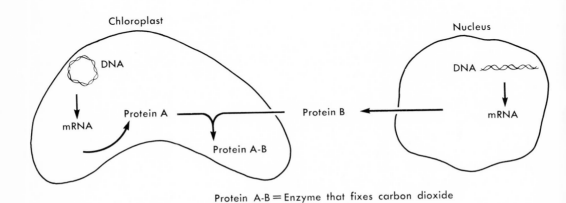

Protein A-B = Enzyme that fixes carbon dioxide

Fig. 18-21. Diagram depicting the cooperation of nuclear and chloroplast DNA in synthesizing the enzyme ribulose-1,5-diphosphate carboxylase, which fixes carbon dioxide in the chloroplast.

lustrated in the example of epistasis, genes seem to act by determining the presence or absence of a particular enzyme. It is very necessary here to understand that the gene is not the enzyme and vice versa. What then is the chemical nature of the gene? To better understand its action and function, one needs to understand the chemical and physical nature of the heredity material. Chromosomes, which carry genes, are found in the nucleus of the cell. The chromosomes consist of protein (40%), DNA (60%), and trace amounts of RNA. Early investigators focused their attention on the two most likely prospects for the genetic material, protein and DNA. Both are very large macromolecules that have the capacity to carry all of the information necessary for an organism. There were certain lines of indirect evidence that supported DNA as the genetic material. DNA is found primarily in the nucleus of a cell, whereas protein is abundant both in the nucleus and the cytoplasm. A direct correlation exists between the amount of DNA in a somatic cell and that in a gamete of the same species, the gamete always having one-half the amount of DNA. The base composition of the DNA (adenine/thymine : guanine/cytosine ratio) is the same for all of the cells of an organism, but it differs from species to species.

Direct evidence for DNA as the genetic material came from three different experiments, the **transformation** of *Diplococcus pneumoniae,* the infection process of bacteria virus (phage T_4), and the mode of infection of **tobacco mosaic virus (TMV).**

Two strains of *Diplococcus pneumoniae* infect mice. One strain is virulent (causes infection, disease, and death), and the second strain is avirulent (enters the organism but does not cause disease and death). When a mouse is infected with the virulent strain, the mouse dies and live virulent bacteria can be isolated from the dead carcass. Infection of a mouse with an avirulent bacterium does not cause the mouse any undue stress or discomfort. When heat-killed virulent pneumococci are injected into the mouse, it does not cause disease and death. However when heat-killed virulent plus normal avirulent bacteria were injected, the mouse died and from the carcass virulent bacteria were isolated. The heat-killed virulent cells contained a transforming principle that changed avirulent to virulent cells. Agents that specifically destroyed protein and RNA did not destroy the **transforming principle,** but agents destructive specifically to DNA did destroy it. It was obvious then that DNA was the transforming agent (Fig. 18-22). Second, the mode of growth of bacterial virus T_4 is illustrated in Fig. 18-23. The bacteriophage is a simple particle that must reproduce in the cytoplasm of a bacterial cell, specifically *E. coli.* The phage is composed of an outer covering (capsid) of protein and an inner core of DNA. It attaches to the bacterial cell tail first and injects the DNA into the bacterial cytoplasm. By the use of radioactive isotopes of sulfur (which label protein) and of phosphorus (which label DNA) it was demonstrated that only phage DNA entered the bacterial cytoplasm and the protein stayed on the surface. Once in the cytoplasm, the phage DNA directs the synthesis of new intact virus particles (millions of them) so that the cell ruptures and releases them to the surrounding medium. The fact that DNA, not protein, directs the synthesis of new virus particles is direct evidence that DNA is the basis of the genetic material. The third line of evidence was work done with tobacco mosaic virus (TMV). TMV is a tobacco plant pathogen. It is a simple particle composed of a protein outer coat and an inner core composed of RNA. The many strains of TMV are distinguished by their protein coats. The virus can only reproduce within the cytoplasm of a tobacco leaf cell analogous to the phage-bacteria relationship. It is possible to synthesize virus particles that contain the RNA of one strain and the protein coat of another (Fig. 18-24): RNA 1/protein 2 or RNA 2/protein 1. When particles of RNA 1/protein 2 constitution are used to infect tobacco leaves, the intact viruses produced as a result of the infection are RNA 1/protein 1. When RNA 2/protein 1 particles were used for infection,

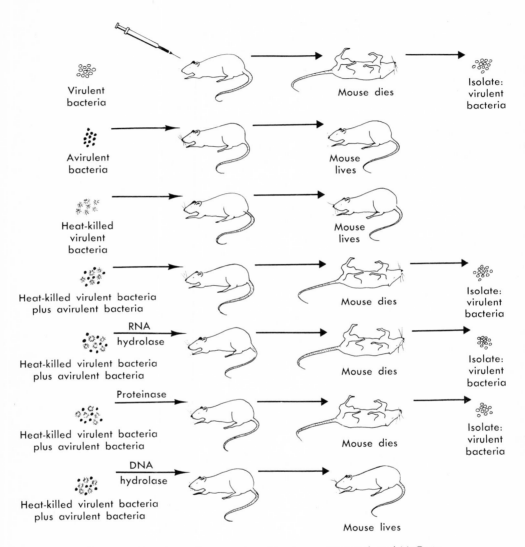

Fig. 18-22. Transformation experiment by Avery, MacLeod, and McCarty.

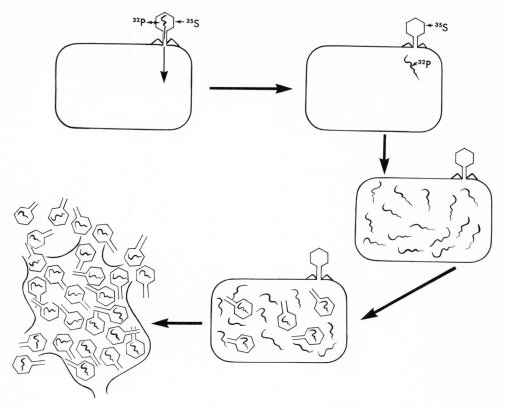

Fig. 18-23. Life cycle of bacteriophage T_2.

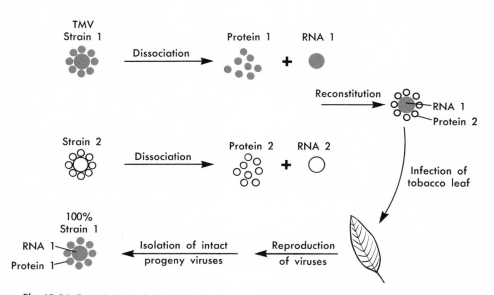

Fig. 18-24. Experiments depicting the control of tobacco mosaic virus structure by RNA.

the particles produced were RNA 2/protein 2 (Fig. 18-24). It is evident that it is the nucleic acid RNA in this case that is carrying the genetic information for TMV. It is clear then, from the direct evidence, that DNA is the molecule that carries the genetic information in DNA-containing organisms and RNA carries it for the rare RNA viruses such as TMV and polio.

STRUCTURE OF THE DNA MOLECULE

Using a technique known as X-ray diffraction, Watson, Crick, and Wilson established the detailed physical structure of the DNA molecule (Fig. 18-25). They found that DNA was a double-stranded structure held

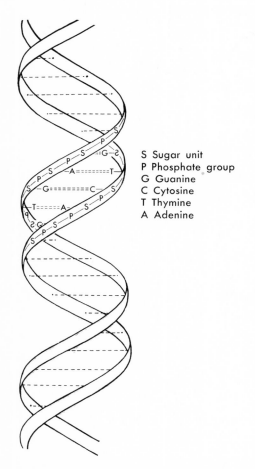

S Sugar unit
P Phosphate group
G Guanine
C Cytosine
T Thymine
A Adenine

Fig. 18-25. Model of the double-helix DNA molecule as described by Watson and Crick.

together by hydrogen bonds. The hydrogen bonds were specifically between the bases adenine and thymine, guanine and cytosine. The sugar deoxyribose phosphate formed the backbone of each strand.

Much experimental evidence supports the concept that genes exert control over cellular function by directing the synthesis of proteins with a specific amino acid sequence. The amino acid sequence in proteins determines their function. In plants there are literally millions of proteins containing about 20 amino acids. How can four bases then code for 20 amino acids? Even the bases taken two at a time (4^2) would be insufficient. Then a minimum of three bases must be necessary to code for the incorporation of all 20 amino acids. This mathematical reasoning proved to be correct as evidenced by many subsequent experiments. A sequence of three bases in DNA encode for the insertion of a single amino acid into cellular protein (Fig. 18-26). The sequence of the three-base triplet is a **codon,** and we will discuss later how it directs the synthesis of proteins.

DNA REPLICATION

Once the nature of the genetic material was firmly established, investigators began to unravel the mode of DNA replication. Based on their model of the DNA molecule, Watson and Crick postulated that the two strands separated and were used as a template for the synthesis of complementary strands (Fig. 18-27). The two new DNA molecules, being exact copies of the parental DNA, consist of one parental strand and one new one. This is **semiconservative replication,** and evidence in support of this concept came from two experiments. First, Arthur Kornberg (1957) isolated the enzyme DNA-dependent DNA polymerase, which catalyzed the synthesis of new DNA using parental DNA as the template. Second, an elegant experiment by Messelson and Stahl (1958) proved that the mode of replication was semiconservative. The experimental organism used was E. coli, the enteric bacillus. It was grown in a synthetic medium using ammonium chloride la-

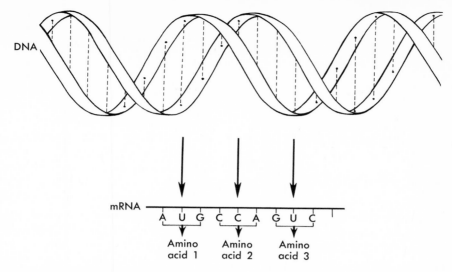

Fig. 18-26. Diagram depicting triplet nature of the genetic code. The base sequence in DNA determines the base sequence in mRNA, which in turn determines the sequence of amino acids.

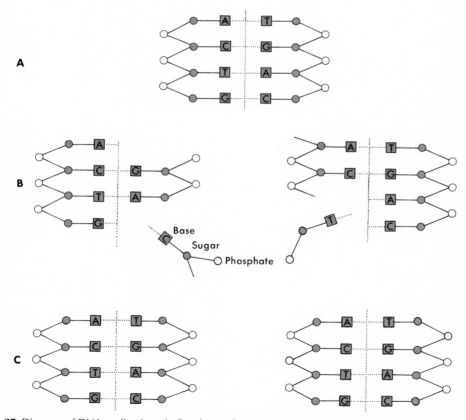

Fig. 18-27. Diagram of DNA replication. **A,** One loop of molecule. **B,** Molecule splits and bonding of nucleotides occurs, through enzyme action, to form new loops. **C,** Two identical helices formed by each half of the original helix, completing replication.

345

beled with the heavy isotope of nitrogen ^{15}N. After many generations of growth in this isotope, almost all of the bacterial DNA was labeled with ^{15}N. Using a density-gradient procedure in the ultracentrifuge, one can identify the DNA labeled with ^{15}N as a function of its density, ^{15}N-^{15}N. After the bacteria had grown many generations in the ^{15}N label, some was removed and the DNA labeled with the isotope was identified in the ultracentrifuge. The remaining cells were transfered to a medium containing the normal nitrogen atom ^{14}N. Now, when the bacteria replicated their DNA, the nitrogen incorporated would be ^{14}N and the DNA would have a lighter density when compared to the parental DNA in the ultracentrifuge. The fact that the parental DNA (heavy density) disappeared completely and an intermediate density DNA appeared was proof that a ^{14}N-^{15}N DNA hybrid was formed, and the mode of replication was semiconservative (Fig. 18-28).

GENETIC CONTROL OF PROTEIN SYNTHESIS

Genes composed of DNA carry the genetic information and transmit that information from the nucleus of the plant cells to the cytoplasm where it directs the synthesis of proteins, either enzymatic or structural units, thus giving the cell a specific observable characteristic that we call a phenotype. Let us trace the genetic control of protein synthesis from the DNA in the nucleus to the protein in the cytoplasm (Fig. 18-29). The DNA in the cell nucleus is a long double helix found associated with protein in the chromosome. The double helix is held together by the specific hydrogen bonding of adenine to thymine and guanine to cytosine. An enzyme, **RNA polymerase,** unzips the strand, breaking the hydrogen bonds and exposing the bases. Using the single strand of DNA as a template, a strand of RNA is synthesized with a base sequence complementary to that of the template DNA. Uracil replaces thymine in RNA so that the pairing would be adenine to uracil and guanine to cytosine.

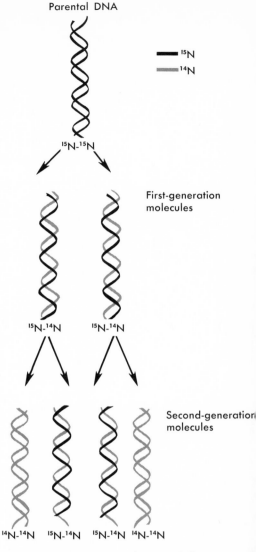

Fig. 18-28. Distribution of parental ^{15}N atoms to daughter DNA molecules. The strands of parental DNA separate at first division and remain separated in subsequent divisions.

The RNA molecule that is synthesized using DNA as a template is **messenger RNA (mRNA).** The specific base sequence of mRNA is determined by the base sequence of the template DNA. The mRNA leaves the nucleus and attaches to the ribosomes in the cytoplasm. The cytoplasm contains **transfer**

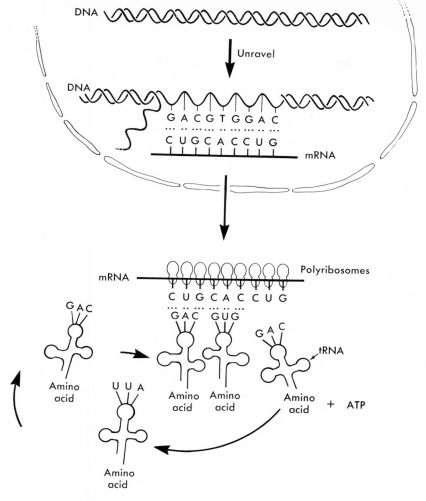

Fig. 18-29. Diagram of the genetic control of protein synthesis.

RNA, a species of RNA that is different from mRNA and ribosomal RNA. Transfer RNA was isolated and characterized by Holley in 1964 (Fig. 18-30). Each transfer RNA contains an anticodon loop that consists of three projecting bases. It also contains a portion that will combine with a specific amino acid. For each amino acid that is to be incorporated into protein, there is a specific tRNA and the reaction is catalyzed by the enzyme amino acyl RNA synthetase. The first step is the reaction of the amino acid with ATP in order to raise its energy level. This produces

an energized amino acid that is attached to its specific tRNA. This complex now moves to the ribosome mRNA complex where it fits into a place on mRNA complementary to the bases on the anticodon loop. Thus the specific amino acids are brought adjacent to one another and peptide bonds are formed between them in a specific sequence.

In summary the sequence of bases in the DNA determines the sequence of bases in the mRNA. The sequence of bases in mRNA determines which tRNAs will be brought adjacent to one another and thus which amino

347

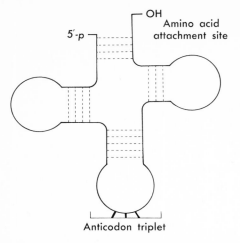

Fig. 18-30. Diagram of the transfer RNA molecule as it was isolated by Holley. 5'-*p*, 5'-Phospho terminus.

acids will be adjacent and form peptide bonds.

REGULATION OF GENE ACTION

The zygote has a specific inherited genotype or the potential for the development of a great many characteristics. This zygote will normally develop from a cell with a total potential. This process of development or differentiation is a gradual one and involves a selective shutdown of genes so that the fully differentiated cell has only about 15% of the total genotype functioning. This is not a random shutdown but a very selective one—a definite regulation. It is obvious that even in a fully differentiated cell not all of the DNA functions all of the time. There are specific molecular regulatory mechanisms in operation, many of which remain obscure.

Jacob and Monod through a series of outstanding experiments with the enteric bacillus *E. coli* have proposed a model for the regulation of gene expression at the molecular level in prokaryotes. This model is called an **operon,** and it consists of two types of genes, structural and controlling. **Structural genes** function by determining the primary structure of protein with a specific function, for example, an enzyme. **Controlling genes** are those that function by controlling the transcription of mRNA along the structural genes. The controlling genes that were characterized were the operator (O gene), the promoter (P gene) and the regulator (i gene). The structural genes are located adjacent to the operator and promoter while the regulator gene is in a site a few units away (Fig. 18-31, *A*). As we have stated earlier in this chapter, to transmit the message in the structural gene to the cytoplasm, mRNA must be synthesized along the DNA in the structural gene. The promoter gene is the site for the initiation of mRNA synthesis. When the promoter gene mutates (Fig. 18-31, *F*), mRNA is not transcribed for the structural genes under its control; thus their message is not sent to the cytoplasm for phenotypic expression. The regulator gene codes for, and causes the production of, a protein molecule called a repressor.

The operator gene functions by combining with the specific **repressor protein** (Fig. 18-31, *B*). When the repressor molecule is combined with the operator gene, mRNA cannot be synthesized along the adjacent structural genes. When a substance called an **inducer** is present in the cytoplasm (Fig. 18-31, *C*), it will combine with the repressor protein, thus

Fig. 18-31. Lactose operon. **A,** Map of the lactose operon showing the location of the structural and O, P, and i genes. **B,** Normal noninduced operon. Regulator gene i produces a repressor molecule, which combines with the operator gene; thus transcription of the structural genes is blocked. **C,** Normal induced system. An inducer combines with a repressor molecule, preventing repressor-operator combination, and the structural genes are transcribed. **D,** Regulator gene mutation. An altered repressor is produced; it does not combine with operator; therefore structural genes are transcribed. **E,** Operator gene mutation. An altered operator gene loses its affinity for repressor, and structural genes will be transcribed. **F,** Promoter gene mutation. Promoter gene has lost affinity for the enzyme that synthesizes mRNA. Therefore transcription will not occur.

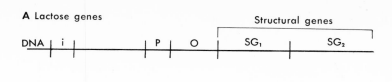

A Lactose genes

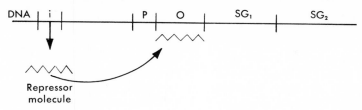

B Normal noninduced system

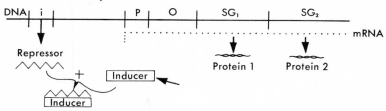

C Normal induced system

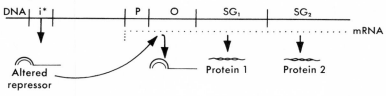

D Regulator mutation

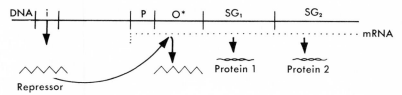

E Operator mutation

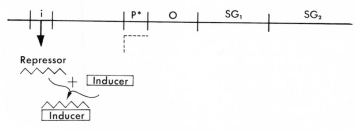

F Promotor mutation

Fig. 18-31. For legend see opposite page.

preventing the repressor from combining with the operator, and the structural genes under the operator's control will be transcribed. This is called an "inducible system." When the regulator gene mutates (Fig. 18-31, *D*) so that a nonfunctional repressor molecule is produced (one that will not combine with the operator), the structural genes are continually transcribed and this is referred to as a "constitutive system". If the operator gene mutates (Fig. 18-31, *E*) so that it can no longer combine with the normal repressor molecule, this is also a constitutive mutant. When the promoter gene mutates, mRNA will not be transcribed and the structural genes will not be expressed under any circumstances (Fig. 18-31, *F*).

PROBLEMS

1. Purple-stem tomato plants are crossed with those having green stems. The F_1 generation have purple stems. The F_2 generation produced a ratio of 3 purple stems to 1 green. What is the genetic basis for the inheritance of stem color in tomatoes? Diagram all crosses, showing gamete formation, genotype, and phenotype.

2. A white, disc-shaped summer squash is crossed with a yellow, sphere-shaped one. The F_1 generation is all white and sphere shaped. The F_2 generation produced the following results:

Phenotype	Number
White sphere	181
White disc	58
Yellow sphere	60
Yellow disc	20

Explain the genetic basis of the above results. Diagram all crosses, showing gamete formation, genotype, and phenotype.

3. In four-o'clocks flower color is inherited as an incomplete dominant, RR = red, Rr = pink, and rr = white. Another segregating gene pair is stem length. Long stems, L, is dominant to its allele short, l. From a cross between a red, short plant and a white, homozygous long one, predict the kinds of phenotypes and their frequencies in the F_1 and F_2 generations.

4. When purple peas are crossed with white ones, the F_1 generation is all purple. The F_2 generation is found in the ratio of 9 purple to 7 white. What is the genetic basis for coloring in peas?

5. Assume three pairs of closely linked alleles, A/a, B/b and C/c, with ABC dominant to abc. Gene A is 10 map units from gene B, and B is 5 map units from gene C.

$$A \xleftrightarrow{\text{10 map units}} B \xleftrightarrow{\text{5 map units}} C$$

The coefficient of coincidence for this region is equal to 1. In a three-point testcross, 1000 progeny were counted. Predict the kinds and frequencies of phenotypes formed.

QUESTIONS AND PROBLEMS

1. Why is meiosis related to the study of genetics?
2. Define Mendel's laws of segregation and independent assortment.
3. What is epistasis? Give an example.
4. What is your concept of a gene?

5. What is the direct proof for nucleic acid as the genetic material?
6. How does DNA control the primary structure of proteins?
7. What is a mutation?
8. Distinguish genotype from phenotype.
9. How does a controlling gene differ from a structural gene?

DISCUSSION FOR LEARNING

1. Discuss cytoplasmic inheritance and the possibility of the mitochondria and chloroplast being autonomous.
2. Discuss the possible use of genetics to improve agriculture.

ADDITIONAL READING

Cohen, S. S. 1973. Mitochondria and chloroplast revisited. Am. Sci. **61**:437, July. (A review of mitochondria and chloroplast genetics.)

Crick, F. H. 1962. The genetic code. Sci. Am. **207**:66, Oct. (An explanation of the nature of the genetic code.)

Levine, L. 1973. The biology of the gene. The C. V. Mosby Co., St. Louis. (A recent elementary textbook.)

Miller, O. L. 1973. The visualization of genes in action. Sci. Am. **228**:34, March. (A comparison of electron micrographs with diagrams of the genetic control of protein synthesis.)

Rich, A. 1963. Polyribosomes. Sci. Am. **209**:44, Dec. (Discussion of the relation of ribosome clusters to protein synthesis.)

Smith, H. J. 1975. Chloroplast and protein synthesis. Nature **254**:13, March 6. (A summary of the recent developments in chloroplast portein synthesis.)

Spiegelman, S. 1964. Hybrid nucleic acids. Sci. Am. **210**:48, May. (Discussion of the formation of mRNA.)

Srb, A. M., Owen, R. D., and R. Edgar. 1965. General genetics. W. H. Freeman & Co., San Francisco. (Textbook that explains basic concepts of classical genetics.)

Yanofsky, C. 1967. Gene structure and protein structure. Sci. Am. **216**:80, May. (A discussion of the relationship between the sequence of nucleotides in DNA and the sequence of amino acids in protein.)

Fig. 19-1. The Galápagos Islands are populated by weird plants and uniquely tame animals. This scene, much as Darwin saw it, shows the typical arid shore vegetation of the island group. Darwin's studies here contributed greatly to his theory of evolution.

CHAPTER 19

Plant evolution

■ The process of evolution provides the only known mechanism for the preservation of life in an ever-changing physical environment. Flexibility tempered by the rigid code of the DNA molecule seems so simple on the one hand and so profoundly complex on the other: simple because of the few components of the double helix of DNA, and complex because of all the diverse organisms of today and times past. Programming at the cellular level through the genetic process determines the survival for reproduction of each member of a population. Those not programmed to cope with their effective environmental influences do not survive to reproduce. This determines the proportion of variants that are able to reproduce and the natural selection of those adapted to the environment. Once gross shifts in the programming have reached the point of reproductive isolation, species "reproduction," or evolution, has taken place.

Charles Darwin gazed upon a scene in the Galápagos islands (Fig. 19-1) that impressed him in two ways: the weird landscape, where common plants and animals were so changed as to be barely recognizable—sunflowers and cactuses became trees—and the continuity of these forms even though so remarkably changed. His thoughts crystallized many years later. His ideas and the discovery of the laws of heredity by Gregor Mendel have led the way in changing biology from a descriptive to a quantitative science.

The details of both the genetic and the evolutionary process are the most intricate of all biological phenomena, and some aspects are still little known. The combined knowledge of all genetic, biochemical, and physiological functions is needed to interpret the phenomenon termed "evolution." There is no doubt that it is the greatest conceptual scheme of science ever devised. In it we see a progression of biological systems from the replication of DNA to cell division, to reproduction of the organism, and finally to the multiplication of species. This seems to be the theme common to all aspects of biology.

EARLY HISTORY OF EVOLUTION THEORY

It took man hundreds, even thousands, of years of thought to untangle the complexities of the evolutionary process. It did not come easy.

Many early writings show that men have long speculated on their own origin and on life around them, in addition to learning how to plant. The search for useful plants, although the beginning of biological science, was not entirely utilitarian. The early development of magic followed by logical reasoning led to a confusing history of human thought. Woven into these thoughts was always the theme of change and development. Contrary to popular belief, the theory of evolution has a history as long as that of all biology. The ancient Chinese, the Babylonians, and the Egyptians speculated about the origin of life on earth. The Chinese in their ancient writings said that five elements were on the earth at the beginning and were the foundation of the universe—water, wood, fire, soil, and gold. Their anthropocentric

353

thinking is obvious. The universe had the elements that man needed. Not recognizing wood as a complex living material but knowing its value to their lives, these writers believed that this substance was so basic that it must have always been present. With these elements they were able to postulate that life came from positive and negative interactions of the primary elements.

The act of water putting out fire is a negative interaction, but the act of fire burning wood is a positive interaction. Water added to the soil produces life, and so on. Thus Confucians and Taoists had some idea of change, descent, complexity, and progress, from the least complex to the more complex.

In their legends of a Creator the ancient Egyptians based their ideas of the origin of life on a first cause. These concepts centered about an abyss of water out of which came life from a state of inertness. These same ideas were proposed in Genesis. The transition from nonliving to living substance was always associated with water.

In their writings the Babylonians allowed mystical powers to permeate the universe. With this as their theme they went on to develop a classification of organisms that was well advanced. Unfortunately, even with the many thousands of examples of Babylonian writing preserved, we know little about the evolutionary thoughts of these people.

The Israelites recognized plants for various reasons—as sources of food, as being harmful, or some as even sacred. Many of these observations were recorded on a plant product, papyrus, obtained from a tall sedge native to Egypt.

During the golden age of ideas man began to use his intellect to toy with abstract ideas and found this much to his liking. Thales (640?-546 B.C.) proposed laws about the origin of the universe, including various mathematical studies on the subject. He recognized the basic nature of water. The whole universe, he said, was composed of water, and he thought of it as but another substance that permits life, believing that the earth floated on water.

Anaximander (611-547 B.C.) proposed the limitlessness of space, which now opposes Einstein's theory of finite space. Anaximander believed that living things arose from mud and recognized an order in them— first plants, then animals, and finally man.

Anaximenes (588-524 B.C.) thought only in terms of spontaneous generation. He agreed with Anaximander on the concept of the limitlessness of space and believed that air was the universal medium. Heraclitus (ca. 540-475 B.C.) proposed the theory that fire produced the changes in the world, and as a result, all things were in constant motion or in a state of ceaseless change. So many of the concepts that we take for granted today were important and new ideas in early times.

Parmenides (born ca. 540 B.C.) thought of the world as immutable and uniform: as it always was, so it will be always. This concept led later thinkers such as Democritus (ca. 460-400 B.C.) to inquire into the nature of change and to propose the term "atom" as the stable element, but he postulated an infinite number in the universe.

Empedocles (495-435 B.C.) is often called the father of the idea of evolution. His elements were reduced to four—fire, air, water, and earth (Fig. 2-1). Change was brought about by two forces, attraction and repulsion. These forces acted in unison upon the four elements and in this way effected change. He also had a concept of adaptation for survival, which was basic to Darwin's theory—once the sexes were separate, species arose. This concept shows that the roots of the idea of evolution were well established, even in those early times. Empedocles also believed in the concept of order in nature—plants, animals, and man.

Anaxagoras (488-429 B.C.), during the same period as Socrates and Democritus, thought of matter as originating from an infinite number of "seeds." Living things sprang from atmospheric germs, not from terrestrial earth.

One reason the theory of evolution took so long to be recognized and accepted was that the basic ideas previously proposed were not absorbed by Aristotle (384-322 B.C.). If they had been, it is likely that the idea of evolution would have developed much sooner because Aristotle's texts were used exclusively for many hundreds of years. His classification of plants is essentially very close to that used today, but he did not directly classify animals and plants. Later compilers arranged his ideas into a classification. He knew that anatomy was necessary for classification or that classification was based on anatomy. He recognized the individual, the species, and the genus as used in the modern logical sense. This concept was, philosophically, part of his four predictables. He considered genera as a category, the members of which have many parts in common, and that it was not necessary to repeat all of these in a description. This was, in effect, the binomial system, but it was not again used consistently until 1753 when Linnaeus restated it. Linnaeus was a great reader of both Aristotle and the Bible and quoted from both extensively. The genus that Aristotle recognized is equivalent to the modern category of "class," but his categories could change, as do those used in logic. Now, under the Linnaean system, the various categories have been stabilized. Aristotle also recognized that species exist and that they have reality in nature, whether this is recognized or not.

Aristotle was a believer in spontaneous generation. His writings may give the idea that he was an evolutionist, but only when they are taken out of context. Certain statements he made might indicate that he had a real concept of evolution, but he did not. He arranged organisms in a seemingly evolutionary order, from the least complex to the more complex, which in modern terms is evolutionary. His relationship was mathematical. Even though he believed in the immutability of the species, he also believed that motion is imparted to the universe by an immaterial first cause. First causality is responsible for all change, he said, thus absolving himself from all credit for evolutionary thought.

DARWIN TO THE TWENTIETH CENTURY

The publication of Darwin's *Origin of Species* in 1859 marked the culmination of 20 years of thought on his part and centuries of slow accumulation of evidence. Charles Darwin (1809-1882) was responsible for changing the field of biology from an organized mass of unrelated facts into a unified science. Although he did not actually propose the theory of evolution, his theories were responsible for what is perhaps now the most unifying concept of biology—the application of the knowledge of the processes of evolution to understanding the significance of biological processes and organisms. This was a turning point for biology, from which issued modern biology based on the principle that all life on the earth has a common origin and operates under the same physical and chemical laws.

The nineteenth century saw the study of the geographical distribution of plants and the development of a knowledge of the relations between climatic and other environmental factors. The presence of a nucleus and chromosomes in a cell was established, and the method of cell division was discovered. The function of plant organs was determined, and so plant physiology as a recognized science was established.

Darwin's theory. In his early life Darwin unknowingly collected the facts he needed, as well as gained the broad experience necessary to see at a conceptual level the processes of life that he was to explain. His voyage around the world from December 1831 to October 1836 as naturalist and geologist on the *Beagle* was the primary source for this experience. During the voyage he took notes on everything that he saw, and these notes later formed the basis for several books published long before his most famous one. Also, he read the works of Reverend Thomas Robert Malthus (1766-1834) on population increase, which suggested to both Darwin and Alfred Russell Wallace (1823-1913) the idea of a struggle to survive. Darwin contributed to the scientific literature enough original

355

research to give him a firm position in the annals of biology even if he had never written the *Origin of Species*. His studies of the formation of coral reefs, of insectivorous plants, and of earthworms are still basic works, although they are seldom remembered because his later work so overshadows them.

At the same time Darwin proposed his theory, Wallace proposed a similar theory. Once again the time was ripe for discovery. One of the great tributes to scientific objectiveness was Wallace's act of withdrawing his paper in preference of Darwin's; thus he gave up a first place in history. The two men collaborated and pooled their ideas to their great credit. Wallace is now best known for his studies of plant and animal distribution.

When Darwin's famous book was published, the effect was immediate and widespread. The fashion of the time became the reading of Darwin, a daring act because to many it denied the existence of God. Surely if monthly book clubs had been in existence at that time, social pressure would have demanded that this special selection be read by all, and it probably would have outsold even such popular novels as our modern classic *Gone With the Wind*. Unfortunately, few understood the book, even those of today. Although it still remains one of the great books to be read by generally educated people, one should not assume that is represents current thinking on the subject.

Darwin lived to see his work scientifically accepted by most, but he felt the brunt of great public controversy at the same time. Nevertheless, Darwin and Newton are buried side by side in Westminster Abbey in London as recognized great men, great scientists, and great contributors to human knowledge. Doubtlessly the theories of both men number among the greatest of scientific achievements.

At the time Darwin's theory was first published, it was known as the "survival of the fittest." It should be noted that he never used the term "evolution." His proposal will be discussed in more detail in this chapter. The following are his main points.

SUMMARY OF THE DARWINIAN THEORY OF EVOLUTION

1. Species change or mutate.
2. There is descent with modification.
3. Variation in individuals permits change by selection.
4. Therefore the best-fitted individuals in a given environment survive; the least fitted perish; "the survival of the fittest."
5. Change is always toward more complex forms, even when parts are reduced or lost through parasitism.
6. Therefore, once a change is made, the new form cannot change back to its original state.

Although each of these concepts had been vaguely presented by others, no one had previously brought them together in one book documented with the detail and logic of Darwin.

In summary, Darwin laid the foundation for all future evolutionary thought. Since then, biologists have been elaborating and refining his theory by adding facts as well as concepts, until today it is recognized as one of the basic biological processes.

EVOLUTIONARY PROCESS

Evolution is the process involving changes in the kinds of organisms from the first life on earth to the present time. It is a continuing process that, by itself, characterizes life on the earth. The actual forces that permit these mechanisms to function is an interaction of chromosomes and the environment.

Assuming, as we must, a common ancestor for all life, it then becomes obvious that all life comes from life. This took a long time to prove, but it is now a scientific fact. It naturally follows then that if life originated only once and if all life comes from life, that all life must be related in an unbroken chain from the present back to the time of its origin. All organisms that ever lived had parents, which in turn had parents, and so on back in an ever-narrowing genealogical line to those first macromolecules that hovered on the brink between living organisms and nonliving chemicals. This is sometimes referred to as a vast sea of protoplasm that forms a network through time to the present. It is described as a network because it is not a series of absolutely integrating units but is organized into supraindividual units called kinds. In the case of bisexual kinds the unit of operation is the breeding population, and in the case of asexual organisms the unit is generally a colony. One or more such morphologically and physiologically similar units capable of producing additional similar units constitutes the species. Based on this is the theory of **homology** by inheritance. Similarity of form and structure is obvious in orderly arranged units of all kinds. Paper clips can be classified (or arranged in orderly units) by grouping like with like. It is validly assumed that parts are homologous, and, therefore, possessors of homologous parts have a common ancestor. This is the theory of homology and the essence of the theory of evolution. Similarity of parts, but not necessarily similarity of function of those parts, indicates relationships. All the information in Appendix B on life cycles in this book is based on this assumption and relates the various stages in the life cycle of plants. Sporophytes, for instance, are homologous throughout, even though in the primitive condition they are limited to single cells producing spores and in the higher plants they form the entire vegetative body. The pitfalls in dealing with the theory of evolution come only in the interpretation of the concept of similarity-by-homology as opposed to the concept of similarity-by-accident. Unrelated similarity must be recognized and accounted for in evolution. So far there is no adequate empirical way to demonstrate whether similarity is related or unrelated.

The true nature of species remains an unsolved biological problem. It seems certain that the term "species" does not necessarily refer always to comparable entities, although our present means of dealing with these phenomena by naming them seems to imply this. We do see species generally as units set apart by a similarity **gap.** That is, there are species' boundaries—morphological, physiological, and geographical. It is this gap concept that delimits species. When any group of organisms are examined, both those now living and those represented by fossils, it is possible to arrange them in an orderly fashion from the least complex to the most complex and from the simple to the specialized. These are first arranged linearly. One of the principal features of this arrangement is the presence of gaps between one group of individuals and another group. Gaps allow the recognition of species. Suppose one were to go into the field and gather all the plants he encountered, taking several specimens of each kind. Upon returning to the laboratory, without consulting a book or making use of any previous knowledge, he sorts and arranges the specimens, in the way most logical from the data at hand, to form an orderly sequence of specimens. It would be obvious at once that some individuals clump in one pile because they have similar features; other individuals clump in other piles for the same reason, and so on. This process is continued until all the specimens are piled, like with like. In other words, the ones that are alike are separated from those that are different. This classification can be made by anyone with any group of objects, regardless of what knowledge the collector may have about them. Any number of characteristics may be used to make this classification, but no matter how it is done, it will usually be possible to make assignments, and there will usually be discrete piles, with degrees of variation within any given pile. It may be necessary sometimes to unite piles and other times to split piles, but regardless, gaps between piles will usually be evident, some narrow and some wide. These groups may be called species, both philosophically and biologically, since the logical process is the same in both fields. Thus far, only the logical process has been discussed.

As one learns more and more about the specimens of plants collected, one will find that some of the things used to separate the piles are significant and others are not. The latter are called "unnatural" because a deeper knowledge of the characteristics shows that the features used conflict with greater evidence in favor of another arrangement. For instance, if the color red is used as a criterion and all the red-flowered plants are put into a single pile, it will soon be noted that there are great differences in the flower structure, the leaves, the size of the plants, and other features. If the plants are rearranged on the basis of flower structure, these other features will agree, and there will be several piles with red flowers, but all of them will be very different otherwise. Therefore the color red is insignificant. The new arrangement will be much more logical. Even though this is a hypothetical case, it embodies all the principles of classification that eventually caused the proposal of the modern theory of evolution. This process and this reasoning led to the discovery of a natural process in biology, speciation or the reproduction of species. Species are first recognized by the presence of gaps and are then defined so that they can be recognized again. The similarities are studied, and each species is assigned to higher categories. Finally, relationship to ancestors is proposed by studying all the data, but particularly the fossil evidence and the genetics. If this is possible, a sound phylogeny can be developed. Most of these studies result in a logical arrangement, however, and such a classification remains as a theoretical proposal changeable as new evidence becomes available.

SPECIATION

Later in this chapter is a brief summary of the actual evolution of the plants, showing the change of species from period to period. All the plant groups mentioned are represented by fossil species, but these are not necessarily ancestors of modern plants. The actual mechanism of speciation, the formation of new species, is a complex process of genetic mutation and hybridization. The new forms may be either slightly different or substantially different from their parents. Changes such as described previously are taking place all the time apparently and result in variation so that no two are exactly alike. This situation obtains in any ecological habitat. Despite the fact that habitats may be characterized in general terms, no two locations are exactly alike, and a given location does not remain constant from year to year. As environmental conditions change, given individuals vary in their adaptability to this environment. At the same time, every individual has a maximum, a minimum, and an optimum tolerance for these factors. Thus, some individuals can survive at lower temperatures than others. During a particularly cold period many members, but not all, of a population may be killed. Those that survive have a genetic aptitude for low temperatures, and some of their offspring will inherit this ability. The gene pool of the population is then shifted toward a tolerance for cold in this case, and this same result follows for every condition that prevails in the habitat. As time passes, the gene pool changes in this manner. If conditions become too severe, however, the population may be entirely destroyed. If the situation is general throughout the range of the species, in other words, in each population, the species then becomes extinct—the usual fate. On the other hand, if some of the many mutations that are always taking place happen to produce a combination of effects that enable the offspring to survive, a new species may then be born.

As a result of this, a great amount of change is going on at the population level. Several factors, such as isolation of a population from the rest of the populations throughout the range of the species, may bring about change. Isolation may be caused by changes in land form, by changes in climate in part of the range of the species, or by invasion by predators in part of the range. As soon as gene exchange is cut off, either temporarily or permanently, the entire **gene pool** is changed because not all the genes available to the species are then available to each population. These breeding inhibitors are termed **isolation mechanisms.** The new gene pool plus the mutations that may have occurred change at varying rates. Eventually there may be change sufficient to be detected in the phenotype, and when it is recognized, a new species has been formed.

The rate of this evolutionary change at the species level varies considerably. It may take a million years or much more, but apparently considerably less when conditions are right. Never, however, has this change to a new species been rapid enough to be observed, except under laboratory conditions. The plants pictured in the most ancient of the herbals look the same as their descendants today. However, under controlled laboratory conditions, these processes can be speeded up to a point at which it is very likely that new species can be created, and in a few cases natural species can be duplicated.

DATA USED TO SHOW THE EVOLUTION OF ORGANISMS

Biologists continue to look for evidence of the evolution of organisms, but the already accumulated mass of information is so overwhelming that most groups can be related and at least a working phylogeny given. Taxonomists use the following kinds of evidence to make a classification and to show the evolution of plants and animals.

Fossils. Before the theory of evolution was proposed, **fossils** were believed to be abortive attempts of the Creator to fashion kinds of plants and animals that were discarded because of imperfections. By the eighteenth century the real nature of these specimens was understood, and fossils were collected, classified, and studied in detail. When these specimens were arranged in a classification sequence in accordance with the classification of existing species and the theory of homology of parts was applied, the ancestry of several groups of plants and animals could be seen. When the rocks in which the fossils were embedded were dated, a time sequence was worked out, and the story of evolution unfolded, making fossil evidence the strongest proof of relationship.

Anatomy. A comparative study of the parts of living organisms is the most common means of showing relationship. Living groups that have a well-preserved fossil series show their pattern of development. If the same patterns of evolution are applied to other groups lacking a fossil history, logical and acceptable hypothetical phylogenies may be constructed. Such reconstructions are called phylogenetic trees and serve a very useful purpose in devising a sound and useful classification based on the principles of evolution. Although a strictly artificial classification can be useful without such procedures, it lacks the meaning and scientific interest of a natural classification.

Genetics and development. The development of the embryo and the plant body may offer important comparative features useful in relating groups that are otherwise very different. Comparative embryology of plants, however, is seldom possible but is used in the flowering plants to a certain extent. Most studies of development in plants are confined to life cycles. Such studies have been mentioned already and may be seen in Chapter 18.

Physiology. Comparative physiology offers evidence in support of evolution in the same manner as does comparative anatomy. Specific physiological processes are under the control of the gene, just as the formation of anatomical structures is controlled by genes. Furthermore, recent work has shown that many genes, perhaps all, have **pleiotropic effects;** that is, a gene will affect a number of characteristics, both structural and physiological. Therefore it is reasonable to believe that if comparative anatomy is a legitimate means of studying evolution, then the comparative study of processes is valid and useful and is especially so as a means of determining the fitness of a plant to its environment and of registering environmental tolerance. Another way is to study the effect of plant juices on the formation of antibodies. Comparative serology has recently become a very active field for both plant and animal studies.

Distribution. If any single form of evidence

for evolution of a group is the most convincing, it is probably the random distribution pattern. Each kind of plant has a specific and particular geographical and ecological (microgeographical) distribution. No organism is universally distributed, despite the frequent use of the word **cosmopolitan** in describing the range of a given species. The so-called cosmopolitan species are restricted to a certain type of habitat and are therefore limited in range. The majority of species have a much more restricted distribution. They are confined to certain very definite areas and to definite conditions in those areas. In evolutionary studies when one species is compared with other species, the range of the species and its adaptation to particular conditions will yield many valuable clues about the origin of a given taxon and about the possible mechanism of evolution in the group because a study of the various members of a taxon will show a variety of distribution patterns. The range of a species, coupled with anatomical changes, furnishes information for speculations on the origin of the taxon and, finally, will show the **radiation** of the group, that is, the development of many species as they evolve and spread. When fossil evidence and knowledge that certain basic distribution patterns exist are available, some idea of the age of the taxon can be determined. These same principles can be applied to extant groups. From these kinds of data a great many very logical phylogenies can be constructed.

Genetics and populations. A change in species can be fixed in a population only by genetic change. Therefore evolutionary process is principally a process of genetics, and it is difficult to separate the two as separate subjects. The concept of the gene pool, that is, the total available genes of a breeding population, is the explanation of the mechanics of evolution. These mechanisms were described earlier in Chapter 18. In all natural populations, that is, populations occupying a place in nature unaffected by man, there is great genetic variation in the individuals that comprise it. However, much of this variation is concealed in the phenotype because it occurs in the form of heterozygotes.

When it is proposed that a species became adapted to a certain situation, the implication is that mutations occurred, which permitted certain individuals to live in a particular place better than can another individual. Because of this, they survived and propagated, whereas the less well-adapted individuals died out. This is what the statement "the survival of the fittest" implies, and this is the process of natural selection.

Polymorphism. Some populations may contain two or more distinct types of individuals and are, therefore, **polymorphic.** Several factors may contribute to this condition. It may be simply a matter of the presence in the population of certain genes as homozygous recessives that are neutral so far as selection is concerned; that is, they are neither beneficial nor unfit. This is known as the Hardy-Weinberg equilibrium. Other cases of polymorphism occur because the rate of mutation is in balance with the rate of selection. Other conditions involving various selective factors may account for polymorphism.

Genetic drift. The Hardy-Weinberg equilibrium applies to random breeding of large populations of plants, especially those that are wind pollinated. In smaller populations, especially those that are isolated, a different phenomenon may take place. Here gene frequency fluctuations may be large, and from year to year the character of the population may show great change. This is termed **genetic drift** and takes place under conditions where a small sample of the genes of the population are drawn upon for the next generation. This might occur among herbaceous plants where a large amount of seed may be produced, but only a very small number of seeds germinate for the following year's plants. Such a condition may obtain until one or the other allele, in the case of a dominant and recessive gene complex, becomes homozygous and fixed. After that, the population will remain stable for the particular character, but other changes may occur through mutation.

Introgressive hybridization. Hybridization

undoubtedly plays an important role in speciation, but it probably is not as important as mutation. One may expect to find in any area a very small percentage of naturally occurring hybrids because isolating mechanisms, especially between species of newly evolved groups, may not be irrevocably fixed. Populations of species that are geographically isolated are said to be **allopatric.** Sometimes such populations may be only ecologically isolated and still be allopatric. Populations of two separate species that live in the same geographical area or ecological habitat are said to be **sympatric.** As one might expect, sympatric species will have stronger isolating mechanisms than do allopatric species. Several things may happen to cause two allopatric, and weakly isolated species to come in contact. They may, for example, enlarge their range because of an optimal growing condition or absence of pests. Or they may come in contact through disturbance of the land by man. When this happens and hybrids form in the area of contact, **introgressive hybridization** may occur. Hybridization in such cases does not result in intermediates between the two species because crosses between the hybrids (F_2) are less likely to occur than a backcross with one or the other of the parent types. In cases of survival of this stock through introgression, the newly formed population may be better fit for the new ecological conditions, and not only is survival possible, but also actual range extension into the new habitat takes place.

SPECIES

Attempts to write a scientific, all-inclusive definition of the category species have been unsuccessful over the past 200 or more years. Since evolutionary theory has emphasized the change and multiplication of species, it seemed imperative that the category be defined because, if the theory proposed that species evolve, then the biologist must know what a species is.

The reason the category has never been defined to the satisfaction of all systematists working in many different groups is simple: the category is an unnatural one and, therefore, cannot have a scientific definition. For example, if it is defined to fit one group of sexually reproducing forms, say the members of the maple tree family Aceraceae, then it will not apply to asexually reproducing bacteria, and so on. Therefore it seems reasonable to think of the category species as a unit that serves only to handle certain types of information about organisms.

The importance of distinguishing between a category and a taxon now becomes apparent. The information reference, species, is a category; therefore, when a name is applied at this level, one understands that the information is limited to facts about certain individual organisms. The organisms themselves form a taxon. The assignment of this naturally existing taxon to the category species is merely a process of information storage and retrieval. This permits writing of descriptions and the identification of specimens in an efficient manner. In fitting these units into classification schemes designed to show relationships and evolution, one must understand that it is the taxon that evolves and not the category.

There are several features of species that are universal. First, all species are composed of individuals collectively referred to as a species. These are the taxa and they are named. Sometimes it may be difficult to determine exactly the limits of an individual, particularly if it exists as a member of a colony, or an individual may be a variant that cannot be immediately assigned to a taxon. These are specific problems, real enough, which can be solved eventually after sufficient study. The individuals form populations, that is, the total number of living individuals in a geographically limited area. All the individuals of a breeding population are related to each other. The area may be microscopic or macroscopic, changing or static, but it is definite at any one moment. The members of this population may or may not be interbreeding, depending on the kind of species and their stage of development. The total area occupied by these living populations constitutes the range of the species. As a final criterion, the former populations, most of which are unknown but which are a part of the species, should be taken into consideration whenever possible. In a few cases, former population representatives may be fossilized, but this is rare.

A species, to be fully understood, should be traced back in the fossil record, but rare is the case where this is possible. However, it is known that fossil series, like extremely variable and close living species, have intermediate forms that fit neither one species nor the other. These may represent transitional populations, or they may be hybrids. In fossil species, because of the time differential, carefully worked out series can be logically pieced together and can truly show the changes that have taken place in the species. All such series have limits. The gaps are the result of the incomplete fossil record. As paleontological research continues, more evidence is gathered to explain the gaps. A three-dimensional concept of species, we can easily see, is necessary, one that includes time, space, and form. These criteria characterize all species. The determination of which populations are species and which are not is the real problem, and a definition of the category species is now of more historical interest than of scientific interest.

Variation

Variation within a particular species of plant is attributable either to environment, hybridization, or mutation. Plants, particularly, often vary as a result of their environment. A plant growing in poor soil that lacks a certain essential mineral will not grow so large or produce so much food or so many flowers and seeds as the same plant growing in fertile soil in which the necessary minerals are present. Changes in plant growth attributable to environment alone are not inheritable and will not be transmitted by the variant plants to their offspring. Organisms can be modified through the action of external influences, but such acquired characters are a modification that arises as a direct response

to an external stimulus and are not inherited because there is no mechanism that will change the coding of the DNA and hence permit transmission of the new trait.

Although the structure of organisms and the way in which their parts function are determined primarily by heredity, the environment is also important in providing an opportunity through selection for modification of inherited characteristics and the way in which they will be expressed. If a plant is grown in the cold, it fails to grow large, the leaves remain small, and the stem stays short. When the plant is exposed to higher temperature, it will develop and increase in size. When grown in the shade, many plants develop larger leaves than when grown in sunlight. The cuticle of the leaf becomes thicker if a plant is growing in dry soil, so that transpiration is reduced. On the whole, plants exhibit great versatility in response to their environment.

Another type of variation is attributable to **hybridization,** which is the result of crossing two varieties of the same species. The offspring of such a cross is likely to have some of the characteristics of each parent and hence is called a **hybrid.** Since they are heritable, the variations resulting from hybridization differ from those induced by the environment.

Because of the ability to produce offspring that resemble the parents, living organisms can be classified into groups of similar and presumably related individuals. At the same time, because offspring are not always identical to the parents, in other words, do exhibit variation, it is possible to study the way differences are inherited. This science is termed "heredity." The physical basis for the laws of heredity is to be found only in the behavior of the chromosomes.

The only physical bridge between two generations of sexual organisms is the sex cell or gamete, and more specific than that, it is the individual chromosomes. The science of genetics deals with this aspect of heredity and is, therefore, concerned with the study of the behavior of genes and chromosomes and their effect on the form and function of plants and animals. Thus we see that heredity deals with the **phenotypes,** or external appearance of an organism, and genetics treats the **genotype,** or internal inherited components.

COEVOLUTION OF PLANTS AND ANIMALS

The obvious dependence of animals on plants because of the net productivity of photosynthesis, that is, excess plant growth, is only part of the ecological relationship of these major groups of organisms. Many other biological features serve to entwine species and species complexes of both groups. So intimate are these systems that the evolution of the two are often in tandem, the changes of one affecting changes in the other. Recent synthesis of data has led to the realization of the importance of the **coevolution** of plants and animals. In fact, it now seems that much of the diversity of many flowering plants and some animals, especially certain insect groups is attributable to this one phase of selection.

Coevolution is adaptive evolution as a result of selection pressures brought about because of the relationship in an ecosystem of two or more species. These relationships may involve plants and animals, host and parasite, two plants, or two animals. As the term is used in this text, it is restricted to plant and animal relationships. Most of these relationships are discussed in various places in the text under the appropriate topic. For example, one of the most obvious coevolutionary relationships deals with insect pollination.

Coevolutionary relationships occur in marine, freshwater, and terrestrial habitats, but they are most obvious in the latter simply because the greatest number of species and the greatest variety of habitats occur on land. Plant-animal relationships are easily fitted into eight classes as shown in the adjacent study box. Each of these are described below.

1. Herbivores, such as cattle, feeding on grass and other foliage may seem to be entirely detrimental to the plants and could in no way benefit these organisms until one remembers that nitrogen, of vital necessity to the plants, is recycled through the dung of the herbivores. However, since this process takes place through the action of bacteria, little change for the benefit of the herbivores can be expected in the plants. The changes that do take place are more likely to be toward the evolution of protective spines or the addition of poisonous alkaloids. Insects feeding on plants are responsible for the evolutionary pressures that increase the poisonous alkaloids in plants. The insects, in turn, not only develop a tolerance for these substances, but also utilize them in their own bodies as a protection against predators.

2. One of the major reasons for the success of insects is that their adaptations permit them to utilize plants for shelter. Many insects bore into plants not just to feed but also to avoid predators and parasites. In some cases plants grow special tissues in which the insects are housed. For example, galls of many kinds are produced by the plant as a reaction to the presence of the insect. The thorn acacias of the tropics provide a site for ant nests by growing special structures at the base of a thorn in which the ants live. Many kinds of tropical plants retain water at their leaf bases where mosquito larvae may breed.

3. Seed- and fruit-infesting insects and

PLANT-ANIMAL RELATIONSHIPS RESULTING IN COEVOLUTION

1. Plants provide food
2. Plants provide shelter
3. Plants provide transportation
4. Plants assist in some way with animal reproduction
5. Animals provide plants with nutrients
6. Animals provide plants with shelter
7. Animals provide plants with a means of dispersal
8. Animals assist plants in pollination

---

other fruit-eating animals, such as snails, may be transported from place to place through the dispersal mechanisms of the plants. Because of these infestations plants have evolved various kinds of seeds and fruits either to resist their attackers or, in the case of fleshy fruits, to attract **frugivores** to aid in plant transportation; this also transports the pest. Protection of the ovary from insect feeding probably is the major cause of the change in position of the ovary in flowering plant evolution.

4. Flying insects need a mechanism for congregating for reproduction. The fragrance of flowers attracts insects that may in turn pollinate the flower. At the same time the insects may copulate, thus receiving this assistance from the flower. In fact, there are certain male bees that are attracted to flowers solely to make use of the flower fragrance as a perfume on their bodies. This attracts the opposite sex so that reproduction may take place. These bees do not feed on the pollen or nectar, but they may act as flower pollinators.

5. Animals may provide plants directly with nutrient material. This is most striking in the carnivorous plants such as the pitcher plant and Venus's-flytrap. The dead insect furnishes nitrogen to the plant. But bacteria and fungi live on and in animals (other than pathogens), especially in the digestive tract where they are fed and sheltered in return for their aid to digestion.

6. Many kinds of plants are provided with shelter by insects through their cultivation or culturing. Subterranean fungus gardens are maintained by ants. These species grow only certain species or strains of fungi and only these fungi are found in these nests. Other species of fungi are transported by wood-boring beetles and inoculated into the wood of the beetle's borrow. The beetles have evolved special pockets or crypts in their bodies in which they store and transport the fungus spores as they move from tree to tree.

7. Insects, birds, and mammals are active seed and plant dispersal agents. This subject is elaborated upon in Chapter 20. In a few cases the vegetative plant is also transported. For example, algae and moss often grow on the fur of the sloth and some other tropical rain-forest animals. Adult beetles carry lichens on their back in the high forests of Papua New Guinea. Other similar relationships are known, but in these cases it is difficult to determine the extent of the coevolution involved.

8. Pollination of many flowering plants is dependent on many species of bees, moths, butterflies, flies, beetles, occasionally bats, and rarely birds. Darwin was among the first to point out that flower structures evolved to attract and accommodate these animal pollinators. The colors of the flowers (usually by carotenoids) serve as visual signals in the plants, with a corresponding development of the visual apparatus and signals in the animals. A superabundance of pollen is produced by the plant and the animals evolve pollen baskets and other suitable equipment for gathering pollen. Nectar requires sucking mouthparts instead of chewing mandibles, and so on. Many complex relationships exist between flower and pollinator. These studies have been made for a long time, but only recently have botanists and entomologists jointly considered the evolutionary effect of these relationships. It now seems very likely that the evolution of both groups is mutually dependent.

PHYLOGENY

The study of evolutionary mechanisms and of the history of the evolution of plants leads botanists to speculate on the phylogeny of plant taxa. They are interested, in other words, in tracing the historical relationships of the plants and, therefore, their evolutionary lives. Our fossil records are far from complete. Therefore any phylogeny proposed is subject to change as new evidence becomes available. All data are used when creating a phylogeny or family tree of plants. Since it is only the higher plants, and particularly the vascular plants, that have left a fossil record, the arrangement of other groups is based primarily on structural similarity. The

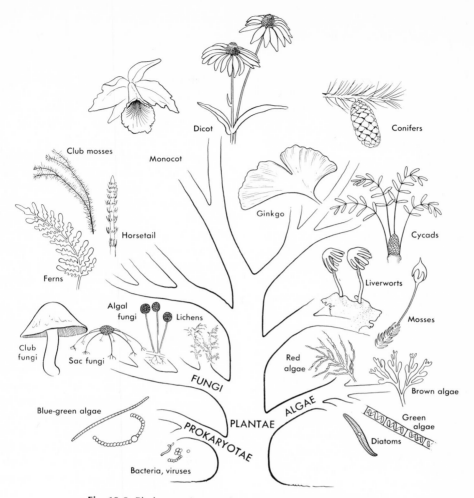

Fig. 19-2. Phylogenetic tree showing the evolution of plants.

current phylogeny of the kingdoms is based on cellular structure and function. The arrangement of the higher divisions of Plantae as given in this text is according to structure and fossil record. Unfortunately there is little paleontological data available to show the arrangement of the orders of the flowering plants, and this group still remains in a state of continued change as research taxonomists continue their attempts to show relationships that are as close to the actual history of the groups as is possible through logical deduction.

Whenever a phylogenetic tree (Fig. 19-2)

is proposed for the evolution of any group, it is based on the supposition that the various species and other taxa evolved in varying degrees into the spectrum of kinds that occur today and that occurred in the past by means of mutations in the gene pool.

SYSTEMATICS

Modern systematic biology concerns itself more with the study of present and past populations than it does with the description of new species. It does not, however, neglect the many thousands of new species of plants and animals remaining to be discovered. But with little doubt, the major features of the classification of plants have been recognized. Changes in classification details can be expected, but it is unlikely that further major changes are forthcoming unless there are radical discoveries in plant physiology and genetics, something that is possible.

Increasing attention is being given to the study of population variations and to the mechanisms that cause these changes in an attempt to find the underlying principles of evolution. This is done by extensive field observations and laboratory and field experimentation.

As mentioned previously, one of the unsolved problems of biology is the true nature of the units called species and how they are formed. The full story of evolution will be told only in the future when this question is answered.

HISTORY OF THE EARTH
Age of the earth

The origin of the earth in the solar system is of concern to astronomers and geologists; it is currently believed to have separated from the sun much more than 4.6 billion years ago because of the age of some moon rocks. However, it is estimated that sedimentation started at about 2 billion years ago. The fact that sediment could form indicates that the earth had cooled enough to permit stratified or layered rock to form. Prior to that time volcanic action was so widespread that all rock was molten and continually changing. Water was present only as steam or vapor. More information about the origin of the earth and the solar system will soon become available as a result of recent space exploration.

Origin of life

We must assume that life originated only once. By this we mean that macromolecules capable of replication are of one type only and that they are the surviving life forms that eventually provided the great diversity of life as we know it today. This is so important to the evolution theory that any other idea

would make all other aspects of the theory logically untenable because the uniformity of cellular biochemistry cannot be coincidental. There is absolutely no scientific evidence against this; also there is little direct evidence to support it. The premise can be assumed only because there is no way to prove it. Not to assume this premise would prevent the organization in any other manner of the available data. Great effort has been expended to explain in a most logical way the origin of life. Attempts have been made to create life in the laboratory and probably has been done to the satisfaction at least of the experimental biologist. With this being done, one way that life could have originated is known. There will be other ways demonstrated as time goes on, but none of these will show how life was created, and none will prove that it was created only once. At the same time, it should be made clear that this does not overrule the possibility of numerous abortive origins. As far as is known at the present time, living substance may be produced every day at that ragged edge between the living and the nonliving. When it was proposed that there was a single origin of life, the supposition was applied to life as it is now on the earth and as it was according to the paleontological evidence at hand. Any other new life appearing now or at any time in the past was either immediately consumed or was unable to maintain itself independently and merged back into the environment.

There is little doubt that life originated in steps as conditions changed on the earth and that these steps took place following the laws of chance. In other words, a succession of events resulted in the formation of primitive life forms as a part of many other chemical formations. These one-time events persisted through natural selection of chemicals and, in the same way that modern species develop, as conditions became suitable. Life as we know it developed under situations no longer present.

The great interest in solving the problem of the origin of life has resulted in the testing of several clever, interesting, and simple systems to show how life might have developed. It seems apparent that the early atmosphere of the earth was reducing in character because of the lack of oxygen. However, the presence of a dense cloud of water vapor produced oxygen at the upper levels through the reduction of the water either by electrical energy or perhaps by cosmic energy. The resulting hydrogen escaped into space, but the heavier oxygen remained and gradually increased. One of the interesting experiments performed to show a possible step involved a "spark-discharge" apparatus into which was placed a mixture of methane, water vapor, ammonia, and hydrogen, the primordial atmospheric gases. When these were exposed to a series of electrical discharges, the gases circulated past the charge by boiling the water on one side and condensing it on the other; small quantities of more complex chemicals were formed, including amino, hydroxy, and aliphatic acids, the three basic types of organic molecules needed for protein structure. In fact, glycine, alanine, aspartic acid, and glutamic acid appeared, all necessary for protein formation. It is not difficult for a biochemist to imagine further steps that might take place, for example, ultraviolet light as an energy source, to continue the building process toward the formation of life particles. As we have pointed out in previous chapters, once the replication of molecules took place, life was present, since no known, naturally occurring system of replication occurs outside of life forms as we know them.

Such steps as these are not the only possibilities for life formation. Very likely many processes actually took place, only one of which was successful in surviving. All the elements of the scientific method are present in the current approach to these studies; therefore it seems logical to believe that those currently discussed may well have been the ones that took place.

Fossil formation

Of the myriad of individuals that have inhabited the earth during its 2 billion year history of life, the chance of even one being preserved as a fossil is so remote that it is almost beyond belief. It seems, therefore, that the hundreds of thousands of fossil specimens available and still being discovered are some indication of the abundance of life during past ages.

Fossils are formed in a variety of ways. One of the more spectacular is petrification. The logs scattered over hundreds of acres in and near the Painted Desert in Arizona are an example of this. Petrification takes place when living forests are covered by soil and rock. The weight of the cover presses the plants. Then by a slow process, mineral-containing water seeps through the covering rock so that, molecule by molecule, the carbon compounds composing the wood are replaced by mineral deposits, and the wood is turned to stone. Every detail of the grain is preserved in an array of colorful rock. Other forms of petrification occur, especially in limestone areas. Sometimes rocks are found that contain perfectly preserved plants or animals, showing every detail of structure. These concretions are often inside round stones, which may be cracked open to reveal the fossil contents within. Still another form of petrification, complete preservation, occurs. Tar and resin deposits reveal nearly perfect plants and animals, the actual body of the organism having been kept intact for thousands of years. Tar pits contain plants that have fallen into this material. The chemical composition prevents decay so that the original structure is preserved. Resinous secretions from trees are sometimes so abundant that they form fossil deposits of amber. Insects, small animals, leaves, and needles are often trapped in this material when it is soft. These are kept in a preserved state comparable to the mounted slides used in laboratory studies.

Most fossils are compressions formed by a covering of rock. The usual process is for silt to cover plants and animals that have either fallen into or live in water. The steady deposits of silt cover these dead bodies, gradually building a rock cover hundreds of feet thick.

After millions of years these rocks are raised up and form land that is eroded away to expose their contents. Coal is formed in a similar way. In this case the vegetation was growing in marshy or swampy areas. After long periods of time a thick mat of vegetation was formed, much like present-day peat deposits. If the land sank and became covered by silt, veins of coal eventually formed because of the great weight of the rock that compressed and turned the vegetation into various types of coal. Under certain circumstances the shape and structure of plants are preserved in such a way that the original plants may be studied.

Finally, impressions of various kinds are found. Leaves that fell into mud left impressions that were later covered by fine silt deposits. Because the two materials, mud and silt, were different, they later separated as stones so that the leaf impressions may be seen in the form of casts.

Geological record

The fossil record of rocks is very incomplete. Obviously, if all the individual organisms that had lived on earth were preserved, this planet would soon have been filled with dead bodies. The fact that only certain kinds of organisms are suitable for fossil preservation and the fact that there is no orderly process of fossil formation prevent any hope that the history of the evolution of all organisms can be obtained from fossils. Because of cellulose in their cell walls, woody plants in particular, as well as many others, are more likely to be fossilized than are animals. The fossil record of plant evolution in general is more complete than that of animals. Even so, there are many gaps in this record so that there is never an unbroken chain of specimens from one stratum to the next.

Geological time

Geologists have worked out a scale, the geological time scale, which names the various periods of time since the beginning of the earth. This scale is based primarily on the layers of stratified rock. Volcanic rock is not dated in this manner because it cannot be measured in time units. It is possible to date both types of rock by radioisotope studies. The original time scales were dated by measurement of the thickness of each layer, and an estimate was then made of the thickness of known silt deposits for a few years. By careful computations, an estimate of the time it took for a given deposit to accumulate could be made. After long study, the various deposits were coordinated and arranged into a time scale. Actually, the geologists used fossils to help date the rocks, and this practice resulted in a paradox because the paleontologist used the rocks to date his fossils! However, when it became possible to date organic matter by ^{14}C and other isotopic tests, the time scale was in need of only minor adjustments. These tests showed that the amount of time for the periods was somewhat underestimated and was actually longer than previously believed.

A revised geological time scale (Fig. 19-3) shows that approximately 2 billion years of sedimentation have occurred on the earth. During the first 1.4 billion years very few fossils were preserved, although the geological record is well known. The geologists divide this long period into various eras, but the biologist generally refers to it as the Precambrian era. Good deposits of Precambrian animals have been found at Ediacara Hills in South Australia, but their exact date has not been determined. The time for known fossils is divided into eras, periods, and epochs. These divisions usually indicate a major geo-

Fig. 19-3. Geological time scale showing the periods during which fossil remains are known. All time is in millions of years. The date 2000 million years ago, at which time this scale starts, represents the beginning of sedimentation and the formation of the first known stratified rock. This scale is based upon new methods of dating that use the radioactive materials in the rock as indicators of the time that has passed since the sediment was deposited. (Modified from Kulp, J. L. 1961. Science **133**:1111.)

Geological time scale

Years ago	ERA	PERIOD	EPOCH	Time passed
0		QUATERNARY	Pleistocene	2,000
			Pliocene	
			Miocene	
	CENOZOIC	TERTIARY	Oligocene	
			Eocene	
50			Paleocene	1,950
100		CRETACEOUS		1,900
150	MESOZOIC			1,850
		JURASSIC		
200		TRIASSIC		1,800
250		PERMIAN		1,750
	PALEOZOIC			
300		PENNSYLVANIAN		1,700
		MISSISSIPPIAN		
350				1,650
		DEVONIAN		
400				1,600
		SILURIAN		
450				1,550
		ORDOVICIAN		
500				1,500
550		CAMBRIAN		1,450
600				1,400
	Precambrian	Ediacara Hills animals?		
		Origin of life?		
2,000	Beginning of sedimentation			0

Fig. 19-3. For legend see opposite page.

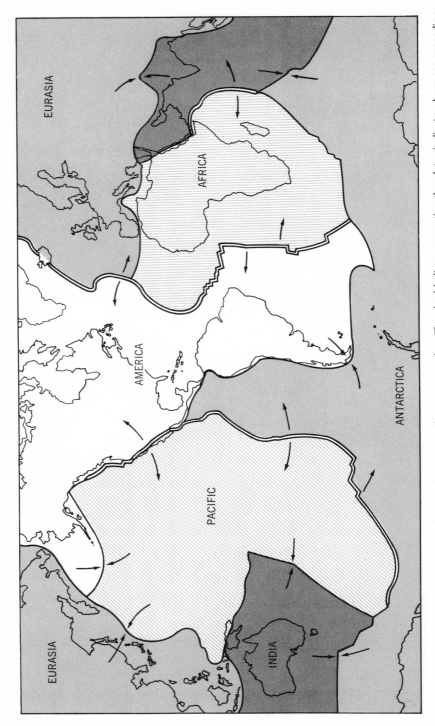

Fig. 19-4. The six major shifting plates of the earth's crust as they appear today. The *double lines* separating the plates indicate where sea-spreading is taking place. The *single lines* indicate zones where the plates are converging and compression is taking place. Earthquake and volcanic activity is found wherever the plates come into contact. (From National Science Foundation. 1972. Patterns and perspectives in environmental sciences, U.S. Government Printing Office, Washington, D.C.)

logical change, some of which will be mentioned.

Plate tectonics

A current theory held by many geologists proposes that the surface of the earth is composed of about 12 gigantic plates of rock floating on an inner, molten core. The six major plates have been roughly mapped (Fig. 19-4). The plates are not stationary, but are moving from place to place, one sliding upon another, responding to internal pressures caused by the continuing processes of geological activity. This theory is termed **plate tectonics** and is used as an explanation for sea-floor spreading.

The continents as we know them, float with the moving plates more or less as islands "frozen" in place on the plates. Their movement is a part of the sea-floor movement; so it appears that the continents are separating and drifting.

Obviously all 12 plates are tightly pressed together so that only very narrow cracks are formed. It is through these cracks that molten lava oozes, causing volcanic activity. This forms chains of volcanic mountains along the sides of the plate, either on land or in the sea. One area between the two contiguous plates is rising, pushing the plates apart. The central rift valley of the Atlantic ocean is an example of this. Another area between two plates is sinking into the plastic mantle below. This edge of the plate is an area of earthquakes as well as volcanic activity. The San Andreas Fault in the western United States is one example of this. So we see that there are two distinct processes taking place, depending on the location on the plate. One edge of one plate moves against another edge and is raised or lowered. The adjacent plate slides under or over, causing earthquake activity. The edge that is lowered may form a deep sea trench. The second process is really the source of power for the movement of the continents. This is the upward thrust of the mantle spreading the edge of two plates apart. This takes place in the ocean and also forms volcanic islands and undersea ridges.

Two major pieces of evidence support this theory: (1) sea-floor mapping, which shows the underwater mountains, valleys, and deep trenches; (2) reading of magnetic fields in the rocks of the sea and the land. As new rock is formed by volcanic lava flows, they adjust to the magnetic fields of the earth at the time they cool. The polarity of the earth reverses periodically and this reversal is recorded in the rocks. So if these fields are measured by suitable instruments as the sea floor is mapped, the fields are also mapped. The edge of one continent is consistent in matching the edge of the adjacent continent of the opposite shore. The sea floor between the continents is composed of newer rock, the most recent being found in the center. This is best seen in the Atlantic Ocean, and the matching of the magnetic fields between South America and Africa consistently agree.

Thus the theory of plate tectonics gives a geological basis for the theory of continental drift.

Continental drift

Over half a century ago the theory of **continental drift** was formally proposed, maintaining that the continents were once all joined together (Fig. 19-5) and that they separated slowly by drifting apart to their present position. But the match of the configuration of the shores on both sides of the Atlantic Ocean had not escaped the notice of geologists for many decades previously. Biogeographers have long postulated **land bridges,** narrow connections of land between continents, as a means of explaining the distribution of plants and animals. The theory of continental drift as originally proposed lacked general appeal, mainly because there was little evidence that such a movement could or had taken place, and although it was hard to believe that the fit of the continents could be entirely coincidental, it still explained very little of the biogeographical problems.

For many years the proposal remained as a controversial, speculative subject. During the past few years the very significant new data offered by the theory of plate tectonics

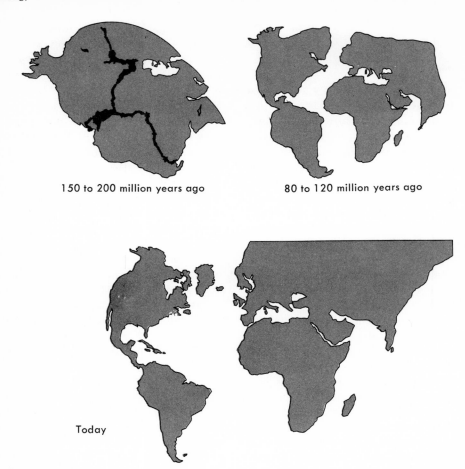

150 to 200 million years ago

80 to 120 million years ago

Today

Fig. 19-5. The original continental land mass, Pangaea, drifted apart to form the continents as we know them today. (From National Science Foundation. 1972. Patterns and perspectives in environmental sciences, U.S. Government Printing Office, Washington, D.C.)

has suddenly given the theory of continental drift not only renewed attention but very widespread acceptance.

Originally it was difficult to imagine an energy source great enough to move entire continents, but as explained in the previous section, the energy is now presumed to come from geological forces inside the earth. Further data obtained through studies that include such items as the precise measurement of minute changes in the distance between two points in Europe and North America, give support to the theory. For example, it is now known that the spreading of the ocean floor increases the distance between London

and New York at the rate of about 1 inch each year. This means that Columbus sailed about 40 feet less in 1492 than he would have to today.

Continental drift theory now proposes that the continents have moved apart and together over long periods of time. The total evidence for this, in summary, now includes the (1) fit of the continents, (2) **paleomagnetic** data, (3) increased age of islands the greater their distance from the midocean ridges, (4) earthquake data, (5) location of volcanoes, and (6) paleoclimatic information, which shows the compatability of the plants with the weather conditions at the location of the land masses at any particular time. The present-day distribution of many plant groups is in accord with these data.

Major events of recent drifting

It is now estimated that the current separation of the continents bordering the Atlantic Ocean began about 200 million years ago, or during the Mesozoic era. This is recent enough to have had considerable effect on the distribution of land plants, and biogeographic studies, as mentioned previously, support this. The major land movements of the Mesozoic (Fig. 19-5) resulted in the subdivision of the single land mass called Pangaea ('all land,' pronounced pan-jee′a). Detailed studies of the distribution of fossil reptiles and early mammals correlate with these land movements. The first of these geological events, according to some geologists, was the union of Europe and Asia, closing the ocean between the two continents and forming the Ural Mountains. Not all geologists agree on the time this took place, but most agree the modern continental distribution started with the original land mass at this time and call it Pangaea. The surrounding ocean is called Panthalassa ('all sea'). As is discussed further on in this text, the vegetation of the late Mesozoic consisted of giant ferns and similar plants; the flowering plants had not yet evolved.

By the Triassic period Pangaea began to split apart forming two major continents, Laurasia in the north, and Gondwana in the south. This important event lead to the differences in the **biota** of the northern and southern hemispheres and resulted in the distinctive Jurassic flora that followed.

It appears that during the early Cretaceous period Africa and South America separated; thus Gondwana was broken up into the present continents of South America, Africa, Antarctica, and Australia. Laurasia at the same time began to split, but retained a land bridge between North America and Eurasia. It was during this period, however, that the first of the flowering plants evolved. This accounts for the wide distribution of many of the more primitive families of these plants. Finally by the end of the period the North American–Eurasian land bridge was submerged. Present-day India was once a part of Antarctica, and it was not until the end of the Paleocene epoch of the Tertiary period that it broke away and floated northward, eventually bumping against Asia. This pressure against the continent formed the new and very high Himalayan Mountains.

A new land bridge was formed probably less than 1 million years ago between eastern Asia and North America as a result of the westerly drift of the North American continent. This bridge accounts for the common distribution of the more recent flowering plant families in Asia and North America.

The theory of continental drift now seems probable, and although it is still theory, these events very likely actually occurred. Nevertheless, certain as yet unexplained contradictory geological evidence exists. More data must be gathered to resolve these problems. The theory will be revised as new data are gathered until a completely logical, corroborated picture of the events is possible. The major aspects, however, seem to be established.

EVOLUTION OF PLANTS

The history of the development and diversification of the Prokaryotae and most of the primitive plants is unknown. Some of the algae, certain mycelia of fungi, and especially spores are preserved as fossils, but little in the way of phylogenetic conclusions may be inferred from these scanty records. The higher plants, especially the vascular plants, have a substantial, but far from complete, fossil record.

Paleozoic era

The evolution of the major groups of plants is fairly well known. The Paleozoic era saw the development of all these groups, except the flowering plants and perhaps the Fungi, as shown in Table 19-1.

The Paleozoic era was the longest period in the development of plants. At the beginning

Table 19-1. Geological history and origin of major plant groups

Era	Period	Prokaryotae and Algae	Meta-phyta	Anthophyta Gymno-sperms	Anthophyta Angio-sperms	Fungi
Cenozoic		Present	Present	Present	Present	Present
Mesozoic	Cretaceous	Present	Present	Present	Present	?
	Jurassic	Present	Present	Present	Origin	
	Triassic	Present	Present	Present		
Paleozoic	Permian	Present	Present	Present		
	Pennsylvanian	Present	Present	Present		?
	Mississippian	Present	Present	Present		
	Devonian	Present	Present	Origin		?
	Silurian	Present	Origin (Psilophyta)			
	Ordovician	Present				
	Cambrian	Present				
Precambrian		Origin				

of the era there was no life on land. The climate was very hot, and there was almost constant rain. Freshwater probably had some simple algae living in it, but the entire evolution of major groups of Algae is generally believed to have taken place in the sea, which at that time still had a relatively low-salt content.

Cambrian. The Cambrian period is characterized by times of flooding of the land forming shallow seas, which resulted in a much different geography from that of today. Continents little resembled the shapes we know them to be today. Undoubtedly the Algae were well developed by the end of the Cambrian period.

Ordovician. The Ordovician period is introduced by the Taconic revolution, a period of land elevation that resulted in the cooling of the land and the diminishing of rain so that various kinds of algae emerged from the sea. In the still very wet condition they were able to gain some footing. Their growth on land caused the first soils so that by the next period, the Silurian, land plants were beginning to develop rapidly.

Silurian. Among the first of the Silurian

plants were woody Psilophyta, distant relatives of the Pterophyta. Many other Silurian plants are known, but their relationships have not been studied very well as yet. These first plants were, of course, all spore producers and were probably generally distributed across this rather barren earth. The climate of the time was uniform, with no seasons.

Devonian. The late Silurian and Devonian periods were characterized by an increasingly arid climate, but certainly the dryness was relative because during the Mississippian and Pennsylvanian periods, known together as the Carboniferous period, the climate was still much wetter than that of present-day tropical rain forests and probably much hotter. There are few known survivors from the Silurian to the Devonian, yet Psilophyta of various kinds remained as the most characteristic plants of the period. Other major groups are present in the Devonian as well, including members of the order Cladoxylales, an early fernlike plant that may have been the start of the Pterophyta (ferns). Plants of the order Hyeniales, also present in this period, may be the ancestors of the Arthrophyta (horsetails). A group of gym-

Fig. 19-6. Carboniferous swamp showing reconstruction of ancient horsetails. (Courtesy Field Museum of Natural History, Chicago, Ill.)

nosperms, the Pityeae, formed the first seed plants, but these were short lived. The dominant plants of the landscape of the period were members of the Psilophyta, Lycophyta (club mosses), and the Arthrophyta. Although these plants developed early, they were temporary and were replaced by other plants in the later half of the Devonian. Much of the Psilophyta flora was replaced by the primitive ferns that became the dominant plants of the Mississippian period. The Devonian period ended, nevertheless, with a well-developed land flora.

Mississippian and Pennsylvanian. The Carboniferous period (Mississippian and Pennsylvanian periods combined) was characterized by a mild, moist climate all over the world. A heavy cloud cover kept carbon dioxide close to the earth and caused the phe-

nomenon known as the "greenhouse" effect, producing high heat and humidity. This and the dense clouds resulted in optimum conditions for luxuriant plant growth, probably the most extensive swamp flora (Fig. 19-6) the earth has ever known if one can judge from the coal deposits that remain. It was during this period that the seed plants were perfected, although throughout these periods the ferns seem to have been well represented. Many of the fronds once believed to belong to ferns are now known to be early Cycadophyta (cycads), the seed ferns. Even though we consider them the most primitive division of the Metaphyta (higher plants) it is not until the Pennsylvanian period that the Bryophyta appear. Their late appearance makes it rather certain that the group is a sideline development and not a contributor to future plant evolution. At the end of the Pennsylvanian period the first conifers appeared. These were the first pollen producers. All the rest of the seed plants of that time were dependent on motile sperm.

Permian. The Permian period was ushered in by a general rise of land masses, which was followed by the building of mountains, the cooling of the climate, the draining of large swamp areas, and the extinction of many plants. South Africa was covered with a glacier. Thus the first great changes in land flora took place. The spore plants, so dependent on an abundance of water for fertilization, were less well fitted in the new climate. The higher ferns, conifers, and cycads developed to become the dominant features of this flora. Because of the climatic diversity, floral regions became evident for the first time. The early conifers were still fertilized by ciliated sperm partly dependent on rain for their dispersal, as are the cycads today. By the end of the Permian period practically all the Carboniferous plants were replaced by modern groups, many of which have survived to the present time. It was also at this time that the continents began the splitting apart and drifting that became much more evident in the Mesozoic era.

Mesozoic era

The Mesozoic era began with the Appalachian revolution, the formation of the Appalachian Mountains in eastern North America, as well as of mountains in other parts of the world, and is characterized by a cooler climate. At this time many of the Paleozoic plants were extinct.

Triassic. Taking the place of the Paleozoic plants were more lycopods, horsetails, and ferns, the last rapidly becoming the dominant plants of the Triassic period. It was during this period that the Araucariaceae developed. Plants of this family have been studied for the distributional evidence they show in support of the theory of continental drift.

Jurassic. Cycadlike plants, the true cycads, and the ginkgos were the dominant plants of this period, and there was a rapid rise in growth of conifers. The angiosperms had their origin in the Jurassic period and developed rapidly. The flora of the Mesozoic era would be recognizable and somewhat familiar today. The whole era is characterized by a climate that became gradually more dry and progressively more severe.

Cretaceous. The Cretaceous period saw the extinction of several groups of plants and the decline of the cycads, horsetails, and ginkgos to minor groups. The flowering plants rose to become the dominant plants, along with the conifers. Unfortunately very little of the early history of the flowering plants has been found in the fossil record; the earliest known examples are fully evolved and can be placed in modern families, and even in genera. The end of the period is characterized by the Laramide revolution, another period of mountain building. It was during this period that high mountains such as the Rocky Mountains were formed and that all ancient forms of plants became extinct.

Cenozoic era

The Cenozoic era is, of course, among the best known. Its vegetation was very modern. All the surviving plant groups now on the earth were present at the beginning of the period so that no new plant divisions or major groups are recorded as originating at this time. There was a gradual evolution of nearly all the land masses that form our present-day geography. All large areas of swamp ground disappeared, except those that remain today. Deserts were extended on the leeward side of high mountain ranges, and extensive grasslands were formed. The various epochs of the Cenozoic era are not characterized by distinctive changes in vegetation types, except for the shifting of biomes. The climate was characterized by distinct seasons, but because of alternate periods of cold and heat, tropical vegetation gave way to temperate and subarctic vegetation, only to have tropical conditions return again. The periods of the era are only the Tertiary and the Quaternary.

Tertiary. The early part of this period is characterized by the spread of tropical vegetation far north because of a general warming trend. This caused the melting of ice and the rise of the oceans, and this in turn provided water for an abundance of more snow at the poles. Therefore a reverse trend took place toward the end of the Tertiary, eventually resulting in the series of glaciations of the last part of the period.

Paleocene. The first epoch of the Tertiary, the Paleocene, is better known from animal fossils than from plants. However, those present show a continuation of a warm climate with a subtropical vegetation extending far into the northern part of North America and Europe.

Eocene. By the Eocene a characteristic tropical flora extended as far north as England, and, strangely, this flora had a strong Indo-Malayan element. These later gave way to forms that may have come from the less tropical Eocene flora of Asia.

Oligocene. The end of the northern tropical flora was reached by the middle of the Oligocene epoch. However, palms, sequoia, and ginkgo reached as far north as Greenland where abundant fossils of deciduous trees indicate that forests once existed. In South America the modern plants of the genera *Nothofagus,* and *Araucaria* developed. These plants are shared with Australia, a phenomenon still not clearly explained by continental drift. The genus *Nothofagus* probably existed on Antarctica from the late Cretaceous to the Miocene and was distributed from this region to its current circumpolar sites. The epoch ended with new mountain building, including such mountains as the Alps.

Miocene. Mountain building reached its climax during this epoch, accompanied by considerable volcanic activity. The Himalayas in Asia and the Cascade range in western North America were formed, and they represent our newest mountains. The flora of the time lost its Indo-Malayan forms and became temperate in character. Tropical plants slowly retreated until today the only remains of Miocene flora in the United States are found in Florida. The epoch closed with the beginning of glaciation in the far North, and Miocene ice still remains there in a few areas.

Pliocene. At the beginning of the Pliocene epoch the temperate regions still retained certain tropical elements in its flora, but by the end, the Ice Age had begun. Most of the flora of the period is known from flowering plants. The trend was certainly toward herbaceous and away from woody plants as shown by the preserved seeds and fruits. This is the final period of the Tertiary of the Cenozoic.

Quaternary. The Quaternary period contains only the Pleistocene epoch, which is characterized by successive ice ages, ages of widespread and successive glaciations. Since this epoch is the most recent, it is not surprising that it is the best known. It is also the period during which man evolved; thus more intensive study has been given to it. Four glacial periods existed, with only about 10,000 years having passed since the last glacial cover of North America. The fascinating history of the Quaternary cannot be dealt with in this introduction. However, it was a period of wholesale extinction of both plant and animal species and the almost explosive development of modern groups. Some northern areas either were not covered by ice or were only lightly covered so that some of the earlier Cenozoic plants persevered in these **refugia.** These areas also provide an isolated and interesting flora in that the study of these plants gives us an insight into preglacial conditions. Many acid and alkali bogs of the northern part of North America are such refugia.

Recent. Our present time period began at the end of the last ice age, and no one knows whether a new ice age is on its way; therefore we cannot determine its exact geological state in relation to other periods and epochs. However, it is known that the Antarctic ice sheet is gradually deteriorating, raising the level of the ocean several inches per century. The flowering plants are now the dominant plants, with conifers as a prominent relic of past ages. The ferns, mosses, bryophytes, and similar spore-producing plants of the Paleozoic and Mesozoic remain on the earth only as a minor part of the flora both in appearance and as a part of the energy-exchange cycles. They are a silent reminder of the past.

QUESTIONS AND PROBLEMS

1. What is meant by a blood relative? Blood lines? And so on?
2. There are two meanings to the word "Related"; one is used in genetics, and the other is used in discussions of evolution and phylogeny. What are these different meanings?
3. Compare a population of plants with a human population. What are the similarities? The differences?
4. Why is not a horticultural variety considered to be a separate species by taxonomists?
5. What is meant by a new species? Is it "new"?
6. Describe how you think a new species might arise, keeping in mind the terms allopatric, sympatric, isolating mechanisms, and gene pool.
7. How would you describe the environment of the earth at the time of origin of life? Of terrestrial life? Of man?
8. How do these primeval conditions relate to the fermentation process described previously?
9. What is meant by mountain building? What are the major periods of mountain building? What effect did this have on climate?
10. Why have the major changes in the development of a plant's anatomy taken place on the earth's surface rather than in the sea?
11. Name the geological eras, periods, and epochs in order and give the approximate dates.

DISCUSSION FOR LEARNING

1. Discuss some possible effects of atomic warfare on the future of plant and animal life on the earth. What kind of changes might take place?
2. Discuss some of the practical applications of genetics; some of the theoretical applications.
3. The idea of evolution has disturbed many people in the past. What are some of the misconceptions that have brought about this fear of scientific knowledge? What are some actual misconceptions or limitations of the current knowledge of the processes of evolution?
4. Discuss the theory of continental drift in relation to the dispersal mechanisms of plants. In other words, how much passive riding on these land masses have plants done?
5. Discuss spore dispersal and pollen dispersal in

terms of the known fossil history. What are the
advantages of one over the other?

6. Discuss the conditions under which the flowering plants developed and those under which
they became the dominant plants of the earth's
surface. Why are they apparently better fit for
present-day conditions than are other kinds of
plants?

ADDITIONAL READING

Ableson, P. H. 1956. Paleobiochemistry. Sci. Am.
195:83-92, July.

Andrews, H. M., Jr. 1947. Ancient plants and the world
they lived in. Cornell University Press, Ithaca, N.Y.

Derek, V. 1963. Principles of paleoecology. McGraw-
Hill Book Co., New York.

Gilbert, L. E., and Raven, P. H. 1975. Coevolution of
animals and plants. University of Texas Press, Austin,
Texas.

Knopoff, L. 1969. The upper mantle of the earth.
Science **163:**1277-1287.

Oparin, A. I. 1957. The origin of life on the earth. 2nd
ed. Academic Press, Inc., New York.

Rutten, M. G. 1962. The geological aspects of the origin
of life on earth. Elsevier Publishing Co., Inc., New
York.

Sullivan, W. 1974. Continents in motion. McGraw-Hill
Book Co., New York.

Walton, J. 1953. An introduction to the study of fossil
plants. Adams & Black, London.

Fig. 20-1. The interrelationship of the giraffes, gazelles, and acacia trees in East Africa is one example of the many complex ecosystems that constitute the ecological web of energy flow.

CHAPTER 20

Plant ecology

■ Ecology is best defined as the interrelationships of organisms with their abiotic and biotic environment. To single out plants and describe their "ecology" would be an illogical exercise at best, because animals and plants are in intimate association in the physical environment. The flow of energy in the biological world takes place not only within the cell, but from cell to cell within the organism, and from organism to organism without distinction as plant or animal. This flow of energy is through interconnections that are a complex web of living activity. This is known as the **ecosystem** (Fig. 20-1).

NATURE OF THE ENVIRONMENT

Only certain features of the environment are effective, that is, cause dynamic, equilibrium-seeking reactions of the organism in a habitat. The other factors, therefore, are inert so far as a particular organism is concerned. For example, the nitrogen gas that constitutes approximately 78% of the atmosphere has no direct effect on the physiological processes of plants. Once converted into a nitrogen compound, however, it is an important and necessary part of the effective environment.

For convenience, we may analize the environment by considering the factors singly. The first dichotomy is between the physical and the living environment. The term "abiotic environment" is useful in classifying environmental features. Thus, the abiotic environment constitutes all of the nonliving factors, whereas the biotic environment is the collective complex of living organisms. These are individually discussed here, following the classification shown in the study box.

THE ENVIRONMENT
Abiotic environment
   Light
   Water
   Atmosphere
   Temperature
   Radiation
   Substrate (soil)
   Chemicals
Biotic environment
   Plants of the same kind
   Plants of a different kind
   Animals

PHYSICAL ENVIRONMENT

Light. The role of light as a source of energy for photosynthesis is obvious. The amount of light available to plants is regulated by latitude, weather, and shade. Shade may be produced by land forms, for example, mountains, by other plants, artificially by buildings or similar structures, and by weather conditions.

The northernmost and southernmost latitudes have the longest growing seasons in terms of length of day but the shortest period in number of days. Therefore plants of these areas are quick to flower and set seed. Because of this and temperature factors (to be discussed later), plants are short and stunted.

The same species growing in temperate latitudes are considerably taller. As the latitude approaches the equator, the length of day becomes more nearly equal throughout the year, with a gradual disappearance of the seasons until at the equator, light conditions are nearly uniform throughout the year, and night and day are equal in length. Consequently, plants of those areas have more uniform growth, and any variation is the result of other factors such as the amount of rainfall. The effects of latitude are apparent but are never independent of other environmental conditions.

Plants growing in mountainous areas may be shaded by land forms. The length of day is shortened by the early morning or afternoon shade of the mountain, depending on the slope exposure. Thus eastern slopes receive early morning sun and western slopes, afternoon sun. Northern slopes receive less light during the growing season than do southern slopes. Plants in canyons may receive relatively little light, whereas plants on mountain tops receive the most light but at the same time are in a cooler climate. The growing conditions in these two types of locations produce different kinds of plants, or at least plants having different characteristics.

Plants growing in the shade of larger plants or buildings are affected in the same way by light conditions. Therefore certain kinds of plants will grow best in shade and others in sun. Neither are able to survive in exchanged situations, showing that they are adapted to their particular habitat. A knowledge of light effects is fundamental in understanding the distribution of plant species.

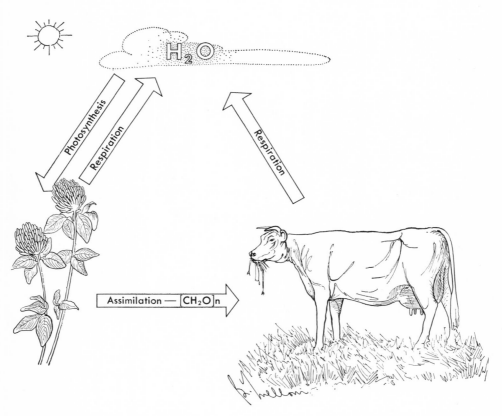

Fig. 20-2. Water-use cycle. Water is necessary for metabolism in all organisms. It is taken in for use in some reactions, and given off as a product, usually as a vapor, in others. The vapor appears as clouds.

The color of light may affect plant growth also. For flower formation, red light is more effective in some species than is blue, and green is less effective than either.

Duration of light during the day, that is, the number of hours of light to which a plant is exposed, evokes certain responses. This is called the photoperiod. Some plants such as corn are long-day plants and will only produce flowers when exposed to daylight for 12½ hours. Other plants such as tobacco are short-day plants and will produce flowers only when exposed to less than 12½ hours of daylight. Still other plants are neither short- nor long-day plants and will produce flowers during either photoperiod.

The number of days of sunshine varies from region to region. The average number for a given area is tabulated and is available in reference books along with other weather data. Areas having many sunny days produce more abundant plant growth than do areas having many cloudy days, unless some other limiting factor is present, for example, water. Some areas are characteristically covered by cloud or fog banks throughout much of the year. Despite an abundance of water, plant growth is limited in such areas by the lack of bright sunshine and by the lower temperature resulting from the cloud cover.

Water. Water is usually the limiting factor in plant growth (Fig. 20-2). Without an abundance of water, photosynthesis cannot take place even if the days are sunny and the soil is rich in minerals. The amount of rainfall (Fig. 20-3), like the amount of sunshine, is

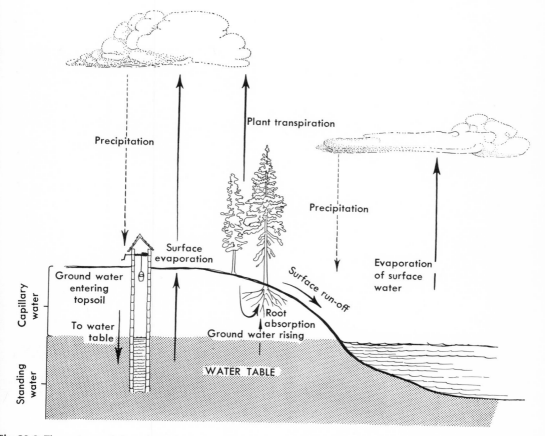

Fig. 20-3. The water cycle is a physical phenomenon and nearly independent of life. However, plants contribute water vapor from transpiration and effectively increase atmospheric humidity in local areas. (See Fig. 20-6.)

Fig. 20-4. Water hyacinths in Arbuckle creek in Florida. This plant chokes out all other life in streams and drainage ditches.

carefully recorded for all areas and is a key factor in agricultural regions. To illustrate the exchange of water, under ordinary growing conditions a square foot of leaf surface transpires 4 ounces of water each 24 hours. At that rate, a corn plant would take up 54 gallons of water during its growing season. An apple tree, however, may lose 95 gallons of water per day by transpiration during the month of July; therefore an orchard of 40 trees would have to take up 480 tons of water during a period of 1 month at the height of the growing season. A rainfall of 4.25 inches would be needed to replenish this water. In addition to this, large amounts of water are used by the plant for processes other than transpiration. Under natural conditions the vegetation found is indicative of the amount of available water. Plants found in a tropical rain forest (Fig. 20-26), where heavy rains occur, are characteristic of that region and are not found in drier areas. Conversely, desert plants, xerophytes, are characteristic of dry areas. The necessary water need not be in the form of rain. Surface or subsurface runoff and marshy soil are other sources of water for plants. A few areas are known as fog

Fig. 20-5. Water lettuce. The roots of these plants are favorite sites for the breeding of certain kinds of pest mosquitoes.

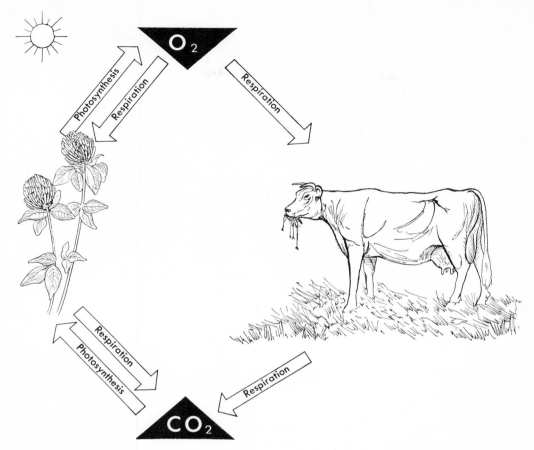

Fig. 20-6. Oxygen and carbon dioxide cycle. Both are dependent on respiration and photosynthesis. (See Fig. 20-9.)

deserts in which rain is rare; yet moderate plant growth is maintained because the moisture in the air is absorbed by the plant. Finally, truly aquatic situations have an obviously unlimited supply of water, and plants are variously modified to live there. For example, air bladders may keep the plant floating on the surface of the water (Fig. 20-4). Plants with floating leaves (Fig. 20-5) have stomas located on the upper surface instead of ventrally. The extent of the root system of plants usually varies with the availability of water.

Atmosphere and oxygen. Oxygen is of primary importance in the growth of plants and in their ecology. The role of oxygen in plant

life is termed the **oxygen cycle** (Fig. 20-6). Usually oxygen is not a limiting factor because plants give off oxygen during photosynthesis. Some plants growing in water have special structures to help aerate their roots. Among the best known are those of the mangrove (Fig. 20-7), with aerial roots. Cypress "knees" (Fig. 20-8), long believed to be aerating structures, are now known to be only a growth reaction to a particular environment.

Atmosphere and wind. Air currents in the lower atmosphere do not often grossly affect plants except by controlling somewhat the rate of transpiration. Wind will blow away water vapor that builds up near the stomas,

Fig. 20-7. Mangroves (*Avicennia* sp.) in a Colombian estuary. The mud in which they grow contains little or no oxygen so that many "breathing" roots grow up into the air.

Fig. 20-8. Cypress knees, a growth adaptation in swampy areas. (U. S. Forest Service.)

and unless they close, the rate of transpiration will increase. In desert areas particularly, this can be an important factor. Winds that are strong, constant, and always from the same direction affect the growth pattern of plants. Trees growing in prevailing winds are bent over, and the branches are pushed to one side. Frequently this effect is evident in coastal or mountainside plants.

Carbon dioxide. The atmosphere contains approximately 0.03% carbon dioxide. However, it varies greatly near the cell's surface, especially if decomposition is occurring, as in a pond or similar area. It has been found that carbon dioxide may measure as much as 0.35% within a few inches of the leaf. Other factors being equal, this increased concentration of carbon dioxide is sufficient to increase the rate of photosynthesis. Increasing the carbon dioxide concentration when other fac-

tors are favorable will increase the rate of photosynthesis, but it is doubtful that carbon dioxide is ever a limiting factor in nature. If one refers to the **carbon dioxide cycle** (Fig. 20-9), it is evident that carbon dioxide fixed in the cell must be returned to the atmosphere, or eventually the atmosphere would become exhausted of carbon dioxide.

The amount of carbon dioxide in the atmosphere may have profound effects on the climate of the earth, since it is this material in the atmosphere that effectively shields the surface from cosmic cold. In fact, through geological history evolutionary changes in the flora have been in response to changes in the carbon dioxide blanket. As the cover increases, the temperature increases (sometimes called the greenhouse effect because it is comparable to the same process that occurs in a greenhouse). Over the past hundred

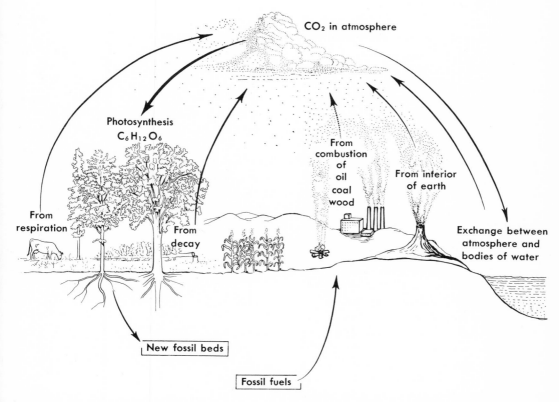

Fig. 20-9. Carbon cycle illustrates the chain of events in the utilization of carbon by plants. The *dotted arrow* indicates one artificial process by which man adds carbon dioxide to the atmosphere.

years, carbon dioxide has increased in the atmosphere by 13%, causing an average increase in temperature of 1° F. (= 0.56° C.). This may account for such events as the Ice Age and the Carboniferous period (Chapter 19). The effect of the increase of carbon dioxide from industry and carbon monoxide from transportation, and other air pollutants, on plant growth has not been determined as yet.

Temperature. Heat, primarily from infrared radiation reaching the earth's surface, affects the rate of photosynthesis and other metabolic processes by increasing or decreasing the speed at which the chemical processes take place. There is little effect other than this, unless the temperature is lowered to the point of freezing or raised to the point at which the protoplasm is damaged. The rate of transpiration increases with higher temperatures. The evaporation of the transpired water vapor will cause a cooling effect that is readily noticeable on hot days in forests and other shaded areas. If water is a limiting factor, the increased transpiration resulting from high temperatures will cause the plant to wilt. Wilting slows down transpiration by allowing the stomas of the leaves to close. If water continues to be a limiting factor for too long a period, wilting will cause permanent damage to the plant, and it may not recover.

Radiation. In addition to radiation of light as we see it, and to infrared radiation, there are other kinds of radiant energy affecting plant life, either already present on the earth, or reaching it from outer space. Ionizing ultraviolet, nuclear, and cosmic rays have sufficient power to damage cells under certain conditions. Fallout from nuclear reactions is now a well-known example of causes of cellular changes of many types. Ultraviolet radiation, usually screened by the atmosphere, may be important at extremely high mountainous habitats. Cosmic rays will be important as space travel develops, and it is now known that the amount of this kind of particulate radiation may be responsible for many mutations. Radioactive minerals in the earth's shell cause changes in plant growth in some areas. After volcanic activity, radioactive lava has the same effect on growth as does fallout. The amount of energy exchange varies with the latitude and altitude, which in turn affects the amount of radiant energy reaching a cell surface.

Substrate (soil). Most of the plants, except some algae and fungi, are terrestrial and are rooted in soil, the substrate. Many plants growing in water are anchored to the bottom soil. All plants are dependent on the moisture and mineral content of the substrate, but completely submerged plants are also dependent on the gases dissolved in the water. In terrestrial plants the soil affects growth in other ways. Most plants depend on the soil to anchor and support the stem and leaves. For these reasons the quality and quantity of the soil are important in plant growth. Contrary to what might be expected, tropical soils are thin and subject to landslides (Fig. 20-10), so that there are many areas in the tropics where plant growth has been disturbed.

Soils of different types have varying moisture-holding qualities. A heavy clay soil will hold moisture because it is composed of very fine particles of rocks. These small particles provide a tremendous surface for holding water. A thin layer of water will cling to each particle by surface tension. Loose soils such as sand hold less water because the particles are larger. The organic matter in the soil also determines its ability to retain water. Heavy clay soils will hold water but make root penetration more difficult because they dry to a hard surface. Such soils also keep out air so that roots may be drowned, eliminating the oxygen needed for their respiration as in other plant parts. Desert soils are generally loose so that the little water available soon drains off. The best soils are loams because they are a mixture of sand and clay, with a considerable amount of humus. This combination holds the proper amount of moisture, drains well, and does not harden at the surface.

Soil is formed principally by the slow cracking of rocks from frost or root penetra-

Fig. 20-10. A landslide in Ecuador. Thin tropical soils do not hold during heavy tropical rains.

tion or by erosion. Humus, organic matter from decaying plant and animal bodies, covers the best soil and provides roots with mineral salts containing nitrogen and phosphorus formed when the final decay process is complete. Humus is a substance added to soil and is not a soil type. Principal soil types are sand, clay, and loam, as mentioned previously. Soil is transported by a variety of means, each of which results in a characteristic deposit. Thus alluvium is formed by receding flood waters that deposit the soil they have held. This is generally a rich soil. **Lacustrine** soils form in lake bottoms that have eventually dried. Such soils are rich if they are high in organic matter, but they may also be poor because of heavy mineral and salt deposits, including sodium salts, borax, and lime. Glacial tills are stony soils deposited at the foot of receding glaciers and are generally poor for agricultural purposes. In desert regions much of the soil is in the form of loess, a deposit of fine yellowish gray loam transported by wind. Actually, loess is rich but porous, causing rapid runoff of the sparse water supply. Usually this soil crusts over to form a hard cover, making plant growth very difficult.

Chemicals. The mineral content of soils is critical for plant growth. Any soil deficient in essential minerals will not support plant growth unless this lack is counteracted either by the use of fertilizers or by water running into the soil from other areas. Even the trace elements must be present. Soils deficient in any of these show characteristic growth deficiencies in the plants they support. Although organic substances are not used directly by the plant, soils high in organic materials support better plant growth by gradually breaking them down into their component parts. The organic matter is also a source of nitrogen through bacterial action.

The type of minerals and the amount of humus regulate the pH (acidity and alkalinity) of soil. Some soils, particularly in humid regions, are acid in reaction because the basic salts are leached out. Others, particularly in desert regions or in certain types of bogs, are alkaline in reaction. Most plants grow best in nearly neutral or slightly acid soils. The salt (NaCl) content of soil, although

Fig. 20-11. Effect of salt on plant growth. The grass in the drum on *right* is growing in soil with a high salt (NaCl) content; the one on *left,* grown under the same conditions but without salt, shows greater growth. (USDA photograph.)

it does not affect the pH, is sometimes so high that plants cannot grow; they are unable to absorb water because of the increase in osmotic pressure outside the root cells (Fig. 20-11). Areas with poor drainage and high evaporation suffer from high salt content. The most notable of such areas is the Great Salt Desert of Utah.

Unstable soils result in a particular kind of dynamic ecology. Among the best examples of this are sandy areas with prevailing winds that blow and shift the soil surface into dunes. Dunes are mounds of sand that shift continually so that momentarily they cover established vegetation. The battle is endless in such regions. Only through man's efforts can the soil be held in place long enough for a firm plant cover to become established. The slightest disturbance from disease, insect pests, or fire will again release the nomadic dunes. They may be man made through faulty agricultural practices, a good example of which is part of the dust bowl.

Other than nitrogen, the minerals essential for plant growth come from the weather-

ing of rocks. Nitrogen, however, does not, since its origin is ultimately the atmosphere. Before atmospheric nitrogen can become available to a plant in soil, it must be combined with other elements. Symbiotic and free-living microorganisms in soil are primarily responsible for combining free nitrogen from the air with other elements. A group of bacteria (*Rhizobium* sp.) that live symbiotically on certain legumes are able to utilize the free nitrogen gas from the air in their metabolism and thus fix the nitrogen in a combined form. These bacteria are called **nitrogen-fixing bacteria.** They are responsible for the formation of nodules on legumes (Fig. 5-11) such as clover, vetch, and the pea. The combined nitrogen, probably in an amino form, is absorbed by the legume and synthesized into its proteins. Other saprophytic bacteria, such as *Azotobacter* and *Clostridium,* are also capable of nitrogen fixation. The organic nitrogen of the legume after death is broken down into nitrates by still other microorganisms in the soil and can then be used by nonleguminous plants. Ani-

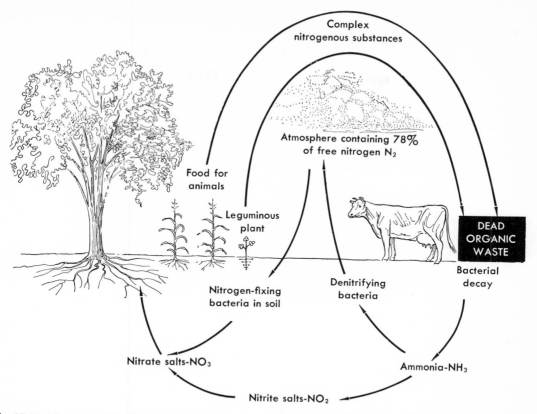

Fig. 20-12. Nitrogen cycle. There are various sources for the nitrogen needed for protein production in both plants and animals. Animals obtain all their nitrogen from either plants or other animals. Plants, however, obtain all theirs from the soil.

mals secure their nitrogen supply from plants or other animals as food, which is used in synthesizing their proteins. The nitrogen of the animal and plant proteins is returned after death to the soil, where it is broken down into nitrates. However, this breakdown occurs in several stages and is aided by the microorganisms in the soil. The organic residue from dead plants and animals is reduced first to ammonia, then by nitrification to nitrites, and finally to nitrates. Nitrates are the only form in which plants can utilize nitrogen. The diagram in Fig. 20-12 illustrates graphically the use and reuse of nitrogen. The transfer of nitrogen from the air to the soil, then to plant and animal proteins, and back again to the soil and air is called the **nitrogen cycle**.

BIOTIC ENVIRONMENT

In addition to the physical factors in the environment, the relationships of individual plants to others of the same kind and of different kinds are of equal importance. Through a study of these relationships the transition from the autecological approach to the synecological approach is apparent. Both these approaches are necessary if the complexities of biological phenomena are to be understood.

Competition with other plants of the same kind. The simplest form of competition in plant life is the need for a certain amount of space for roots in the soil and for exposure of leaves to light. As any plant grower knows, well-spaced plants grow best. The more crowded the conditions, the greater will be the reduction in the size of the plants and the number of seeds. On the other hand, beneficial effects from crowding are also possible. The colony of a single species of plant, once it occupies all available space in a given area, presents a united front against the invasion of other species. Dioecious plants must be close enough together so that pollen can easily fertilize the female plant. In many monoecious plants this is necessary also because the anthers mature at a different time from the pistils or some structural device prevents cross-fertilization; therefore fertilization may not occur in widely separated plants.

Competition among plants of different kinds. This is well known to the practical biologist. Weeds penetrate any cultivated area, and the struggle between these two types of plants is as active and dynamic in its slow, persistent way, as is competition between animals. Some plants have developed special structures to aid them in their invasion techniques. Reproduction by runners is one, the more vigorous part of the plant colony contributing to the growth of the invading shoot until it becomes established. Vines, such as poison ivy or the strangling fig, can completely choke out large trees and occupy the space once held by them. A vine growing over smaller plants can block out the sunlight that would otherwise reach them, and these plants must give way to the new growth. Any pest or disease that weakens one species and leaves another unharmed will give the latter the advantage, and it will soon occupy the space held by the former. Certain plants have the ability to colonize quickly an area destroyed by fire. However, once they have become established, other plant species will invade the area and by some superior growth habit push out the original colony to take over. Thus there is a continuous succession of plants until a climax condition is reached. Once this is achieved, conditions remain more stable until the area is again disturbed. Weeds are particularly tenacious in this respect. Some, for example, plantain (Fig. 20-13), will flower and fruit, using nutrients stored in their leaves, after they have

Fig. 20-13. Plantain fruiting even after being uprooted. (USDA photograph.)

been removed from the soil by the gardener.

Parasites of plants fall into two main classes, bacteria (Fig. 20-14) and fungi that cause disease, and worms that cause nematode diseases by invading the plant tissues. External feeders such as insects and other herbivores cannot be considered parasites in the usual sense of the word because their effect on the plant is immediate. Plant diseases caused by fungi are many, and because of their importance to agriculture, their study and control form a separate field of biology, plant pathology. Some of these are discussed in Chapter 7. Today, nematode parasites of plants have become increasingly important. They have been brought to the attention of the lay public by importation of the golden nematode disease of potatoes. There are many other diseases of this kind, all causing serious loss and influencing competition for space by plants.

Certain symbiotic relationships among plants occur. One of the best known is the relationship between certain algae and fungi to form lichens. This is the relationship between two plants of different kinds for their

Fig. 20-14. Crown gall bacterial disease affecting plants. (Courtesy A. S. Riker, Department of Plant Pathology, University of Wisconsin, Madison, Wis.)

mutual benefit and is called **symbiosis.** Other relationships of this nature are to be found, such as the denitrifying bacteria that grow in root nodules.

The term "symbiosis" is now used in a much broader sense than is implied in the preceding paragraph. It includes almost any mutual and constant relationship between two species, including parasitism. This is the more generally used meaning of the word, but a few moments' reflection will show how difficult it is to limit the term and to assign a given case to this phenomenon.

Animal relationships. Plants of many species, but not all, are eaten by many kinds of animals. Without this relationship, there would be no animals. It seems obvious,

therefore, that the growth of plants is sufficient not only to maintain the plants themselves but also to provide for the entire animal kingdom, including man. This is done primarily by overproduction of the vegetative and reproductive parts of the plant in the form of accessory tissue such as fruits and nuts. Secretions such as nectar and sap also provide food for animals. These relationships are quite obvious. Less apparent is the fact that feeding on plants by animals has become an established pattern so that, in the normal balance of nature, the plant species are not harmed by this predatory action. It is only when the balance is momentarily upset that the plant population is harmed. If the plant is under cultivation by man, animal predators are called pests, and steps are taken to eliminate them. This view is anthropocentric and not meaningful in a true biological understanding of the phenomenon. Plants not cultivated or otherwise useful also are held in check by predators. So far as the plant species is concerned, being preyed upon by animals is only one factor to prevent the species from occupying the maximum possible space for which it is fitted. This means that there may be a constant struggle for the greater area occupied. Likewise, from the point of view of the animals, if the plants grew better, the animals would have more food and could occupy more space. Therefore, in any given case, the relationship is one of dynamic equilibrium, sometimes with great fluctuations of individuals. Once host specificity such as this is firmly established, it will result in the extinction of both. Only man and a few other animals, for example, some ants, have learned to profit by the cultivation of plants. When this happens, the relationship is no longer one of host and predator but becomes symbiotic in the broadest sense.

Actually, the kinds of animals feeding directly upon plants are relatively few. Most animals feed on other animals as a part of a food chain that only eventually gets back to the plants. These food chains may involve many kinds of plants and animals, one eating the other until either man or some animal

393

dominant in a given community is reached.

Most notable of the animals that are **phytophagous** (that is, feed on plant parts) are many insects and the herbivorous mammals. Both of these groups have many species that feed directly upon the plants. Aquatic animals, particularly some arthropods and some fish, are the herbivores of the sea. Most of these are filter feeders; that is, they feed by straining seawater through their mouths and eat the filtrate of microscopic plants, as well as of animals. Together, the floating plants and animals form the **plankton.** The extent of this feeding is apparent when one sees the thousands of tons of fish removed from the sea each year or when one realizes that some of the whales are plankton feeders and grow to tremendous sizes on a diet of plankton.

The following outline briefly sketches "plant" and "animal" relationships.

I. Unicellular plants

Consumed entirely by macrozoa such as sponges, jellyfish, anemones, corals, flatworms, rotifers, entoprocts, bryozoans, sea worms, lamp shells, mollusks, worms, crustaceans, starfish, and vertebrates.

II. Multicellular algae

A. Marine algae—eaten mainly by crustaceans and vertebrates.

B. Freshwater algae—eaten mainly by crustaceans, fish, reptiles, and birds, but rarely by mammals.

C. Terrestrial algae—eaten by small insects and rarely by some other terrestrial arthropods.

III. Vascular plants

A. Roots—insects, particularly the larvae of beetles and moths, are obligate root feeders, and a number of larger mammals, for example, pigs, are sometimes root feeders. The insects are primarily borers in roots, which causes considerable loss of vitality and frequently death to the plants.

B. Stems—this part of the plant is usually the least palatable, partly because of the hard, nonnutritious supporting tissue. The majority of stem-feeders are boring insects or ruminate mammals, for example, cow and horse.

C. Leaves—the greatest variety of plant-feeding animals are leaf feeders. Many mammals depend entirely on the consumption of leaves of vascular plants. Unless the plant is well protected by heavy spines or poisonous secretions, it may be stripped of its leaves by these animals. In addition, the adults and larvae of beetles, moths, and some bees feed on or in the leaves of plants. These number into the tens of thousands of species, many of which are host-specific. Because of the variety of leaf-feeding mechanisms, these must be further elaborated.

1. Feeders on the entire leaf—insect larvae and mammals.

2. Leaf-skeletonizers—insect larvae that feed only on the softer parts of the leaf, leaving untouched the skeletons of vascular tissue.

3. Leaf-miners—larvae of moths, flies, midges, sawflies, and beetles. The larvae and sometimes the adults of these insects bore into the leaf and feed on the parenchyma tissue, leaving a tunnel or mine in the leaf. The mine grows larger as the insect grows, producing a characteristic winding tunnel pattern on the leaf that is different for each species of miner.

4. Leaf-cutters—some species of insects cut away parts of the leaf, using it to build nests, or as their food supply in subterranean fungus gardens.

5. Leaf-rollers—a few insects, though not feeding on the leaf, cut and roll the leaf to form a nest held together by silk.

D. Fruit- and seed-feeders—mammals, birds, and insects consume a large part of the fruit and seeds produced by many species of plants.

E. Flower-feeders—animals consuming floral parts before they mature into fruit and seeds are relatively rare. However, the sugary secretion of the flower, the nectar, is consumed by birds and insects. Usually, while these animals are feeding, they come in contact with pollen, which is also gathered, either by accident or on purpose, and transferred to other plants so that pollination results. A large number of flowering plants are entirely dependent on this method of pollination. A few insects are pollen feeders, but do not pollinate.

F. Gall-makers—a number of insects, particularly flies and wasps, and also some fungi produce galls, that is, growths on leaves, stems, and roots of plants. The shape and characteristics of the gall depend on the insect or fungus guest living within and are specific for each guest and each host. The galls are different from those resulting from disease in that the normal function of the plant is little affected.

G. Sap-feeders—insects with sucking mouthparts, particularly the true bugs such as plant lice and some others, are able to pierce the tissues of plants to feed directly on the fluid within the vascular tissue of the leaves and stems of the plant. This not only weakens the plant but also may stunt or even kill back the plant if feeding takes place at the growing tip. The scars left by such feeding also destroy the esthetic value of flowers and fruit.

IV. Fungi

The larger fungi such as the sac fungi and club fungi are eaten by some mammals, rarely by birds, and by a great number of insects, many of which are host-specific. Several families of flies and beetles have many species that are obligatory feeders on fungi of this type.

Other more specialized animal-plant relationships are properly reserved for detailed study.

PLANT HABITATS

The environment of plants, combined with their adaptive structures and physiological latitude, determines where they live, their habitat. This exceedingly complex web of functions programmed in the genes of all organisms results in a changing world of great variety. The few plants living in the Chihuahuan Desert near Big Bend National Park, Texas, struggle for existence in a severe habitat (Fig. 20-15). A closer look at habitats reveals certain patterns of distribution that make it possible to describe large areas of the earth's surface in terms of the flora and fauna found there. Each species, no matter how

Fig. 20-15. Typical xerophyte (a cactus) in the Chihuahuan Desert in Texas. Note the sparse vegetation in this extremely arid situation.

abundant or how rare it may be, has a range that may be circumscribed in geographical and environmental terms.

The habitats of both plant and animals are difficult to consider separately because each area investigated combines many specialized habitats. It is customary for ecological studies to be made from one of two approaches. One deals with the study of the individual organism, species, or group of related species. This organismic approach may be termed **autecology.** For example, a single plant may be studied under varying growth conditions. Its environment and its relationships to the environment may be described for a single species. From data obtained for each of a group of species, generalized statements may be made about the ecology of a genus, a family, or even an order. The larger the taxon, the less precise any of these statements may be. More recent studies have been concentrated on local breeding populations of a single species, and interest in the comparative autecology of separate populations leads to information about speciation pathways.

A second approach, termed **synecology,** deals with the study of **communities,** or associations, of different kinds of organisms functioning more or less as a unit, an ecosystem. Communities may be described according to the dominant (most obvious) plants in the association, such as an oak-hickory-magnolia association. Such studies are often called population ecology, community ecology, or ecosystem ecology.

The tendency of several fields or branches of biology to concentrate on populations, that is, several different biological approaches, has led to some confusion. At the present time the term "population" is used by three separate biological disciplines in three different ways. The systematist uses the term to describe an interbreeding group of individuals that form one part of the total living members of a species. The geneticist is concerned with populations according to their genotypes and studies them only as a sample of the species. The ecologist treats a population as a segment of a community of species with interrelated functions.

The mathematics and details of the energy cycles in ecosystems is beyond the scope of this text. The field of ecology is vast, and many important topics cannot be mentioned in an introductory text, for should we attempt to do so, we would become involved in a complex terminology that would tend to magnify this part of botany out of due proportion. All the basic principles of botany are involved in the study of ecology so that the student is assured of the necessary foundation for more specialized studies if desired.

As we have shown, each species of plant has not only a range but also a relationship with the physical and the biological environment. It is possible, therefore, not only to map the geographical range of the species but also to give a habitat description of each. Obviously, no attempt will be made here to do this for even the most dominant of the flowering plants. Certain general statements are made about the distribution of the various divisions of the plants in Chapters 6 to 11.

A close study of the range of any species shows that it is not continuous throughout the geographical limits of its distribution. Instead it is divided into **populations,** each with a range that is circumscribed by the limits of interbreeding. The populations fall into roughly three major types. The first is characteristic of a common widespread species. Here there are broad areas of overlap in the range of a population so that the gene flow is limited only under unusually adverse conditions, for example, long periods of foul weather, cold, or dry condition in part of the range of the species, or similar restrictions to free crossbreeding. The second situation is intermediate and characterizes a great many plants that occupy somewhat specialized habitats. Here the populations are more or less isolated, but there is considerable gene flow through wind or insect pollination. Again, during severe weather periods, portions of the range may become temporarily isolated from the rest of the species. The third condition is characteristic of those species often described as rare. They are composed of a series of isolated populations, usually with specialized habitat requirements. Each population may show sufficient variation from the others to exhibit a succession of characters following a geographical pattern. This is known as a cline. The gene flow between these populations is very limited and miscegenation may not take place except after a lapse of several years.

With this concept of a population in mind, combined with our knowledge of gene action, it is now possible to understand the mechanisms of isolation and evolution described in the next chapter. It is for this reason that the systematist is very much interested in the distribution and ecology of plants.

PLANT DISTRIBUTION IN RELATION TO ENVIRONMENT

The distribution of a species, the geographical area and ecological habitat occupied by any particular plant species, is determined by the anatomical and physiological characteristics of each member of the population. The physical and biological environments of a given geographical area are not suited equally well to all species. Therefore any particular area is inhabited by characteristic species.

Two major factors determine the distribution of plants. Generically speaking, these are heredity and environment. The hereditary composition of a plant determines the degree of adaptability to a given environment; that is, the organism can live in a particular environment only to the extent of its genetic latitude. Within this framework the individual is affected by environmental factors.

NICHE

The analysis of a plant community shows not only those complex interrelationships already described in this chapter as ecosystems, but also that each species has a functional place in the ecosystem known as its **niche.** Niche differs from habitat, as function differs from place or site. No two species function biologically exactly the same within the ecosystem. Thus some species of plants produce the food for the community and others have the role of decomposers and occasionally predators (parasites—bacteria and fungi). The slot of each species is clearly defined and any tendency to leave this slot and invade another results in competition. No two species can occupy the same niche at the same time (known as **"Gause's principle"**) and the struggle for the niche results in one or the other species being eliminated. Many examples of this have been shown experimentally. One of the most dramatic is an experiment by Harper,* who showed that one of the two species of duckweed (small floating aquatic flowering plants) grown in mixed cultures would soon dominate the other as they competed for space simply because one species was able to float better than the other. Often some plants (for example, a sage) produce aromatic materials that prevent the germination of seeds of grasses. This sets up a zone around these plants where grass does not grow. It seems that always one species is a little better adapted than the other so that it wins the slot. This explains much of what happens in succession. This is not to deny that two closely similar plants can function within the same community—it depends on their demands. Following the same ways of life in equilibrium is possible so long as the resources of the community are sufficient to support them both. Nevertheless there are always subtle differences between the two. All species are in constant danger of aggression from other species utilizing a different niche. Thus a true climax situation, implying equilibrium, cannot exist for long, especially with the influence of man on the ecosystems.

PLANT FORMS IN RELATION TO ENVIRONMENT

Plants, like animals, have definite shapes and forms according to the habitat in which they live. Animal form is usually geometrical, whereas plant form is randomly organized. Even so, it is not difficult to see habitus types in plant species and to correlate them with

---

*Harper, J. L. 1961. Symposia of the Society for Experimental Biology **15:**1-39.

their habitat. Some of the more obvious of these types are listed and described.

Hydrophytes. Plants growing in water are characterized by lanceolate leaves if they are submerged, with stems reaching nearly to the surface and stem length depending on the depth of the water, or with floating leaves having dorsal stomas for direct contact with the atmosphere. Hydrophytes may have air bladders to keep the plant afloat. All such plants are turgid since water is never a limiting factor, and the tissues are usually thick, with large alveolar spaces.

Freshwater habitats are often separate climax situations, with many cyclic activities going on independently of land plants and animals. One such situation was created at the Panama Canal when several river valleys were flooded for the canal. Large mats of floating vegetation (Fig. 20-16) are formed that not only are discrete ecological niches but also that provide vast areas for the breeding of larvae of anopheline mosquitoes, the adults of which are carriers of malaria.

Mesophytes. Most terrestrial species are mesophytes, the body form of which varies according to the closeness of an adequate water supply for the roots and the method of obtaining maximum exposure to light. This type of plant is characteristic of most of the earth's surface, and their particular variations

and characteristics can only be described in a lengthy systematic floral catalog.

Xerophytes. Desert and arid-land plants provide a series of limited habitus features familiar to any desert traveler. The spiny cactus is characterized by lack of leaves, and the chlorophyll is confined to thick, water-filled stems that form a storage system capable of holding the large amounts of water absorbed during the limited period of rainfall. The stomas of these plants are usually located within pits on the stem surface, and the body of the plant is covered by a waxy secretion, making the plant nearly impervious to loss of water by transpiration. Other desert plants may resemble mesophytes in many respects but are often leafless during much of the year, or the leaves are reduced to small leaflets. The former produces leaves during the rainy season only and remains nearly dormant throughout the rest of the year. The root system of xerophytes either extends very deep into the ground to reach deep, subsurface water or is very shallow but widespread, as in cactuses, to absorb as much of the surface water as possible during the brief rains. This action is called **hydrotropism** (Fig. 14-17).

Aerial plants. A few plants live in aerial habitats completely independent of soil and are therefore without a root system. Spanish

Fig. 20-16. Large mats of floating vegetation in Gatun Lake in Panama.

Fig. 20-17. Field of pitcher plants (three species intermixed) in a swampy area in Alabama. These plants are adapted to soils of a low nitrogen content and high acidity.

moss in the southeastern United States is an example. These plants are dependent upon a humid climate and absorb water through their stomas directly from the atmosphere. Other plants, such as certain bromeliads of the tropics, are epiphytes, some saprophytic and some not.

Other types. Saprophytes, commensals, and parasitic plants are likewise characterized by a particular body form, easily recognized in all cases. The Indian pipe is an example of a seed-producing saprophyte. Certain lichens form a type of Spanish moss on trees in California (not to be confused with the Spanish moss bromeliad of the southeast). Orchids are also examples of this, but are capable of extensive photosynthesis as well. Mosses and fungi may take on particular forms in special habitats. There are many other examples.

Sometimes plants have devised elaborate mechanisms to aid in obtaining food. There are many species of insectivorous plants (Fig. 20-17) with trapping devices for the capture of insects, the decaying bodies of which are a supplementary source of minerals and salts needed for the metabolism of the plant. Plants such as Venus's flytrap and the pitcher plant are, by their insect-capturing abilities, capable of living in areas deficient (for them) in nitrogen. These are usually boggy areas that are low in nutrients because of the downward movement of water that prevents the accumulation of minerals in the soil.

DISTRIBUTION MECHANISMS OF PLANTS

All plants have some kind of dispersal mechanism, most of which are discussed elsewhere in this book. Therefore the following outline will serve merely as a summary. Although plants are usually thought of as stationary organisms, at some time during the life history of every species locomotion is possible and usually extensive.

1. Floating distribution
   a. Entire vegetative plant—unicellular and multicellular algae, bacteria, and some vascular plants are dispersed by both freshwater and saltwater currents.
   b. Vegetative parts—storm action or partial decay sometimes results in the dispersal of plant parts that will continue to live and grow into a new plant by rooting, once a new location is reached.
   c. Reproductive parts—spores and zygotes of many plants are dispersed by floating in water.
   d. Seeds—seeds are dispersed by water currents.
2. Swimming
   A few of the algae are flagellated in the vegetative stage and are dispersed by active swimming. Many algae have flagellated gametes and zygotes that allow this type of dispersal.
3. Ground distribution
   Some vegetative dispersal is possible by the growth of runners, either as stems or roots. New plants become established in this manner, finally breaking away from the parent plant. A few species become established in new areas by the rooting of growing leaves after becoming detached from the parent.
4. Aerial distribution
   a. Vines—the growth of vines over large areas is the simplest means of aerial distribution.
   b. Spores—the terrestrial cryptogams depend almost entirely on the dispersal of spores formed vegetatively, rarely by sexual reproduction. These light, microscopic plants are distributed by wind currents and often have worldwide distribution. In past geological ages, before the evolution of seed plants, this means of dispersal was nearly universal among terrestrial plants.
   c. Seeds—many seeds have, as accessory parts, fine hairs or winglike projections that are caught by wind currents, to be carried long distances. Most are lost—but

even with the establishment of one in a new locality, a new population is started. It is, of course, advantageous to a plant species to have the widest possible seed dispersal because the seed stage is precarious. Many things can happen to prevent the germination of seeds; therefore great numbers are produced by the plant. The greater the number of seeds, the better they are protected; and the more generally they are distributed, the better are their chances for germination.

5. Distribution by animals
   a. Fruit—many plants produce a fleshy fruit that may be eaten by animals who may carry it for some distance before it is consumed. If large, the seeds are removed and tossed aside to germinate. Small seeds are eaten with the fruit and passed through the digestive tract unharmed by chewing or digestive juices, finally to be deposited on the soil along with the fertile feces, providing an ideal situation for germination and growth. Many barren islands remote from continental land masses have been populated by plants brought to them in the intestines of birds.
   b. Seeds—nuts are gathered and stored by rodents and similar animals. A few roll away and may germinate in new ground. Other seeds may be equipped with spines that catch in the fur of animals or clothing and are thus transported. Sometimes complicated mechanisms are devised; for example, the seeds of parasitic mistletoe are transported by the mistle thrush. This bird is especially fond of mistletoe berries but does not eat the seeds. The plant is dependent on finding a suitable branch of a juniper tree to parasitize. The mistle thrush rubs the seeds of the berries off against the branch on which it perches, and the seeds become fixed to the branch by the viscid internal flesh of the berry. A large number of woodland plants are dispersed by ants that pick out seeds made attractive to them by special oil bodies or **elaiosomes.** The desire of the ants to imbibe this oil causes them to gather the seeds and carry them off. They do not eat the seeds, and they are cast away after the oil is consumed, thereby dispersing them.

c. Pollen—dispersal of pollen by animals (bees, etc.) serves as a means of mixing genetic traits throughout the range of the species and does not contribute directly to the extension of the range. However, when these genes are thus mixed, the amount of variation may be higher and the chance of adaptation to new areas somewhat greater.

This summary covers only the principal dispersal mechanisms. Other special features are discussed in books on ecology and plant distribution.

CLASSIFICATION OF BIOTIC COMMUNITIES

No entirely satisfactory system of divisions of the world's flora can be devised except as a purely arbitrary descriptive device. Many widely different systems are used, but certain ones are commonly encountered in the literature. The three means of classification are based on (1) major or dominant species;

(2) the physical habitat, for example, aquatic, terrestrial, and marine; (3) function, for example, the energy systems, or ecosystems.

The term **biota** refers to the entire assemblage of plants and animals, in all their stages, living on the earth at any given period of time. Thus it is the biota that we classify into various levels of communities based on their kind and their distribution.

Realm

Realms deal with the gross distribution patterns. Several systems are available to describe the division of the world into floristic regions or realms. Each of these is rather complicated and is based both on climate and major or dominant plants. They have not been as widely accepted or as useful as the one used by zoogeographers. These regions are given here exactly as used by zoologists (Fig. 20-18). Generally botanists do not use this system of realms but prefer to use geographical regions instead. Thus one would

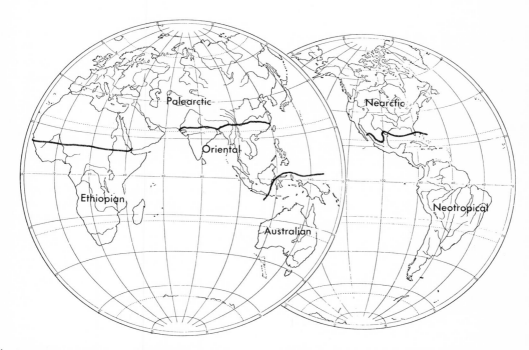

Fig. 20-18. Zoogeographical realms. These are the classical regions used by zoologists but serve no more than a convenient reference for the discussion of the distribution of most organisms.

study the flora of North America rather than the flora of the Nearctic region. There are many floras of states or geographical regions.

Biomes

It is primarily the physical environment that determines the distribution of plants. Rainfall, sunlight, and temperature variations of the earth's surface are reflected in the vegetation of each area. These conditions bring about rather distinct areas known as **biomes** (Fig. 20-19), the major vegetation zones of the continents. Each biome may appear in several areas in the world. These regions will be composed of a distinctive set of plants so that the separate biomes of each kind, although superficially resembling each other, differ greatly in their species communities.

Several classifications of biomes are possible, but there is general agreement on these areas, except some are occasionally lumped together for simplicity of treatment. The 10 biomes treated here are found in many or most of the realms.

Ocean. Only recently biologists have turned their interest to studies of the ocean as a biome. Its complexities form almost a separate subdivision of ecology. Certainly little is known yet about the distribution of plants and their interrelationships in the ocean environment. This vast area, 70% of the earth's surface, compared to only 3% occupied by freshwater, has been referred to as the last frontier. It is not a single barren biome, but rather, one characterized by a very diversified fauna and flora. Most of its surface is occupied by microscopic plants able to withstand the constant motion of the waves. Such environmental factors as light, temperature, salinity, and pressure, in harmony and disharmony with many animals, form an important site of biological energy exchange. As one approaches the shore, the environment becomes more severe. It is well known that the interface (**ecotone**) of a biome or a more restricted habitat is an area of tension, and so, one biome may abruptly change into another.

Reef. One such oceanic habitat is the reef. Many tropical islands and offshore areas near continents in tropical regions have formations called barrier reefs. Algae as well as

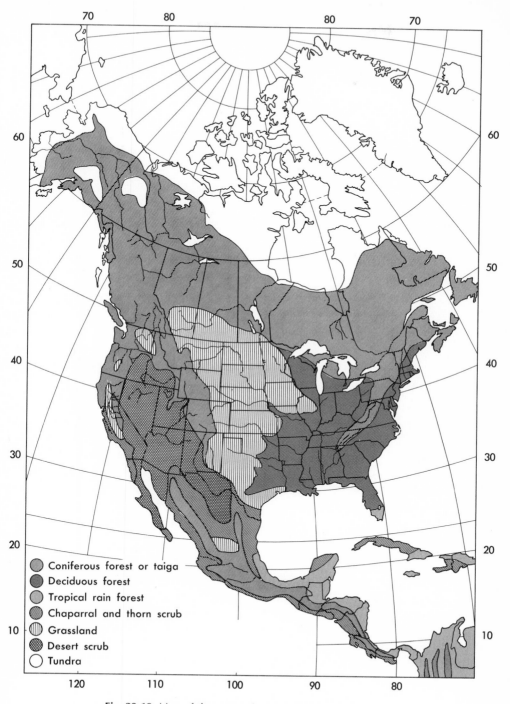

Fig. 20-19. Map of the vegetation types of North America.

Legend:
- Coniferous forest or taiga
- Deciduous forest
- Tropical rain forest
- Chaparral and thorn scrub
- Grassland
- Desert scrub
- Tundra

Fig. 20-20. Tundra. Note particularly the absence of trees except at the interface, where they are dwarfed (left edge of scene). (U.S. Forest Service.)

Fig. 20-21. Oat field in Alaska where the growing season is short, but the sunlight hours each day are long. (USDA photograph.)

coral (Coelenterata) form intricate habitats, the complexities of which are the subject of a great amount of literature.

Shore. True shore areas, especially in the intertidal region, are subjected to a wide range of environmental changes, and here one finds various uniquely adapted plants such as fucus and other typical intertidal plants.

Tundra. Tundra areas are located on the fringe of the northern and southern limits of the continent and on the tops of tall mountains, usually at elevations over 12,000 feet (Fig. 20-20). Such areas (Fig. 20-21) are characterized by long cold winters and short rapid growing seasons, which produce stunted plants. Characteristic plants are mosses, lichens, grasses, and sedges. True

404

Fig. 20-22. Western grassland area with trees only along permanent stream beds, and foothill mountains at the edge of the grassland. (USDA photograph.)

tundra has permanently frozen soil (permafrost) beneath a shallow surface. This biome merges into the extreme northern and southern eternal snow fields, which are entirely devoid of all plant life except for occasional patches of unicellular algae and sometimes on exposed rocky surfaces, a few lichens.

Grasslands. The great central plains of the United States, Canada, Asia, and limited areas of Africa and Argentina, form the grasslands (Fig. 20-22). These regions are treeless or nearly so, except for small areas along the banks of streams. Characteristic plants are grasses and composites such as asters, goldenrod, and sunflowers. Rainfall is sufficient to support grass but not trees, so it is the controlling factor. Unfortunately much of the world's grassland has been destroyed by overgrazing and is now virtually a desert. In tropical regions with abundant rainfall, but long dry seasons, a special type of a grassland termed "savanna" is found. Scattered clumps of trees may occur in these areas. Extensive areas of central and southern Africa, parts of Australia, and scattered areas in South America have this kind of grassland. It is here that much of the big-game herbivores are found. Such areas are subject to widespread fires that serve to remove the dead grass stems and make room for the new growth.

Chaparral. Areas with heavy winter rains, and long, hot, dry summers, especially on mountain slopes, may develop a climax of trees and shrubs with thick, hard, evergreen leaves. These areas are called chaparrals, probably the most poorly defined biome. Such areas are widely scattered but are best known from California, Mexico, along the shores of the Mediterranean Sea, and along the southern coast of Australia. Again, fire plays an important part in the growth and decline of this climax.

Deserts. Areas that receive an average of less than 10 inches of rainfall per year are considered deserts. Associated with these are semiarid areas that receive more rainfall but that remain arid because of the rapid runoff of rain. The deserts of North America occur in eastern Oregon, western Idaho, Utah, and Nevada, extreme western Colorado, New Mexico, the eastern part of southern Califor-

Fig. 20-23. Typical desert scene in western Texas. This area is nearly bare of vegetation. Similar areas with wind-eroded rock finely ground into sand may form dunes as in parts of the Sahara desert.

nia, Arizona, western Texas (Fig. 20-23), and much of northern Mexico. Cactus and similar xerophytes, yucca, mesquite, creosote bush, and grasses of various sorts characterize these areas. The desert regions of South America are found west of the Andes Mountains along the coast of Ecuador, Peru, Chile, and central Argentina. Extensive deserts are found in Africa, Asia, and Australia. The barren areas of the Sahara desert may have wind-blown sand dunes of fine sand, but much of the area is rocky or gravelly, with an occasional oasis surrounding an artesian well. Southwest Africa has a large area of arid land somewhat similar to the desert of southwestern United States. The Arabian peninsula, eastern Pakistan, parts of Siberia, and much of central Australia complete the list of deserts of the world. Each of the major deserts is subdivided into characteristic regions, and each has a unique flora. Most deserts lie in the leeward shadow of high mountain ranges that receive most of the water dropped by the clouds as they pass over the mountains.

True desert in North America is typified by the Sonoran desert in Arizona (Fig. 20-24). The presence of creosote bush (*Larrea* sp.) and saguaro characterize much of this region. Cactus may be an indicator of deserts, but many cactuses are found outside of desert regions, and some deserts have cactuslike plants of the family Euphorbiaceae.

Coniferous forests. Coniferous trees, especially pine, spruce, and juniper, often grow in almost pure stands that create a distinctive biome. However, different regions have different compositions. Those of North America are described separately. Coniferous forests also occur across northern Europe and Siberia as well as at higher altitudes of most mountain ranges.

Northern evergreen forests. The area south of the tundra or below the tundra on moun-

Fig. 20-24. Sonoran desert. This type of desert maintains a variety of distinctive plants and offers interesting and rather simple associations of plants and animals, each with special adaptations for survival in an area having less than 10 inches of rainfall annually. Note the giant cactus. (U.S. Forest Service.)

Fig. 20-25. Southern evergreen forest of pure stands of pine. Note the characteristic ground cover of palmetto. (U.S. Forest Service.)

tains is composed mainly of conifers and forms the northern evergreen forests (also known as taiga, pronounced ty'ga). This area extends across Canada, where the climate is less severe than that of the tundra region but where also the winters are cold and the summers cool. Pine and similar conifers predominate, and there are a few deciduous and herbaceous plants.

Southern evergreen forests. The coastal plains from Virginia through Florida to eastern Texas form another type of evergreen forest (Fig. 20-25). These areas are often swampy, and much of the region is covered with longleaf or pitch pine. The wet areas grow live oak, water oak, bald cypress, gum, and magnolia. Spanish moss, an **epiphyte,** grows on the live oaks throughout most of the area.

Western evergreen forests. The Rocky Mountains and Pacific Coast areas at altitudes above 7000 feet are covered with ever-

greens of one kind or another, even when deserts are found at lower altitudes in the same area. There is an abundance of rainfall at the higher altitudes. The plants vary according to the altitude, with ponderosa pine (Fig. 20-26) at the lower altitudes and Douglas fir, white fir, quaking aspen, alpine fir, and spruce in ascending order to the higher levels.

Deciduous forests. Forests with trees that drop their leaves in winter are found in many temperate regions throughout the world (Fig. 20-27) south of the northern evergreen forests. These deciduous forests begin in North America in southern Ontario and western and central New York State and extend southwest along the Appalachian Mountains to Louisiana and eastern Texas and west to Oklahoma and southern Minnesota. Oak, hickory, maple, chestnut, elm, ash, birch, rhododendron, mountain laurel, and ferns of various species are typical. Higher rainfall

Fig. 20-26. Ponderosa pine in Kaibab Forest, Arizona. High rainfall occurs in the high mountains, which creates islands of forest surrounded by the lower desert. (U.S. Forest Service.)

Fig. 20-27. Eastern deciduous forest. A great variety of trees occurs in these forests, creating variable habitats. (U.S. Forest Service.)

Fig. 20-28. Tropical rain forest. Note the large tree ferns and the epiphytes not found outside these forests. (U.S. Forest Service.)

is characteristic of these regions, much of which has been severely lumbered. Almost no natural deciduous forests are left because of land clearing, lumbering, the growth of cities, and the building of roads.

Tropical forests. True tropical rain forests occur only in Central and South America. However, parts of Mexico, the lower Rio Grande region of Texas, the southern tip of Florida, the Florida Keys, and the West Indies (Fig. 20-28) have many of the elements characteristic of tropical forests. Palms, lianas, epiphytes, and orchids are characteristic plants of tropical rain forests, many of which are found in the less dense stands in North America. Other tropical rain forests occur in South America, in the Amazon Basin, in central and southern Africa, India, Burma, and the vast tropical regions of Indonesia, New Guinea, and northern Australia.

Community

Although both realms and biomes are in the broad sense communities, they are far too vast to be studied and described in enough detail to demonstrate the variety of ecosystems established in these areas. Therefore further division of the biome is necessary, even though such divisions are not always clearly defined. The term "community," as previously stated, is often applied to these aggregations of plants and animals. So many communities have been described for the terrestrial habitats of the earth that no attempt will be made to describe them here.

Communities are ever changing, so that even when they may appear at first to be uniform, close examination will reveal stages, or **sere,** through which portions pass in the development of a **climax** situation. A climax is the last sere of community development, and the most stable. As portions of a climax community die or are destroyed, a new series of **successions** take place, usually following an established order until the climax is again reached. Therefore certain plants, such as aspen or fire birch, are indicators of a sere, whereas others, for example, the ponderosa

pine, indicate a climax. A recent treatment of the vegetation types of the United States (Küchler, 1964) lists and maps well over 100 distinctive communities. A similar treatment by Shelford (1963) describes these and others in detail.

ECOLOGY, POLLUTION, AND ENVIRONMENTAL SCIENCE

One aspect of ecology of great concern to modern society is the ecology of human life. Man has greatly disturbed many, perhaps most, of the ecosystems of the earth. Land management, air and water pollution, and conservation of wildlife are now major problems of great concern to many biologists. A whole new field, environmental science, has suddenly achieved considerable importance. Each of its many branches are of direct concern to ecologists, and this applied ecology is attracting many students.

Two of these problems are of concern to the botanist: fire and fire exclusion, and air pollution. Botanists are involved in many other environmental problems, but these have been discussed elsewhere in this text.

Fire and fire exclusion

Certain seeds depend on the heat from a ground cover fire to initiate germination. Grass foliage dies back during the winter months and seeds lie dormant. The burning of the dead vegetation stimulates the production of auxin in the tip of grasses, which in turn promotes the growth of new leaves. Most fires are started by lightning striking a forest or grassland; very few relative to this cause (approximately 2%) are started through carelessness. All fire is the rapid oxidation of organic matter composed almost entirely of plant material. These natural fires have occurred over the earth's surface for as long as plants have occupied the land. They play an important role in the balance and succession of plant species in many ecosystems. The effect of these fires is to eliminate some of the dead wood and leaves and certain plants that cannot tolerate fire. Those plants that remain uninjured after a fire are particular to a cer-

Fig. 20-29. Controlled burning, developed at Tall Timbers Research Station near Tallahassee, Florida, stabilizes the habitat and prevents both destructive fires and succession from the coniferous forest to a deciduous forest.

tain stage in the succession of plants toward a climax. Both in forests and in grasslands these plant associations usually represent an early stage in the succession toward a climax.

Probably since the beginning of agriculture man has noticed the effect of ground fires on the growth of vegetation, and he has intentionally set fire to the dead vegetation to achieve the natural effect of encouraging plant growth. Most primitive agriculturalists today set fire to their fields at the beginning of the growing seasons. However, through the efforts of foresters and rangeland managers the practice of spring burning has been discouraged. Preventing fires is called fire exclusion, a practice that has brought about new problems. Forests kept free of fire produce a dense undergrowth that slowly causes the change of the forest ecosystem and eventually leads to an entirely different type of forest. To the detriment of the commercial forest, this undergrowth, once ignited by lightning or carelessness, builds up an intense heat, which causes crown fires in the valuable timber trees. The elimination of range or grassland fires either by prohibiting them or putting them out when they start permits the growth of plants that are not fire tolerant and at the same time greatly changes the nature of the grassland by the presence of woody shrubs and small trees. It is now believed that many areas profit greatly by the practice of controlled burning (Fig. 20-29). This stabilizes the ecosystem at the desired point and prevents further succession toward a climax. One of the best examples of this is in the southeastern United States where it is necessary to burn each year or at least once every 2 or 3 years in order to keep the pine forest stable. These valuable forests are a

Fig. 20-30. An example of the worldwide problem of air pollution, in this case, in a remote area in Arizona where a chemical plant contributes to the problem.

major source of wood for paper and plastics. In grassland areas the effects of burning are several. It chiefly (1) stimulates the growth of certain grasses, (2) keeps out woody plants, (3) increases the proportion of desirable range plants, and (4) cuts down or eliminates many of the imported and undesirable weed plants. Certainly recent research on fire and fire exclusion clearly shows the desirable aspects of controlled burning in many parts of the world.

Air pollution, acid rain, and plant growth

Air over and near cities becomes polluted with smoke (carbon particles), carbon monoxide, hydrocarbons of various kinds, sulfur dioxide, and metallic oxides. The carrying capacity of the air is limited, and if the wind does not dilute the polluted air, smog and other pollution problems arise. Chemical changes take place in the air by the action of sunlight on hydrocarbons, carbon dioxide, sulfur dioxide, and other pollutants to form photochemical smog. This material is very deleterious to both plant and animal tissues. Leaves are particularly affected because the pollutants enter through the stomas and circulate within the air space of the leaf. There the caustic action of the smog kills many of the cells. The damaged plant then has the appearance of those affected by fungal diseases. Different air pollutants cause different-appearing lesions on the plants. An expert may be able to determine the kind of pollutant causing the damage; so these plants may be indicators of the amount of air pollution in an area.

Exactly how widespread (Fig. 20-30) and subtle the effect of air pollutants is on plant growth is not known. Air pollution is not confined to cities, however. The increase of sulfur dioxide in the atmosphere from these pollutants is causing more noticeable damage to plant life, buildings, paint, and metal. A large amount of sulfur is contained in crude oil. The refining process could remove most of the sulfur but the cost involved would greatly increase the price of gasoline and so much of it is left in. When this is burned as fuel in automobiles and trucks, sulfur dioxide escapes into the atmosphere from the exhaust. It is not removed by the new antipollutant devices now required on automobiles. Sulfur dioxide combines with water in the atmosphere to form a weak solution of sulfuric acid only to return to the earth's surface as acid rain. Obviously this is very widely distributed and causes much damage. But because it cannot be seen in automobile exhaust, it is largely ignored.

Some plants are more sensitive to these pollutants than are others. It should be no surprise that some of our most important crop plants, such as alfalfa, cotton, wheat, field beans, and soybeans are the most sensitive. Garden flowers, shade trees, vegetables, and even some weeds are also sensitive. Once the chronic or acute markings to the extent of about 5% of the leaf surface appear, the crop yield is affected. This effect soon becomes a major problem in areas near large cities, and unless corrective measures are taken, a major source of food and fiber is lost in those areas. Current solutions using stack scrubbers and catalytic converters (as discussed previously) are far from satisfactory, and so the problems remain.

PLANTS FOR CONSERVATION AND LAND DEVELOPMENT

More subtle uses of plants include water conservation and the prevention of land erosion. Conservation may be defined most simply as the prevention of the disturbances of natural balances or the restoration of these balances. These disturbances are many; some are natural, but most are man created.

Natural communities of plants form a cover over the soil, which holds back large quantities of water and prevents rapid runoff of rain water. This not only prevents erosion but also maintains a water reserve, which in turn prevents drought conditions. When the natural vegetation cover is disturbed through cultivation and landscaping or highway, airport, and community development, this water-retention effect is destroyed. Floods and erosion result.

Erosion of soil is one of the greatest conservation problems. Water and wind move vast quantities of soil from place to place, much of which ends up in the sea where it is of little use. Wind erosion is responsible for much of the configuration of land in arid regions. Gorges and mountains are the result of water erosion. Both types may be prevented by stopping the useless removal of trees that hold back both wind and water. The current practice of flood prevention, rather than control, has cost this nation dearly. The practice of cutting down trees and digging straight channels simply removes water in a hurry, along with millions of tons of topsoil, and leaves behind a water shortage of ever-increasing magnitude. Planting trees and providing for backwaters as a reservoir for flood waters are solutions to the problem in all but montainous areas where the damage is already too great (Fig. 20-31). There, only expensive dams can keep back the water. Wind erosion can be prevented by planting windbreaking trees (Fig. 20-32). In irrigated areas these trees should be grown along the edges of all fields, and a plant cover should then be maintained at all times. The practice in the Midwest of early harvest and immediate plowing to prevent the growth of weeds, followed by late planting, is a continuing folly that has turned vast areas into dust bowls in the past.

Leaching of the soil by the removal of humus and plant cover causes a deficiency in mineral content, which soon renders such areas barren of all possible plant growth. The solution for the prevention of this loss is obvious. Such areas may be reclaimed only by extensive fertilization, as well as by water control and planting of erosion-resisting grass (Fig. 20-33).

Fig. 20-31. The anchoring ability of the root system of plants serves a two-way purpose—to anchor the plant and at the same time to hold the soil, which also helps to prevent erosion. In this photograph young pine trees are planted in badly eroded mountain areas. Some will survive and eventually hold back the soil, and thus further erosion will be prevented. (U.S. Forest Service.)

Fig. 20-32. Windbreaks along the edge of irrigation ditches surrounding cultivated fields of young plants help to prevent wind erosion of the soil. (U.S. Forest Service.)

414

Fig. 20-33. Holy grass is planted by soil conservationists in barren areas like this to anchor the soil. This grass grows naturally in windswept alpine areas. Here it holds the soil until other plants can become established. (USDA photograph.)

Trees affect local weather conditions in several ways. They increase the relative humidity by transpiration and, at the same time, produce a cooling effect in the shade by evaporation of the water transpired. This in turn contributes to cloud formation from the rise of water vapor, thus making more moisture available for rain (Fig. 20-3). Trees, herbs, and grasses all provide food for wildlife, which in turn carry on complex food chains, and some of these animals, especially the game animals, are of direct interest to man. Many food chains, however, only remotely affect man's economy and unless he is interested in and appreciates the esthetic value of understanding these complexities of nature, he does not concern himself with them.

In agricultural areas a highly organized soil conservation program is sponsored by governmental agencies. Farmers participating in this program are required to grow windbreaks, terrace their land, practice contour plowing, that is, plow across slope lines, add required fertilizers, and practice proper crop rotation. Following the latter practice, a natural building of the soil can be effected by first planting soil-building plants such as legumes, after which may be planted crops that tend to deplete the mineral content of

the soil. Some plants, for instance tobacco, rapidly deplete the mineral content of soil. This may be easily seen by looking into any ash tray. The soil of the tobacco land of the nation rests there in the form of the high ash content of the burnt tobacco leaves. The ash is nearly pure mineral salts. Eventually the conservation practices just described will bring back the land, but a long program of proper education lies ahead before the lands of the nation are finally stabilized.

The development of new agricultural land is becoming more and more important. As urban areas squeeze out more farmland, arid areas are developed by means of irrigation. All too frequently, however, this results in the exploitation of both land and water, and after a short period the land is abandoned in worse condition than before. There is great need for research in the proper development and management of new lands. Unfortunately, politics often enters into this problem, and strong lobbies prevent the proper control of such areas. As an example of this, the cattle interests have prevented proper legislation that would control grazing in western arid areas (Fig. 20-34) and in our national forests. Unless these selfish interests are soon controlled, irreparable damage will be done. As it is now, the once extensive west-

415

Fig. 20-34. *Right and in background,* Effect of overgrazing. *Left,* Luxuriant growth in an area in which no grazing has occurred. Much of our western land was once grassland, but through improperly controlled grazing it has been turned into unproductive desert. (U.S. Forest Service.)

ern grasslands will take many years to restore.

Equally important to future generations is the conservation of natural areas. In these days of rapid transportation and relative abundance of leisure time, millions of people flee the urban areas in search of recreation in natural areas of this country. The natural areas are rapidly diminishing. Man already has exterminated many animals such as some kinds of quail and the passenger pigeons and nearly has exterminated many others. These he knows about, but many other interesting animals as well as plants have been exterminated unknowingly. National parks have been set aside, plus many natural areas, as wildlife refuges to preserve as much of the natural country as possible. Much of the interest in these areas centers around the preservation of game animals, with little regard, however, for the conservation of all wildlife, including plants, insects, and other forms of life that are of interest for themselves, not as something to shoot or fish so that a man can assert his superiority. A policy that will fully

preserve these areas is urgently needed. Visitors to natural areas should be educated to appreciate fully the delicate ecological balances and the widespread effect of disruption of even the most subtle of these balances. This we owe to future generations.

QUESTIONS AND PROBLEMS

1. Where would you expect to find the following plants: maple, ponderosa pine, bald cypress, cactus, spruce, grass, willow?
2. Describe the dominant plant community of your particular areas.
3. What are the three different meanings to population? What is meant, biologically speaking, by overpopulation?
4. What are the controlling factors in plant distribution, and how do these factors operate?
5. From the discussion so far and from what can be observed of life about you, list the conditions under which life can exist as we know it.
6. What are the chances of finding life like ours on other planets? In other regions of space?
7. Explain how a greenhouse is a controlled growth chamber.
8. Under what conditions of plant growth can

there be, and often is, an oxygen deficiency?

9. How would you set up a balanced aquarium? A terrarium?

10. Considering the nature of protoplasm, why is extreme cold less injurious to plants than is extreme heat?

11. When is wind a factor in the effective environment? Name some other factors that are a part of the effective environment only under certain circumstances.

DISCUSSION FOR LEARNING

1. Many tropical plants have very large leaves, and most desert plants have very small leaves, yet cactus grows equally well in both habitats. How do you explain this? Include in your discussion the effect of watering desert plants.

2. Plants in arctic and alpine habitats grow very rapidly. At the same time, they require a very long photoperiod. What happens when such plants are grown in more southern, warmer climates? Consider in your discussion the factors that induce flowering.

3. Discuss the reasons why aquatic plants cannot grow on land and why land plants cannot grow in water. Are there exceptions?

4. Discuss the differences between the flora of a region and the vegetation of the area.

5. Describe and discuss some obvious succession of plants in an area known to you.

6. What is the relationship between ecology and the evolution process? Discuss this in terms of your study so far in this textbook.

7. Discuss the relationship of the cell wall to the effective environment. What does it protect? What processes are slowed down as a result?

8. Considering all the factors involved in the effective environment, discuss the following agricultural practices: (a) spacing of plants; (b) weed control; (c) insect control; (d) irrigation. What factors are not ordinarily controlled?

9. Which one of the environmental factors is most likely to lead to overpopulation if it is in overabundance?

ADDITIONAL READING

Black, C. A. 1968. Soil-plant relationships. 2nd ed. John Wiley & Sons, Inc., New York.

Darlington, P. J., Jr. 1965. Biogeography of the southern end of the world. Harvard University Press, Cambridge, Mass.

Daubenmire, R. F. 1974. Plants and environment. 3rd ed. John Wiley & Sons, Inc., New York.

Gleason, H. A., and A. Cronquist. 1964. The natural geography of plants. Columbia University Press, New York.

Good, R. 1953. The geography of the flowering plants. 2nd ed. Longmans, Green & Co., Ltd., London.

Kellman, M. C. 1975. Plant geography. Methuen & Co., London.

Küchler, A. W. 1964. Potential natural vegetation of the conterminous United States. Amer. Geographical Soc. Sp. publ. no. 36 (map and book), New York.

Odum, E. P. 1971. Fundamentals of ecology. 3rd ed. W. B. Saunders Co., Philadelphia.

Odum, E. P. 1963. Ecology. Holt, Rinehart & Winston, Inc., New York.

Oosting, H. J. 1956. The study of plant communities. 2nd ed. W. H. Freeman & Co., Publishers, San Francisco.

Platt, R. B., and J. Griffiths. 1964. Environmental measurement and interpretation. Reinhold Publishing Corp., New York.

Richards, P. W. 1952. The tropical rain forest. Cambridge University Press, New York.

Shelford, V. E. 1963. The ecology of North America. University of Illinois Press, Urbana.

Walters, H. 1973. Vegetation of the earth. Springer-Verlag. New York.

Went, F. W. 1955. The ecology of desert plants. Sci. Am. **192**:68-75, April.

Fig. 21-1. Man, beast, and plow created civilization. The continuation of human life now depends on constant improvement of agricultural techniques.

CHAPTER 21

Plants and man

■ The triad of man, beast, and plow (Fig. 21-1) was millions of years in the making. Man took at least 2 million years to change from the little creature hardly distinguishable from any of several species of primates, except for the stone in his hand, to the erect man capable of plowing and building. As we have already discussed in Chapter 1, the modern history of man has been directly the result of his learning to cultivate. The earliest evidence of cultivation is only about 12,000 years ago, or at the end of the Ice Age. Once started, civilization developed rapidly in some parts of the world, and slowly in others. The turning point did not involve plants alone, however. It was necessary to domesticate animals as well as the plants—exactly which came first is difficult to determine. It seems certain that plants were cultivated to feed man and his domestic animals concurrently. Once this was possible, it was no longer necessary to follow the herds to new feeding ground. The nomadic way of life could cease and permanent settlements were established. As we look about us now and see large cities and rapid transportation, we marvel at the results of simple botanical knowledge. Equally incredible is the uneven development of this civilization that leaves some people in the world still living as nomads.

Primitive agriculture began, without doubt, simply by protecting stands of useful plants. Removing unwanted plants to make room for natural reseeding of the desirable plants began the actual cultivation and "created," the new group of plants now known as "weeds." The following steps, so obvious to us now, were major discoveries in the evolution of civilization: gathering of the best seed, planting in the best soil, preparing the soil for the seed, and soil improvement by watering and fertilizing. All these practices were slowly acquired, and even though the agriculturalists through the ages knew how to improve plants through selection, it was not until this century that we learned why it happened.

ORIGIN OF CULTIVATED PLANTS

Plants have been under cultivation by man for so long that most of them are taken as a matter of course. The following lists show that the isolation of the Old World and the New World resulted in separate cultivation of food plants. A surprising mixture of plants after voyages of Columbus has promoted a more varied diet, which is evident by the New World plants that have taken Old World names (Jerusalem artichoke) or that are now grown more extensively in the Old World (pineapple), or by the Old World plants (orange) that are now grown more extensively in the New World.

New World plants not grown in the Old World before the time of Columbus:

corn	squash
lima bean	pumpkin
potato	peanut
sweet potato	papaya
kidney and other beans	avocado
tomato	pineapple
green pepper	guava
Jerusalem artichoke	chocolate
sunflower	cashew

Old World plants not grown in the New World before the time of Columbus:

wheat	carrot	plum
rye	breadfruit	cherry
barley	mangosteen	alfalfa
oats	onion	wine grape
millet	garlic	apricot
sorghum	spinach	peach
rice	eggplant	olive
buckwheat	lettuce	fig
turnip	endive	almond
cabbage	celery	quince
rutabaga	asparagus	pomegranate
chard	peas	watermelon
mustard	soybean	cucumber
radish	yam	mango
beet	apple	clover
parsnip	pear	

The first cultivated plants were selected from wild stock because they were the best tasting and the easiest to grow. This natural selection process changed many of our plants so that they bear little resemblance to their ancestors. It is obvious that this selection is possible through the genetic mechanisms inherent in the plants. Only in this century have systematic attempts been made to develop new crops through crossbreeding. However, almost all our modern crop plants are hybrids (Fig. 21-2) developed for higher yields, greater food value, or as more attractive ornamentals.

Most of the original wild forms of our modern cultivated plants are known, and these forms can be traced back easily to a certain wild species. In fact, many of these plants will revert to the wild form if allowed to grow unattended, the very existence in their present form depends on the continued attention of man.

Fig. 21-2. Close-up of Wichita wheat, an example of increased food production from hybrids. These grains of wheat are much larger than those of wild wheat or earlier cultivated varieties. (USDA photograph.)

Fig. 21-3. Constant selection for larger flowers and improved color has changed simple wild flowers into large, complex, polyploid plants. (Courtesy Ferry-Morse Seed Co., Detroit, Mich.)

CHANGES IN PLANTS THROUGH CULTIVATION

Cultivated plants are usually larger than their wild relatives. They have broader and thicker leaves, sturdier sprouts, stems, and stalks, fleshier roots, and larger flowers (Fig. 21-3 and 21-4). Their fruit and seeds show these same characteristics. This gigantism is attributable to **polyploidy,** the multiplication of the **chromosome** complement in the cells (see Chapter 18 for an explanation). The study of these phenomena in plants has led to the science of plant breeding. Further consideration of this topic is beyond the scope of this text.

FOOD PLANTS

Food plants provide us with all of the essential elements needed for growth and energy. Carbohydrates (Figs. 21-5 and 21-6) are almost entirely obtained from plants. Protein, fat, minerals, and vitamins are all obtained from plants. Since all animals are dependent directly or indirectly on the food stored in plant cells as the result of photosynthesis, it follows that these same materials from animals is an indirect source. Some plants provide us with one kind of food, whereas others are a better source for other kinds of food. The cereals (grasses such as corn, rice, and wheat) are the principal source of carbohydrates used in bread and similar food produces. Corn is rich in fat, yielding corn oil, as well as protein. Soybeans (a legume) are used as a source for oil, and in addition, they are an excellent supplier of protein, including the so-called textured protein, an all vegetable "meat." Other legumes such as beans, peas, and peanuts are also an important source of protein.

Forage plants, for example, clover and grasses, indirectly provide us with food by feeding livestock. Many plants provide us with food seasonings or spices, the common names of which are familiar to all (sage, garlic, bay leaves, cinnamon, thyme, rosemary, savory, cloves, anise, and wintergreen to name a few). Most of our vegetable plants are primary sources of essential minerals in our diet, provided that they are properly prepared. Corn, cauliflower, carrots, spinach, and potatoes are high in food value. Others, such as eggplant and squash, help to round out our diet. Plants by means of their fruits (for example, apples, cherries, and peaches) provide us with vitamins, essential elements of our diet that we cannot manufacture ourselves. Citrus fruits and tomatoes are important sources of these vitamins. Some plants provide us with the essential ingredients of

Fig. 21-4. Fruits, too, in this case garden peas, are improved through selection and crossbreeding.

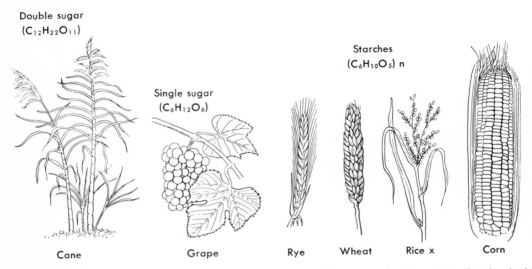

Double sugar
$(C_{12}H_{22}O_{11})$

Single sugar
$(C_6H_{12}O_6)$

Starches
$(C_6H_{10}O_5)\ n$

Cane   Grape   Rye   Wheat   Rice x   Corn

Fig. 21-5. Various kinds of plants yield different carbohydrates, as these examples show. Note that the double sugar from the sugarcane sap is found in the stem; the simple sugar is in the liquid of the fruit; the starches are stored in the dry seeds.

Fig. 21-6. Grapes are a source of not only sugar, but also, through the work of nature and human hands, wines of many delicate flavors. (USDA photograph.)

Fig. 21-7. Fibers of the cotton boll, a form of fruit, supply much of the clothing and other fabrics used throughout the world. (USDA photograph.)

beverages, such as coffee, tea, and cola, and with the hops, malt, and wine grapes, the raw materials for alcoholic drinks. In addition, distilled spirits are prepared from rye, corn, wheat, juniper berries, and sugar cane. A few of the lower groups, fungi, molds, and bacteria, are essential to cheese production, and some can be eaten directly, such as mushrooms, puffballs, truffles, and others.

FIBER PLANTS

Fiber plants are essential to many manufacturing processes. Cotton (Fig. 21-7) linen (made from flax), jute, Indian hemp, Manila hemp, sisal hemp, kapok, and other plants produce long cells or closely packed groups of cells with tough cell walls to form fibers used in the making of cotton cloth, rope, burlap, thread, string, and packing materials. After suitable chemical treatment these plants are also sources of rayon, paper, cellulose, lacquers, cellophane, and many other products used in modern industry.

LUMBER TREES AND FOREST PRODUCTS

Lumber trees (Fig. 21-8) are among the most useful plants known to man. The use of wood for building materials is as old as civilized man. Some of the best lumber is cut from oak, mahogany, ash, and maple (hardwoods) and from redwood, cedar, and pine (soft woods). Oak, hickory, and pine are used for fuel, pitch pine being used for the manufacturing of charcoal. Posts, mine timbers, poles, and marine pilings are cut from oak, ash, and Osage orange. Veneers and plywood are made from pines, maple, and mahogany. Railroads use red oak for ties. White oak is used in the manufacture of barrels, but the cooperage trade has become almost entirely extinct, since cardboard containers made from pine and redwood scraps are now in general use. Excelsior for packing material is made from aspen and cottonwood. The few wooden shingles still used come principally from red cedar. Differences in the cell struc-

423

Fig. 21-8. This scene in northern Arizona tells the story of man's need for lumber and his greed. Cutting ponderosa pine in Arizona creates an erosion problem, turns much of the land into a semidesert, and, as can be seen from this photo, pollutes the air. If man is to survive, he must learn to reap the benefits of plant growth without destroying the land that produces it.

ture and in the amount of gum and resins in the wood determine the quality of wood and the use to which it is put. Wood is also used for the manufacture of tars, resins, alcohol, acetone, wood gas, and turpentine. Chicle for chewing gum, rubber, cork, and some spices are additional wood products. A few secondary forest products deserve mention—cork from cork oak, Christmas trees from pine, fir, spruce, and hemlock, and woods for inlay and other ornamentation from hardwoods such as locust, persimmon, ebony, and mahogany.

DEVELOPMENT OF NEW PLANTS

World food shortages, caused mainly by overpopulation of many areas and poor or unequal distribution of available food, have spurred interest in finding a near perfect, cheap source of all of the nutritional requirements of man. Most wheat, corn, and rice is high in carbohydrate and low in protein. A diet consisting mainly of these grains will

Fig. 21-9. A new high-lysine corn, discovered by Purdue University scientists, known as opaque-2. This corn has twice as much lysine, a protein building block, as normal corn and therefore greater food value than other corn hybrids. (Courtesy Purdue University Agriculture Experiment Station Information Service.)

result in serious dietary deficiencies, sometimes causing the disease known as kwashiorkor. This is the leading cause of death among infants and children in many parts of the world.

A few years ago, almost by accident, a mutant corn (Fig. 21-9) was discovered at Purdue University. The corn, upon testing, showed a much higher percentage of the amino acid lysine. This is one of the essential amino acids that the body cannot produce from other foods, and so it must be supplied with this material directly. The new corn, known as high-lysine corn, meets the requirements of a nearly balanced diet, which is also relatively cheap. It is a corn with an opaque, chalk-white kernel (therefore known as **opaque-2**). This corn plus an inexpensive vitamin and mineral supplement provides an adequate diet for about 10 cents a day.

The different appearance of this corn has prevented its immediate widespread acceptance. So further research has resulted in still further improvement of the corn so that this high protein is now combined with a hard, flinty-textured kernal that will be purchased at the market. Still another problem had to be solved by genetic manipulation: the corn that was produced must grow well under tropical conditions. Tropical regions suffer most from high-population levels and low-protein diets. Now a corn combining all of the required features has been produced and is being planted in many of these areas.

High-lysine corn is the most dramatic solution to the increased yield of food plants needed to counteract famine caused by overpopulation. Many other crop plants have been improved steadily over the years so that yields per acre have just about kept pace with population increases. Very likely this will continue to some extent, but most scientists are fearful that the end will come long before zero population growth is reached.

INSECTS AND INSECTICIDES

The role of insects as pollinators has been discussed briefly in Chapters 16 and 20. Although many insects serve in this important way, including a coevolutionary process only recently recognized, by far a larger number of insects destroy plants through excessive feeding and by the transmission of diseases. Elm trees, once the stately tree of cities and towns, are dying or dead from the fungus causing Dutch elm disease, transmitted by sap-sucking insects. Foliage damage by the codling moth, and damage to the growing tip of conifers by the spruce budworm are recent problems brought about by the cutback in the use of dangerous **insecticides.** Many other examples of damage to crops, ornamental plants, and forests can be cited.

The use of many kinds of organic and inorganic **insecticides** to combat and control insect damage to plants has resulted in a new problem: pollution of the environment. The bane on insecticides recreates such insect problems as the codling moth and the spruce budworm. The continued use of insecticides poisons the land, water supplies, and eventually kills wildlife and endangers human life. Part of the problem has been the result of overuse, or nonprofessional use of these materials. Safer insecticides and their proper use, along with various biological control practices using parasitic and predatory insects to bring about integrated control, is apparently the only solution to this problem.

WILD PLANTS FOR FOOD

Tomatoes, once called loveapples, were first believed to be poisonous, as were believed many of the plants now under cultivation. Food shortages, high cost of food, and a desire on the part of many people to "return to nature" has resulted in a new search for edible wild plants. This can be a dangerous as well as an interesting pastime. Some plants are extremely poisonous; so those who desire to try new foods must take the time to learn how to identify plants. Once they have mastered their local flora they will be in a better position to judge the danger of eating certain plants. All the species of some plant families are known to be poisonous. Others have a few that are poisonous and some that are edi-

Fig. 21-10. Water hemlock, *Cicuta douglasii* shown here, is probably the most poisonous plant in the United States; the poison is found principally in the roots, which are often mistaken for parsnips. (USDA photograph.)

ble, and so on. The poisonous substances of some plants are destroyed by cooking and may be safely eaten. Many kinds of plants are eaten locally and are unknown elsewhere. Residences of the area often eat these plants more or less routinely. Newcomers can rely on this as a safe practice, provided that they learn to *correctly identify* the plant. A knowledge of plants may provide new eating experiences from all parts of the world and offers a pleasant adjunct to the study of botany.

POISONOUS PLANTS

Snakes and other animals are well known as a source of fatal poisons. Among the most poisonous substances known is that produced by the mushroom *Amanita phalloides*. The toxic agents, two similar substances known as phallotoxins and amatoxins, are amino acids linked together by peptide bonds to form a continuous ring. When eaten, these substances not only cause severe intestinal distress but also damage the cell membranes of liver cells, the result of which is fatal sev-

eral days after the eating of the mushrooms.

Other plants are almost as poisonous. The water hemlock (Fig. 21-10) is well known for its poisonous roots. Several species occur in the United States. Also the famous poison hemlock (said to be taken by Socrates to commit suicide) has been introduced from Europe and occurs throughout United States and Canada.

A long list of plants contain poisonous substances. Many of these plants are commonly eaten, such as potato, asparagus, horseradish, tapioca, rhubarb, and many others under certain conditions of use and according to the degree of sensitivity of the individual may be poisonous. A very long list of species has been compiled that are poisonous to stock. One well-known example is arrow grass (Fig. 21-11).

Other plants exude substances that cause severe skin lesions. Poison ivy (Fig. 21-12) is well known for this. Related species in this group of plants are irritating also. Some plants, such as the common weed called

Fig. 21-11. Arrow grass grows in wet areas, as shown in this view, where it is sometimes eaten by sheep and cattle with fatal results because the poisonous substance is hydrocyanic acid.

Fig. 21-12. Poison ivy causes irritation to the skin of most people. The shiny leaves in groups of three make it easily recognizable.

Fig. 21-13. Manchineel, a small tree native to the beaches of Central America, the West Indies, and Florida. The milky sap of the fruit and leaves causes severe irritation and even blindness.

Fig. 21-14. Digitalis used in the treatment of some heart ailments is manufactured from foxglove plants such as these shown under cultivation. (Courtesy Merck Sharp & Dohme, Inc., Division of Merck & Co., Inc., Rahway, N.J.)

stinging nettle, have poison spines that cause temporary irritation. Many plants have milky sap containing poisonous materials, some of which are skin irritants. One causing very serious illness is the manchineel (Fig. 21-13).

DISEASE-PRODUCING PLANTS

Many plants cause diseases of animals and man. The bacteria and fungi are the greatest offenders of this kind. Bacterial diseases are discussed in greater detail in Chapter 5.

Fungal diseases, such as those of the skin and lungs, are briefly mentioned in Chapter 7. Some bacteria and fungi help to limit animal populations and thus are useful to man in the control of certain pest animals. For example, the fungus causing milky disease in the Japanese beetle is a practical control of this pest. It is now being spread artificially to control the beetle, thereby preventing serious damage to fruit crops and ornamental plants.

Molds such as the genera *Alternaria, Hormodendrum, Penicillium, Aspergillus, Mucor,* and *Rhizopus* are major wind-borne spore allergens, second only in importance to the pollens in causing asthma and rhinitis. Spores of fungi originating in the midwestern states can be wind borne to eastern cities in 24 hours. Western states have the least incidence of the disease. July, September, and October are the important months for this allergy. Pollen from ragweed, goldenrod, and other plants cause the disease known as "hayfever."

MEDICINAL PLANTS

Medicinal plants have contributed greatly to the welfare of man. Among the drug-producing plants and their products are belladonna, a dilator extracted from the belladonna plant; foxglove (Fig. 21-14), the source of digitalis for the treatment of heart disease; cascara and senna, yielding cathartics; and morphine and opium, powerful drugs from the poppy. Many drugs formerly came from plants, but they are now made synthetically from petroleum.

PLANT DISEASES

Diseased plants cause great economic loss each year and often famine in many parts of the world. Losses may run as high as 25 billion dollars in a single year. Most plant diseases are caused by fungi and bacteria. Many examples of species causing plant diseases have been cited in this text. These diseases are the subject of research and experimentation by federal and state agricultural agencies. **Fungicides** are applied by agriculturalists much the same as insecticides.

Treatment of seeds with fungicides before planting has become routine. However, these diseases are very difficult to control, primarily because of their universal transmission by spores, usually airborne, but sometimes insect borne, as discussed previously. Therefore, each species of crop plant and ornamental is exposed to many pathogens. But not all plants react the same to the disease. Certain individuals, just as in humans, are more resistant to the disease than are others. When resistant strains are discovered, these plants are selected for cultivation, or sometimes, as in the case of wine grapes, for grafting of one part of the plant to a resistent root and stem of another plant. This may solve the problem. But the complexity of the life cycles of most of the fungus diseases further complicates any attempts to control these diseases. A very good example is the involved host relationship of wheat rust (see Appendix B, life cycle 15), involving both the grain and the wild barberry bush. Further discussion of this intricate subject is beyond the scope of this text.

WEEDS AND HERBICIDES

Weeds are not a particular kind of plant, as many people suppose. Any plant growing in a lawn, garden, or cultivated field that is not desired by the grower is considered to be a weed. Among the many kinds of weeds of cultivated crops, all of which are obnoxious because they reduce the yield per acre, are mustard, crabgrass, pigweed, dock, wild garlic, locoweed, and plantain. Additional species are continually being introduced inadvertently, usually through commerce. For example, some introduced pests of current interest are water hyacinths and milkweed vine in the southern United States.

A great many such plants are more tenacious of life than are the cultivated plants protected by man. For this reason there is a struggle between the two kinds of plants, and the cultivated plant soon loses if the grower falters in the battle. In many cases it would be better to join the weeds than to fight em. Traditionally, many of the so-called weeds have been excluded from our list of desirable plants, even though many are attractive and useful. For instance, the common dandelion has all the beauty of a chrysanthemum and is a delicious green with the food value of similar cultivated plants. It would be easier, no doubt, to grow dandelions instead of a lawn!

Much time and expense is spent on weed control. No longer is it possible to economically hire the bodily removal of weeds. Instead, machine cultivation and the use of herbicides, usually caustic chemicals such as sulfuric acid or organic compounds, are used. The attempt to eliminate the undesirables has resulted in problems similar to those caused by the use of insecticides. Herbicides contribute to our ever-growing list of pollutants. Weed control, therefore, like insect and plant disease control, must incorporate, in addition to the use of chemicals, certain biological control techniques. One way currently used with some success is to find a phytophagous insect specific to the weed. These insects, which are usually native to the same region as the weed, are reared in large numbers and introduced in the area where the plant is a pest. Other attempts have been made to introduce specific plant diseases to control the weed, but with little success. More effort needs to be spent in selecting plants for cultivation, particularly along roadsides and for landscaping. These should be hardy plants that will crowd out the weeds, grow to a desirable height without the need to mow or prune, and otherwise take minimum care.

PLANTS AND AIR POLLUTION

The increase of sulfur dioxide in the atmosphere from the burning of fuels of various kinds has affected the vegetation in areas of sulfur dioxide concentration. The leaf lesions caused by sulfur dioxide pollution resemble various plant diseases; so it is not always possible to immediately state that the plant is affected by the pollutant. Also, several other materials in the air may affect plant growth. See further discussion on p. 413.

PRACTICAL USE OF BOTANICAL KNOWLEDGE

Knowledge of the basic principles of the structure and physiology of plants is indispensable to practical botanists such as horticulturists, floriculturists, and orchardists. One of its applications is in the pruning of trees and shrubs. The chief purpose of pruning is to remove dead and diseased branches to prevent the entry of parasites and their subsequent spread to the main trunks. In addition, pruning is necessary to give the desired shape to ornamental plants, to increase side branches, or to induce flowering. Unless it is properly done, it may result in improper healing of the cut surface and encourage rot or parasitic attack.

Grafting is another practical application of stem structure and physiology. It is a process in which two freshly cut stem surfaces are bound together so that the cambium layers of both are in contact. The usual types of grafting involve the union of a base stem having a root (the **stock**) and a cutting (the **scion**) (Fig. 21-15). There are many ways to do grafting. Depending on the way that scion and stock are united, the types of grafting are called saddle grafting, cleft grafting, tongue grafting, or side grafting (Fig. 21-16).

Grafting is of use in propagating seedless varieties of plants such as navel oranges and hybrid plants, in inducing more rapid fruiting since plants may produce fruits earlier if grafted than if propagated from seeds, in accommodating a species of plant to a new environment, and in checking or eliminating parasites (particularly of roots).

Plants are sometimes propagated through stems by a process called **layering.** A stem is bent down and buried partly under the ground. That part of the stem under the ground will develop adventitious roots; after it has started to grow, it can be separated from the parent plant (Fig. 21-17). Another example of plant propagation is the use of cuttings. A short stem is removed from the parent plant and its cut end placed in moist sand or soil (or in water for some plants). It will usually develop adventitious roots at this

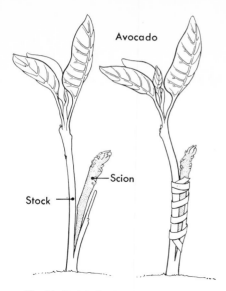

Fig. 21-15. Method of side grafting.

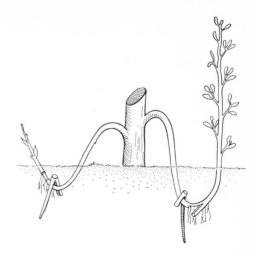

Fig. 21-17. Vegetative propagation of plants by the layering method (see text).

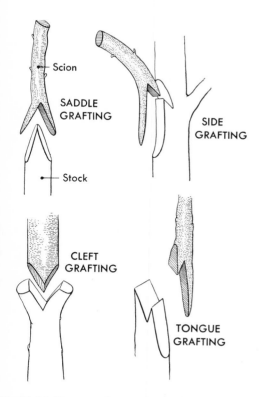

Fig. 21-16. Diagram of several methods of grafting.

lower end. Many horticultural plants are propagated in this way.

BOTANY AS A HOBBY

Growing plants in the home or in a greenhouse is extremely popular. Probably no other aspect of science is of as great an interest to the general public. Many people pride themselves in having a "green thumb." Gardening in and around the house requires a basic knowledge of plant growth requirements, a knowledge usually gained by word of mouth, by trial and error, and through the medium of a multitude of garden books and magazines.

Unfortunately, some of this information is not based on scientific fact, such as the recent "best seller" book ascribing a sensitivity of plants to man's emotions, prayers, and conversation. The fact that such a book is accepted by the public shows both a lack of responsibility on the part of those offering such misinformation to the public, and the lack of a sound scientific background on the part of those who accept as fact this kind of duping. Those interested in home gardening and planting should be urged to consult with their local county agricultural agents, the state university, and similar reliable information sources.

In addition to growing plants, some people like to collect and press wild plants to make a systematic collection to study plant systematics. Directions for these studies follow.

PREPARATION OF PLANT SPECIMENS

Because a detailed study of plants cannot be conveniently made in the field, many techniques for laboratory study have been developed. These processes include the proper method of collecting, preparation, and storing of specimens. Preparation for the study of microscopic anatomy demands that special reagents be used in the field so that these parts can be preserved in as near their lifelike state as possible. Special treatment must be given to particular groups of plants so that their parts may be properly preserved for later identification. Details of such methods can be readily demonstrated in the laboratory, and reference to them is made in several manuals that describe these techniques.

Students of taxonomic botany usually begin by making a collection of plants. The plants are gathered in the field and are placed in cellophane bags within a **asculum,** a metal can that is easily carried. After the day's collection, the specimens must be pressed for preservation. This is done by means of a plant press consisting of two frames of standard size, 12 by 18 inches. Between these frames are placed a packet of felt blotters and corrugated cardboard ventilators cut to the size of the frames. The plants are carefully arranged between sheets of folded newsprint or newspaper, which are next arranged between two blotters, one plant per pair of blotters. These are alternated with cardboard ventilators until a package about 2 feet thick is obtained. Then the frames are laid over and under the package and drawn tightly together with straps. These presses are placed over low heat so that the water absorbed by the blotters from the plants is removed through the channels in the cor-

rugated cardboard. After about 24 hours the presses are opened and the now limp plants are arranged for the best display of parts. They are then relocked in the press and dried for another 24 hours. Then the dried and flattened plants are removed and are ready to be glued to the herbarium sheets made of high-quality paper. Attached to each of these sheets is a label identifying the plant and, most important, field notes telling where each of the plants was growing and any other information about the plant and its immediate environment.

If chromosome counts are to be made, it is necessary that growing tips or developing parts of the flower be preserved in the field in some specific fixing fluid, such as Newcomer's,* which will preserve the cells in various stages of mitosis or meiosis so that the chromosomes may be studied later in the laboratory.

Sometimes in the study of various kinds of plants it is desirable to observe the microscopic structures of the stem, roots, or other parts. The preparation of these parts requires a long and involved **microtechnique.** The cross section of a stem, for example, must first be fixed in a solution like alcohol or formalin, dehydrated through a series of graduated steps of increased alcohol and decreased water, and, finally, after complete dehydration, placed in xylol, which is miscible with paraffin. Once the specimen is embedded in paraffin, it can be cut into sections only a few microns thick by means of a microtome. The paraffin sections are then fixed to a glass slide, the paraffin is removed by xylol, and the xylol is replaced by a gradual series of alcohols until the specimen is again in water. The section is stained, the whole process of dehydration repeated, and a cover glass fixed to a slide with a material like Canada balsam. Because of this procedure, the specimen may now be easily and safely examined under the high magnification of a compound micro-

---

*Science **118:**161, 1953. The formula is 6 parts isopropyl alcohol, 3 parts propionic acid, 1 part ether (petroleum), 1 part acetone, and 1 part dioxane.

scope without danger of breaking the specimen.

IDENTIFICATION

Plant scientists must identify wild plants before they can report certain kinds of research data, particularly studies that involve distribution, variation, certain kinds of genetic studies, and investigations on the processes of speciation and evolution. The process of identification involves several steps, each of which lead to greater accuracy.

1. Spot identifications—often botanists are able to recognize common local plants on sight. They learn these plants by comparing them with pictures and herbarium specimens and, through practice, are able to recall the names, often with a high percentage of accuracy. This method has the disadvantage that plants from other regions are unrecognized or incorrectly identified, and even some local plants may not be separated as distinctive because of the lack of critical appraisal of characters.

2. Botanists, fortunately, have a wealth of regional manuals, termed "floras," to use for accurate identification of the flowering plants. These contain keys for the identification of the taxa, descriptions of the plants, their distribution, and, frequently, illustrations of the species. The lower plants are not well covered.

3. Specimens collected for research purposes may need to be identified with the use of technical monographs. To find these, one must search the available literature using the various library tools.

4. Specimens identified with the aid of manuals should be verified by comparison with correctly identified herbarium specimens. Research material is often sent to specialists for vouchering, that is, verification of the identification.

QUESTIONS AND PROBLEMS

1. Considering the many synthetic products manufactured from petroleum, what are still the most important cultivated plants producing useful products?
2. Name one or more cultivated plants that have similar wild relatives.
3. Name one or more cultivated plants that are very different from their wild relatives.
4. There are many plants in the world that are grown locally but are never shipped to markets in other sections of the country or world; what are some of these that are found in your area? Can you name others seen during your travels?
5. What factors enter into the growth of "weeds," the cause of plant diseases, and damage from insect pests? One might best answer this question by making a table showing the factors on one side, the problems across the top, and the discussion in the body of the chart.

DISCUSSION FOR LEARNING

1. Discuss the possibilities of continuing to feed the world's expanding population through the development of greater yields of plant products. What types of plants might be most useful?
2. Many drug-manufacturing companies are sending "plant explorers" throughout the world in search for new drug plants. Discuss the reason for this interest and why it continues despite the rapid development of synthetic products.
3. Conservation reforms are critically needed. The popular press is full of attempts to prevent air pollution and water poisoning by insecticides and herbicides, but very little about other urgent problems such as prevention of land erosion and the proper management of grassland. Discuss some of the things each private individual might do to hasten these needed reforms.

ADDITIONAL READING

Healey, B. J. 1975. The plant hunters. Charles Scribner's Sons, New York. (The search for new plants is exciting and hazardous. This is a popular account of the discovery of plants that are now grown in gardens around the world.)

Heiser, C. B., Jr. 1969. Nightshades: the paradoxical plants. W. H. Freeman & Co., San Francisco. (The origin, history, and uses of several important plants of the potato family.)

Kingsbury, J. M. 1964. Poisonous plants of the United States and Canada. Prentice-Hall, Inc., Englewood Cliffs, N.J. (The taxonomy and biochemistry of plants poisonous to man and livestock, now the standard reference for this topic.)

Mangelsdorf, P. C. 1974. Corn: its origin, evolution, and improvement. Harvard University Press, Cambridge, Mass. (A nontechnical monograph on this very important food plant.)

Montgomery, F. H. 1964. Weeds of the northern United States and Canada. Frederick Warne & Co., Inc., New York. (A nontechnical guide to the common weeds with illustrations useful for identification.)

Russell, H. R. 1975. Collecting and cooking wild foods. Thomas Nelson & Sons, New York. (This book gives accurate and detailed information about the use of many wild plants for food.)

Schery, R. W. 1972. Plants for man, 2nd ed. Prentice-Hall, Inc. Englewood Cliffs, N.J. (A standard textbook on economic botany.)

Schwanitz, F. 1966. The origin of cultivated plants. Harvard University Press, Cambridge, Mass. (Aimed at the nonspecialist, this book provides a clear and concise account of the transformation of wild species into cultivated plants.)

Wagner, R. H. 1974. Environment and man, 2nd ed. W. W. Norton & Co., Inc., New York. (This is a balanced treatment of most of the world's major environmental problems, including an extensive section on insecticides and herbicides.)

APPENDIX A

Plant classification

■ This list of plants summarizes the major group names used throughout the text. By reference to this list the student will see in one place all of the divisions, and a detailed list of families and common species of the flowering plants. The list is indexed by both scientific and common name. Frequent reference to this appendix during study of the text will be helpful in fixing in one's mind the placing of the diversity of plant life in our current classification sceme. The life cycle numbers refer to the life cycles discussed in Appendix B.

Kingdom PROKARYOTAE

DIVISION 1. Virulenta—viruses (life cycle 1)
DIVISION 2. Schizophyta—bacteria
    (Nineteen parts, including gliding bacteria, mycelial bacteria, and spirochetes—see p. 78)
DIVISION 3. Cyanophyta—blue-green algae (life cycle 2)

Kingdom PLANTAE
Algae

DIVISION 4. Chlorophyta—green algae (life cycles 3 to 6)
DIVISION 5. Chrysophyta—yellow-green and golden brown algae and diatoms (life cycle 7)
DIVISION 6. Euglenophyta—euglenoid algae
DIVISION 7. Pyrrhophyta—flagellated algae
DIVISION 8. Phaeophyta—brown algae (life cycles 8 and 9)
DIVISION 9. Rhodophyta—red algae (life cycle 10)

Fungi

DIVISION 10. Myxomycota—slime molds
  Class Myxomycetes—plasmodial slime molds
  Class Acrasiomycetes—cellular slime molds
  Class Labyrinthulomycetes—cell-net slime molds
  Class Plasmodiophoromycetes—endoparasitic slime molds
DIVISION 11. Oomycota—oosphere fungi (life cycle 11)
  Class Oomycetes—oosphere fungi
  Class Chytridiomycetes—true chytrids and related fungi
  Class Hypochytridiomycetes—hypochytrids
DIVISION 12. Zygomycota—conjugation fungi (life cycle 12)
  Class Zygomycetes—conjugation fungi
  Class Trichomycetes—commensal fungi

DIVISION 13. Ascomycota—sac fungi (life cycles 13 and 14)
DIVISION 14. Basidiomycota—club fungi (life cycles 15 and 16)
DIVISION 15. Deuteromycota—imperfect fungi

Bryophytes

DIVISION 16. Hepatophyta—liverworts and hornworts (life cycle 17)
DIVISION 17. Bryophyta—mosses (life cycle 18)

Primitive vascular plants

DIVISION 18. Psilophyta—whisk fern and allies
DIVISION 19. Lycophyta—club mosses and allies (life cycle 19)
DIVISION 20. Arthrophyta—horsetails and allies (life cycle 20)
DIVISION 21. Pterophyta—ferns (life cycle 21)

Conifers

DIVISION 22. Cycadophyta—cycads
DIVISION 23. Ginkgophyta—ginkgos
DIVISION 24. Coniferophyta—conifers (life cycle 22)
DIVISION 25. Gnetophyta—Gnetales

Flowering plants

DIVISION 26. Anthophyta—flowering plants (life cycle 23)
  Class Angiospermae
    Subclass Dicotyledones
      *1. Order Magnoliales

1.1 Magnolia family—Magnoliaceae	
Tulip tree	*Liriodendron tulipifera*
Magnolia	*Magnolia soulangeana*
Sweet bay	*Magnolia virginiana*
1.5 Custard-apple family—Annonaceae	
Papaw	*Asimina triloba*
1.6 Nutmeg family—Myristicaceae	
Nutmeg, mace	*Myristica fragans*

      3. Order Laurales

3.9 Laurel family—Lauraceae	
Camphor	*Cinnamomum camphora*
Cinnamon	*Cinnamomum zeylanicum*
Mountain laurel	*Kalmia latifolia*
Avocado	*Persea americana*
Sassafras	*Sassafras albidum*

      4. Order Piperales

4.2 Pepper family—Piperaceae	
Black pepper	*Piper nigrum*

      5. Order Aristolochiales
        5.2 Old World pitcher-plant family—Nepenthaceae
      6. Order Nymphaeales

6.1 Water-lily family—Nymphaeaceae	
Common water lily	*Nymphaea odorata*

      7. Order Ranunculales

7.1 Crowfoot family—Ranunculaceae	
Anemone	*Anemone coronaria*

---

*The numbering system followed here is that used by Stebbins, 1974 (see Chapter 11). Since not all orders and families are listed, there are gaps in the numbers.

Columbine	*Aquilegia vulgaris*
Cowslip, marsh marigold	*Caltha palustris*
Larkspur	*Delphinium ajacis*
Hellebore	*Helleborus niger*
Peony	*Paeonia suffruticosa*
Buttercup	*Ranunculus acris*

7.3 Barberry family—Berberidaceae

Japanese barberry	*Berberis thunbergii*
Barberry	*Berberis vulgaris*
May apple	*Podophyllum peltatum*

8. Order Papaverales

8.1 Poppy family—Papaveraceae

Prickly poppy	*Argeomone platyceras*
Bleeding heart	*Dicentra spectabilis*
Poppy	*Papaver somniferum*
Bloodroot	*Sanguinaria canadensis*

9. Order Sarrenceniales

9.1 Pitcher-plant family—Sarraceniaceae

Pitcher plant	*Sarracenia purpurea*

11. Order Hamamelidales

11.3 Plane-tree family—Plantanaceae

Sycamore	*Platanus occidentalis*

11.5 Witch-hazel family—Hamamelidaceae

Witch hazel	*Hamamelis virginiana*
Sweet gum	*Liquidambar styraciflua*

13. Order Leitneriales

13.1 Corkwood family—Leitneriaceae

Corkwood	*Leitneria floridana*

14. Order Myricales

14.1 Sweet-gale family—Myricaceae

Sweet fern	*Comptonia peregrina*
Sweet gale, bog myrtle	*Myrica gale*
Bayberry	*Myrica pensylvanica*

15. Order Fagales

15.2 Beech family—Fagaceae

White oak	*Quercus alba*
Water oak	*Quercus nigra*
Post oak	*Quercus stellata*
Cork oak	*Quercus suber*
Live oak	*Quercus virginiana*
American chestnut	*Castanea dentata*
Beech	*Fagus grandifolia*

15.3 Birch family—Betulaceae

Alder	*Alnus serrulata*
Paper birch	*Betula papyrifera*
Hazelnut	*Corylus americana*
Filbert	*Corylus aveliana*
Hornbeam, ironwood	*Ostrya virginiana*

17. Order Caryophyllales

17.2 Four-o'clock family—Nyctaginaceae

Four-o'clock	*Mirabilis jalapa*

17.3 Pink family—Caryophyllaceae

Sweet william	*Dianthus barbatus*

17.4 Cactus family—Cactaceae
   Prickly-pear cactus   *Opuntia chlorotica*
   Cholla   *Opuntia versicolor*
   Organ-pipe cactus   *Cereus thurberi*
   Saguaro, giant cactus   *Carnegiea gigantea (=Cereus giganteus)*
   Barbados gooseberry   *Piereskia aculeata*
17.5 Mesembryanthemum family—Aizoaceae
   Pigweed (in Arizona)   *Trianthema portulacastrum*
17.8 Purslane family—Portulacaceae
   Bitterroot   *Lewisia rediviva*
   Portulaca, purslane   *Portulaca oleracea* var. *sativa*
17.10 Goosefoot family—Chenopodiaceae
   Saltbush   *Atriplex polycarpa*
   Beets   *Beta vulgaris*
   Chard, Swiss chard   *Beta vulgaris* var. *cicla*
   Spinach   *Spinacia oleracea*
17.11 Amaranth family—Amaranthaceae
   Tumbleweed   *Amaranthus albus*
   Pigweed, lamb's quarters   *Chenopodium album*
19. Order Polygonales
   19.1 Buckwheat family—Polygonaceae
      Buckwheat   *Fagopyrum esculentum*
      Bindweed   *Polygonum cilinode*
      Rhubarb   *Rheum rhaponticum*
      Dock   *Rumex crispis*
22. Order Thaeles
   22.4 Tea family—Theaceae
      Camellia   *Camellia japonica*
      Tea   *Thea sinensis*
   22.10 Saint-John's-wort family—Hypericaceae
      Mangosteen   *Garcinia mangostana*
      Saint-John's-wort   *Hypericum perforatum*
23. Order Malvales
   23.5 Linden family—Tiliaceae
      Jute   *Corchorus capsularis*
      Basswood, linden   *Tilia americana*
   23.6 Stercula family—Sterculiaceae
      Cacao, cocoa plant, chocolate tree   *Theobroma cacao*
      Cola   *Cola acuminata*
   23.7 Bombax family—Bombacaceae
      Ceiba tree (produces kapok) *Ceiba pentandra*
   23.8 Mallow family—Malvaceae
      Hollyhock   *Althaea rosea*
      Cotton   *Gossypium arboreum*
      Hibiscus, rose of Sharon   *Hibiscus syriacus*
      Mallow   *Malva neglecta*
24. Order Urticales
   24.1 Elm family—Ulmaceae
      Elm   *Ulmus americana*
   24.3 Mulberry family—Moraceae
      Breadfruit   *Artocarpus altilis*
      Hemp, marijuana   *Cannabis sativa*
      Fig   *Ficus carica*

Hop	*Humulus americanus*
Osage orange	*Maclura pomifera*
Mulberry	*Morus alba*

24.5 Nettle family—Urticaceae

Nettles	*Urtica dioica*

26. Order Violales

  26.4 Violet family—Violaceae

Violet	*Viola odorata*
Pansy	*Viola tricolor* var. *hortensis*

  26.14 Ocotillo family—Fouquieriaceae

Ocotillo	*Fouquieria splendens*
Boogum tree	*Idria columnaris*

  26.16 Rockrose family—Caricaceae

Papaya	*Carica papaya*

  26.18 Begonia family—Begoniaceae

Tuberous begonia	*Begonia tuberhybrida*

  26.20 Gourd family—Cucurbitaceae

Watermelon	*Citrullus vulgaris*
Cantaloupe	*Cucumis melo*
Cucumber	*Cucumis sativus*
Gourd	*Cucurbita argyrosperma*
Squash	*Cucurbita maxima*
Pumpkin	*Cucurbita pepo*
Squirting cucumber	*Ecballium elaterium*

27. Order Salicales

  27.1 Willow family—Salicaceae

Cottonwood	*Populus deltoides*
Lombardy poplar	*Populus nigra* var. *italica*
Quaking aspen	*Populus tremuloides*
Pussy willow	*Salix discolor*

28. Order Capparales

  28.3 Mustard family—Brassicaceae

Golden tuft	*Alyssum saxatile*
Horseradish	*Armoracia lapathifolia*
Mustard	*Brassica campestris*
Kohlrabi	*Brassica caulorapa*
Rutabaga	*Brassica napobrassica*
Kale	*Brassica oleracea* var. *acephala*
Cauliflower	*Brassica o.* var. *botrytis*
Cabbage	*Brassica o.* var. *capitata*
Brussels sprout	*Brassica o.* var. *gemmifera*
Broccoli	*Brassica o.* var. *italica*
Turnip	*Brassica rapa*
Radish	*Raphanus sativus*

29. Order Ericales

  29.5 Heath family—Ericaceae

Wintergreen	*Gaultheria procumbens*
Huckleberry	*Gaylussacia baccata*
Indian pipe	*Monotropa uniflora*
Red laurel	*Rhododendron catawbiense*
Azalea	*Rhododendron arborescens*
Rhodora	*Rhododendron canadense*
Blueberry	*Vaccinium corymbosum*
Cranberry	*Vaccinium macrocarpon*

31. Order Ebenales
   31.2 Ebony family—Ebenaceae
      Ebony   *Diospyros ebenum*
      Persimmon   *Diospyros virginiana*
32. Order Primulales
   32.3 Primrose family—Primulaceae
      Fairy primrose   *Primula malacoides*
      Moneywort   *Lysimachia Nummularia*
33. Order Rosales
   33.13 Saxifrage family—Saxifragaceae
      Hydrangea   *Hydrangea paniculata* var. *grandiflora*
      Gooseberry   *Ribes grossularia*
      Currant   *Ribes sativum*
      Venus's-flytrap   *Dionaea muscipula*
   33.14 Rose family—Rosaceae
      Cockspur   *Crataegus crus-galli*
      Hawthorn   *Crataegus intricata* var. *straminea*
      Quince   *Cydonia oblonga*
      Strawberry   *Fragaria chiloensis* var. *ananassa*
      Apple   *Malus sylvestris*
      Almond   *Prunus amygdalus*
      Apricot   *Prunus armeniaca*
      Sweet cherry   *Prunus avium*
      Plum, prune tree   *Prunus domestica*
      Peach   *Prunus persica*
      Nectarine   *Prunus persica* var. *nectarina*
      Pear   *Pyrus communis*
      Rose   *Rosa cathayensis*
      Blackberry, blackcap *Rubus occidentalis*
      Raspberry   *Rubus strigosus*
      Loganberry   *Rubus ursinus* var. *loganobaccus*
      Bridal wreath   *Spiraea prunifolia*
34. Order Fabales
   34.1 Pea family—Leguminosae
      Peanut   *Arachis hypogaea*
      Senna   *Cassia senna*
      Redbud   *Cercis canadensis*
      Beggarweed   *Desmodium tortuosum*
      Honey locust   *Gleditsia triacanthos*
      Soybean   *Glycine max*
      Kentucky coffee tree *Gymnocladus dioica*
      Alfalfa, lucerne   *Medicago sativa*
      Sensitive plant   *Mimosa pudica*
      Locoweed   *Oxytropis splendens*
      Lima bean   *Phaseolus limensis*
      Kidney bean   *Phaseolus vulgaris*
      Pea   *Pisum sativum*
      Mesquite   *Prosopis juliflora*
      White clover   *Trifolium repens*
      Wisteria   *Wisteria floribunda*
37. Order Myrtales
   37.3 Mangrove family—Rhizophoraceae
      Mangrove   *Avicennia nitida*

440

37.7 Water-chestnut family—Trapaceae
   Water chestnut   *Trapa natans*
37.9 Myrtle family—Myrtaceae
   Myrtle   *Myrtus communis*
   Allspice   *Pimenta dioica*
   Guava   *Psidium guajava*
   Clove   *Syzygium aromaticum*
37.10 Pomegranate family—Punicaceae
   Pomegranate   *Punica granatum*
37.11 Evening-primrose family—Onagraceae
   Evening primrose   *Oenothera biennis*
38. Order Cornales
  38.2 Sour-gum family—Nyssaceae
   Swamp tupelo   *Nyssa aquatica*
   Sour gum   *Nyssa sylvatica*
  38.4 Dogwood family—Cornaceae
   Dogwood   *Cornus florida*
39. Order Proteales
  39.1 Oleaster family—Elaeagnaceae
   Buffalo berry   *Shepherdia argentea*
40. Order Santalales
  40.5 Sandalwood family—Santalaceae
   Sandalwood   *Santalum album*
  40.6 Mistletoe family—Loranthaceae
   Mistletoe   *Phoradendron juniperinum*
42. Order Celastrales
  42.3 Staff-tree family—Celastraceae
   Bittersweet   *Celastrus scandens*
  42.7 Holly family—Aquifoliaceae
   American holly   *Ilex opaca*
43. Order Euphorbiales
  43.2 Spurge family—Euphorbiaceae
   Sapodilla (produces chicle)   *Achras zapota*
   Candelilla   *Euphorbia antisyphilitica*
   Poinsettia   *Euphorbia pulcherrima*
   Rubber tree   *Hevea brasiliensis*
   Manchineel   *Hippomane mancinella*
   Cassava   *Manihot esculenta*
   Castor-oil plant   *Ricinus communis*
   Mexican jumping bean*   *Sebastiania pavoniana*, also *Sapium biloculare*
44. Order Rhamnales
  44.1 Buckthorn family—Rhamnaceae
   New Jersey tea   *Ceanothus americanus*
   Cascara buckthorn   *Rhamnus purshiana*
   Indian jujube   *Zizyphus mauritania*
   Chinese date   *Zizyphus jujuba*
  44.3 Grape family—Vitaceae
   Virginia creeper   *Parthenocissus quinquefolia*
   Boston ivy   *Parthenocissus tricuspidata*
   Grape   *Vitis vinifera*

---

*The bean "jumps" because it is infected by the larva of a moth, *Carpocapsa saltitans*, family Tortricidae.

45. Order Sapindales
  45.8 Horse-chestnut family—Hippocastanacea
    Horse chestnut, buckeye   *Aesculus hippocastanum*
  45.9 Maple family—Aceraceae
    Silver maple   *Acer saccharinum*
  45.10 Bursera family—Burseraceae
    Frankincense   *Boswellia carteri*
    Myrrh   *Commiphora myrrha*
  45.11 Cashew family—Anacardiaceae
    Cashew   *Anacardium occidentale*
    Mango   *Mangifera indica*
    Poison ivy   *Rhus radicans*
    Staghorn sumac   *Rhus typhina*
  45.13 Quassia family—Simaroubaceae
    Tree of heaven   *Ailanthus altissima*
  45.16 Rue family—Rutaceae
    Lemon   *Citrus limon*
    Grapefruit   *Citrus paradisi*
    Orange   *Citrus sinensis*
  45.17 Mahogany family—Meliaceae
    Mahogany   *Swietenia mahogani*
  45.18 Caltrop family—Zygophyllaceae
    Creosote bush   *Larrea tridentata*
46. Order Juglandales
  46.2 Walnut family—Juglandaceae
    Swamp hickory   *Carya cordiformis*
    Pecan   *Carya illinoensis*
    Walnut   *Juglans regia*
47. Order Geraniales
  47.2 Flax family—Linaceae
    Flax (produces linen)   *Linum usitatissimum*
  47.3 Coca family—Erythroxylaceae
    Coca, cocaine plant   *Erythroxylon coca*
  47.4 Geranium family—Geraniaceae
    Lemon geranium   *Pelargonium crispum*
  47.8 Balsam family—Balsaminaceae
    Touch-me-not   *Impatiens biflora*
49. Order Umbellales
  49.2 Carrot family—Umbelliferae
    Dill   *Anethum graveolens*
    Celery   *Apium graveolens* var. *dulce*
    Caraway   *Carum carvi*
    Water hemlock   *Cicuta douglasii*
    Poison hemlock   *Conium maculatum*
    Carrot   *Daucus carota*
    English ivy   *Hedera helix*
    Parsnip   *Pastinaca sativa*
    Anise   *Pimpinella anisum*
50. Order Gentianales
  50.1 Logania family—Loganiaceae
    Nux-vomica tree (produces strychnine)   *Strychnos nux-vomica*
    A woody vine that produces curare   *Strychnos toxifera*
  50.2 Gentian family—Gentianaceae
    Fringed gentian   *Gentiana crinita*

50.3 Dogbane family—Apocynaceae

Dogbane	*Apocynum androsaemifolium*
Indian hemp	*Apocynum cannabinum*
Milkweed	*Asclepias syriaca*
Periwinkle	*Vinca minor*

50.4 Olive family—Oleaceae

Forsythia	*Forsythia suspensa*
Ash	*Fraxinum americana*
Privet	*Ligustrum vulgare*
Olive	*Olea europaea*
Lilac	*Syringa vulgaris*

51. Order Polemoniales

51.2 Nightshade family—Solanaceae

Belladonna	*Atropa belladonna*
Pepper (bell or green)	*Capsicum frutescens* var. *grossum*
Tomato	*Lycopersicon esculentum*
Tobacco	*Nicotiana tabacum*
Petunia	*Petunia hybrida*
Chinese lantern plant	*Physalis alkekengi*
Horse nettle	*Solanum carolinense*
Eggplant	*Solanum melongena* var. *esculentum*
Black nightshade	*Solanum nigrum*
Potato	*Solanum tuberosum*

51.4 Morning-glory family—Convolvulaceae

Sweet potato	*Ipomoea batatas*
Morning glory	*Ipomoea purpurea*

51.7 Phlox family—Polemoniaceae

Wild blue phlox	*Phlox divaricata*
Phlox, annual	*Phlox drummondii*

51.9 Borage family—Boraginaceae

Hound's-tongue	*Cynoglossum officinale*
Forget-me-not	*Myosotis scorpioides*

52. Order Lamiales

52.1 Verbena family—Verbenaceae

Verbena	*Verbena hybrida*

52.2 Mint family—Labiatae

Hyssop	*Hyssopus officinalis*
Lavender	*Lavandula officinalis*
Peppermint	*Mentha piperita*
Spearmint	*Mentha spicata*
Bells of Ireland	*Molucella laevis*
Basil	*Ocimum basilicum*
Marjoram	*Origanum vulgare*
Rosemary	*Rosmarinus officinalis*
Sage	*Salvia officinalis*
Savory	*Satureja hortensis*
Thyme	*Thymus vulgaris*
Vetch	*Vicia sativa*

53. Order Plantaginales

53.1 Plantago family—Plantaginaceae

Plantain	*Plantago lanceolata*

54. Order Scrophulariales

54.3 Snapdragon family—Scrophulariaceae

Snapdragon	*Antirrhinum majus*

| Foxglove | *Digitalis purpurea* |
| Mullein | *Verbascum thapsus* |

55. Order Campanulales
   55.2 Bellflower family—Campanulaceae

| Canterbury bells | *Campanula medium* |
| Bluebell | *Campanula rotundifolia* |

56. Order Rubiales
   56.1 Madder family—Rubiaceae

| Cinchona, chinchona | *Cinchona ledgeriana* |
| Coffee | *Coffea arabica* |

57. Order Dipsacales
   57.1 Honeysuckle family—Caprifoliaceae

| Honeysuckle | *Lonicera sempervirens* |

   57.4 Teasel family—Dipsacaceae

| Teasel | *Dipsacus sylvestris* |

58. Order Asterales
   58.1 Aster family—Compositae

Ragweed	*Ambrosia trifida*
Burdock	*Arctium minus*
Aster	*Aster laevis*
Sticktight, beggar-tick	*Bidens tripartita*
Chrysanthemum	*Chrysanthemum morrifolium*
Endive	*Cichorium endivia*
Chicory	*Cichorium intybus*
Thistle	*Cirsium arvense*
Yellow cosmos	*Cosmos sulphureus*
Artichoke	*Cynara scolymus*
Dahlia	*Dahlia pinnata*
Sunflower	*Helianthus annuus*
Lettuce	*Lactuca sativa*
Goldenrod	*Solidago speciosa* (and many others)
Marigold	*Tagetes lucida*
Dandelion	*Taraxacum officinale*
Salsify (= vegetable oyster)	*Tragopogon porrifolius*
Cocklebur	*Xanthium spinosum*
Zinnia	*Zinnia elegans*

Subclass Monocotyledones
  59. Order Alismatales
   59.3 Water-plantain family—Alismataceae

| Arrowhead | *Sagittaria graminea* |

  61. Order Najadales
   61.3 Arrow-grass family—Juncaginaceae

| Arrow grass | *Triglochin palustre* |

   61.4 Najas family—Najadaceae

| Naiad | *Najas marina* |

   61.5 Pondweed family—Potamogetonaceae

| Pondweed | *Potamogeton natans* |

  63. Order Commelinales
   63.4 Spiderwort family—Commelinaceae

| Spiderwort | *Tradescantia virginiana* |

  66. Order Poales
   66.1

| Wild oat | *Avena fatua* |

Oat	*Avena sativa*
Bamboo	*Bambusa bambos*
Bermuda grass	*Cynodon dactylon*
Crabgrass	*Digitaria sanguinalis*
Sweet grass, holy grass	*Hierochloe alpina*
Barley (produces malt)	*Hordeum vulgare*
Millet	*Pennisetum glaucum*
Rice	*Oryza sativa*
Sugarcane	*Saccharum officinarum*
Rye	*Secale cereale*
Sorghum	*Sorghum vulgare*
Wheat	*Triticum aestivum*
Corn	*Zea mays*

68. Order Cyperales
   68.1 Sedge family—Cyperaceae

Sedge	*Carex morrowii*
Papyrus	*Cyperus papyrus*

69. Order Typhales
   69.2 Cattail family—Typhaceae

Cattail	*Typha latifolia*

70. Order Bromeliales
   70.1 Pineapple family—Bromeliaceae

Pineapple	*Ananas comosus*
Spanish moss	*Tillandsia usneoides*

71. Order Zingiberales
   71.4 Banana family—Musaceae

Plantain (not to be confused with *Plantago*)	*Musa paradisiaca*
Banana	*Musa paradisiaca* var. *sapientum*
Manila hemp	*Musa textilis*
Traveler's palm, traveler's tree	*Ravenala madagascariensis*

72. Order Arecales
   72.1 Palm family—Palmae

Betel palm	*Areca cathecu*
Coconut palm	*Cocos nucifera*
Date palm	*Phoenix dactylifera*
Saw palmetto	*Serenoa repens*

74. Order Pandanales
   74.1 Pandana family—Pandanaceae

Pandana	*Pandanus pacificus*

75. Order Arales
   75.1 Arum family—Araceae

Jack-in-the-pulpit	*Arisaema triphyllum*
Skunk cabbage	*Symplocarpus foetidus*

   75.2 Duckweed family—Lemnaceae

Duckweed	*Spirodela polyrrhiza*
Wolffia	*Wolffia punctata*

76. Order Liliales
   76.2 Pickerelweed family—Pontederiaceae

Water hyacinth	*Eichhornia crassipes*

   76.3 Lily family—Liliaceae

Sisal hemp	*Agave sisalana*
Century plant	*Agave americana*
Onion	*Allium cepa*

Leek	*Allium porrum*
Wild garlic	*Allium sativum*
Asparagus	*Asparagus officinalis* var. *altilis*
Queen-cup	*Clintonia uniflora*
Autumn crocus	*Colchicum autumnale*
Lily of the valley	*Convallaria majalis*
Adder's-tongue, dogtooth violet	*Erythronium americanum*
Lily	*Lilium canadense*
Narcissus	*Narcissus tazetta*
Greenbrier	*Smilax herbacea*
Tulip	*Tulipa gesneriana*

76.4 Iris family—Iridaceae

Crocus	*Crocus susianus*
Gladiolus	*Gladiolus gandavensis* (hybrids)
Iris	*Iris versicolor*

76.14 Yam family—Dioscoreaceae

Yam	*Dioscorea sativa*

77. Order Orchidales

77.4 Orchid family—Orchidaceae

Orchid	*Cattleya labiata*
Lady's slipper	*Cypripedium acaule*

APPENDIX B

Plant life cycles

■ Life cycles of plants have long been fundamental to the understanding of plant evolution. At the same time, students have struggled to master these cycles. With this in mind, 23 life cycles are set apart in this appendix and presented in a comparative sequence. Student and instructor may select those most useful for their studies.

A fixed format is followed regardless of the variation in the life cycle described. Illustrations of the cycle followed, the parts involved, and a habitus drawing of the species described enable one to follow the changes from group to group. The classification of the example chosen is given; then the vegetative characteristics of the plant are described. A discussion of asexual reproduction then follows, accounting for the vegetative plant, sporophyte, sporangium, and spore; next sexual reproduction is described, accounting for the gametophyte, antheridium, **archegonium** (or oogonium in lower plants,* sperm, egg, zygote, and embryo in each case. The means of dispersal is also described because reproduction not only provides for the formation of new individuals but also for the dissemination of the species. If one or more of these stages is missing, this is so stated in the proper place by the word "none." Remember, however, that in many cases the relationship between these stages from plant to plant is functionally analogous rather than structurally homologous. One should never assume that, because a phase of the life cycle is described after one of these headings, it is equivalent homologously to that structure in other plants.

---

*When the structure is an oogonium, the heading "archegonium" is put in quotation marks to indicate that a true archegonium is *not* present.

LIFE CYCLE 1—T-EVEN PHAGE (Fig. 1)
DIVISION 1. Virulenta
Characteristics

The simple growth cycle of a T-even phage (under the old classification, a T_2, T_4, and T_6 phage, now members of the family Phagoviridae) will serve as a generalized example (Fig. 1). The virion attaches to the surface of the bacterium. (In an animal virus the virion is engulfed or absorbed by the host cell.) The process is termed absorption, and the attachment is irreversible. Shortly thereafter, penetration takes place, a process consisting of the injection of the DNA contents of the virion into the host cell. The capsid remains outside and is no longer needed. Next is the latent period, so-called only because nothing is observed in the culture. However, inside the bacterium two processes take place. First, there is a period of protein synthesis known as the eclipse period. Enzymes, capsid coats, and DNA are produced from raw material provided by the bacterial cell. A second part of the latent period is termed the maturation period, during which time condensation of the DNA takes place and the capsids are added. About 200 to 1000 virions are released, ending the cycle with the final, or release, phase.

During the recombination phase there is an exchange of genetic material. This is dependent on mixed infections that require the absorption of two or more virions by the cell. There are three types of such infections: (1) phenotypic mixing, which is caused by the combining of capsids of two different forms of virions during the growth process, resulting in a different external appearance of the newly formed virions; (2) complementation, which involves two different defective viruses that help each other by supplying each others missing genes (differs from helper viruses); (3) phage recombination, during which time portions of the DNA molecule recombine with others so that there is an actual exchange of genetic material that follows the same laws of heredity found in the higher plants.

Members of the Virulenta are from 10 to 200 nm in diameter and capable of growth and reproduction only when they have penetrated a living cell. Their dispersal form, the virion, functions as does the spore of higher plants.

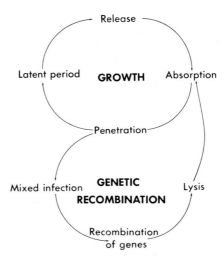

Fig. 1. Life cycle of a T-even phage.

LIFE CYCLE 2—NOSTOC Sp. (Fig. 2)
DIVISION 3. Cyanophyta
CLASS Myxophyceae
Characteristics

The vegetative body of *Nostoc* sp. consists of short filaments of rounded cells embedded in a firm matrix of gelatinous material (Fig. 3). These filaments are referred to as trichomes, and each may have one or more enlarged cells, the **heterocysts.** The filaments are grouped together into colonies, which at first are microscopic but which grow into balls 30 to 60 mm in diameter. These balls are jellylike, sometimes hollow, and may be irregularly shaped. The species of this genus are either aquatic or terrestrial. Many occur as floating colonies in sunny ponds and others in swift streams attached to stones. Some live as deep as a meter in the soil, and still others have a symbiotic association with certain fungi to form a lichen.

Asexual reproduction

Vegetative reproduction. The breaking apart of the colony may take place as a result of storms and similar disturbances, but otherwise there is no usual vegetative propagation.

Sporophyte. There is no true sporophyte; however, some of the cells are differentiated into **hormogones** forming a **hormogonium.** The hormogones are sporelike in function. They can germinate to produce new colonies.

Sporangium. None.

Spore. None. Thick-walled **akinetes** are formed from any vegetative cells in mature colonies that act as spores and that are capable of germinating to form a new filament in the colony.

Sexual reproduction

No form of sexual reproduction has been observed in *Nostoc* sp. or related genera.

Means of dispersal. The floating balls of filaments provide the only dispersal mechanism for these plants.

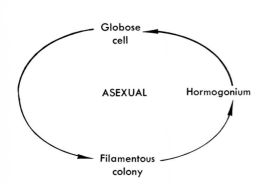

Fig. 2. Life cycle of *Nostoc* sp.

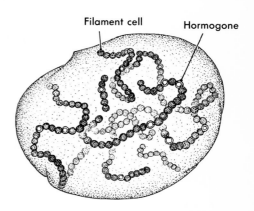

Fig. 3. *Nostoc* sp. Vegetative stage.

LIFE CYCLE 3—CHLAMYDOMONAS Sp.
(Fig. 4)
DIVISION 4. Chlorophyta
CLASS Chlorophyceae
Characteristics

The genus *Chlamydomonas* sp. is probably one of the most primitive of the green algae; it is found in standing water, in wet soil, and in swimming pools. The cells may become so numerous as to color the water green. It is a unicellular alga, the cell shape of which varies from spherical to ovoid. One large chloroplast almost completely fills up the cell. Within the chloroplast is a single, round pyrenoid that appears to be the center of starch formation. The cell also contains a centrally located nucleus, an eyespot sensitive to light, and two contractile vacuoles that probably function as organelles of excretion. At the anterior end of the cell are two flagella that pull the cell through the water. The entire cell is bounded by a conspicuous wall composed of cellulose.

Asexual reproduction

Vegetative reproduction. At certain times, 2, 4, or 8 zoospores may be produced within the parent cells, which they resemble except that they are smaller.

Sporophyte. None.

Sporangium. None.

Spore. Daughter cells swim out of the parent cell through the dissolved wall to form new plants.

Sexual reproduction

Gametophyte. There is no structural differentiation of a gametophyte plant as distinct from a sporophyte plant. However, the parent cell may divide into 8, 16, or 32

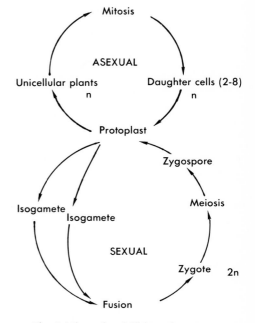

Fig. 4. Life cycle of *Chlamydomonas* sp.

smaller cells that resemble miniature *Chlamydomonas* sp. plants. These are the gametes or sex cells that are released into the water.

Antheridium. None.

"Archegonium." None.

Gamete. All are exactly alike; therefore, they may be called **isogametes.** They are not differentiated into egg and sperm. The gametes in some species look like the **zoospores** except that the former are smaller. Apparently in this life cycle there are sex cells and sex, but in a primitive type such as *Chlamydomonas* sp. the fusing gametes are isogametes; therefore no indication of maleness or femaleness is found. In other species there may be heterogametes.

Zygote. The gametes fuse in pairs (Fig. 5); ordinarily each fusing member of the pair comes from different parent cells. Since the gametes that fuse are alike, the resulting fusion process is called isogamy. However, the fusion cell formed is a zygote. A thick wall forms around the zygote and it becomes a resting cell that is adapted to withstand adverse conditions (such as drying).

Although there is no sporophyte, the zygote is a single cell containing one diploid nucleus that undergoes reduction division to produce 4 nuclei that eventually become incorporated into 4 zoospores. Each released zoospore develops into a *Chlamydomonas* sp. plant.

Embryo. None.

Means of dispersal. The flagella of the plant cells are the mechanism for the zoospores' dispersal; they allow the species to spread as far as physiologically suitable conditions permit.

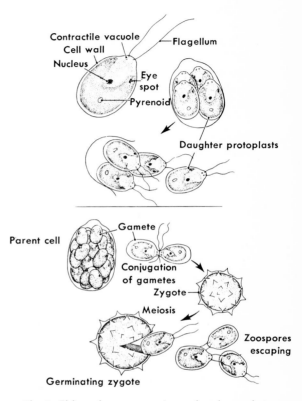

Fig. 5. *Chlamydomonas* sp. Asexual and sexual stages.

LIFE CYCLE 4—VOLVOX AUREUS (Fig. 6)
DIVISION 4. Chlorophyta
CLASS Chlorophyceae
Characteristics

The cells composing the vegetative body of *Volvox aureus* are individual organisms grouped together to form an organized colony (Fig. 7). Each cell is connected by a thin protoplasmic strand to the adjacent cells. The whole macroscopic colony slowly moves and rotates by the coordinated efforts of the projecting flagella. The colony is usually spherical or nearly so, is hollow within, and varies in size according to the number of cells. Mature colonies have from 500 to over 60,000 cells, depending on the species. *Volvox aureus* occurs in freshwater pools, both temporary and permanent types, and often becomes abundant enough to give a greenish color to the water.

Asexual reproduction

Vegetative reproduction. The vegetative growth of the colony takes place until it reaches the size normal for the species. Further growth of the colony is very limited or stops entirely.

Sporophyte. None. The colony acts as the sporophyte. Daughter colonies are produced by special cells, the **gonidia,** which by rapid cell division produce hollow, spherical daughter colonies. These project into the center of the parent colony. When the inside is nearly filled with the smaller colonies, the parent breaks apart and liberates the new forms.

Sporangium. None. The gonidium is analogous to the sporangium of higher plants.

Spore. None.

Sexual reproduction

Gametophyte. Vegetative colonies may at times become gametophytes, producing either sperms or eggs or both. These develop from special cells.

Antheridium. Sperm-producing cells enlarge and undergo mitosis, producing hundreds of sperm that collect in a small pocket, the antheridium.

"Archegonium." A few of the cells in a colony enlarge, lose their flagella, and form the egg cells, which are held in a small pocket, the oogonium.

Sperm. The sperm cell or **antherozoid** is smaller than the vegetative cells of the col-

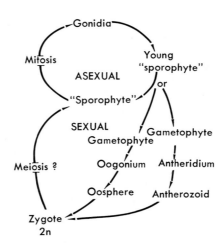

Fig. 6. Life cycle of *Volvox aureus.*

452

ony, spindle shaped, and biflagellated, but it has poorly developed chloroplasts.

Egg. The egg or **oosphere** is larger than the vegetative cells, is rounded, and lacks flagella. If this cell is not fertilized, it may develop into a new colony parthenogenetically. It is then known as a **parthenospore.**

When meiosis takes place is not clear. In some members of this order, alternate generations of haploid and diploid cells are known.

Zygote. The antherozoids are released from the antheridium as free-swimming cells. They swim to the oosphere and fuse with it. After fertilization the zygote produces a thick covering and becomes an **oospore.** This is released from the egg pocket and remains in the spore condition for some time before germination. During this resting period meiotic division of the nucleus may take place. Once the oospore wall is broken down, cell division proceeds and a new colony is formed.

Embryo. None.

Means of dispersal. Floating, combined with slow swimming of the colony, permits the dispersal of the species.

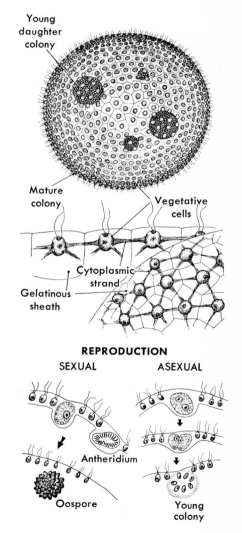

Fig. 7. *Volvox aureus.* Vegetative, sexual, and asexual stages.

453

LIFE CYCLE 5—ULOTHRIX Sp. (Fig. 8)
DIVISION 4. Chlorophyta
CLASS Chlorophyceae
Characteristics

Species of *Ulothrix* are common in freshwater situations, where they form bright green, unbranched filaments having a small diameter (0.025 mm). These filaments are composed of a single row of small cells, each surrounded by a cellulose wall (Fig. 9). At the base of the filament there is a well-developed **holdfast cell** that anchors the plant to some substrate. Although the holdfast lacks a chloroplast, it is otherwise not unlike the rest of the cells. In all the other cells the chloroplast forms a broad, transverse, ring-shaped band lying near the cell wall. This plant is truly multicellular; there is differentiation in both structure and function. The plant filaments increase in length by mitosis of the individual cells.

Asexual reproduction

Vegetative reproduction. None, other than fragmentation.

Sporophyte. None.
Sporangium. Zoosporangia.
Spore. Zoospores. Anywhere from 2 to 4, 8, 16, or 32 zoospores are produced within unicellular zoosporangia. The zoosporangia are transformed vegetative cells. The zoospores each bear 4 flagella; they form new filaments by cell division after coming to rest in a suitable location.

Sexual reproduction

Gametophyte. The cells making up the filaments are the haploid generation. Gametes are produced in cells morphologically like those that produce zoospores.

Antheridium and "archegonium." None as such. The vegetative cells serve as unicellular gametangia that bear 8, 16, 32, or 64 gametes. These gametes are similar to the zoospores, except smaller, but bear only 2 flagella. *Ulothrix* sp. is considered to be het-

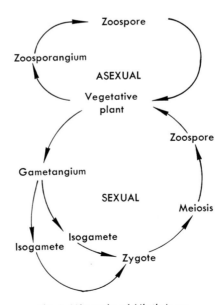

Fig. 8. Life cycle of *Ulothrix* sp.

erothallic, since the fusing gametes are produced by different filaments.

Sperm and egg. None. However, the isogametes act as sperm and egg and probably differ biochemically. Fertilization occurs between isogametes and is termed isogamy.

Zygote. A diploid cell is formed by the fusion of haploid gametes and then undergoes a period of rest prior to meiosis and the formation of 4 haploid zoospores. Once the zoospores are formed, they attach themselves and produce new filaments by cell division.

Embryo. None.

Means of dispersal. *Ulothrix* sp. is disseminated by the actively swimming zoospores.

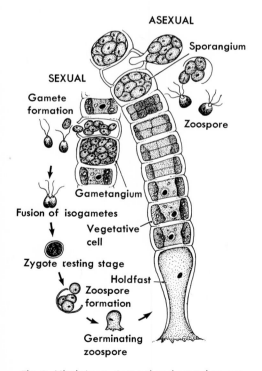

Fig. 9. *Ulothrix* sp. Asexual and sexual stages.

LIFE CYCLE 6—CHARA Sp. (Fig. 10)
DIVISION 4. Charophyta
CLASS Charophyceae
Characteristics

These plants are members of a single order, Charales. They are a group of green plants that grow submerged and are attached by rhizoids to the mud at the bottom of pools or slow-moving water. The chief feature of the class is a series of internodes with narrow, leaflike branches at the nodes (Fig. 11). Since plants of the genus *Chara* are inhabitants of waters especially high in calcium, they frequently become encrusted with lime—hence their common name stonewort. Their continued growth in water and their eventual accumulation result in a deposit of marl on the bottoms of ponds.

The cells of these multicellular plants contain more than one nucleus, and there are numerous small chloroplasts embedded in a peripheral layer of cytoplasm. The vegetative plants may have both male and female fructifications attached at the nodes (Fig. 11).

Asexual reproduction

Vegetative reproduction. The only type of reproduction known to occur is by the detachment of vegetative outgrowths that develop into new plants.

Sporophyte. None.
Sporangium. None.
Spore. None.

Sexual reproduction

Gametophyte. The vegetative plant is the haploid generation and bears separate sex organs at the nodes. Both the male **globule** and the female **nucule** consist of a gamete-producing part surrounded by a multicellular sheath. Most species are homothallic.

Antheridium. The globule in *Chara* sp. is borne below (Fig. 11) the female nucule. Motile sperm or antherozoids are produced by the globule.

"Archegonium." The oogonium of *Chara* sp. is developed above the globule and is called a nucule. The initial cell of this structure divides to form a row of cells, the apical one of which is the oogonial mother cell. The rest of the cells grow around it while the apical cell enlarges to become the egg.

Sperm. Motile sperm or antherozoids are produced in large numbers within the globule to escape as free-swimming sperm.

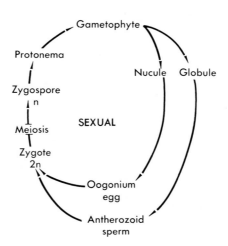

Fig. 10. Life cycle of *Chara* sp.

Egg. The egg is nonmotile and remains within the nucule.

Zygote. Fertilization takes place within the nucule. The fusion of the nonmotile egg and motile sperm forms the zygote, around which is secreted a thick wall. Eventually, the thick-walled zygospore along with other parts of the nucule drop to the bottom of the pond or stream.

Embryo. None. Before germination the zygote nucleus divides by meiosis into 4 nuclei. At the time of germination the 4 nuclei present in the zygote contribute to the initial development of the multicellular plant. However, the 3 nuclei in the basal cell soon disintegrate, and the new plant is formed by the continued division of this unincleate cell to form the protonema and eventually the gametophyte.

Means of dispersal. Dispersal of the species is entirely attributable to the passive scattering of the zygospores. The movement of the water causes their scattering as well as that of the detached vegetative outgrowth.

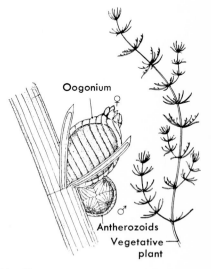

Fig. 11. *Chara* sp. (stonewort). Vegetative plant (with reproductive organs on *right*). Male and female fructifications in detail on *left*.

LIFE CYCLE 7—VAUCHERIA Sp. (Fig. 12)
DIVISION 5. Chrysophyta
CLASS Xanthophyceae
Characteristics

This genus is not really representative of the class because it is one of the most advanced forms. It occurs in both freshwater ponds and moist or muddy land areas, where it may form a dark green, feltlike mat. It is filamentous, with very few crosswalls, and forms large, multinucleated cells (Fig. 13). The filaments may extend for a foot or more within the mat. The absence of crosswalls makes these filaments a syncytium. Thousands of small, disc-shaped chloroplasts occur in the cytoplasm. If a branch of the filament is injured, the opening may be sealed off by the generation of a crosswall. Food is stored only as oil droplets. This is a commonly available species and therefore is used as an example of the yellow-brown algae.

Asexual reproduction

Vegetative reproduction. There are no vegetative structures for asexual reproduction. However, the vegetative plant produces zoospores as well as gametangia.

Sporophyte. None.

Sporangium. Zoospores are produced by a terminal division of a filament, the zoo-sporangium. This area is separated from the remainder of the filament by a cell wall but is still multinucleated.

Spore. The spore is in the form of a multinucleated, flagellated zoospore of relatively large size. It is liberated under suitable conditions through a crack in the wall of the sporangium. It swims slowly away to a new site, where it settles down to form another vegetative thallus.

Sexual reproduction

Gametophyte. Sexual reproduction is effected by the formation of adjacent sex organs on the same filament. This area, then, is analogous to the gametophyte.

Antheridium. The antheridium is a special branch with a crosswall at the tip of the branch separating and isolating a small number of nuclei with the surrounding protoplasm. This antheridium branch is elongate and somewhat narrower than the main filament of the plant. The apex of the branch is tapered, and when the branch is mature, a small pore opens for sperm release.

"Archegonium." The oogonium is formed

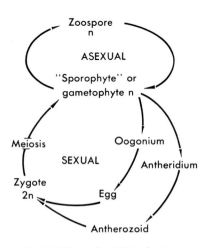

Fig. 12. Life cycle of *Vaucheria* sp.

from a branch near the antheridium. It is separated from the rest of the plant by a crosswall at the base. It swells to form a large rounded oogonium. When the branch is mature, a pore similar to that of the antheridium is formed.

Sperm. Uninucleated, biflagellated antherozoids are formed within the antheridium. These are released through the pore and are free swimming.

Egg. The single egg cell produced in the oogonium has only one nucleus and occupies the entire space within the oogonium.

Zygote. The antherozoids swarm to the pore of the oogonium. Even though several may enter, only one of them penetrates the egg. It does not fertilize the egg in the usual way, however; instead, a single sperm nucleus penetrates the ovum and enlarges considerably before it fuses with the nucleus of the ovum. This then becomes a central nucleus in the zygote. As soon as fertilization has taken place, a multilayered wall is formed, and the zygote remains dormant for several months.

Embryo. The germinating zygote develops directly into a new sporophyte. However, there is evidence that the nucleus divides meiotically before germination; therefore the sporophyte may be haploid.

Means of dispersal. The zoospores, by slow but free-swimming movement, provide the principal means of dispersal for this plant.

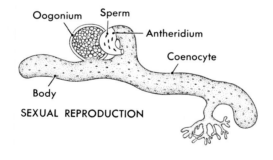

SEXUAL REPRODUCTION

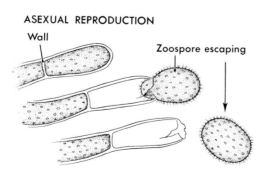

ASEXUAL REPRODUCTION

Fig. 13. *Vaucheria* sp. Sexual and asexual stages.

LIFE CYCLE 8—FUCUS Sp. (Fig. 14)
DIVISION 8. Phaeophyta
CLASS Phaeophyceae
Characteristics

Fucus sp. is so commonly available that it is used as an example of the brown algae even though it is not really representative of the type of life cycle of the class. The thallus is a dichotomously branching shoot arising from a flattened, disc-shaped holdfast (Fig. 15, A). The "branches" are ribbon shaped but short and blunt with evident midribs. Hollow, bladderlike air sacs are scattered along the thallus and serve to buoy up the plant when it is submerged. These plants are widely distributed throughout the cold coastal waters of the northern hemisphere.

Asexual reproduction

Vegetative reproduction. No special structures. The only known method is by fragmentation. Broken pieces may live as floating plants. This is specially true of the genus *Sargassum*.

Sporophyte. The *Fucus* sp. plant is diploid—hence the sporophyte or spore-producing plant.

Sporangium. Since the spores are of two different sizes, the sporangia wherein they are produced are called **microsporangia** and **macrosporangia**. These sporangia are produced in small cavities, or **conceptacles**, which are found on the tips of branches in swollen, cone-shaped structures called receptacles (Fig. 15, B).

Spore. Meiosis of the nucleus occurs in the microsporangium as well as in the macrosporangium to produce 4 meiospores in the sporangia. Each of the 4 meiospores within the microsporangium undergoes 4 mitotic divisions to produce 64 haploid cells. These haploid cells eventually become microgametes or sperm. In the macrosporangium each meiospore undergoes a single mitosis to produce a total of 8 nuclei. A small amount of cytoplasm eventually is massed around each to separate them into spherical, nonmotile macrogametes or eggs.

Sexual reproduction

Gametophyte. The 16 haploid cells formed from each meiospore within the microsporangium may be regarded as the male gametophyte; similarly, the 8 haploid nuclei formed from the meiospores within the macrosporangium may be regarded as the female gametophyte.

Antheridium. Each of the 64 haploid cells within the microsporangium develops into a microgamete or sperm with the haploid chromosome number.

"Archegonium." The oogonium arises from the megagamete already described.

Sperm. Each sperm has 2 lateral flagella and is motile. When mature they escape from the microsporangium and swim to the eggs.

Eggs. The 8 nonmotile eggs develop within the megasporangium from the haploid cell masses. When mature, they are set free to float in the water.

Zygote. Fertilization of the egg by the sperm occurs in the water. Each floating egg becomes surrounded by many minute sperm. The sperm attach themselves to the egg by one **flagellum** and lash the water with the other, causing the egg to rotate. Finally, a sperm penetrates the egg, thus fertilizing

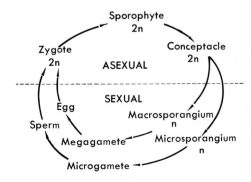

Fig. 14. Life cycle of *Fucus* sp.

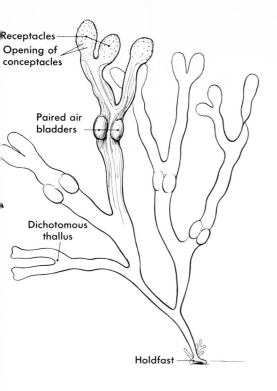

it. This union of egg and sperm results in a diploid cell, the zygote.

Embryo. None as identified in the higher plants. However, the zygote immediately grows and develops into the sporophyte plant. It attaches itself to some suitable support and becomes the mature *Fucus* plant.

Means of dispersal. By means of fragmentation and at the time of fertilization the egg is dispersed by the ebb and flow of the tide.

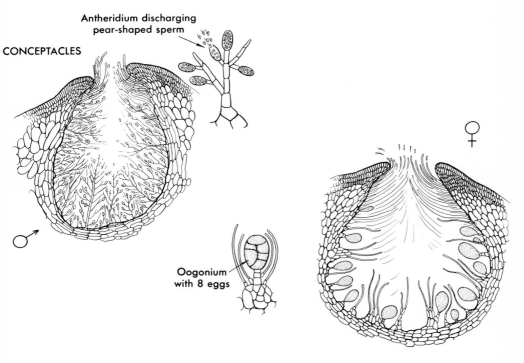

Fig. 15. *Fucus* sp. **A,** Vegetative stage. **B,** Reproductive stage.

LIFE CYCLE 9—ECTOCARPUS Sp. (Fig. 16)
DIVISION 8. Phaeophyta
CLASS Phaeophyceae
Characteristics

The common brown alga *Ectocarpus* sp. is found in cold seawater around the world. It is attached to other kinds of algae as an epiphyte and consists of a main filament from which arise small branches (Fig. 17). The cells are uninucleated and contain many small plastids. The plants form small tufts several inches long on other algae, both in the tidal zone and below low tide. The life cycle of this species exhibits an alternation of generations that is characteristic of some of the higher forms of plant life. However, all the plants look similar but can be distinguished by their fruiting bodies. Hence, the plant is heterothallic.

Asexual reproduction

Vegetative reproduction. None.

Sporophyte. The sporophytes are indistinguishable from the filamentous gametophyte except as they produce zoosporangia (Fig. 17). The cells of the filament are diploid.

Sporangium. The sporangia are most typically unilocular. They are formed from a single uninucleate cell at the tip of a filament. The nucleus in this cell divides many times producing, as a result of meiosis, anywhere from 32 to 128 small, biflagellated, and haploid zoospores.

The sporophyte of *Ectocarpus* sp. may produce on the same filament in addition to the unilocular sporangia, structures called **plurilocular** or neutral sporangia. They resemble the gametangia of the gametophyte filaments but produce zoospores instead of gametes. However, these zoospores are not produced by meiosis. Therefore they are diploid filaments like the parent plant.

Spore. Both haploid and diploid zoospores are produced. The former produce gametophyte filaments, whereas the latter produce sporophytes. Both are biflagellated and free swimming.

Sexual reproduction

Gametophyte. The zoospores produced within the unilocular sporangia are haploid

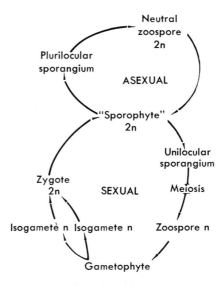

Fig. 16. Life cycle of *Ectocarpus* sp.

and develop into the haploid filament (gametophyte generation). These filaments are morphologically like those of the sporophyte filaments but are obviously different physiologically.

Antheridium. None.

"Archegonium." None.

Sperm and egg. None. However, this alga does produce isogametes that have been demonstrated to be physiologically different. Those from a single gametangium will not fuse together. They must come from 2 different gametangia. Should they not fuse, a single gamete may develop parthenogenetically into a new gametophyte filament.

Zygote. Two haploid isogametes from different filaments fuse to produce a diploid zygote. The zygote formed germinates and produces a new filament (the sporophyte filament) that bears zoosporangia. Thus there is an alternation of a sporophyte and a gametophyte generation.

Embryo. None.

Means of dispersal. Dispersal is almost entirely dependent on the free-swimming neutral zoospores and the zygotes.

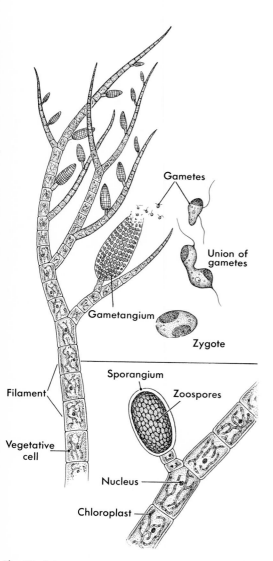

Gametes

Union of gametes

Gametangium

Zygote

Sporangium

Zoospores

Filament

Vegetative cell

Nucleus

Chloroplast

Fig. 17. *Ectocarpus* sp. Vegetative and reproductive stages.

LIFE CYCLE 10—NEMALION Sp. (Fig. 18)
DIVISION 9. Rhodophyta
CLASS Rhodophyceae
Characteristics

Nemalion sp. produces a narrow, branching, wormlike thallus, reddish brown in color and composed of a single row of cylindrical cells, each with several disc-shaped chloroplasts (Fig. 19). The reproductive organs are inconspicuous at the apices of some of the branches. The plants are small, are several inches in length, and are attached to rocks. It is a marine species growing in the intertidal zone during the summer months and is exposed during low tide.

Asexual reproduction

Vegetative reproduction. None. If the plants become detached and floating, they will usually die.

Sporophyte. The sporophyte is represented by the carposporophyte, a small parasitic plant located on the gametophyte. This portion of the gametophyte is referred to as the **cystocarp** and is developed from the **procarp.**

Sporangium. The terminal cell of the lateral filaments enlarges to become a carposporangium. The haploid carpospore is liberated from the carposporangium to become attached to some substratum.

Spore. The carpospore is liberated from the carposporangium as a naked spore. It attaches itself to some substratum and becomes enclosed within a thickened wall. Some time later it will germinate to produce a filament that becomes organized into the plant body.

Sexual reproduction

Gametophyte. The gametophyte is vegetative and is developed from the carpospore. It produces both antheridia and carpogonia, since it is homothallic.

Antheridium. The spermatangia are produced on the same plant as the carpogonia. Many spermatangia are produced on the terminal ends of the spermatangial branches.

"Archegonium." The oogonium is present on the carpogonial filaments. Each carpogonial filament consists of 3 to 5 cells. The terminal cell has an elongated protuberance, the trichogyne, which functions during fertilization.

Sperm. The male gametes are known as

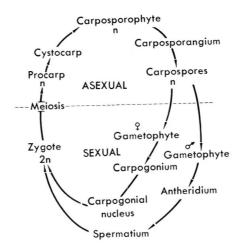

Fig. 18. Life cycle of *Nemalion* sp.

spermatia. They are nonmotile, and a single spermatium is formed within each spermatangium. The single nucleus of the spermatium divides to form two male nuclei after contact with the trichogyne.

Egg. None. The single carpogonial nucleus serves as the female gamete and is found at the base of the carpogonium.

Zygote. The nonmotile spermatia rest against the trichogyne, and both male nuclei pass into it. They migrate toward the base of the carpogonium, and one of the male nuclei fuses with the female or carpogonial nucleus to form the zygote nucleus, which undergoes meiosis. Continued divisions of the daughter nuclei with the formation of cell walls produce lateral filaments.

Embryo. None as found in higher plants.

Means of dispersal. The liberation of the carpospores results in dispersal of the plants.

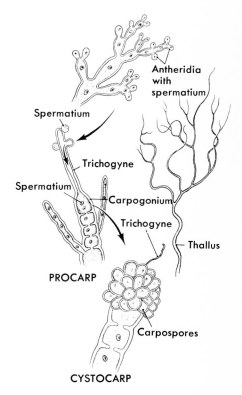

Fig. 19. *Nemalion* sp. Reproductive stages.

LIFE CYCLE 11—SAPROLEGNIA Sp.
(Fig. 20)
DIVISION 29. Oomycota
CLASS Oomycetes
Characteristics

Members of this genus are common water molds, and some are known to occur also in soil. Most species are saprophytic on plant and animal debris in fresh water, but a few are parasitic on algae and one on fish. The plant body is small, forming a mass of white hyphae upon the food source (Fig. 21). The hyphae are coenocytic and branching. Note the similarity between this life cycle and that of *Vaucheria* sp. (life cycle 7).

Asexual reproduction

Vegetative reproduction. Growth by continuous hyphae permits extensive growth, and fracture of parts of the plant allows for some additional spreading and the formation of new plants.

Sporophyte. None.

Sporangium. Special branches of the hyphae project outward from the substrate. The apical portion of this branch develops a cross-wall that separates the sporangium from the remainder of the plant. The cytoplasm of this area darkens as zoospores are formed. When the branch matures, an apical pore opens from which the zoospores may escape. Spore production results after a cell wall appears around the nuclei in the sporangium and by the mitotic division of the already haploid nuclei.

Spore. The spores are biflagellated. Upon maturity, they swim from the sporangium, swarm for some time and then withdraw their flagella and secrete a cyst wall. After remaining dormant for some time, they germinate by producing another zoospore of a somewhat different shape. These secondary zoospores again swim about until they find a suitable location, after which they settle down and germinate into new sporophytes. The phenomenon of having two periods of motility of zoospores is known as **diplanetism,** but the effect of this can only be assumed to be a means of dispersal and at the same time a means of providing a mechanism for passing through adverse conditions.

Sexual reproduction

Gametophyte. The vegetative plant produces well-developed oogonia and antheridia. This plant is **heterogamous.**

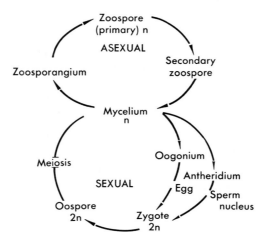

Fig. 20. Life cycle of *Saprolegnia* sp. (water molds).

Antheridium. The antheridia are produced on hyphae located around the oogonia or from an entirely different hypha. They contain several nuclei, and the antheridia grow until they come into contact with an oogonium. On contact, the antheridium produces a tube that pierces the wall of the oogonium.

"Archegonium." There are produced unicellular oogonia that arise as swellings on terminal hyphae separated by a crosswall. Most of the nuclei of the oogonium degenerate, but one remains to form the egg nuclei of each egg.

Sperm. None. Only "males" present are the sperm nuclei of the antheridium.

Egg. One or more eggs is present in each oogonium.

Zygote. The fertilized eggs develop into thick-walled oospores. Several months later they will germinate to develop hyphae that terminate in zoosporangia, in which are produced zoospores.

Embryo. None.

Means of dispersal. Dissemination is most effective by means of the motile zoospores. The oospores are also means for dispersal.

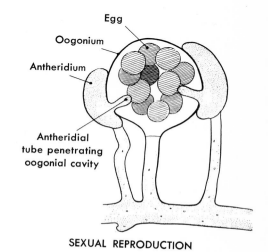

SEXUAL REPRODUCTION

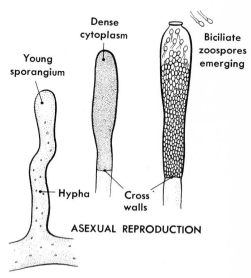

ASEXUAL REPRODUCTION

Fig. 21. *Saprolegnia* sp. Sexual and asexual reproduction.

LIFE CYCLE 12—RHIZOPUS STOLONIFER
(Fig. 22)
DIVISION 12. Zygomycota
CLASS Zygomycetes
Characteristics

The common black bread mold, *Rhizopus stolonifer*, has long been a familiar example of the Zygomycetes. Members of this group are of minor economic importance only because of their damage to stored fruits and vegetables. Bread mold is characterized by white mycelia, prominent black zygospores, and black fruiting bodies (Fig. 23). The growth may be extensive and readily visible to the unaided eye.

Asexual reproduction

Vegetative reproduction. Stolons grow out from a group of rhizoids to penetrate the substrate. This is a means of vegetative growth and enlargement of the plant body. If these stolons become broken, separate plants are formed.

Sporophyte. None.

Sporangium. The sporangia are visible as small, black globules on the mycelium. They develop as terminal swellings on upright hyphae (Fig. 23). These swellings become filled with the multinucleated protoplasm, which then is separated from the rest of the hypha by a dome-shaped crosswall called the columella.

Spore. The many nuclei in the sporangium collect a small mass of cytoplasm around them and become transformed into uninucleated spores. The sporangium wall ruptures when the spores are mature. On reaching a suitable medium and under conditions of proper temperature and moisture they will germinate and give rise to a new mass of hyphae.

Sexual reproduction

Gametophyte. Under certain circumstances the formation of gametes takes place. This occurs when two separate strains of *Rhizopus stolonifer* are growing near each

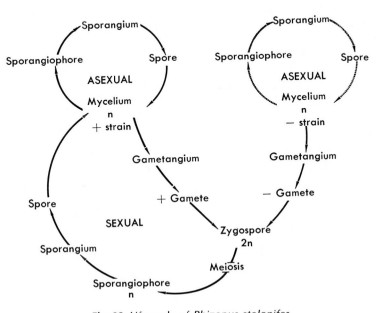

Fig. 22. Life cycle of *Rhizopus stolonifer*.

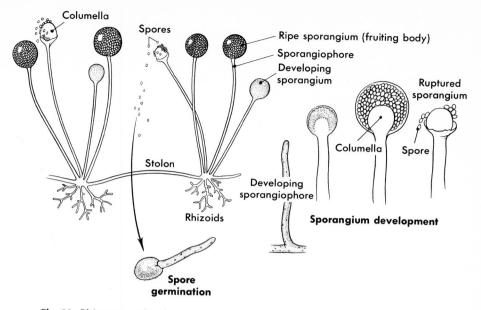

Fig. 23. *Rhizopus stolonifer* (black bread mold). Vegetative and asexual stages.

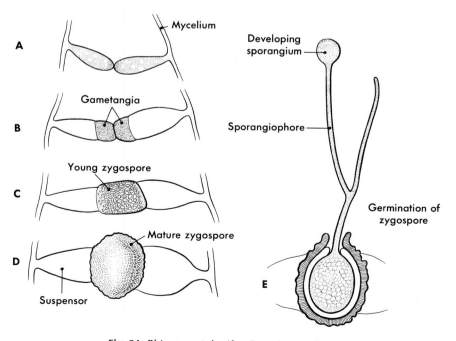

Fig. 24. *Rhizopus stolonifer.* Sexual reproduction.

other (Fig. 24). These have been termed plus and minus strains and are equivalent to the separate sexes of other plants.

Gametangium. When the mycelia of two plants of opposite strains come near each other, lateral lobes are developed. These are the progametes, and this portion is darker than the mycelium proper because of thickening of the wall of the gametangia in the vicinity of their union. The apical portion is separated by a crosswall to form the gametangia. These are equivalent in function to the antheridium and archegonium of the higher plants.

Gamete. The gametes are multinucleated. When the gametangia come into contact with each other, the cell wall between them dissolves, and their nuclei pair off to form diploid nuclei. Unfused nuclei probably disintegrate.

Zygote. The multinucleated zygote formed at the time of copulation of the gametangia enlarges to form a **zygospore.** This is held in place by the original lobes of the respective mycelia. The zygospore remains protected by a heavy covering for some time. Many such zygospores give the growth a black appearance along the line of fusion. If conditions remain favorable, the zygospore will germinate and produce new mycelia and develop **sporangiophores** and sporangia while remaining in its original location.

Embryo. None.

Means of dispersal. The spores from the asexual stage are the principal means of dispersal. That these spores are present everywhere can by readily proved by leaving homemade* bread in a moist, warm place, where it soon develops a luxurious growth of mold.

---

*Bakery bread has preservatives added that prevent molding for several days.

LIFE CYCLE 13—
SCHIZOSACCHAROMYCES Sp. (Fig. 25)
DIVISION 13. Ascomycota
CLASS Ascomycetes
Characteristics

The yeasts are typically unicellular and microscopic in size (Fig. 26), with oval cells, large vacuoles, and scattered inclusions of reserve food. The nucleus is obscure except when the plant is budding. The nucleus divides under favorable conditions, and when rapid budding is taking place, small chains or clusters of cells may be seen.

Asexual reproduction

Vegetative reproduction. The most common kind of asexual reproduction is by budding. A small protuberance forms on the side of the yeast cell. The bud grows for a time, after which it may become detached from the parent cell. The cells may separate, although in some instances they cohere and continue to bud, producing a long chain.

Sporophyte. None.

Sporangium. A yeast plant may become a kind of spore-bearing structure called an **ascus.**

Spore. By parthenogenesis 8 ascospores may develop within the ascus, which in turn develops into new yeast cells.

Sexual reproduction

Gametophyte. The yeast plant (each single cell) is the haploid generation.

Gametangium. None.

Gamete. The haploid yeast cells form an ascus and behave like sex cells inasmuch as 2 of the cells are of one strain and 2 of another. They may pair and fuse to form a diploid cell.

Zygote. After fusion of 2 yeast cells the fused cell can be considered a zygote, since it is diploid. The fused cell nucleus undergoes meiosis, producing 8 haploid ascospores within the ascus. The ascospore, as a spore, develops into the new yeast cells.

Embryo. None.

Means of dispersal. Ascospores are produced as a result of both sexual and asexual reproduction. Dissemination of yeast is accomplished upon their release from the ascus.

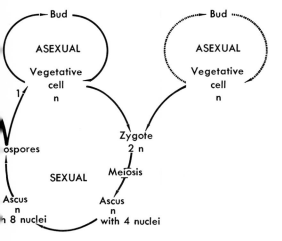

. 25. Life cycle of *Schizosaccharomycetes* sp. (yeast).

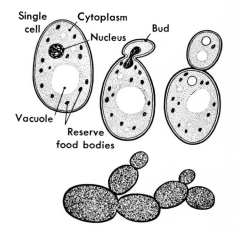

Fig. 26. *Schizosaccharomycetes* sp. (yeast). Vegetative and asexual stages.

LIFE CYCLE 14—MICROSPHAERA Sp.
(Fig. 27)
DIVISION 13. Ascomycota
CLASS Ascomycetes
Characteristics

The powdery mildews represent a group of plants that produce diseases in other plants. They get their common name from the **conidia** that are produced and that give the surface of the host plant a powdery appearance (Fig. 28). The mycelium of *Microsphaera* sp. is not extensive, since it infects only the epidermis of the leaf. The **cleistothecium** containing the asci is a small body attached to the epidermis. This changes color as it matures—it is white at first and then becomes darker until it is finally black.

Asexual reproduction

Vegetative reproduction. The growth of the mycelium as it covers a greater area of the leaf is the only form of vegetative reproduction. The vegetative plant consists only of a short **haustorium,** a special hypha that penetrates the cell for absorption of nutrients and that is rhizoid in nature, the interconnecting stolons, and the conidiophores, short, erect hyphae producing the conidia.

Sporophyte. None.

Sporangium. The conidiophore is the equivalent of the sporangium, producing the conidia in rows.

Spore. The conidia, therefore, are the equivalent of the spores of other fungi. These small bodies are released from the apex of the conidiophore and are capable of germinating into new plants, thus completing the asexual cycle.

Sexual reproduction

Gametophyte. The mycelium, composed of septate hyphae, represents the gametophyte generation.

Antheridium. Late in the season the production of conidia ceases, and the mycelium

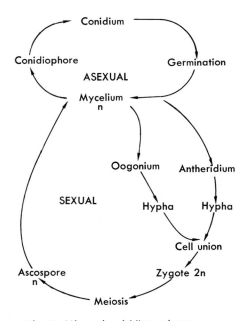

Fig. 27. Life cycle of *Microsphaera* sp.

develops poorly differentiated sex organs. From adjacent hyphae two very short branches arise side by side. One of these bears the single-celled antheridium. The antheridial cell becomes closely applied to the female reproductive organ or oogonium.

"Archegonium." No true archegonium, however, an oogonium termed ascogonium is produced adjacent to the antheridial cell–bearing hypha. After an opening forms between the antheridium and the ascogonium, the nucleus of the antheridial cell moves into the ascogonium, but nuclear fusion is delayed.

Sperm. None as such, but the oogonial nucleus acts as an egg.

Zygote. The union of the antheridial and oogonial nuclei produces a diploid cell or zygote. After fertilization, a vegetative cover of haploid hyphae, the **perithecium,** grows around the developing ascospore to enclose it completely. The **ascus mother cell** forms within the cleistothecium and then goes through 3 successive nuclear divisions to form 8 haploid ascospore nuclei, and eventually ascospores. These develop new mycelia.

Embryo. None as found in the vascular plants.

Means of dispersal. Dispersal is by the conidia shed from the conidiophore and also by the ascospores produced within the ascus.

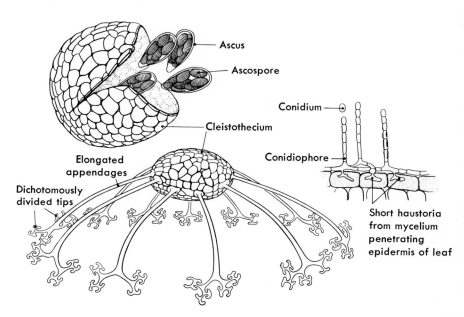

Fig. 28. *Microsphaera* sp. (powdery mildew). Spores are produced as a result of both asexual (conidium) and sexual (ascospore) reproduction.

LIFE CYCLE 15—PUCCINIA GRAMINIS
(Fig. 29)
DIVISION 14. Basidiomycota
CLASS Basidiomycetes

Characteristics

For the most part, the wheat rust fungus is a noxious parasite living on its host, the wheat leaves and stems. However, only part of its life cycle (Fig. 30) is spent on the wheat plant. To complete its life cycle, a part of its growth period must occur on the American barberry bush. Therefore it alternates between the wheat and barberry bush but does the most serious damage to the wheat. The most effective control of the ravages of the wheat rust has been accomplished by the elimination and destruction of the barberry bush in areas where wheat is grown. The only source of infection of wheat plants each spring is a spore (aeciospore, Fig. 31) produced by the fungus on the barberry leaf.

The wheat rust fungus produces a number of different kinds of spores that develop into filaments or hyphae that are found between the cells of the leaves and stems. This is called a rust because of the reddish brown color of the spores that are found chiefly upon the surface of the host leaves and stems.

Asexual reproduction

Vegetative reproduction. The mycelium (mass of hyphae) grows between the cells of the host plant, frequently rupturing through the surface to produce the rust spots characteristic of this parasitic fungus.

Sporophyte. None as described for the higher plants. The binucleated mycelium growing between the cells of the barberry leaf leads to the production of the binucleated cells of the **aecidia.** Aecidia are the cuplike formations found on the barberry leaf (Fig. 31), which finally rupture through the leaf epidermis to release the binucleated **aeciospores** (Fig. 31). The aeciospores are shed in the spring and are wind borne to the young wheat plants that they infect. Upon germination of these spores a mycelium is produced between the cells of the wheat leaf and stem, which finally ruptures through the epidermal surfaces as "red rust" spots. The mycelium on the wheat produces **uredospores,** one-celled, orange-colored spores (the "summer spores") within a sporangium called a **uredium.** Again, the uredospores are wind borne and are capable of infecting other wheat plants. In this way the disease may be spread throughout an entire field. In late summer the mycelium on the wheat will form a **telium** and produce two-celled "winter spores" or **teliospores** (Fig. 32). Each of the two cells contains two nuclei, which initially fuse as the teliospore matures. Since the teliospore is a dark brown spore producing what appears as black rust spots on the surface of stems and leaves of the wheat, this stage is called the "black rust." It is the most destructive stage because it weakens the stems and causes them to break, thus preventing the harvesting of the wheat.

The teliospore has a thick, protective wall and is found lying in the wheat stubble or on the soil surface, where it remains all winter. Teliospores remain dormant until spring, when they germinate on the ground. Each teliospore produces a short hypha that develops a basidium (the basidial stage, the only stage having no host). Four **basidiospores** (Fig. 32, A) are produced on the basidium after meiosis. They form two strains. These are carried by the wind; if they fall on a barberry leaf, they will germinate and give rise to an intercellular mycelium.

Sporangium. The aecidium, basidium, and uredium are all sporangia producing their respective spores.

Spore. There are binucleated aeciospores, uredospores, and teliospores, which are produced on their respective host plants, as well as the haploid basidiospores.

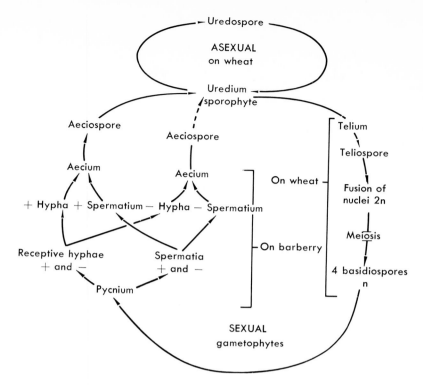

Fig. 29. Life cycle of *Puccinia graminis* (wheat rust).

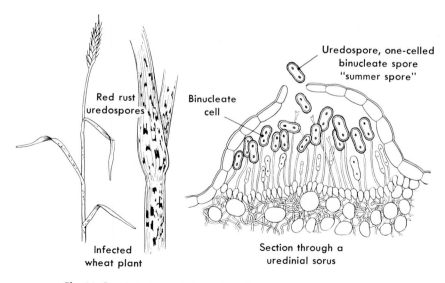

Fig. 30. *Puccinia graminis* (wheat rust). Uredial stage (asexual).

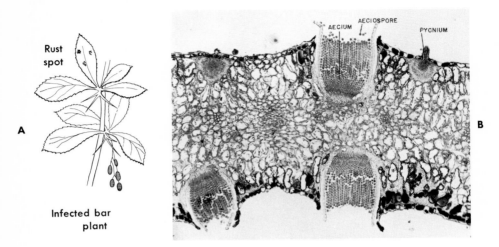

Fig. 31. *Puccinia graminis*. Sexual stage. **A,** Infected barberry plant. **B,** *Puccinia graminis* on barberry leaf showing at least one median pycnium and one median aecium. (X100.) (**B** courtesy George H. Conant, Triarch Products, Ripon, Wis.)

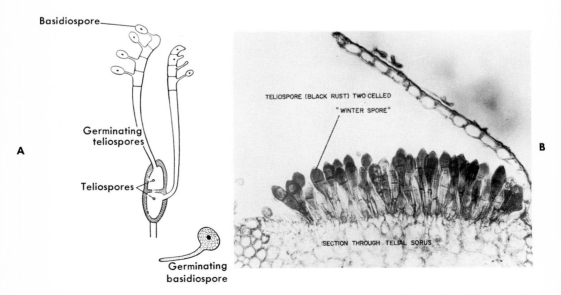

Fig. 32. *Puccinia graminis*. **A,** Telial stage showing teliospores on wheat stem. (X300.) **B,** Basidial stage. (**B** courtesy George H. Conant, Triarch Products, Ripon, Wis.)

Sexual reproduction

Gametophyte. None as such. However, the haploid basidiospores develop into haploid hyphae of two strains mixed together on the leaf surface of the barberry leaves. Eventually small, flask-shaped structures, pycnia, are formed, primarily on the upper surface of the leaf (Fig. 32, *B*). These produce hyphae called spermagonia. Small cells called **spermatia** are produced at the ends of the hyphae. The spermatia ooze out on the leaf surface in a sweet liquid that is attractive to insects.

Gametangium. None as such. However, the pycnium also produces either plus or minus receptive hyphae depending on the basidiospore that produced the **pycnium.**

Gamete. Sperms are not produced, but plus and minus spermatia serve the same purpose. The receptive hyphae act as eggs.

Zygote. None as such. When a plus spermatium is carried, by an insect visiting the sweet spermatia mass, to a minus hypha, or vice versa, the plus and minus cells fuse. Following this fusion a binucleated mycelium will develop on the underside of the barberry leaf; this eventually leads to the production of aecidia. Aeciospores are formed after nuclear meiosis to form binucleated hyphae cells. This completes the life cycle of the wheat rust fungus.

Embryo. None.

Means of dispersal. This dispersal of the fungus is accomplished by the windborne basidiospores transported to the barberry leaf, from the barberry leaf to the wheat plant by the wind-borne aeciospores, and from one wheat plant to others by the wind-borne uredospores.

LIFE CYCLE 16—AGARICUS CAMPESTRIS
(Fig. 33)
DIVISION 14. Basidiomycota
CLASS Basidiomycetes
Characteristics

Mushrooms are well known to everyone. The portion that is eaten, however, is merely the fruiting body, which is composed of a **stipe** or stem and a **pileus** or cap (Fig. 34). When the fruiting body is mature, the underside of the pileus exposes the **gills.** Each surface of the gill bears spores. The annulus is part of the covering of the pileus remaining on the stalk after it has broken to expose the gills so that the spores may be shed. The vegetative plant is more extensive than the fruiting body, since the former consists of mats or masses of subterranean hyphae that feed on organic matter. Mushrooms grow best around decayed trees or in fertilized soil such as a pasture. They form the "fairy rings" of literature because they often grow in a circle around the area of the roots of trees long since decayed.

Asexual reproduction

Vegetative reproduction. The extensive growth of the hyphae forms interconnecting plants that send up fruiting bodies.

Sporophyte. The sporophyte is composed of uninucleate cells forming extensive underground growths. Their interwoven and matted mycelial strands are called **rhizomorphs.** At maturity the rhizomorphs form buttons that are pushed above the ground to form the fruiting body.

Sporangium. The gill region of the fruiting body is the sporangium. The gill proper is composed of interlaced hyphae. Two kinds of structures develop along the sides of each gill, the paraphyses or sterile filaments and the basidia or fertile filaments. The basidium is binucleated.

Spore. As the basidium develops, the two

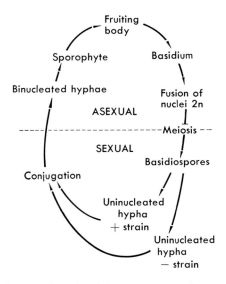

Fig. 33. Life cycle of *Agaricus* sp. (mushroom).

nuclei, one plus and the other minus, fuse to form a delayed diploid cell. Meiosis immediately follows, producing four nuclei. Meanwhile the basidium develops four lobes into which the new haploid nuclei migrate. In these lobes they develop basidiospores subtended by a short stalk or **sterigma.** Basidiospores are shed when they are ripe.

Sexual reproduction

Gametophyte. The reproductive cycle of this common plant takes place in the soil. The basidiospores germinate to produce small uninucleated hyphae. Hyphae formed from basidiospores are either plus or minus.

Gamete. Hyphae from germinating basidiospores of opposite strains conjugate to form binucleated cells. It is these cells that develop into the sporophyte.

Zygote. None as such.

Embryo. None.

Means of dispersal. Spores shed from the fruiting body are blown about by the wind and germinate when they come into contact with a suitable substrate. These spores are gathered by commercial growers and used as seed in mushroom farms in caves.

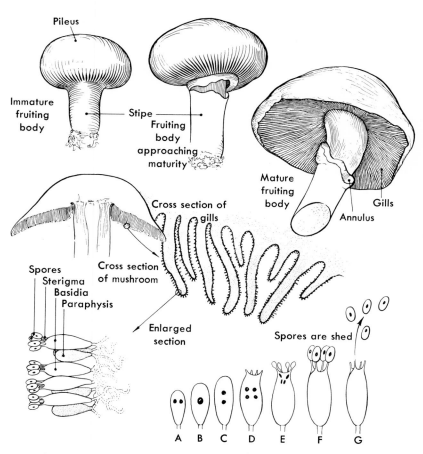

Fig. 34. *Agaricus* sp. (mushroom). Fruiting body and spore formation.

LIFE CYCLE 17—MARCHANTIA Sp.
(Fig. 35)
DIVISION 16. Hepatophyta
CLASS Hepaticae
Characteristics

The vegetative body is the gametophyte, composed of a prostrate dorsiventral thallus typical of the class (Fig. 36). Small unicellular rhizoids project into the substrate. The dorsal cells are green. Growth is radial from the point of origin of the thallus, which develops from the protonema. The surface of the thallus appears to have a protective, scalelike covering. Epidermal tissue surrounds the thallus, but only the upper surface has stomas. Beneath the epidermis the tissue differentiation is slightly more complex than that of the algae. The middle portion of the thallus is filled with chlorenchyma, interspersed with parenchyma storage cells. Two gametophytes of opposite sex are produced, and these can be readily distinguished by the nature of the **gametophores;** those of the male are stellate and those of the female have an umbrellalike shape (Fig. 36).

Asexual reproduction

Vegetative reproduction. When conditions are such that sexual reproduction is delayed, vegetative reproduction may take place by the production of structures called cupules located on the upper surface of the thallus. Within the **cupules** appear small buds, the gemmae. These break off from the cupules and are washed away. When they find a suitable growing site, they develop a new gametophyte thallus. If mechanically broken, the thallus will also regenerate into a new plant.

Sporophyte. The sporophyte is a small diploid plant growing from the undersurface of the archegoniophore of the female plant. It consists of a foot, a stalk or seta, and the sporangium. It is partially parasitic upon the gametophyte and is surrounded by a perianth produced by the gametophyte.

Sporangium. When the sporophyte is mature, the apical sporangium develops within the archegonium (see discussion of embryo that follows). The spore mother cells divide by meiosis and mitosis to produce simple spores. Within the sporangium are developed peculiar springlike elater cells that aid in spore dispersal.

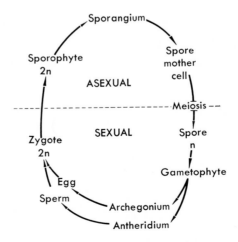

Fig. 35. Life cycle of *Marchantia* sp. (liverwort).

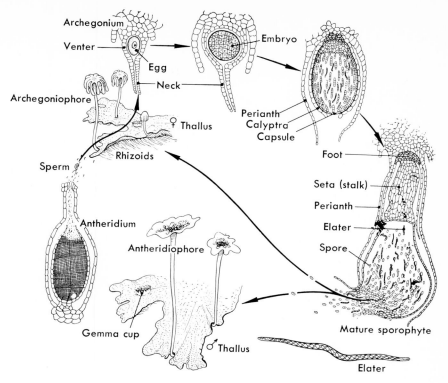

Fig. 36. *Marchantia* sp. (liverwort). Gametophyte and sporophyte generations.

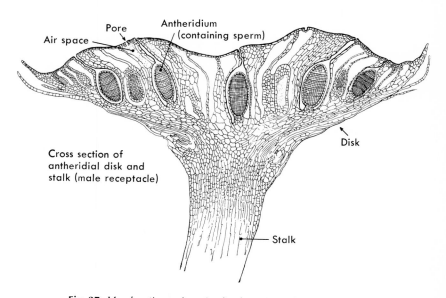

Fig. 37. *Marchantia* sp. Longitudinal section of antheridial disk.

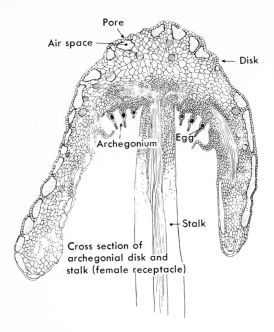

Air space
Pore
Disk
Archegonium
Egg
Stalk
Cross section of archegonial disk and stalk (female receptacle)

Fig. 38. *Marchantia* sp. Longitudinal section of archegonial disk.

Spores. As soon as the spores become mature, the covering or perianth of the sporangium and the archegonium ruptures, and the microscopic spores are ejected. Upon reaching a suitable habitat, they germinate to produce a small protonema, a filament-like, embryonic thallophyte.

Sexual reproduction

Gametophyte. The vegetative and dominant plant in this species is the gametophyte, which has just been described. It develops from a small protonema. Upon reaching maturity, it produces either **antheridiophores** or **archegoniophores,** depending on the sex of the thallus. These stalks, several per plant, extend a few millimeters above the surface of the thallus.

Antheridium. The antheridia (Fig. 37) are developed in cavities in the expanded disc on the antheridiophore. They consist of small sacs of developing sperm.

Archegonium. The archegoniophore develops archegonia (Fig. 38) on the undersurface of its disc. These small structures are saclike, with an entrance channel or canal, the **neck canal.** When the egg is mature, this canal opens.

Sperm. True single-celled, flagellated sperm cells are produced in the antheridium. They are haploid, being produced from the haploid tissue of the gametophyte. Since they must swim to the egg, they are released only when the plant is flooded with water.

Egg. A true haploid egg, much larger than the sperm, is produced in the archegonium. This remains in the center at the base of the neck canal.

Zygote. A diploid zygote is formed when a sperm reaches the egg and fuses with it. The sperm enters the archegonium by way of the neck canal and swims to the egg. Both cytoplasm and nuclei fuse.

Embryo. The zygote grows by mitosis into an embryo, which eventually develops into the mature sporophyte described. Neither the egg, embryo, nor sporophyte leaves its original position within the center of the archegonium.

Means of dispersal. Spores and gemmae provide the distribution mechanism of these plants.

LIFE CYCLE 18—POLYTRICHUM Sp.
(Fig. 39)
DIVISION 17. Bryophyta
CLASS Musci
Characteristics

This typical moss forms matlike colonies on soil or soil-covered rocks (Fig. 40). The gametophytes are fairly large, with numerous erect "stems" and small "leaves," giving them a bushy appearance.

Asexual reproduction

Vegetative reproduction. Several gametophytes develop from the protonema of a germinating spore. This structure persists and may be responsible for a small original colony of gametophytes. Gemmae and bulbils occur in mosses, and fractured plants are capable of regeneration.

Sporophyte. The embryo develops into the sporophyte, which is almost completely parasitic upon the female gametophyte. It extends dorsally above the archegonial branch of the thallophyte, however, and is readily seen in sporulating moss. It consists of the following parts: (1) a foot, which penetrates the tissue of the gametophyte from which nu-trients are absorbed, (2) a stalk or seta, which arises from the foot, and (3) the sporangium, which is formed at the apex of the stalk.

Sporangium. The sporangium is a smaller, somewhat lanceolate **capsule** with a central **columella** of sterile tissue surrounded by the spore-bearing layer. The apex of the capsule is fitted with a small caplike deciduous cover or **operculum,** which opens to release the mature spores.

Spore. Spores are produced by meiosis of the spore mother cells. Each spore mother cell produces four haploid spores. The microscopic spores are stored within the capsule until conditions are dry enough for their release and dissemination. After the spores are released, they may remain dormant for long periods of time. When conditions of temperature and moisture are optimum, the spore absorbs water, and the cell within divides. It soon bursts open the spore case. Continued division of the spore forms a protonema typical of the group.

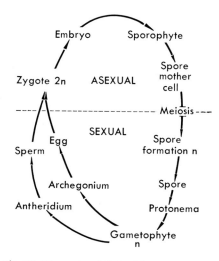

Fig. 39. Life cycle of *Polytrichum* sp. (moss).

Sexual reproduction

Gametophyte. The protonema, which is somewhat flattened and resembles an alga, continues to grow, producing rhizoids that anchor the plant to the substrate. After a short growing period, "buds" composed of a compact group of cells are formed. These buds develop into "stems" and soon put forth green "leaves," thus forming the young, vegetative moss plant or gametophyte.

When this plant reaches maturity, sex organs are developed, usually at the apex of a thallus stem. In some species the male and female organs are borne on the same plant, whereas in others they develop on separate plants but in the same colony.

Antheridium. The sperm are produced by mitosis in a small capsule, the antheridium. When the sperm mature, a pore opens at the apex of the antheridium, allowing the sperm to leave.

Archegonium. The female structures are similar to those of *Marchantia* sp. and, as in the antheridia, are borne at the apex of the fertile shoots. A short stalk supports the **venter** in which the egg is formed. A neck is produced that hollows out to form a neck canal when the archegonium is mature.

Sperm. The male sex cells are biflagellated. They swim from the antheridia only when the plant is drenched in water.

Egg. The egg cell is large like that of *Marchantia* sp. and remains in the venter of the archegonium.

Embryo. The zygote develops shortly after fertilization and grows into a sporophyte.

Means of dispersal. Like *Marchantia* sp., these plants depend on dissemination of spores for dispersal and formation of gemmae.

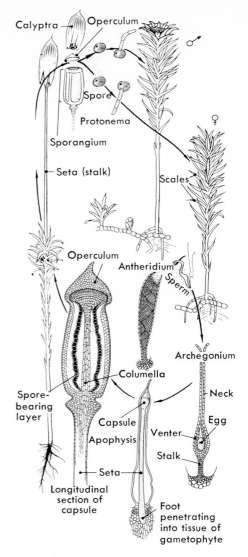

Fig. 40. *Polytrichum* sp. (moss). Gametophyte and sporophyte generations.

LIFE CYCLE 19—SELAGINELLA Sp.
(Fig. 41)
DIVISION 19. Lycophyta
CLASS Lycopsida
Characteristics

Selaginella sp. is a small bushy plant resembling some mosses (Fig. 42). The vegetative plant is the sporophyte, producing loose and inconspicuous strobili at the apices of some of the branches. The leaves are of two types arranged in rows along the length of the stem, two rows of small leaves and two rows of large leaves.

Asexual reproduction

Vegetative reproduction. No special structures are present for vegetative reproduction; however, portions of the plant may run along the ground so that several clumps may be produced.

Sporophyte. The sporophyte has already been described.

Sporangium. The sporangia are of two types, both of which are found on the same strobilus (Fig. 43). The strobilus is characterized by having, in addition to the sporophyll leaves, a small tonguelike scale, the **ligule,** which fits into a pit at the base of the sporophyte. The basal portion of the strobilus produces megaspores and the apical portion microspores. Both types of sporangia are borne in the axil of the leaves, forming the strobilus. These are the **sporophylls** and are either **megasporophylls** or **microsporophylls,** according to the type of sporangia they subtend.

Spore. Each megasporangium produces four megaspores by meiosis of the diploid spore mother cell. The mature spore is cov-

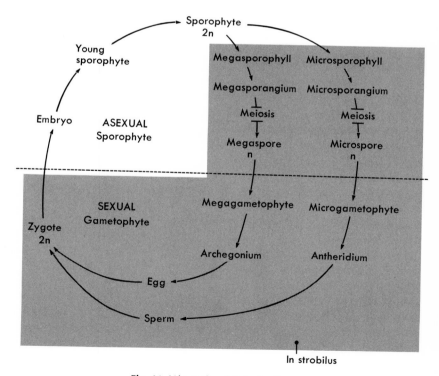

Fig. 41. Life cycle of *Selaginella* sp.

485

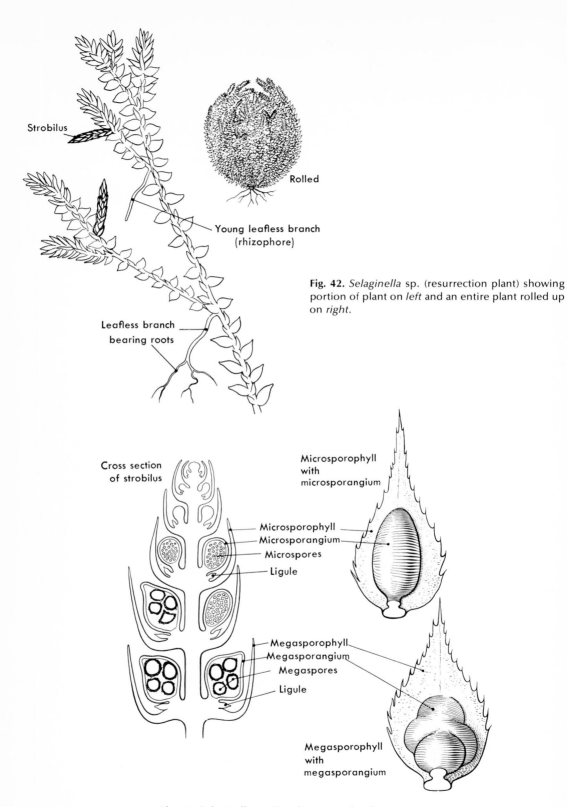

Strobilus

Rolled

Young leafless branch
(rhizophore)

Fig. 42. *Selaginella* sp. (resurrection plant) showing portion of plant on *left* and an entire plant rolled up on *right*.

Leafless branch
bearing roots

Cross section
of strobilus

Microsporophyll
with
microsporangium

Microsporophyll
Microsporangium
Microspores
Ligule

Megasporophyll
Megasporangium
Megaspores
Ligule

Megasporophyll
with
megasporangium

Fig. 43. *Selaginella* sp. Development of male gametes.

ered by a rough coat, which protects the spore until germination. The microspores are produced in a similar fashion, except that the many microspore mother cells in the microsporangium produce hundreds of minute microspores.

Sexual reproduction

Gametophyte. The megaspores remain in the megasporangium where they develop into a megagametophyte. This small plant, though not parasitic on the sporophyte, foreshadows parasitism in the higher plants. As the megagametophyte develops, a layer of cells is produced to form a female **prothallus.** The rest of the spore serves as a reserve food supply. Several archegonia with a venter and a venter canal cell are produced on the prothallus. An egg develops in the venter, awaiting fertilization.

Meanwhile the microspore in the microsporangium develops into microscopic **microgametophytes** or prothallial cells (Fig. 44). While the antheridium is still in the spore stage, ciliated sperm are produced.

Antheridium. The antheridium is attached to the sporophyte.

Archegonium. The archegonium is attached to the sporophyte.

Sperm. The spores mature into the gametophyte and act as pollen grains. They are released from the microsporangium and are either washed from the strobilus or are wind borne. Some eventually land on the female portion of the strobilus. When moist, they burst open, and the ciliated sperm leave the old spore case.

Egg. The venter canal of the archegonium opens when the megagametophyte is ma-

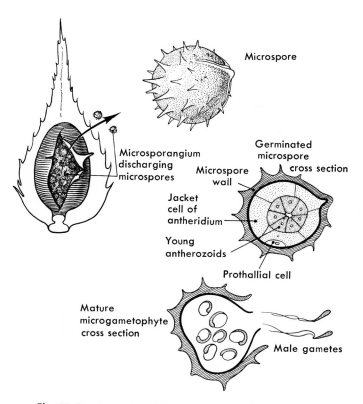

Fig. 44. Development of the microspore of *Selaginella* sp.

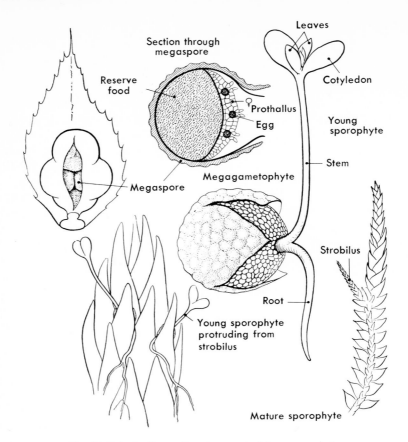

Fig. 45. *Selaginella* sp. Development of the sporophyte.

ture. The egg is a small cell located at the base of this canal.

Zygote. Several of the sperm enter the ruptured spore case of the megagametophyte and find a venter canal. The two cells fuse, restoring the nucleus to the diploid condition.

Embryo. As soon as fertilization takes place, an embryo develops, obtaining its food from the reserve remaining in the spore.

Soon a young sporophyte is developed (Fig. 45), eventually drops from the strobilus of the parent sporophyte, and becomes established in the soil.

Means of dispersal. The lack of spores producing independent gametophytes has prevented the wide dispersal of these plants so characteristic of other cryptogams. Most dispersal is the result of water washing the young sporophyte to new locations.

LIFE CYCLE 20—EQUISETUM Sp. (Fig. 46)
DIVISION 20. Arthrophyta
CLASS Sphenopsida
Characteristics

The vegetative and reproductive characteristics of the horsetails fit the characteristics of the class; they are described in Chapter 9.

Asexual reproduction

Vegetative reproduction. The rhizome or subterranean stem may run under the ground for several feet, putting forth aerial branches. If they become detached, the separate branches will continue to grow and produce new individuals.

Sporophyte. The sporophyte is the vegetative plant, many branches of which are sterile (Fig. 47). The spore-producing branches are separate and colorless and receive their nourishment from the underground rhizome connecting these branches with the vegetative ones.

Sporangium. A well-differentiated strobilus is borne at the apex of the fertile branch. This is a compact structure that darkens as it matures. It is a series of short stalks, the sporangiophores containing sporangia. These are probably modified sporophylls.

The apex of the sporangiophore is six sided, giving a geometrical appearance to the strobilus before it ripens. A sporangium consists of six bracts grouped around the sporangiophore (Fig. 48). Inside, the spore mother cells divide by meiosis to produce small spores.

Spore. The spores are all alike in this plant. They are unique in that as accessory structures they have coiled appendages, termed **elaters.** As the sporangium dries, these elaters expand and force open the strobilus, thus aiding in release and dissemination of spores.

Sexual reproduction

Gametophyte. The spores are released and are blown about. After about 12 hours they will usually germinate. They will each grow into a small thalluslike green gametophyte partly suberged in the soil. When mature, the gametophyte produces gametangia. The thallus may persist for some time after fertilization.

Antheridium. Antheridia are produced at

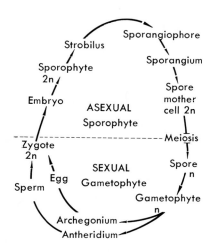

Fig. 46. Life cycle of the *Equisetum* sp. (horsetail).

489

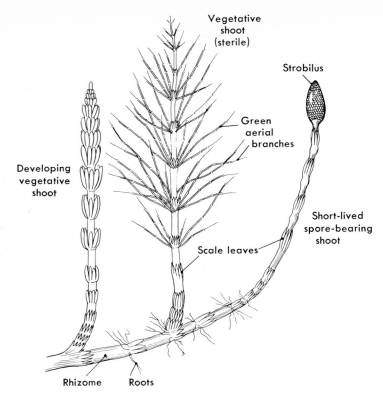

Fig. 47. *Equisetum* sp. (horsetail). Vegetative plant.

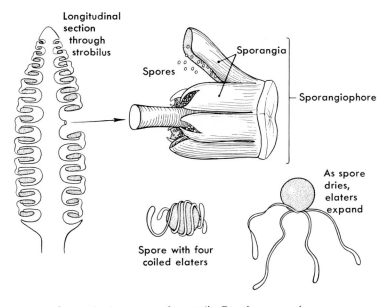

Fig. 48. *Equisetum* sp. (horsetail). Development of spores.

the upper end of the prothallus, the gameto-phyte.

Archegonium. Archegonia are produced at the base of the short branches of the prothallus.

Sperm. The sperm are coiled and multicilated and are capable of swimming to the archegonium.

Egg. The eggs are located in a venter at the base of a fertilization canal.

Zygote. Fertilization takes place during rainy weather, the sperm swimming from one gametophyte to another to fertilize the ripened eggs.

Embryo. A young sporophyte develops within the fertilized archegonium. It has primary roots, stem, and leaves in addition to a small "foot" by means of which it obtains its first food from the gametophyte. Eventually it grows into a sporophyte and replaces the parent gametophyte, which gradually dies.

Means of dispersal. Spore dissemination by the sporophyte provides the means for the distribution of the species.

LIFE CYCLE 21—POLYPODIUM Sp.
(Fig. 49)
DIVISION 21. Pterophyta
CLASS Filicopsida
Characteristics

These ferns are fairly large sporophytes with long pinnate fronds (Fig. 50). They are common in wooded areas throughout North America and are often grown as ornamentals. The sporangia are arranged in sori along the margin of the **pinnae** of the vegetative fronds. The plants of this genus sporulate in the summer and fall and remain green throughout the year.

Asexual reproduction

Vegetative reproduction. Growth of the rhizomes may produce several clusters of fronds interdependent on each other. If these become separated, however, the plants persist and develop new clones, which grow up in the spring by uncoiling (circinate vernation).

Sporophyte. The spore-producing plant is vegetative. The frond is composed of many pinnae arranged laterally along the petiole or rachis. Each pinna is also pinnate.

Sporangium. The sporangia are located in small clusters along the margin of the undersurface of the pinnae of the fronds. These clusters are called sori and are partly covered by a leaf fold, the **indusium.** When the spores are ripe, this area appears dark brown. The sporangium is suspended from the frond by a short petiole. Each sporangium has a remarkable device for the release of the spores. Along one side is a row of cells with thick inner walls and thin outer walls. This structure is called the **annulus.** Along the other edge is a row of thickened cells, two of which are called **lip cells.** The cells of the annulus are normally filled with water so that turgor pressure keeps the cell inflated and in position. Upon drying, the thick inner walls spring together and rupture the sporangium at the area of the lip cells, releasing the spores. If moist conditions return, the sporangium will again close because the cells of the annulus become reinflated, forcing the walls apart. Eventually the sporangium ruptures beyond closing. This mechanism ensures the release of spores during dry weather when they are most likely to be widely distributed. The spores are produced by meiosis of the spore mother cells.

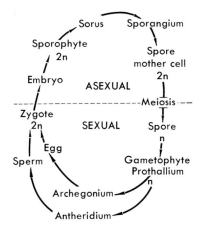

Fig. 49. Life cycle of *Polypodium* sp. (fern).

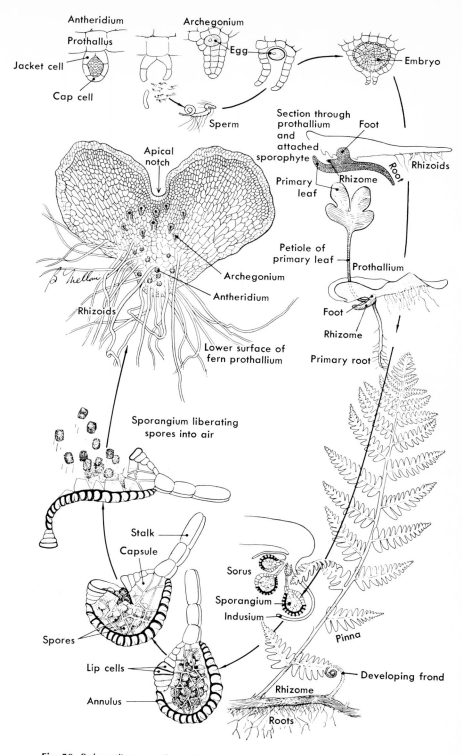

Fig. 50. *Polypodium* sp. (fern). Sporophyte and gametophyte generations.

Spore. The spores are microscopic in size, are usually haploid, although they are diploid in this genus, are covered with a resistant coating, but are without elaters or other structures for dissemination. They are small and light enough to be carried great distances by air currents.

Sexual reproduction

Gametophyte. Upon reaching a suitable habitat, the spores germinate, usually during the following spring, and grow into a rhizoid-bearing green filament. However, the filament is not a true protonema. This filament eventually develops into a small heart-shaped prothallium about ½ to 1 inch in diameter, and this prothallium, when mature, develops the gametangia. The prothallium, which is green in color, is capable of limited photosynthesis. It is composed of a thin layer of cells but lacks a vascular system and is therefore devoid of true roots, stems, and leaves.

Antheridium. The male or sperm-producing structures develop as small lobes on the lower surface of the prothallium among the rhizoids.

Archegonium. The female structures are also located on the lower surface of the prothallium, usually near the apical notch some distance from the antheridia. The archegonium consists of a venter harboring a single egg cell and the venter canal.

Sperm. The sperm are irregularly shaped cells with many cilia, enabling them to swim freely.

Egg. The egg is nonmotile and remains in the archegonium.

Zygote. During rain the sperm are released from the antherida, swim to the archegonium, enter the venter canal, and fertilize the egg. The zygote thus formed remains in the archegonium.

Embryo. An embryo soon develops at the expense of the gametophyte. It first puts forth a primary root and a primary leaf. A small foot remains in the archegonium to absorb nutrients. As growth continues, a rhizome develops, and the new sporophyte appears. As the sporophyte grows, the prothallium is destroyed and disappears.

Means of dispersal. The species, like all ferns, depends on spores for wide dispersal.

LIFE CYCLE 22—PINUS Sp. (Fig. 51)
DIVISION 24. Coniferophyta
CLASS Gymnospermae
Characteristics

Members of the genus *Pinus* are well known as valuable trees generally distributed in the northern hemisphere of both the Old World and the New World. The trees have a central stem or trunk with lateral branches and branchlets. The leaves are in the form of needles, usually in groups of two to five, and are uniform in length, the length depending on the species. Pine trees are evergreen and monoecious, producing both male and female pendulous **cones.** The female cones have **scales** and **bracts** in spirals and are held tightly closed, except for a slight opening at the time the megagametophyte matures to permit pollination. Until the seeds mature, they remain closed.

Asexual reproduction

Vegetative reproduction. None.

Sporophyte. Like all seed plants, the sporophyte is the vegetative plant, and it produces microsporangia and megasporangia on the same plant.

Sporangium. The reproductive parts are in cones and represent a strobilus formed from modified leaves. Each cone is a cluster of sporophylls with a short central stem. The two types of cones are distinguished according to the sex of the gametophyte they produce. Microsporophylls are the staminate cones and are the smaller of the two; megasporophylls are the carpellate cones.

Microsporophyll. Each microsporophyll is

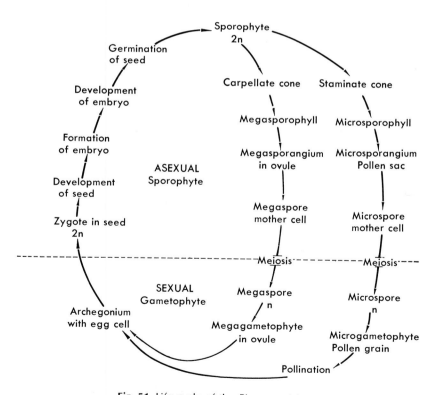

Fig. 51. Life cycle of the *Pinus* sp. (pine).

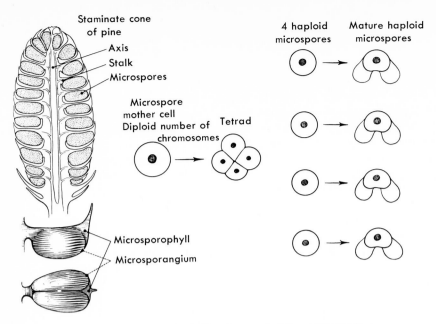

Fig. 52. *Pinus* sp. (pine). Development of the microspore.

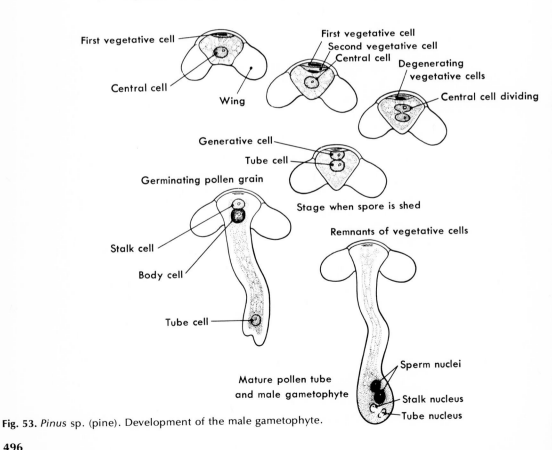

Fig. 53. *Pinus* sp. (pine). Development of the male gametophyte.

a scale attached to the axis of the cone by a short stalk and produces two microsporangia on its lower surface.

Megasporophyll. Each megasporophyll is also a scale subtended by a bract much larger than that of the staminate cone. It bears 2 ovules near its base, each containing a megasporangium. The bracts are conspicuous only in young cones, but in some conifers the bracts may exceed the scales in the mature cone.

Microsporangium. The tissue of the microsporangia gives rise to numerous diploid cells, microspore mother cells, which pro-

duce a tetrad of haploid cells by meiosis. They separate into four microspores.

Megasporangium. The megasporangia, like the microsporangia, produce diploid generative cells, the megaspore mother cells, which divide by meiosis to form, in this case, 4 haploid megaspores, 3 of which eventually degenerate. Unlike the cryptogams, these megaspores are produced in megasporangial tissue modified to form an ovule or nucellus, a structure similar to the venter of the cryptogam gametophyte. A section through one of these ovules shows that it consists of a single integument with an opening, the mi-

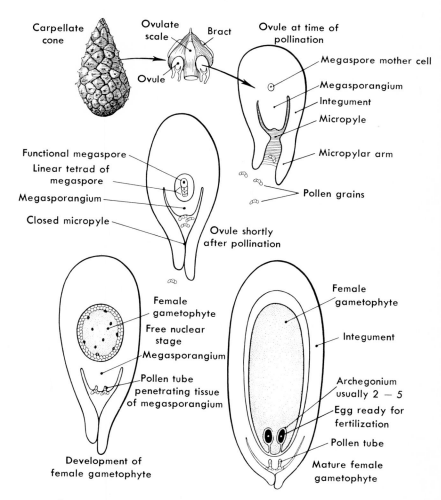

Fig. 54. *Pinus* sp. (pine). Development of the female gametophyte.

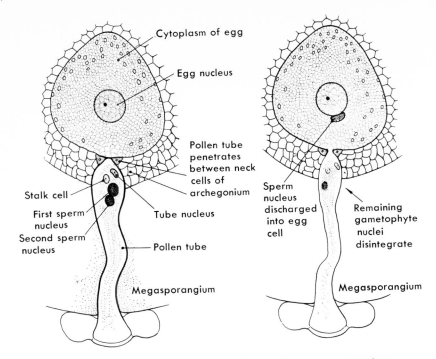

Fig. 55. *Pinus* sp. (pine). Fertilization of the egg by the sperm nucleus.

cropyle, directed toward the axis of the cone. This structure is equivalent in function to the venter canal of the archegonium of the cryptogams but is not homologous in structure because it is formed by the sporophyte and not by the gametophyte.

Microspore. As soon as microspores (Fig. 52) are produced, they develop lateral projections or wings. They remain in the microsporangia, however, for further development into a microgametophyte.

Megaspore. The megaspore remains in place in the ovule for further development into the megagametophyte.

Sexual reproduction

Gametophyte. The gametophytes of all seed plants are greatly reduced to microscopic parasites on the sporophyte. The megagametophyte never leaves the parent, but the entire microgametophyte is in the form of a pollen grain, which is usually wind borne. One of the sperm or male nuclei fertilizes the egg of an archegonium.

Microgametophyte. The haploid microspore cell divides by mitosis to form a large central cell and a smaller vegetative cell (Fig. 53). The vegetative cell, now equivalent to the prothallium of the cryptogams, again divides to form a first and a second vegetative cell. Then both cells degenerate and remain as remnants throughout the further development of the pollen grain. This is the extent of the vegetative part of the microgametophyte and the end of what was, in the lower plants, the dominant vegetative plant.

Megagametophyte. Further development of the megaspore does not take place until a few pollen grains enter the micropyle, after which the latter closes, preventing entrance of more pollen (Fig. 54). However, before actual pollination, the megaspore mother cell divides meiotically to produce 4 megaspores, 3 of which degenerate. The remaining functional megaspore enters the free-nuclear stage; that is, the haploid nucleus divides several times by mitosis to form the megagametophyte.

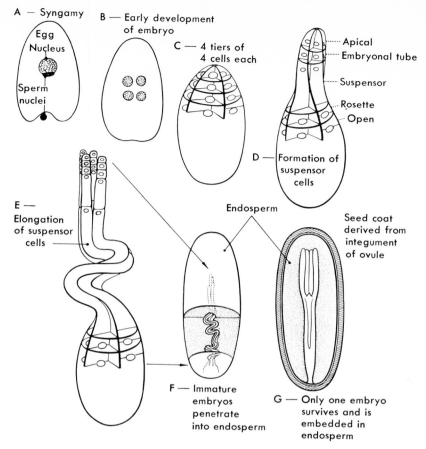

Fig. 56. *Pinus* sp. (pine). Development of the embryo in the ovule.

Antheridium. The pollen grain does not develop a true cellular antheridium. However, the central cell divides to produce a generative cell and a tube nucleus. It is in this stage that it is shed from the microsporophyll and then carried by the wind to the **carpellate cone.**

Archegonium. The megagametophytes within the two ovules of every megasporophyll develop archegonia, each with an egg, an archegonial or neck canal, and an endosperm, which is the vegetative body of the gametophyte and which will become the food supply of the seed. At the same time the sporangium is enlarging to form the seed coat and wings of the seed.

Sperm. Sperm nuclei in pollen tube.

Egg. Egg nucleus within the megasporangium.

Zygote. Once it is inside the megasporangium, the pollen grain is attached near the micropyle by a sticky fluid exuded at this stage by it. The pollen grain now produces a pollen tube. The generative cell divides to produce a stalk cell and a body cell. These cells then migrate down the pollen tube, where the body cell again divides to form 2 sperm nuclei.

Fertilization (Fig. 55) takes place about a year after pollination. The pollen tube penetrates between the neck cells of the archegonium, and one of the sperm nuclei enters the cytoplasm of the egg to fuse with the egg nucleus to form a diploid zygote. The

499

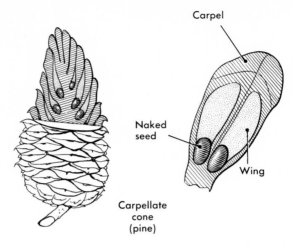

Fig. 57. Arrangement of seeds in gymnosperms showing how they are exposed in the axil of the bract.

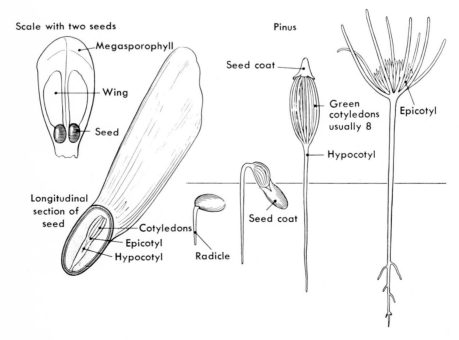

Fig. 58. *Pinus* sp. (pine). Germination of a seed into a seedling.

remaining nuclei of the microgametophyte degenerate along with the rest of the structure.

Embryo. As soon as fertilization has occurred by **syngamy** (Fig. 56, *A*), a proembryo is formed. The egg nucleus divides into 4 separate nuclei geometrically arranged (Fig. 56, *B*) within the zygote. These in turn divide twice, walls form between them, and 4 tiers of 4 cells each are formed (Fig. 56, *C*). Each cell of the lower tier forms an embryo, which now begins to elongate (Fig. 56, *D*) so that there is a bottom rosette, a suspensor in the middle, and an embryonic tube with an apical growing region. The suspensor cells elongate, and the apical region develops (Fig. 56, *E*) until the immature embryos or **proembryos** penetrate into the endosperm of the seed (Fig. 56, *F*). Shortly thereafter all but one of the proembryos disintegrate, and the true embryo survives embedded in the endosperm (Fig. 56, *G*). By continued growth making use of the nutrients of the endosperm, this embryo differentiates into a hypocotyl, an epicotyl, and cotyledons. The embryo is surrounded by the endosperm and this in turn by the seed coat (Fig. 57). Thus the ovule has become transformed into a seed that develops wings that aid in its dispersal when it is released from the pine cone.

By absorption of water from the soil, the embryo enlarges and splits open the seed coat (Fig. 58). A radicle or primary root grows down into the soil for the absorption of more water. The remainder of the endosperm is used for further growth of the embryo, which soon appears above ground. The hypocotyl elongates to form a stem. Cotyledons develop chlorophyll, push off the remainder of the seed coat, and spread out. The epicotyl develops into the true leaves, which by continued growth through photosynthesis produce a pine tree.

Means of dispersal. When ripe seeds are formed in the carpellate cone, it loosens and opens to release them. By means of the wings that are attached, these seeds are blown about by the wind until they reach suitable soil, usually not far from the parent sporophyte.

LIFE CYCLE 23—LILIUM Sp. (Fig. 59)
DIVISION 26. Anthophyta
CLASS Angiospermae
SUBCLASS Monocotyledoneae

Characteristics

The genus *Lilium* has a very large number of species. The tiger lily is a common example. The lily family, Liliaceae, is easily identified by its flower, which has a perianth composed of 6 parts, and a superior ovary with 3 carpels and 6 stamens (Fig. 16-3). The lily flower is widely used to illustrate the flower of angiosperms, the dominant plant life of the current geological era. The life cycle of these plants developed after a long period of evolutionary change, which has resulted in a highly specialized reproductive structure known as the flower. The steps leading to this are seen in many of the preceding life cycles (Fig. 60).

Asexual reproduction

Vegetative reproduction. Various means of vegetative reproduction can be found in the flowering plants, all of which have been discussed in Chapter 16. The lily is usually propagated by means of bulbs instead of seeds.

Sporophyte. A very large and independent (except for a brief and early period) sporophyte is present. It consists of roots, stems, leaves, and flowers—the plant body previously discussed. Angiosperms differ from gymnosperms in that flowers instead of cones are present; the ovules and seeds are covered, that is, enclosed by the carpel (megasporophyll) rather than naked as in the pine; the carpel or carpels constitutes the fruit when mature.

Sporangium. A microsporangium (pollen sac) in the anther and a megasporangium in the ovary form a part of the flower. A cross section of a very young anther reveals 4 pollen sacs or microsporangia in an anther (Fig. 61). Two of these lie on either side of the supporting tissue through which runs a single vascular bundle (Fig. 61). In each are many microspores or pollen mother cells. The anther is an elongate structure, and pollen mother cells are found throughout its length. When the pollen mother cells are first

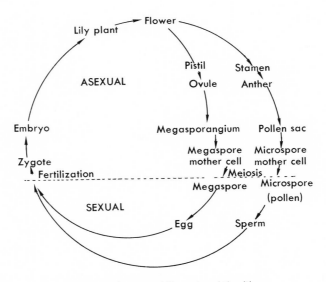

Fig. 59. *Lilium* sp. Life cycle of the lily.

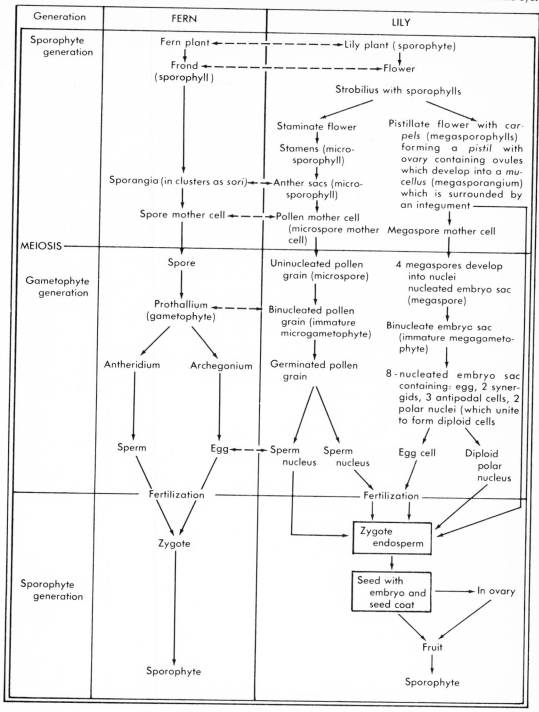

Fig. 60. Comparison of the life cycle of two plants, a fern and a lily. The fern has two independent generations, each living as a separate plant, whereas the lily has comparable stages, but the gametophyte is parasitic on the spermatophyte. This diagram shows that the two cycles are essentially the same, but that the angiosperm cycle is a refinement of the primitive cryptogamic cycle of the fern.

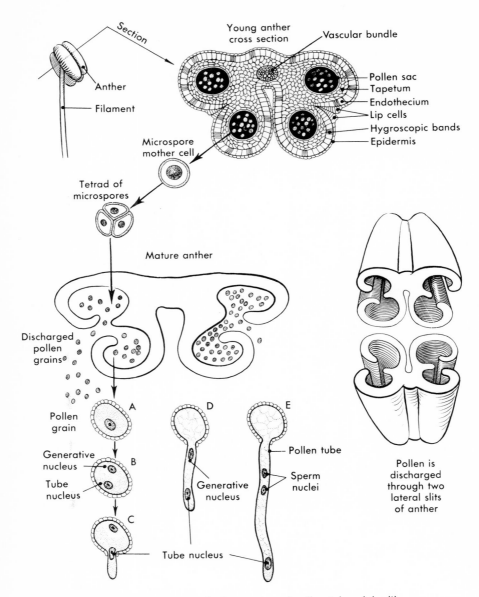

Fig. 61. Development of microspore and pollen tube of the lily.

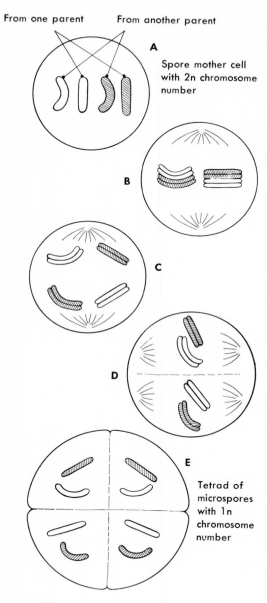

From one parent   From another parent

A

Spore mother cell
with 2n chromosome
number

B

C

D

E

Tetrad of
microspores
with 1n
chromosome
number

Fig. 62. Diagram of sporogenesis.

formed, they are closely packed together. Later, however, they separate and become spherical. As the growth of the anther progresses, the nucleus of each pollen mother cell divides by a special process termed **sporogenesis** (Fig. 62). The chromosomes line up during metaphase in pairs (tetrads), and, after each has duplicated itself, individual members of the pair will separate and move to opposite poles, eventually producing 2 daughter nuclei that have one half the number of chromosomes of the mother cell. The chromosomes of each daughter nucleus undergo a second and equatorial division, and the half chromosomes separate into 4 nuclei, each with the haploid or reduced number of chromosomes. Cell walls form, giving rise to a group of 4 cells known as microspores. This cluster of 4 spores is designated a tetrad of microspores. These spores may remain together in a tetrad arrangement. Usually they separate and develop into pollen grains. This is the only male gametophyte.

The development of the megasporangium usually occurs concurrently with that of the pollen. The carpels of the angiosperm pistil are spore-bearing leaves that have ovules and bear the megasporangia. Within the megasporangium is a megaspore mother cell that undergoes 2 nuclear divisions, resulting in 4 haploid nuclei. The production of these nuclei is essentially like that of the microspores inasmuch as both have the haploid number of chromosomes. At first they are arranged at the micropylar end of the cell, but later, 3 migrate to the opposite (**chalazal**) end (Fig. 63, *D*). Next the 3 chalazal nuclei divide by mitosis to form 2 large triploid nuclei. Meanwhile the haploid nucleus at the micropyle divides by mitosis, forming 2 haploid nuclei. Finally, a fourth mitosis takes place in all 4 nuclei, resulting in 8 nuclei, 4 triploid and 4 haploid (Fig. 63, *E*). This is a functional female gametophyte.

Spore. Both megaspores (Fig. 63) and microspores (pollen) are produced by the floral parts as a result of meiosis.

Microspore. The conversion of a micro-

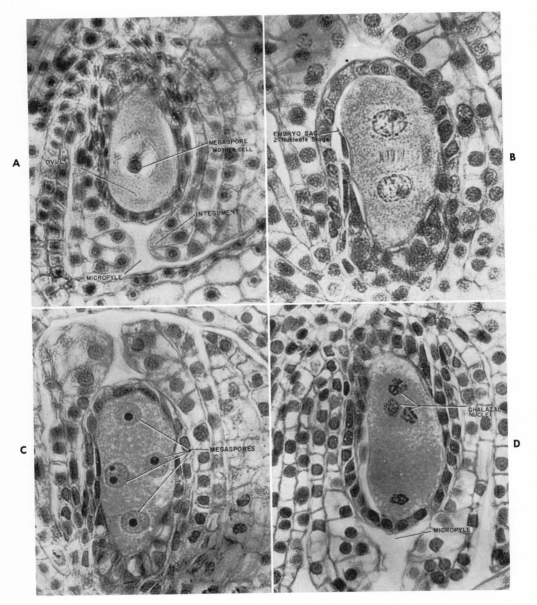

Fig. 63. A-H, Development of megaspore and female gametophyte in an angiosperm. **A,** Cross section of a lily ovary showing an embryo sac with a megasporocyte about to undergo division. (X500.) **B,** Lily embryo sac showing the 2-nucleate stage; remains of spindle fibers can be seen between the nuclei (secondary megasporocytes). (X500.) **C,** First 4-nucleate stage with the nuclei (megaspores) arranged in the diamond-shaped pattern, typically preceding migration of the three chalazal nuclei. (X500.) **D,** Same lily embryo sac showing a later migration. (X500.) (Courtesy George H. Conant, Triarch Products, Ripon, Wis.)

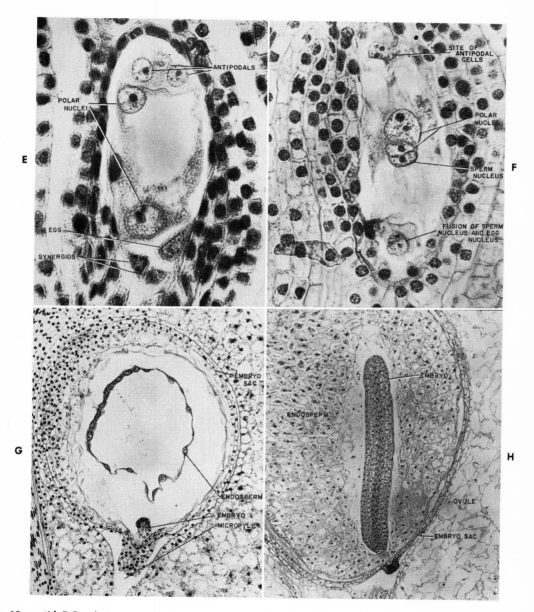

Fig. 63, cont'd. E, Rarely occurring true 8-nucleate stage, with all nuclei clear and distinct. (X500.) **F,** "Double fertilization"; two innermost nuclei of the 8-nucleate sac migrate to the center and are referred to as "polar nuclei"; these fuse together with one of the male nuclei from the pollen tube; the other male nucleus fuses with one of the micropylar nuclei (egg), while the rest of the micropylar nuclei (with synergids) disintegrate or remain nonfunctional. (X500.) **G,** Cross section of a lily fruit showing a young embryo sac and the free-nucleate endosperm. (X80.) **H,** Cross section of a lily fruit showing an older embryo. (X35.) (Courtesy George H. Conant, Triarch Products, Ripon, Wis.)

spore into a pollen grain is the result of the division of the microspore nucleus into 2 nuclei, the **generative nucleus** and the **tube nucleus.** The wall of the microspore thickens, and its outer surface may become covered with ridges, spines, or other characteristic markings of the species producing it. Thus the wall of the microspore becomes the wall (**exine**) of the pollen grain and is one of the most resistant natural substances known. Each pollen sac is eventually filled with a powdery mass of mature pollen grains. As the anther matures during the development of pollen, the pollen sac splits longitudinally, and the pollen is free to be dispersed.

Megaspore. The 8 nuclei are embedded in a common mass of cytoplasm, without separation into individual cells. This is called the embryo sac in the flowering plants. The number of nuclei in the embryo sac varies according to the species, and although 8 is the common number, there may be as few as 4.

Sexual reproduction

Gametophyte. There are 2 gametophytes—male and female—as in the pine. But both gametophytes are much reduced in size, the pollen grain and the embryo sac; both are completely dependent on the sporophyte generation, since neither gametophyte has chlorophyll.

Antheridium. Only the 2 sperm nuclei of the pollen grain represent the male sex organ.

Archegonium. The only reminders of the female sex organ are the **egg nuclei** (Fig. 63, *E*). One nucleus of each group of 4 migrates to the center of the embryo sac to become **polar nuclei.** Each ovule is often bent back upon itself and enveloped by one or more **integuments,** except for a small opening at the tip called the micropyle. The integuments form as the female gametophyte develops. The entire ovule is attached to the placenta by a basal stalk or funiculus.

Egg. After the migration of the polar nuclei, 3 haploid nuclei remain at the micropylar end of the embryo sac. Two are called **synergids,** and the other is the egg. The 3

nuclei remaining at the end of the embryo sac opposite, chalazal, to the micropyle are known as the **antipodals** (Fig. 63, *E* and *F*).

Sperm. Two male gametes or sperm are produced as a result of the division of the generative nucleus of the pollen grain (Fig. 63, *F*).

Pollination. The pollen grain settles on the stigma. The tube nucleus forms the pollen tube that must grow through the tissues of the stigma, style, and ovary to the ovule instead of merely landing in or near the micropyle of the ovule (Fig. 63), as in the gymnosperms.

The development of the egg sac within the ovule, the presence of the sperm in the pollen, and the transfer of pollen to the stigma are all events leading up to fertilization in angiosperms (Fig. 16-4). The stigma of the pistil is covered with a sticky mixture of water, sugar, and other substances. Once the pollen falls on this stigma, it is held there, and in this environment the pollen immediately starts to grow or germinate. It is possible to induce many kinds of pollen grains to germinate in a sugar solution on a slide so that the development of the pollen tube can be observed.

The pollen grain bulges out to produce a **pollen tube** that grows down through the style. As the protoplasm flows out into the tube, the pollen grain is left empty on the stigma. The pollen tube produces digestive enzymes that decompose the tissues of the style as it makes its way toward the ovule. Usually the tube nucleus is near the tip of the elongating tube, and the generative nucleus is somewhat behind it (Fig. 61). The generative nucleus divides so that 3 nuclei may be present in the tube. Eventually the pollen tube arrives at and penetrates the ovule, either through the funiculus or through the micropyle, as indicated in Fig. 16-4. Once the pollen tube reaches the embryo sac, the 2 sperm nuclei, which have been derived from the division of the generative nucleus, are discharged into the sac. The tube nucleus and pollen tube are left to die and degenerate. One of the sperm nuclei fuses with the egg and the other with the fused polar nuclei. The fused egg and sperm have the diploid (2n) number of chromosomes. The fused polar nuclei result in a 5n chromosome number. The fertilized egg marks the beginning of the new sporophyte or diploid generation and will develop into the embryo of the seed by a series of mitotic divisions.

Zygote. In the life cycle of the angiosperms water is not necessary for fertilization. Unlike the pine, the endosperm of the lily is formed by the fusion of 2 polar nuclei and 1 of the sperm nuclei (Fig. 63, E and G). Hence the endosperm is 5n, which, by a series of mitotic divisions becomes the food storage tissue of the seed. In the pine the endosperm was formed from the female gametophyte; hence it was haploid (n). The final union of egg and sperm (nuclei) results in a fertilized egg, or zygote. This fusion of egg and sperm nuclei is fertilization.

Embryo. After fertilization, the zygote undergoes a series of divisions leading to the development of the embryo. The female gametophyte becomes the functioning embryo sac (Fig. 63, H), within which the embryo is found. The embryo is attached to the embryo sac by a suspensor at its basal end. The apical cell produces the embryo proper. In many plants, such as the bean, the endosperm degenerates before the seed is mature (Fig.

14-4). When this occurs, food is stored in the embryo seed coat, or the nucellus, which is the megasporangium of the ovule enclosing the megagametophyte. The union of one sperm nucleus with the egg nucleus and the other with the fused polar nuclei is frequently called **double fertilization** and is characteristic only of angiosperms. The presence of a polyploid tissue for food storage has no apparent advantage over a haploid or diploid one.

Although fertilization in most angiosperms is required for the production of seeds, there are some in which this is not true. The production of seeds without fertilization is known as **apomixis.** In apomictic plants it is not unusual for more than one embryo to be present inside a seed, although normally fertilized plants have only one.

Means of dispersal. During the growth of the embryo and the endosperm there is also rapid growth of the other tissues of the ovule. The ovule increases in size, the integument may harden, the embryo becomes dormant, and the ovule has thus ripened into a seed. The ovary usually increases in size, and its tissues become differentiated into those parts involved in the production and dissemination of the seeds. The mature ovary containing the seeds, together with any accessory parts, is now called a fruit.

Glossary

■ The technical names in boldface type in the text are defined in the glossary that follows. The term is indicated in this manner only once in the text, usually the first time it is used.

Å (see *Ångström unit*).

abscisic acid a plant-growth inhibitor that induces abscission, stomatal closure, dormancy, and growth inhibition.

abscission zone the separation layer of cells formed at the base of some leaves and other organs that permits these parts to be shed.

accessory fruit a fruit composed of parts of the flower in addition to the ovary.

acetyl coenzyme A (acetyl CoA) a compound formed from the condensation of pyruvic acid with coenzyme A, responsible for passing carbon atoms into the Krebs cycle.

acetylcholine a chemical compound responsible for the transmission of nerve impulses in animals.

achene a small dry, one-seeded, indehiscent fruit with the seed coat and fruit wall unfused.

acid a susbstance containing hydrogen whose solution in water increases the number of hydrogen ions present.

actinomycin D an antibiotic that combines with DNA to prevent the formation of messenger RNA.

active absorption the taking of ions into a cell against diffusion pressure as the result of the expenditure of energy.

active secretion the exchange of ions when the principal ions in the cell are at a higher concentration than the solutions outside.

adenosine diphosphate (ADP) a chemical substance present in active cells during metabolic activity and involved in the metabolic economy of the cell; the end product along with inorganic phosphate that results in energy release from adenosine triphosphate (cf. *ATP*).

adenosine triphosphate (ATP) an energy-yielding compound with high-energy bonds involved in the metabolic release of energy within the cell.

adnate two or more series of parts united in a flower.

ADP (see *adenosine diphosphate*).

adventitious bud a bud that arises on a part of the stem other than its axil or apex.

adventitious root a root that develops from parts other than the primary root or one of its branches.

aeciospore a spore formed within an aecium; a type of spore produced by the rusts.

aecidium (plural, *aecidia*) a group of binucleated hyphal cells at one stage in the life cycle of a rust that produce spore chains.

aerial stem a stem that grows above the ground.

aerobe an organism that uses free oxygen during respiration.

aggregate fruit a fruit composed of a cluster of ripened ovaries from a single flower.

akinete vegetative cells of a filament or trichome that are capable of generating new filaments.

allele one gene of a pair of alternate genes found in the same position on homologous chromosomes.

allopatric populations of two species occupying separate and distinct ranges.

allopolyploid polyploid organism with more than two sets of homologous chromosomes that arise through hybridization of different species.

alternate leaves leaves occurring at different levels successively on opposite sides of the stem.

alternation of generations alternation of a spore-producing stage with a gamete-producing stage in the life cycle of a plant.

ameboid like an ameba, i.e., a cell without a firm cell wall and capable of movement by the flow of its cell protoplasm.

amino acid organic compounds with the general formula $NH_2\text{-}\overset{\text{H}}{\underset{\text{R}}{\text{C}}}\text{-COOH}$ that combine to form proteins.

amyloplasts plastids that store large amounts of starch.

amylum stars star-shaped aggregates of cells filled with starch found in certain bryophytes; these cells are capable of asexual reproduction.

anabolism constructive stages in metabolism, e.g., assimilation; opposed to catabolism.

anaerobic organism an organism that does not use free oxygen in its respiratory process.

anaerobic respiration respiration without the use of free oxygen; fermentation.

anaphase stage of mitosis in which the chromosome halves separate and begin to move to opposite poles.

androecium the stamens, collectively; male reproductive organs of a plant.

Ångström unit (Å) a unit of length, equal to one tenth of a millimicron, or one tenth of a nanometer, or one ten millionth of a millimeter.

Animalia the division of living organisms comprising the animal kingdom.

annual a plant that grows from seed and produces seed in one growing season.

annual ring the layer of wood or xylem formed during one year's growth of cambium.

annulus a row of specialized cells in the wall of a sporangium that aid in the release of spores.

anther the pollen-bearing part of a stamen; a pollen sac or microsporangium.

antheridiophore a disk-headed stalk that bears antheridia.

antheridium a male gametangium in which sperm are produced.

antherozoid motile male sexual cells in antheridia.

anthocyanin one of several substances of a blue or violet pigment found in flowers, leaves, and stems.

antibody a globulin type of protein produced within an organism that opposes the action of an antigen.

antigen a foreign substance, usually a protein, that causes the production of specific antibodies within an organism.

antipodal one of the three nuclei opposite the micropyle of the embryo sac.

antitoxin a type of antibody that neutralizes a toxin.

apetalous without petals on the flower.

apical meristem the embryonic tissue at the tip of a stem (or root); also called promeristem.

aplanospore a nonmotile spore of algae; an encysted spore of fungi.

apomixis asexual reproduction, of which there are many types, that substitutes for or replaces sexual reproduction.

apothecium a saucer-shaped or cup-shaped ascocarp that bears asci on its extended or exposed surface; characteristic of the lichens and certain fungi (discomycetes).

archegoniophore the portion of the thallus of some cryptogams that develops the archegonium.

archegonium a multicellular female gametangium of higher plants.

ascocarp sporocarp of ascomycetes, the sac fungi.

ascogonium the female gametangium of ascomycetes, the sac fungi.

ascospore a spore produced by the ascomycetes within an ascus.

ascus a saclike structure that contains ascospores.

ascus mother cell a cell that produces an ascus.

asepalous without sepals.

asexual not involving the fusion of sex cells or the fusion of their nuclei; without sex organs.

atom the smallest particle of an element that can exist alone or in combination with other atoms and elements.

atomic weight a relative measure based on that of carbon with a weight of 12 and computed by the summation of the protons and neutrons in the atomic nucleus.

ATP (see **adenosine triphosphate**).

autecology the study of an individual organism or individual species in relation to the environment.

autogamy the asexual fusion of nuclei.

autopolyploids polyploid organisms with more than two sets of homologous chromosomes.

autotrophs organisms capable of producing their own food and exemplified by plants, which can use completely inorganic nutrients.

auxin growth-regulating substance found in plants; several types.

axial placentation the placenta on a central axis in arising from it.

axile placentation the placenta on a central axis in the ovary.

axillary bud the bud located in the axil of the leaf; also called lateral bud.

bacillus a rod-shaped bacterium.

backcross crossing a hybrid with one of its parents or parental forms.

bacteria a large group of typically prokaryotic unicellular microscopic organisms widely distributed in air, soil, water, the bodies of living plants and animals, and dead organic matter.

bacteriophage a virus parasitic on bacteria.

bark the tissue that lies outside the vascular cambium or the periderm and tissues external to it in woody stems or roots.

base a substance containing hydroxyl groups (OH) whose solution in water increases the number of hydroxyl ions present.

basidiocarp a sporocarp of the club fungi.

basidiospore a spore produced by a basidium, by the basidiomycete fungi.

basidium the spore mother cell of the basidiomycete fungi.

berry a fleshy fruit with the entire pericarp fleshy.

biennial a plant that grows a taproot and a rosette of leaves during the first growing season and produces flowers and seed during the second growing season.

binominal system a system of nomenclature in which the name of a species is composed of two words, the generic name and the specific name.

biochemistry the study of the chemicals and chemical processes found in living organisms.

bioluminescence emission of light by living organisms as the result of internal oxidation of cellular substances.

biome a large ecological habitat, such as tundra, desert, or evergreen forest; a major life zone of plants and animals.

biota all living organisms, as individuals, living at one time.

bisexual having the male and female sex organs or cells in separate individuals.

bivalent chromosome homologous chromosomes when the two are associated during synapsis in meiosis.

blade the thin, expanded portion of a leaf.

botany the study of all the organisms belonging to the kingdoms Prokaryotae, Fungi, and Plantae, including their evolution, function, and classification.

bract small, modified leaves or scales subtending a flower or its pedicel.

Brownian movement the random movement or agitation of small particles in suspension.

bud an embryonic or young shoot that has the capacity to produce leaves, stems, flowers, or roots.

bud scales protective leaflike coverings of buds.

bud scar the scar left after a bud has fallen off.

buffer a substance that resists a change in the pH of a solution.

bulb an underground storage leaf bud composed of enlarged and fleshy scalelike leaves.

bulbil a thickened axillary bud that may be detached and produce a new plant; also an aerial bulb.

calyx collective term for the sepals of a flower.

cambium layer of meristem that is responsible for the lateral growth of a stem or root.

capillitium a network of threads in the capsule of the fruiting body, especially in the slime molds.

capsid the protein coat of a virus.

capsomere the subunit proteins from which the capsid of the mature virus particle is assembled.

capsule the sporangium of a bryophyte; the dry dehiscent fruit of an angiosperm; the slimy covering around certain bacterial cells.

carbohydrates organic compounds composed of carbon, hydrogen, and oxygen, with the general formula, $C_nH_{2n}O_n$.

carbon dioxide cycle the circulation of carbon dioxide during food synthesis and respiration.

carotene reddish orange pigment found in many plant cells.

carpel female part of megasporophyll of an angiosperm flower.

carpellate cone megasporophylls of conifers.

carpogonium part of the procarp; cells resulting in the sporocarp after fertilization.

carpospores the spores produced by the carpogonium after fertilization.

carposporophyte a small parasitic plant located on the gametophyte of certain algae.

catalyst a substance that alters the rate of a chemical reaction without itself being changed during the reaction.

category a hierarchical arrangement of levels of taxa, each level a category; e.g., the taxon Rosaceae belongs to the category family.

cell small mass of protoplasm, usually containing a nucleus or nuclear material; the fundamental unit of structure in plants.

cell division (see *mitosis*).

cell membrane the outer living and limiting boundary of protoplasm in a cell.

cell wall a cell secretion covering plant cells, usually composed of the carbohydrate cellulose and certain minerals.

cellulose the carbohydrate forming the walls of plant cells.

central placenta a stalk that extends from the base of the ovary into the locule.

centrioles small cylinders found in the cytoplasm of animal and some algal cells that are associated with the formation of cilia and flagella.

centromere the constricted part of the chromosome to which the spindle fiber is attached.

chalazal the end of the megaspore mother cell opposite the micropyle.

chemosynthesis the formation of a more complex compound from simpler compounds by chemical action; the formation of organic compounds from inorganic compounds by some plant cells.

chiasma the point at which two chromosomes cross and fuse during meiosis.

chitin a characteristically animal substance found in the cell walls of many fungi; a polysaccharide usually mixed with a protein.

chilling a period of exposure to cold temperatures prior to growth or germination.

chlamydospore a thick-walled resting spore.

chlorenchyma a tissue composed of parenchyma cells that contain chloroplasts.

chlorophyll the green pigment in plants necessary for the process of photosynthesis; a methylphytol ester of the chlorophyllins.

chloroplast a cell organelle that is the site of photosynthesis.

chromatid one of the two parallel halves of a fully contracted chromosome as it appears during the late prophase of mitosis.

chromatin the nucleic acid of a nucleus as it shows up as threads or rods after staining.

chromatography a process used to separate and show the component amino acids of a protein.

chromatophores cells containing pigment substances.

chromoplast a body (plastid) in the cell that contains a pigment.

chromosome characteristic and deeply staining body in the nucleus composed primarily of DNA; bears genes or determiners of heredity.

circinate rolled on its axis with the apex as its center; coiled inwardly.

circinate vernation unrolling of the developing fern fronds, often called "fiddleheads"; see *vernation*.

clamp connections a special structure found in some fungi during cell division to ensure the proper distribution of the set of two nuclei for each cell.

cleistothecium an ascocarp in which the asci and ascospores are developed within a closed envelope.

climax a steady community reached at the peak of development for a particular area.

clone a group of plants developed asexually from a single plant.

closed system a process maintained and concluded without input or output after its original construction.

coalescent fused parts, particularly in reference to flower parts.

coccus a spherically shaped bacterium.

coefficient of coincidence a ratio comparing the actual frequency of double crossovers with the expected frequency in a three-point testcross.

coenocyte a group of protoplasmic units, each with a nucleus, enclosed in a common cell wall.

coenzyme a nonprotein substance, usually a vitamin, that is necessary for enzyme activity.

coevolution the concurrent evolution of a plant species and an animal species through selection and adaptation of features involving both species.

cohesion the molecular attraction of water molecules causing sap flow.

colchicine an alkaloid obtained from the autumn crocus, *Colchicum* sp., which has the property of arresting cell mitosis at the metaphase stage.

coleoptile a protective growing sheath covering the plumule of the seeds of grasses and similar plants during germination.

collenchyma a supporting tissue made up of somewhat elongated cells, the walls of which are thickened with cellulose, particularly at the corners; like parenchyma cells, they are living.

colloid the state of a substance composed of particles that are intermediate in size and that will neither settle out nor form a true solution; opposed to crystalloid.

colony a group of contiguous cells held together by cell cement, usually derived from a single ancestor, growing on a solid surface.

columella the axis or central part within a sporangium, capsule, or carpel.

community all the living and nonliving components of the environment of a particular area, usually as a part or division of an ecosystem.

companion cells elongated parenchyma cells adjoining sieve tubes in the phloem tissue.

complete flower a flower that bears sepals, petals, stamens, and pistils.

complex permanent tissue xylem and phloem.

complex tissue tissue consisting of two or more kinds of cells that perform the same function.

compound leaf a leaf in which the blade is divided into a number of leaflets.

compound pistil a pistil formed by the fusion of two or more carpels.

computer an electronic apparatus used to carry out

repetitious and highly complex mathematical operations at high speed.

conceptacle a hollow structure containing sex organs.

cone a specialized branch that bears sporophylls, each subtended by a bract.

conidiophore the organ that bears conidia.

conidium a propagative cell, asexually produced and separated from the parent.

conjugation a process involving the fusion of isogametes to form a zygote.

continental drift a theory proposing that there is and has been movement of the continents by actual drifting apart.

controlled system an experimental system where one may change a controlling factor to study the effect on the entire system.

controlling gene a gene that controls the expression of other genes.

converter gene a gene capable of changing a color-determining gene permanently to one determining another color.

core the central part of certain fruits, as in the apple; the true fruit, exclusive of accessory tissue.

cork a tissue composed of cells whose walls are impregnated with a fatty substance called suberin.

cork cambium the layer of meristematic cells that produces cork; also called phellogen.

corm a short, stout, erect underground stem in which food is stored.

corolla collective term for the petals of a flower.

cortex the tissue lying between the epidermis and the endodermis of a stem or root.

cosmopolitan generally distributed; ubiquitous.

cotyledons seed leaves or food storage parts of a seed.

cristae infoldings of the inner mitochondrial membrane to form pouches or discs; the site of Krebs cycle enzyme action.

crossing-over an exchange of chromatin material between homologous chromatids during pairing of the chromosomes at meiosis.

cross-pollination transfer of pollen from the anther of a flower of one individual to the stigma of the flower of another individual.

crystalloid crystallike; opposite of colloid.

cupule a reproductive structure producing gemmae; the cup of such fruits as the acorn.

cuticle the outermost skin or external pellicle containing the epidermal layer of cells.

cutin a waxy substance secreted by epidermal cells.

cystocarp the portion of the gametophyte of certain algae that bears the carposporophyte.

cytokinin a class of plant growth hormones important in the regulation of cell division, organ initiation, and delaying of senescence.

cytology the study of cells.

cytoplasm the protoplasm of the cell other than the nucleus.

cytoplasmic inheritance inheritance by genes located in the mitrochondrion or chloroplast, as opposed to nuclear inheritance.

data facts or statistics, either historical, or derived from calculations or experimentation; data may be processed in a computer and retrieved in a form that differs from that stored.

data processing the storage and retrieval of data.

deciduous falling at the end of a season of growth, as leaves and flower petals; usually refers to trees other than evergreen trees.

defective virus a virus that cannot reproduce except in combination with a helper virus.

dehiscent fruit a dry fruit that splits open on maturity for release of seeds.

deoxyribonucleic acid (DNA) one of two nucleic acids composed of nucleotides formed from a nitrogenous base, the sugar 2-deoxyribose, and phosphoric acid.

determinate flower an inflorescence in which the first flowers to open are at the apex or the center of the cluster.

diakinesis the last stage of the first prophase of meiosis in which the definitive chromosome has been formed and the nuclear membrane is still intact.

diatomaceous earth earth composed of shells of dead diatoms.

dictyosome flattened curved discs forming the subunits of the Golgi apparatus of the cell.

differentiation the development of a cell or an organism into a mature or permanent cell or mature organism.

diffusion the process by which molecules or other particles intermingle as a result of their thermal motion.

diffusion pressure the force or pressure developed when particles or ions move from a region of greater concentration to one of lesser concentration.

dihybrid a hybrid involving two characteristics or two pairs of alleles.

dikaryotic containing two separate and genetically different nuclei within the same cell.

dioecious presence of staminate and pistillate flowers on different plants.

diplanetism the phenomenon involving the production of zoospores that swarm and then encyst, followed by the production of a second type of zoospore that again swarms before germination.

diploid (2n) with two sets of homologous chromosomes; the 2n number.

diplotene the stage during first prophase of meiosis when the chromosomes separate.

disaccharide a carbohydrate, $C_{12}H_{22}O_{11}$, in the form of a crystalline solid, soluble in water, sweet to the taste; a sugar.

disease any abnormal condition that interferes with the physiological functioning of the organism.

diversity the variation of the species from group to group.

DNA (see *deoxyribonucleic acid*).

dominant refers to a characteristic dominant in inheritance, one which overshadows some contrasting characteristic, the recessive.

Donnan equilibrium When a simple electrolyte such as NaCl establishes equilibrium across a permeable membrane, the concentration of both ions is the same on each side of the membrane. If a nondiffusible ion is present on one side of the membrane, it inhibits the flow of ions bearing an opposite charge; therefore, the concentration of the diffusable ions at equilibrium will not be the same on both sides of the membrane.

dormancy a period in the plant life cycle during which growth is suspended or retarded to the extent that it cannot be visibly detected.

dormin another name for the plant-growth inhibitor abscisic acid.

double fertilization union of one sperm nucleus with the egg nucleus and another sperm nucleus with the fused polar nuclei.

drupe a fruit with a thin exocarp, a thick fleshy mesocarp, and a stony endocarp.

dry fruit a fruit with a leathery, papery, or woody pericarp.

dyad two chromatids lying parallel to each other during meiosis.

dynamic equilibrium a system in which the interior is in balance and kept in balance, by physiological processes, with the exterior.

ecology the study of the relationship of organisms in their environment.

ecosystem all the living and nonliving components of the environment of a particular kind of habitat as a pond, desert, or similar system.

ecotone the interface between biomes.

egg female gamete or sex cell.

egg nucleus one of the nuclei of the embryo sac.

elaiosome (pronounced e-lye′o-, or e-lay′o-) an oil-secreting body on the outer surface of seeds; the oil is an attractant to ants.

elater an elongated structure produced in the sporophyte of the liverwort and on the spores of horsetails that aids in the dispersal of spores.

electron minute negative particles found in the atom.

element a substance that cannot be broken down into a simpler one by ordinary chemical means.

embryo a plant produced as the result of gamete union and development but before rapid growth at time of germination.

embryo sac the cells in the ovule in which the embryo is formed.

embryonic region a region of actively dividing cells.

emulsion the suspension of droplets of a liquid within another liquid.

endergonic reaction a chemical reaction that absorbs heat.

endocarp the inner layer of a pericarp.

endodermis the layer of cells just inside the cortex, particularly in a root.

endoplasmic reticulum a fine network of vesicles and canals found in the cellular cytoplasm.

endosperm the food-storage portion of many seeds; a part of the female gametophyte.

energy the property of something that enables it to do work.

entropy the energy lost from a system and no longer capable of performing useful work.

environment the nonliving and living surroundings of an organism.

enzyme a protein that acts as an organic catalyst.

epicotyl that part of the embryo in the seed that produces the shoot.

epidemic a rapid spread and increased prevalence of a disease or organism among people.

epidermis the tissue that covers an external surface such as that in leaves, young roots, and stems; usually one layer thick and covered with cutin; primarily a protective tissue.

epigynous a flower with the ovary beneath the flower parts.

epiphyte plant that lives on the surface of other plants.

epiphytotic said of an infectious plant disease that tends to occur sporadically over a wide area and to affect large numbers of susceptible plants whenever present.

epistasis the interaction of nonallelic genes that constitutes a singular phenotypic trait.

ergotism a diseased condition of animals or man produced from eating grasses or grains infected with an ergot fungus.

ethylene a gas that affects the geotropic responses of plants and is produced by ripening fruits.

etioplasts colorless plastids that differentiate into chloroplasts.

eukaryotic cell cell with definite nucleus surrounded by a nuclear membrane.

exergonic reaction a chemical reaction during which heat is given off.

exine the outer coat of the pollen grain.

exocarp the outer thin layer of a fruit; the "skin."

experimental design the plan for the procedure to be used to investigate a scientific problem.

experimentation to test by comparing through measurement or controlled processes to discover information or gather data; valid experiments must be repeatable.

F_1 the first filial generation produced as a result of two individuals mating.

F_2 the second filial generation produced from the mating of two F_1 individuals.

fascicular cambium the vascular cambium within the vascular bundle.

fat organic compound composed of carbon, hydrogen, and oxygen in which the hydrogen and oxygen are not in a ratio of 2 to 1 as in carbohydrates; 3-carbon compound, called glycerol, combined with fatty acids; solid at ordinary temperature; one of the lipids.

fatty acid even-numbered carbon chains, either saturated or unsaturated.

fermentation process whereby organisms extract energy from carbohydrates and other compounds in the absence of oxygen.

fibrous root root composed of many slender roots, most of which are the same size; typical of monocots.

filament a series of cells linearly arranged in a continuous series, e.g., spirogyra plants; the stalk of a stamen.

flagellum (pl. **flagella**) a whiplike thread attached to some cells (spores, sperms, and bacteria) that makes motility possible.

fleshy fruit the fruit in which all or most of the pericarp is soft and juicy.

florigen a flowering hormone that is formed in induced leaves and causes flowering. It is not yet chemically isolated.

flower a group of modified leaves specialized for reproduction and borne on the end of a short stem in most cases.

flowering factors the plant regulators that affect flowering.

flowering response the growth and opening of flowers as a response to hormone action.

fossils the remains or impressions of plants and animals preserved in the various strata that compose the earth's crust; usually extinct species are represented.

free energy the energy of a system capable of performing useful work.

frond the leaf of a fern or similar large leaves.

frugivores organisms that eat fruit.

fruit a fully developed and ripened ovary or ovary cluster.

funiculus attachment of the ovule to the placenta.

gametangium a special structure producing gametes.

gamete a sex cell; reproductive cell.

gametophore the stalk of the gametophyte in mosses on which the antheridia and archegonia are formed.

gametophyte a sexual generation; gamete-producing generation; haploid (n) generation.

gamopetalous a flower with a fused or continuous calyx.

gamosepalous a flower with a fused or continuous calyx.

gaps the unfilled variation possibilities between species; the differences between species leave gaps between one series of individuals belonging to a species and a second series.

Gause's principle no two species may occupy the same niche at the same time.

gel the solid state of a colloid.

gemma an asexual or vegetative growth from a parent body (usually liverworts) that is capable of growing into a new plant.

gene the unit of heredity.

gene pool the total available genes in a breeding population.

generative nucleus one of the nuclei of the pollen grain concerned with fertilization.

generic name the name of a genus to which a species is assigned.

genetic drift changes in the gene frequency attributable to the survival of small samples of genes from year to year.

genetics the study of the mechanism of heredity.

genotype the genetic makeup; the gene composition of an individual.

genus (pl. **genera**) see **generic name**.

geotropism growth movement stimulated by gravity.

germination growth or development of a plant from a seed or spore.

gibberellin growth regulator inducing increased stem elongation, particularly in dwarf mutants.

gill flat, spore-producing plate on the underside of the pileus of agarics, e.g., the gill of a mushroom.

globule the sperm-producing structure in chara and similar plants.

glucan a polysaccharide found in fungus cell walls.

glycerin part of the composition of fat; 3-carbon chain compound.

glycerol part of the composition of fat; also called glycerin.

glycolysis the metabolic pathway that converts 1 mole of glucose into 2 moles of pyruvic acid in the absence of oxygen.

Golgi apparatus organelle of the cell that stores and secretes glycoproteins and mucopolysaccharides.

gonidia special cells, asexually produced and nonmotile, that separate from the parent gametophyte to produce daughter plants.

grana small granules embedded within the matrix of the chloroplast of a cell.

ground meristem a tissue that will develop into the pith and cortex of stems.

growth an increase in mass either through cell enlargement or through increase in the number of cells by mitosis.

guard cells the cells that surround the stomas of leaves.

guttation the accumulation of droplets of water at the hydathodes; formation of water droplets on leaves caused by root pressure.

gynoecium the pistils collectively.

haploid having one set of chromosomes; the *n* number characteristic of the gametophyte.

hapteron a discoid outgrowth, highly adhesive, by which a peridiole is fixed to some substratum.

Hatch-Slack pathway a photosynthetic pathway in which carbon dioxide is fixed directly into a 4-carbon compound.

haustorium a specialized hypha of the fungi that is absorptive; a modified, suckerlike, adventitious root found in some higher plants.

helper virus a virus that combines with a defective virus for reproduction.

herbaceous stem a stem that is composed of soft tissue, usually green and not woody.

herbarium a collection of preserved, mounted, identified, and systematically arranged plants.

heredity the study of the inheritance of variations.

heterocyst a large, apparently dead cell that causes the separation of sections of an algal filament; responsible for dividing algal filaments into hormogonia.

heterogametes gametes of two sexes showing morphological differences.

heterogamous producing two kinds of gametes on the same vegetative plant.

heteroploidy having different chromosomes of the set present in different numbers; unbalanced polyploidy.

heterosis a characteristic that is the result of a cross between two varieties to produce a new variety that exceeds both parents in desirable qualities; hybrid vigor.

heterothallic with two different kinds of filaments or thalli.

heterotrophs organisms that require organic nutrients.

heterozygous involving zygotes of more than one kind.

hilum seed scar; point of attachment of the seed to the ovary wall.

histology the study of tissues.

histone a basic protein found associated with DNA and can yield a high proportion of basic amino acids on hydrolysis.

holdfast cell the anchoring cell of an alga, usually without chlorophyll.

holozoic feeding on solid food particles.

homology the study of equivalents of certain parts in different organisms believed to have the same evolutionary origin, although they may have the same or different function.

homozygous having allelic pairs that are alike.

hormogone a cell of a hormogonium.

hormogonium small segments of cells broken from certain filamentous algae capable of growing into new filaments.

hormone a growth substance.

hybrid the result of a cross between two individuals that differ in one or more characteristics.

hybrid vigor (see **heterosis**).

hybridization the result of a crossing of two organisms that differ in one of more characteristics.

hydathodes specialized cells in leaves that are responsible for the formation of droplets of water or other fluid not associated with stomas.

hydroponics the art of growing plants in nutrient solutions without soil.

hydrotropism growth movement in response to water.

hypanthium a cup-shaped mass of tissue at the sides of a developing flower above the receptacle.

hypertonic solution a solution with a greater osmotic pressure than an isotonic solution.

hypha the filaments of the fungi.

hypocotyl embryonic stem portion of an embryo of a seed just below the cotyledon attachment.

hypogynous a flower with the calyx, corolla, and stamens arising beneath the ovary.

hypothesis the organization of available data and information into a system for testing and verification.

hypotonic solution a solution in which the osmotic pressure is less than an isotonic solution.

imbibition the process by which solids (especially colloids) absorb water and swell.

imperfect flower a flower that lacks either stamens or pistil.

incomplete dominance the dominance of one gene over the other not complete so that the phenotype shows features of them both.

incomplete flower a flower that lacks either sepals, petals, stamens, or pistil.

indehiscent fruit a fruit that does not open or split upon maturity.

indeterminate flower an inflorescence in which the tip continues to grow so that the first flowers to open are at the base or outside the cluster.

indoleacetic acid the principle auxin found in plants.

inducer a substance that combines with the repressor protein of an operon thus allowing the structural genes to be transcribed.

indusium the membranous covering of a fern sorus.

inflorescence a cluster of flowers on the same peduncle.

information documents communicated or received concerning a particular fact or circumstance; information is stored and retrived without change.

information retrieval system (IRS) a process involving computer machinery used to code and file documents and references to documents so that they may be rapidly retrieved for use.

inorganic compound a chemical compound usually not containing carbon in the form found in living organisms or their products.

integument the covering of the ovule, usually developing into the seed coat.

interfascicular cambium the vascular cambium that is found between the vascular bundles.

internode the area of the stem between two successive nodes.

interphase the resting and growth stage of the cell between mitoses.

introgressive hybridization incorporation of genetic material from one population into that of another by repeated backcrossing of two distinct species.

ion a charged atom.

irregular flower a flower the floral parts of which are not symmetrical, e.g., the sweet pea.

isogamete a gamete that can fuse with another of the same size, structure, and behavior to form a zygote.

isolating mechanism any of several factors that inhibit breeding or the circulation of genes in the gene pool.

isotonic solution a solution in which the osmotic pressure is the same on either side of the membrane.

isotope any of the atoms of an element with a different atomic mass than those normally found in a molecule.

isotope tracer an element of radioactive properties used to tag a molecule so that its metabolic pathway may be traced.

kelp a large brown marine alga, e.g., *Fucus*.

Krebs cycle a sequence of aerobic, enzymatic reactions that effects the breakdown of the acetyl group (derived from pyruvic acid or some other source) to carbon dioxide and water with a release of energy that is used in the formation of the high-energy phosphate bonds of ATP; also called the citric acid or tricarboxylic acid cycle.

lacustrine a type of soil characterized by a gray, silty sediment found on the bottom of lakes.

land bridge a connection of land or isthmus of former existence over which animals and plants migrated to new areas.

lateral conjugation conjugation or union between two adjacent cells of the same filament.

lateral root (see *secondary root*).

layering the rooting of stems by bending them down and covering them with soil.

leaf axil angle between the petiole of the leaf and the stem.

leaf mosaic pattern formed by the leaves of a plant.

leaf scar the scar left by a leaf after it has fallen off.

lenticel a pore in the bark or epidermis of the stem

through which air enters the living tissues beneath.

leptotene stage in the first prophase of meiosis in which the chromosomes are threadlike.

lethal gene a gene that causes death in the homozygous state or frequently an abnormal effect in the heterozygous state.

leukoplast colorless plastid.

liana a vine or climbing stem.

lignin an organic material associated with cellulose in cell walls of plants, especially sclerenchyma cells.

ligule tonguelike appendage at the top of a leaf sheath in grasses.

lip cell thickened cells along the edge of the annulus of the fern that are the first to separate on dehiscence.

lipid one of many organic compounds that have the common property of being variably soluble in organic solvents and only slightly so in water (see *fats*).

literature information in printed form and distributed to libraries and individuals.

locules cavities in the ovary in which ovules are found.

lysis breaking down or decomposition.

lysogenic bacterium a bacterium harboring certain latent viruses.

lysosome small storage vessels in some cells.

macronutrients the 10 elements found to be essential for the nutritional needs of plants and required in large quantities compared to micronutrients (cf.).

macrosporangium a cell producing macrospores found in the conceptacles of some algae.

map unit a realtive measure of the distance between genes based on the frequency of crossing over between them. One map unit is equal to a 1% frequency of crossing over.

matter anything that occupies space and has weight.

medullary ray vascular ray.

megaspore a spore that develops within a megasporangium and produces a megagametophyte.

megasporophyll a leaflike organ that bears one or more megasporangia.

meiosis (pronounced my-o′sis) reduction division in the germ cell chromosomes at maturation; reduction from the diploid (2*n*) to the haploid (*n*) number of chromosomes.

Mendelism the study of heredity as proposed by Gregor Mendel.

meristem embryonic tissue of plants undergoing cell division and differentiation giving rise to additional tissue.

mesocarp the middle layer of a fruit wall.

mesophyll the serveral layers of cells containing chloroplasts that are found between the two epidermal layers in the leaves.

messenger RNA (mRNA) RNA produced using DNA as a template, the first step in the synthesis of proteins.

metabolism both the constructive and destructive chemical processes (anabolism and catabolism) that occur in a living organism.

metaphase stage in mitosis in which the chromosomes become arranged in the center of the cell (on the equatorial plate).

metaxylem xylem that matures last from procambium.

microbiology the study of microorganisms such as bacteria, protozoans, and viruses.

microgametophyte the plant that develops from a microspore.

micronutrients elements essential for plant nutrition but only in minute quantities; trace elements.

micropyle opening in the ovule through which the pollen tube enters.

microsporangium a sporangium that produces microspores.

microspore a spore that develops into a male gametophyte; a young pollen grain.

microsporophyll a leaf or similar structure that bears a microsporangium.

microtechnique the methods of preparing material for microscope study.

middle lamella a thin layer of material formed between adjacent plant cells during division.

midrib the largest vein in the net-veined leaf.

mitochondria a double-membrane organelle present in the cytoplasm; the site of cellular ATP production.

mitosis a type of cell division in which the daughter cells have the same number of chromosomes as the parent cell; usually differentiated into stages called prophase, metaphase, anaphase, and telophase.

modifying gene a gene having only a slight effect by itself but which modifies the expression of other genes.

mole the gram molecular weight of a molecule.

molecular biology the study of the molecular architecture of the cell, and cellular processes at the molecular level.

molecular weight the relative weight of a molecule when the weight of carbon is taken as 12.

molecule two or more atoms of one or more elements combined to form the smallest single particle of a chemical substance.

monoecious possessing both male and female reproductive organs on the same plant but in separate flowers.

monohybrid offspring resulting from a cross of two individuals involving only one pair of genes.

monosaccharide a simple sugar, soluble in water, e.g., glucose, fructose.

monosomic loss of one chromosome from a set because of nondisjunction during meiosis.

morphogenesis the process that involves the origin, development, and differentiation of plant tissues and organs.

morphology the study of form and structure.

mRNA, see **messenger RNA.**

multicellular the entire organism composed of a few to many cells, usually more than 12.

multiple fruit cluster of several ripened ovaries produced by many flowers on the same inflorescence.

multiple gene gene that acts with another gene to produce a single characteristic.

mutagenic agent a substance that causes mutations.

mutation a characteristic different from any in the ancestry of the individual possessing it and capable of being transmitted to offspring.

mycelium a mass of hyphae forming a network that produces the fungus body.

Mycota the kingdom of organisms containing the viruses, bacteria, and blue-green algae. See **Prokaryotae.**

mycorrhiza a symbiosis of the mycelium of a fungus with the roots of conifers, beeches, and orchids.

n number an indication of the chromosome number without specifying the exact number of chromosomes; indicates the haploid number.

naked flower a flower with neither calyx nor corolla.

neck canal the narrow canal that extends the length of the neck of the archegonium.

nectar sugary secretion produced by the nectaries of the petals.

nectaries glandular hairs at the base of petals that secrete nectar.

neutron an uncharged elementary particle with nearly the mass of a proton.

niche the functional place of the organism in the ecosystem.

nitrogen cycle the series of steps in the fixing and release of nitrogen in the environment.

nitrogen-fixing bacteria the bacteria (*Rhizobium* spp.) living symbiotically with certain legumes or freely in the soil and able to use free nitrogen gas from the air in their metabolism.

node the area of the stem at which the leaves or buds are attached.

nonhistone protein protein associated with DNA that does not contain a high content of basic amino acids upon hydrolysis (cf. **histone**).

nucellus the megasporangium of an ovule located inside the integument and enclosing the female gametophyte.

nucleic acid a long-chain polymer composed of repeating nucleotide subunits.

nucleolar organizing region a site on chromosomes that causes the synthesis of nucleoli.

nucleolus a spherical or ovoid body in the nucleus that disappears during mitosis; usually associated with specific chromosomes.

nucleotides substances made up of sugar phosphates combined with either purine or pyrimidine bases, which are the building blocks of nucleic acids.

nucleus a differentiated portion of the protoplasm within a cell that generally controls the metabolic and reproductive activities of the cell.

nucule the egg-producing structure in the Charophyceae.

nullisomic loss of both chromosome pairs of the set.

oidium a thin-walled hyphal cell formed from the fragmentation of a hypha or from an oidiophore.

oogonium a unicellular female gametangium in which one or more eggs are produced.

oosphere an unusually large, nonmotile female sex cell.

oospore a thick-walled spore that develops from an oosphere usually after fertilization.

opaque-2 a strain of mutant corn that contains a high level of the amino acid lysine (high-lysine corn).

operculum the lid that covers the moss capsule.

operon a genetic unit composed of structural genes and regulator genes.

opposite leaves two leaves (or buds) present at the same node on opposite sides of the stem.

organ a group of tissues combined to perform a specific function.

organ system a group of organs interconnected structurally and functionally, e.g., the flower of a plant.

organelles the divisions or parts of a cell.

organic compound substance that contains carbon, except carbon dioxide and the carbonates that have a mineral origin.

organism any individual plant or animal.

osmosis the diffusion of water from a region of greater water concentration to one of lesser water concentration through a selective membrane.

osmotic pressure the pressure resulting from the resistance offered by a cell to the diffusion of water during osmosis.

ovary the enlarged base of a pistil that bears the ovules.

ovule portion of the ovary that develops into a seed.

oxidation the removal of an electron, usually from hydrogen by the addition of oxygen.

oxygen cycle the circulation of oxygen molecules through plants during photosynthesis and respiration.

pachytene the stage during first prophase of meiosis when crossing-over takes place.

paleomagnetism the direction of the magnetic poles of the earth shift from time to time as shown by the record in ancient rocks by the direction of their magnetized particles.

palisade layer one or two, sometimes more, rows of mesophyll cells found just beneath the epidermis.

palmate palmlike or hand-shaped, with a number of veins radiating from the petiole of the leaf.

palynology the study of pollen, particularly ancient pollen, and its distribution.

parenchyma a simple tissue composed of large thin-walled cells undifferentiated in shape and containing conspicuous intercellular spaces.

parietal on the wall of the ovary.

pars amorpha the structureless central core region of the nucleolus.

parthenospore an unfertilized oosphere that develops a new organism.

passive absorption absorption of any substance by simple diffusion.

pathogen an organism capable of producing a disease in another organism.

pectin a substance found in the middle lamella between cells and acts as a cement.

pedicel the stalk of an individual flower of an inflorescence.

peduncle the stalk that bears a single flower or the main stalk of an inflorescence.

pentose-shunt pathway a metabolic pathway that produces ribose and reduced NADP.

peptide secondary amide linkages between the α-carboxyl and α-amino functions of adjacent amino acids.

perennial a plant that continues to grow for several to many years, producing flowers and seed after the first few years; a woody plant.

perfect flower bearing both stamens and pistil in the same flower.

perianth the calyx and corolla, collectively; covering of an archegonium.

pericarp the wall of a mature ovary.

pericycle a layer of parenchyma cells found between the phloem and the endodermis of roots; the outermost boundary of the stele.

periderm a term that includes cork, cork cambium, and phelloderm.

peridiole a free chamber containing basidiospores, as in the fungi.

perigynous a flower in which the sepals, petals, and stamens encircle the ovary but are detached from it by a cup-shaped receptacle that rises above the base of the ovary.

perithecium a spherical or flask-shaped ascocarp with an apical pore.

permeation the passage of a substance through the cell membrane.

petal one of the floral parts just inside the calyx, usually white or colored.

petiole a distinct stalk supporting the blade of a leaf.

pH a logarithmic measure of the hydrogen-ion concentration of a solution.

phage another name for a bacteriophage.

phanerogams a name sometimes applied to the seed-producing plants.

phelloderm the secondary tissue formed by the cork cambium at the inner surface.

phellogen the cork cambium.

phenotype visible hereditary characteristics; the appearance of the individual.

phloem a complex vascular tissue of a plant; the product of the vascular cambium; chief function is to conduct food down the stem.

photomorphogenesis the response of tissue morphology to light.

photon a quantum of electromagnetic radiation.

photoperiodism growth and development responses of plants to different periods of exposure to light.

photosynthesis the manufacture of carbohydrates from carbon dioxide and water in the presence of chlorophyll, with light as the energy source.

photosynthetic capable of carrying on photosynthesis; manufactures food in the presence of light.

phototropism growth movement induced by light stimulus.

phycocyanin a blue pigment present in some of the algae.

phycoerythrin a red pigment found in some of the algae.

physiology the study of the life processes and functions of the entire plant, as well as of the individual organs and tissues.

phytochrome a bluish protein that functions as a photoreceptor system in leaves.

phytohormone any plant hormone.

phytophagous organisms that feed on plant parts.

pileus the umbrella-like cap of the mushrooms and allies.

pinna a primary division of a pinnate leaf; a leaflet.

pinnate divided or arranged like a feather.

pinnately bicompound the primary leaflets are compound.

pinnately compound a series of leaflets arranged on opposite sides of the leaf rachis.

pistil ovule-producing part of a flower consisting of a single carpel (simple pistil); composed of a stigma, style, and ovary.

pistillate flower having pistils but no stamens.

pith parenchyma tissue located in the center of a stem surrounded by xylem.

pith ray the same as a medullary ray.

placenta the organ of the ovary that bears ovules.

plankton free-floating aquatic organisms.

Plantae multicellular or organisms with walled and frequently vacuolate nucleate cells and almost always with photosynthetic pigment in plastids.

plaques open spaces in a bacterial colony where lysis of bacteria has taken place through viral action.

plasma membrane the interfacial membrane of a cell.

plasmodium a multinucleated mass of naked protoplasm; the vegetative stage in the life cycle of the slime mold.

plasmolysis shrinking of the protoplasm from the cell walls as a result of loss of water through osmosis.

plastid organelle in the cytoplasm of the cells of most plants (Plantae), some Mycota, and rarely Animalia, usually containing chlorophyll.

plate tectonics the study of the geological plates that make up the surface of the earth.

pleiotrophic effect plural phenotypic effects of a single gene.

pleuropneumonia-like organisms (PPLO) a small viruslike organism, actually a small bacterium.

plumule the bud of the embryo, composed of apical meristem, that will develop into the shoot.

plurilocular with more than one locule.

polar nucleus one of the nuclei of the embryo sac that unites with the other polar nucleus to form the primary endosperm nucleus.

poles the opposite ends of a cell during mitosis between which the spindle appears.

pollen a microspore and young male gametophyte of seed plants.

pollen grain pollen.

pollen sac microsporangium, containing pollen, located within the anther.

pollen tube tube that contains the sperm nucleus and extends down through the style.

pollination transfer of pollen from stamen to pistil.

polymorphic more than one form or habitus in a single species.

polypeptide a compound that is composed of amino acids but has a lower molecular weight than a protein.

polyploidy some multiple of the haploid number of chromosomes other than the diploid.

polysaccharide the combination of several monosaccharide units, e.g., cellulose, starch.

polysomic a diploid individual with one chromosome represented three or more times.

pome fruit in which the seed cavities are surrounded by fleshy floral tube and receptacle.

population a breeding population is a unit of reproduction held together by bonds of fertilization and hereditary descent with a continuity in space and time; the individuals are variable within the population; populations differ one from another and may or may not be connected in space.

PPLO (see **pleuropneumonia-like organism**).

prickles sharp, pointed, weak outgrowths of the stem not associated with vascular tissue.

primary permanent tissue permanent tissue formed by the apical meristem.

primary root the main root.

primary structure the sequence of amino acids in a protein.

procarp a stage in the development of the carposporophyte of certain algae.

proembryo immature embryo.

Prokaryotae kingdom of organisms containing the viruses, bacteria, and blue-green algae.

prokaryotic cells in which the nucleus is not delimited by a membrane.

promeristem a region in which growth takes place by increase in number of cells rather than by increased cell size.

prophase the stage in mitosis in which the chromosomes become distinct and organized.

protein organic compound composed of carbon, hydrogen, oxygen, nitrogen, and usually sulfur; composed of amino acid chains.

prothallium (see prothallus).

prothallus the gametophyte stage of ferns and other plants resulting from the germination of the spore; bearing sex organs; a prothallium.

proton a positively charged particle in the atomic nucleus.

protonema the threadlike growth resembling an alga from which the moss gametophyte develops.

protoplasm a term applied to the entire contents of a living cell.

protostele a stele, in which pith is absent, formed by xylem surrounded by phloem.

protoxylem primary xylem or the first xylem to mature at any level of the root or stem.

protozoa unicellular organisms that lack chlorophyll and are usually considered to be animals.

pycnium in rusts, a pocket at the surface of the host in which pycniospores are produced.

quantum very small increment of energy

rachis a part of the leaf or flower to which the leaflets are attached; main axis of a compound leaf; petiole of a fern frond.

radiation of species the adaptation of many closely related species to a variety of habitats so that they have become reproductively isolated, e.g., the species of *Iris*.

radicle embryonic root.

raphe a ridge on a seed leading away from the hilum or scar of the funiculus.

realm a geographical division of the earth based on the faunal components.

receptacle the expanded end of a stem to which the floral parts are attached.

recessive one of contrasting characters not apparent in the presence of a dominant gene.

record any durable packet of information used in communication of information.

redox potential a measure of the ability of a substance to either receive or give electrons.

reduction the gaining of an electron.

refugia isolated areas harboring plants and animals not exterminated by a previous glaciation or simular factor.

region of elongation that portion of a growing root or stem in which cells elongate.

region of maturation in a root, the same as root hair region.

repressor protein a protein produced by the i gene of the lactose operon; combines specifically with the operator gene.

reproduction the process of duplication of individuals.

research in science, the search for unknown facts by observation and experimentation.

respiration the production of chemical energy from the organic components of living protoplasm within organelles.

rhizoids roots of nonvascular plants.

rhizome a horizontal underground stem.

rhizomorphs a mass of twisted hyphae behaving as an organized unit such as is found in the mushroom.

ribonucleic acid (RNA) one of the two nucleic acids composed of nucleotides formed of a nitrogenous base, the sugar ribose, and phosphoric acid.

ribosome small bodies composed of ribonucleic acid and protein, the site of protein synthesis.

RNA (*see* ribonucleic acid).

RNA polymerase the enzyme that catalyzes the formation of mRNA using DNA as a template.

root a primary organ of vascular plants the functions of which are anchorage and absorption.

root cap the group of protective epidermal cells surrounding the root tip.

root hair slender outgrowth of the epidermal cells in the zone of differentiation of roots.

root hair region the zone of differentiation in the root tip where the epidermal cells become specialized as absorbing cells; area in which root hairs develop.

root pressure the pressure caused by the movement of water into a root.

root tip the meristematic apex of a root in which mitosis is occurring.

runner a long, slender, nearly leafless horizontal stem.

salt combination of a positively charged ion with a negatively charged ion, neither ion being hydrogen or hydroxyl.

samara the particular type of seed characteristic of maple trees; winged seed pair.

sap water and minerals or water and sugar flowing in the vascular tissue of a plant.

saprophyte an organism that obtains its food from nonliving organic matter.

scalariform conjugation conjugation between cells of two different filaments giving the appearance of a ladder.

scales small appressed leaves or bracts, usually vestigial.

science knowledge that has been accumulated by the use of the scientific method and organized under the direction of a scientist.

scientific method the system used to investigate a problem or answer a question by gathering data and information, coding this for processing by forming a hypothesis, repeating to gather more data or confirm the information, and presenting the results for use in other investigations.

scientific name (see **specific name**).

scientific notation a method for expressing a given quantity or number by showing the digits as multiples of 10.

scion the shoot or stem that is detached and then attached to the stock in grafting.

sclerenchyma supporting tissue of dead cells, the walls of which are uniformly thickened with lignin.

secondary growth an increase in the amount of vascular tissue in a stem or root.

secondary permanent tissue permanent tissue formed by the cambium and other meristematic tissue.

secondary root a root lateral to the primary root.

secondary structure a structure assumed by protein molecules that may be either a helix or a sheet.

sedimentation velocity coefficient a measurement of the size and density of a particle by use of ultracentrifugation.

seed embryonic sporophyte in the dormant condition, the endosperm or food storage in the embryo, and the seed coat.

seed coat the modified integument of the seed, usually composed of the integument and the nucellus.

selectively permeable said of a membrane when it is able to allow the passage of some substance but not others, often regardless of molecular size.

self-pollination transfer of pollen from the anther to the stigma of the same flower or another flower on the same plant.

semiconservative replication the mode of replication of the DNA double helix that produces new DNA molecules consisting of one parental strand and one new one.

semipermeable permeable to some substances usually of small molecular size but not to others usually of large molecular size.

senescence a period during which growth ceases and the death phase begins.

sepal the outermost floral part; a division of the calyx.

septate divided into cells; having crosswalls in reference to hyphae.

sere a stage in succession before reaching a climax.

sexual a type of reproduction involving the fusion of gametes or sex cells and their nuclei.

shoot the stem and the leaves of a plant, especially those newly developed.

shrub woody plant that usually branches profusely at or near the ground level.

sieve tube a series of phloem cells with perforated end walls arranged end to end to form a tube.

simple fruit the result of the maturation of the ovary of a single pistil.

simple pistil a pistil with only one carpel.

simple tissue a tissue in which the cells are essentially alike in structure and function.

sol the fluid state of a colloid.

soredium a reproductive granule of a lichen composed of both the algal and fungal elements.

sorus a cluster of sporangia on a fern frond.

source document any representation of information that becomes input to a data processing operation.

species (pl. **species**) kind of organisms composed of interbreeding individuals with more characteristics in common than with any other organism; genetically related organisms.

specific name the latinized, or Latin, name of an organism composed of the generic and the specific names; the name of the species.

sperm male gamete or sex cell.

spermatium a nonflagellated male gamete or gametangia in the red algae and some fungi.

spermatophytes the seed-producing plants.

spindle threadlike condensation of the cytoplasm formed during mitosis and to which the chromosomes become attached.

spine modified leaf.

spirillum corkscrew-shaped bacterium.

spongy layer the layer of cells of the mesophyll of a leaf that contains large air spaces.

sporangiophore a structure that bears one or more sporangia.

sporangium a capsule or structure in which spores are produced.

spore an asexual reproductive structure, usually unicellular, that is capable of giving rise to a new plant.

sporocarp a special frond of a fern containing only the sori.

sporogenesis the formation of spores (see *sporulation*).

sporophyll a spore-bearing leaf.

sporophyte an asexual generation; spore-producing generation; diploid (2*n*) generation.

sporulation the formation of spores.

stamen pollen-producing organ of a flower; usually composed of an anther and a filament.

staminate cone microsporophylls of conifers.

staminate flower flower with stamens but no pistil.

starch sheath the inner layer of cortex when it contains large amounts of starch.

statospores resting spores characteristic of diatoms.

stele the conducting region that occupies the center of the root, stem, and leaves and is composed of xylem, phloem, pericycle, endodermis, and pith.

sterigma the short stalk that subtends the basidiospore, the conidium, or the sporangium.

sterol complex ring compound that is chemically related to fats but resists saponification, e.g., cholesterol.

stigma the part of the pistil that receives the pollen, usually located at its apex.

stipe a stalk such as that in mushrooms or algae.

stipule small leafy structure at the base of the petiole of a leaf.

stock in grafting, the basal portion of the stem to which another piece, the scion, is attached; in genetics, a race or group of genetically similar organisms.

stolon a horizontal rooting shoot in higher plants (also called a runner); in fungi, a horizontal hypha.

stoma (pl. *stomas, stomata*) opening in the epidermis surrounded by two guard cells.

stone cell sclerenchyma cell that is similar to a fiber but much shorter; form the outer stony parts of the pits of cherries and the gritty portion of pear fruit.

strobilus (pl. *strobili*) a conelike collection of sporophylls borne on a stem or axis.

structural gene a gene that determines the sequence of amino acids in a protein.

style the stalklike portion of the pistil that bears the stigma.

suberin waxlike waterproof substance present in the walls of cork cells.

succession the passing through seres toward a climax.

swarm spores zoospores freely swimming after release from the cell chamber.

symbiont an organism forming a part of a symbiotic relationship.

symbiosis two kinds of organisms living together with some mutual benefit.

sympatric population of two species with overlapping ranges.

synapsis the pairing of homologous chromosomes during first prophase of meiosis.

syncytium a tissue or structure containing many nuclei but not divided into separate cells.

synecology the study of groups of organisms that are associated together as a unit.

synergid one of the two functionless nuclei of the embryo sac.

syngamy the fusion of the sperm and egg nuclei in the cytoplasm in sexual fertilization.

system a combination of parts forming a complex whole.

systematic a study of classification, identification, evolution, genetics, ecology, and distribution of plants.

taproot root in which the primary root is enlarged and the secondary roots are few, as in many vegetables, i.e., biennials.

taxon a group of organisms named and assigned to a category; e.g., Rosaceae is a taxon of plants of the category family.

taxonomy the classification and identification of plants; the cataloging of the species.

teliospore a thick-walled resting spore found in the rusts and smuts in which nuclear fusion occurs.

telium the group of binucleated cells that produce teliospores.

telophase stage in mitosis in which the chromosomes are regrouped in the daughter nuclei and the nuclear membrane reappears.

tendrils may be either modified stems, leaves, or leaflets, or stipules that are capable of twining around some object.

terminal bud the bud located at the very tip or apex of a stem.

tertiary structure a complex configuration formed by protein molecules usually related to their function.

testcross the cross of an F_1 individual with the homozygous recessive parent.

tetrad a group of four cells, usually spores, produced by two divisions of a spore mother cell; also a group of four chromosomes formed at meiosis.

tetraspore one of the four spores in a tetrasporangium that will develop into the gametophyte.

tetrasporophyte a plant that resembles the gametophyte in the red algae and produces tetraspores in the tetrasporangium.

thallus a simple plant body lacking true roots, stems, and leaves; characteristic of algae, fungi, liverworts, and so on.

theory the result of the testing and verification of hypotheses by the addition of more data and information; a concept.

thermodynamics the study of energy changes in either chemical or physical systems.

thigmomorphogenesis the response of plant growth and development to mechanical stimulation.

thorn a modified stem forming a sharp-pointed woody structure of various sizes and shapes.

thylakoid disc a part of chloroplast.

tissue a group of similar cells having the same function.

tonoplast the membrane of a plant vacuole.

TMV tobacco mosaic virus.

toxin poisonous secretion of a living system.

trace elements elements necessary for plant growth but used in extremely small amounts; micronutrients.

tracheid the fundamental type of xylem cell that is elongate with tapering ends and, when mature, is nonliving.

transfer RNA (tRNA) RNA molecules that combine with amino acids and sequence them on the mRNA template.

transformation transforming the genotype of an organism.

transforming principle DNA

translocation the movement of water, minerals, and metabolites from one place to another within the plant.

transpiration the loss of water vapor from a leaf by evaporation and diffusion.

tree a woody plant whose main stem rises above the ground before it branches.

trichogyne (pronounced trik'o-jĭn) the distal receptive end of the ascogonium.

trichome a hair or bristle; a filament of globose cells.

trifoliate three leaflets attached to the petiole of the leaf.

tRNA (see **transfer RNA**)

trumpet hyphae filaments with large cells inflated at ends that serve in a vascular capacity in some brown algae.

tube nucleus one of the nuclei of the pollen grains not involved in sexual fusion.

tuber a thick, fleshy stem modified for food storage.

turgid a cell is turgid when the pressure resulting from osmosis is great enough to exert a force against the inner surface of the cell membrane, as in a hypotonic solution; distention as a result of turgor pressure.

turgor pressure the pressure developed as a result of diffusion of water into a living cell; pressure exerted against the cell wall.

ultrastructure cell components smaller than the usual range of the light microscope, or from about 2000 to 2 Å.

unicellular the entire organism composed of one cell.

unilocular with a single chamber.

unit membrane concept of the plasma membrane.

uredium a short filament of binucleated cells producing uredospores.

uredospore a binucleated spore produced by the uredium.

vacuoles organelles of the cell that serve as storage and communication channels.

valence the degree of combining power of an element or a radicle.

vascular bundle a group of conducting tissues (xylem and phloem) found in a plant.

vascular cambium the meristematic tissue that produces phloem toward the outside and xylem nearer the inside; responsible for the lateral growth of stems and roots.

vascular ray thin radial layers of xylem and phloem extending from the pith to the cortex.

vascular system the arrangement of xylem and phloem cells, the fluid-conducting tubes of the higher plants.

vascular tissue conducting tissue of stems, roots, and leaves; xylem and phloem.

vasculum a metal can in the shape of a flattened cylinder with a side door and a shoulder strap, used to store plants in the field that are gathered for pressing later in the laboratory.

vegetative body the part of the plant concerned with all functions except reproduction.

veins vascular bundles in the leaf that can be seen as ridges on the surface.

venter the enlarged end of an archegonium within which the egg develops.

vernalin a plant hormone, not yet isolated, which causes vernalization.

vernalization the chilling of a flowering plant and thereby the slowing or stopping of the growth process before the plant will produce leaves or flowers or before seeds will germinate.

vernation the arrangement of the leaves within the bud.

vessel a conducting tube in the xylem.

vine a climbing stem.

virion the mature virus outside a cell.

virus a noncellular organism-like unit composed of a nucleic acid core and a protein shell functioning only within a host cell.

weed an undesirable plant, no particular kind.

whorls three or more leaves (or buds) located at the same node.

woody stem a stem that is usually thick and composed chiefly of xylem or wood.

xanthophyll yellow pigments present in plant cells and sap.

xanthoxin a plant-growth inhibitor similar to abscisic acid.

xylem complex vascular tissue of a plant; the product of the vascular cambium; wood; chief function is to conduct water up the stem.

zeatin a natural cytokinin.

zone of differentiation the area of the stem or root just behind the zone of elongation in which the cells are developing into the primary permanent tissues.

zone of elongation the area of the stem or root just behind the apical meristem in which the cells are undergoing some modification, particularly increase in size.

zoogametes gametes of algae capable of free swimming.

zoology the study of some or of all the organisms belonging to the kingdom Animalia.

zoospore a motile asexual spore.

zygospore a spore formed by the fusion of two gametangia that enters a resting stage.

zygote a fertilized gamete.

zygotene the stage during first prophase of meiosis when synapsis takes place.

Index

A

Å (Ångström), 28
Abscisic acid, 270, 301, 310, 311, 317
Abscisin II, 310
Abscission, 303, 316
   layer, 249, 318
   zone, 249, 316
Absorption, 253, 257, 258
   passive, 258
Acacia, 380
Acer negundo, 311
Acer saccharinum, 174, 442
Aceraceae, 442
Acetabularia, 88
Acetyl CoA (coenzyme A), 191, 195
Acetylcholine, 313
Achene, 291, 293
Achras zapota, 441
Acid, 37, 38
   fatty, 38, 39
   fumaric, 195
   indoleacetic, 301
   indolebutyric, 311
   indolepyruvic, 302
   2-ketoglutaric, 195
   lactic, 194
   lunalaric, 310
   malic, 195
   naphthaleneacetic, 303, 304
   nucleic, 38, 40, 41
   palmitic, 38
   3-phosphoglyceric, 207
   rain, 412
   ribonucleic, 40, 41
   ribulose-1,5-diphosphate, 207
   saturated fatty, 39
   succinic, 195
   tricarboxylic cycle, 194
   unsaturated fatty, 39
Acrasiomycetes, 105, 106
Actinomycetes, 78
Actinomycin D, 308
Active absorption, 258
Active secretion, 49

Active transport, 57, 258, 259, 260, 267
Adansonia digitata, 167
Adder's tongue, 446
Adenine, 42, 341, 345
Adenosine diphosphate (ADP), 190
Adenosine triphosphate (ATP), 190
Adnate, 283
Aecidium, 474
Aeciospore, 474
Aerobes, 77
Aesculus hippocastanum, 442
Agar
   plate, 74
   slant, 75
Agaricaceae, 119
Agaricus campestris, 117, 479
   life cycle, 478
Agavaceae, 187
Agave americana, 186, 445
Agave sisalana, 445
Age of earth, 366
Agriculture, primitive, 419
Ailanthus altissima, 442
Air pollution, 412
Aizoaceae, 438
Akinetes, 449
Alder, 164, 437
Alexander the Great, 17
Alfalfa, 440
Algae, 1
   biological significance, 100
   blue-green, 1, 69
   branching filaments, 91
   brown, 96
   classification, 435
   filamentous, 91
   flagellated, 96
   freshwater, 86
   golden brown, 93
   red, 98
   yellow-green, 93
Alismataceae, 178, 444
Alismatales, 178, 444
Allele, 325

Allium
cepa, 445
porrum, 445
sativum, 446
Allopatric, 361
Allopolyploid, 337
Allspice, 441
Almond, 440
Alnus serrulata, 437
Alternaria sp., 120
Alternation of generations, 87
Althaea rosea, 438
Alyssum saxatile, 439
Amaranth family, 438
Amaranthaceae, 438
Amaranthus albus, 438
Ambrosia trifida, 444
Ameboid, 104
American holly, 441
Amino acid, 39, 40
Amino group, 39
Amylase, 308
Amyloplasts, 59
Amylum stars, 93
Anabolism, 190
Anacardiaceae, 442
Anacardium occidentale, 442
Anaerobe, 77
Anaerobic respiration, 192
Analogy, 26
Analysis, 24
Anaphase, 298-300
Anaxagoras, 354
Anaximander, 354
Anaximenes, 354
Anderson, Edgar S., 8
Androecium, 281
Anemone, 436
Anemone coronaria, 436
Anethum graveolens, 442
Angiospermae, 161
classification, 436
Ångström unit (Å), 28
Animal
distinct from plants, 2
relationships, 393
Animalia, 1, 5
Anise, 175, 442
Annona, 309
Annonaceae, 436
Annual, 213, 339
Annual rings, 239
Annulus, 492
Anther, 281
Antheridia, 88, 98
Antheridiophore, 482
Antherozoid, 452
Anthoceropsida, 130
Anthocerus sp., 129

Anthocyanin, 251, 316, 318
Anthophyta, 160
Antibiotics, 112
Antibody, 80
Anticodon loop, 347
Antigen, 80
Antipathogens, 108
Antipodals, 508
Antirrhinum majus, 443
Antitoxin, 80
Apetalous, 283
Apical dominance, 302
Apical meristem, 223
Apium graveolens var. *dulce*, 442
Aplanospore, 108
Apocynaceae, 443
Apocynum androsaemifolium, 443
Apocynum cannabinum, 443
Apomixis, 509
Apothecium, 110
Apple, 440
May, 163, 437
thorn, 177
Apricot, 171, 440
Aquifoliaceae, 441
Aquilegia vulgaris, 437
Arabs, 18
Araceae, 180, 445
Arachis hypogaea, 440
Arales, 180, 445
Araucaria, 377
columnaris, 146
Archegoniophore, 482
Archegonium, 447
Arctium minus, 444
Areca cathecu, 445
Arecales, 180, 445
Argeomone platyceras, 165, 437
Arisaema triphyllum, 184, 445
Aristolochiales, 162, 436
Aristotle, 16, 18, 355
Armoracia lapathifolia, 439
Arrow grass, 444
family, 444
Arrowhead, 444
family, 178
Arthrophyta, 139
Artichoke, 178, 444
Artocarpus altilis, 438
Arum family, 180, 445
Asci, 109
Asclepias syriaca, 175, 443
Ascocarps, 110
Ascogonium, 110
Ascomycota, 109
Ascospore, 109
Ascus mother cell, 473
Asepalous, 283
Ash, 175, 443

Asimina triloba, 436
Asparagus, 446
Asparagus officinalis var. *altilis*, 446
Aspen, 170
  quaking, 170, 437
Aspergillus sp., 110, 112
Aster, 178, 444
  family, 178, 444
Aster laevis, 444
Asterales, 177, 178, 444
Asthma, 109
Athlete's foot, 103, 120
Atmosphere, 385
Atom, 33, 34
Atomic nucleus, 34
  structure, 34
  weight, 34, 35
Atomic power, 27
ATP, 59, 195
Atriplex polycarpa, 438
Atropa belladonna, 443
Australia grass tree, 185
Autecology, 396
Autogamy, 96
Autopolyploid, 337
Autotrophs, 1
Autumn crocus, 446
Auxin, 257, 270, 301, 302, 317
  effect, 304-306
  synthesis, 304
Avena fatua, 444
Avena sativa, 445
Avicennia sp., 386
  nitida, 440
Avirulent, 341
Avocado, 436
Axial placentation, 285
Axil, 247
Axillary, 242
Azalea, 439
Azotobacter, 390

B

Babcock, E. B., 8
Babylonians, 353
Bacillariophyceae, 93, 95
Bacillus, 75
Backcross, 326
Bacon, Roger, 18
Bacteria, 1, 69, 73
  beneficial, 81
  classification, 68
  conjugation, 76
  culture, 74
  gliding, 78
  iron, 76
  lysogenic, 72
  nitrogen-fixing, 77

Bacteria—cont'd
  phototrophic, 77
  sulfur, 76
  types, 74
Bacteriochlorophylls, 204
Bacteriophage, 71, 341, 343
Bakanae disease, 305
Balsam family, 442
Balsaminaceae, 442
Bamboo, 179, 180, 445
Bambusa bambos, 445
Bambusa vulgaris, 180
Banana, 180, 182, 445
  family, 445
Banyan, 167, 168
Baobab, 165, 167
Barberry, 118
  family, 437
  Japanese, 437
Bark, 238
Barley, 445
  grain, 308
Base, 37, 38
Basidia, 112, 116
Basidiocarp, 116
Basidiomycete, 117
Basidiomycota, 112
Basidiospores, 112, 474
Basil, 443
Basswood, 167, 438
Bayberry, 437
Bean, 172
  kidney, 440
  lima, 440
  Mexican jumping, 173
  string, 315
Beech, 164, 437
  family, 164, 437
Beets, 165, 438
Beggar-tick, 444
Beggarweed, 440
Begonia, 170
  family, 439
  tuberous, 439
Begonia tuberhybrida, 439
Begoniaceae, 170, 439
Belladonna, 177, 443
Bellflower family, 177, 444
Bells of Ireland, 443
Benson-Calvin pathway, 208
Berberidaceae, 163, 164, 437
Berberis vulgaris, 437
Bermuda grass, 518, 445
Berry, 289
Beta vulgaris, 438
  var. *cicla*, 438
Betula papyrifera, 437
Betulaceae, 164, 437

Bidens tripartita, 444
Biennial, 213
Bindweed, 438
Binominal system, 8
Biochemistry, 27
Biological Abstracts, 23
Biological clock, 312
Bioluminescence, 204
Biomes, 402
　plant, 402
Biota, 373, 401
Biotic communities
　classification, 401
Birch, 164
　family, 164, 437
Bird-of-paradise, African, 180, 181
Bisexual, 326, 330
Bitterroot, 438
Bittersweet, 441
Black nightshade, 443
Black pepper, 436
Blackberry, 171, 440
　blackcap, 440
Blade, 247
Bleeding heart, 164, 437
Bloodroot, 164, 437
Blue grama, 179
Blue-green algae, 1
　description, 81
Blue-light receptor, 312, 313
Bluebell, 444
Blueberry, 170, 439
Bolting, 312
Bombacaceae, 167, 438
Bombax family, 438
Boogum tree, 167, 439
Borage family, 177, 443
Boraginaceae, 177, 443
Boron, 254, 257
Borrelia sp., 80
Boston ivy, 173, 441
Boswellia carteri, 442
Botanical gardens, 25
Botany, 2
　division, 14
　history, 16
　study of, 14
Bouteloua gracilis, 179
Bract, 495
Brassica
　campestris, 439
　caulorapa, 439
　napobrassica, 439
　oleracea
　　var. *acephala*, 439
　　var. *botrytis*, 439
　　var. *capitata*, 439
　　var. *gemmifera*, 439

Brassica—cont'd
　oleracea—cont'd
　　var. *italica*, 439
　rapa, 439
Brassicaceae, 439
Breadfruit, 438
Bridal wreath, 440
Brisbane box tree, 213
Broccoli, 170, 439
Bromeliaceae, 180, 445
Bromeliales, 180, 445
Brownian movement, 45
Brunfels, Otto, 8
Brussels sprouts, 170, 439
Bryophyta, 130
Bryophytes, 127, 253
　classification, 436
Bryopsis sp., 91
Buckeye, 442
Buckthorn family, 441
Buckwheat, 438
　family, 438
Bud, 242
　adventitious, 242
　scale, 242
　scar, 243
　terminal, 242
Buffalo berry, 441
Buffers, 37
Bulb, 243, 245
Bulbil, 93
Burning, controlled, 411
Bursera family, 442
Burseraceae, 442
Burdock, 444
Buttercup, 162, 437

C

Cabbage, 170, 439
Cacao, 438
Cactaceae, 165, 438
Cactus
　family, 438
　organ-pipe, 166, 438
　prickly pear, 166
Calcium, 254
　absorption, 256
Calcium pectate, 256
Callus tissue, 302
　wound, 303
Caltha palustris, 163, 437
Caltrop family, 442
Calvin, Melvin, 202
Calvin-Benson pathway, 210
Calyx, 160
Cambium, 227, 264, 297
Cambrian, 374
Camellia, 165, 438

Camellia japonica, 438
Campanula medium, 444
Campanula rotundifolia, 444
Campanulaceae, 177, 444
Campanulales, 177, 444
Camphor, 161, 436
Candelilla, 173, 441
Canlerpa sp., 91
Cannabis sativa, 438
Cantaloupe, 439
Canterbury bells, 177, 444
Capillitium, 105
Capparales, 170, 439
Caprifoliaceae, 178, 444
Capsicum frutescens var. *grossum*, 443
Capsid, 70
Capsomere, 70
Capsule, 483
Caraway, 175, 442
Carbohydrate, 38
  metabolism, 272
Carbon, 253
  cycle, 387
Carbon dioxide
  cycle, 387
  in environment, 387
  fixation, 207
Carboniferous, 375
Carboxyl group, 40
Carex morrowii, 445
Carica papaya, 169, 439
Caricaceae, 170, 439
Carnegiea gigantea, 166, 438
Carotenes, 88, 205
Carotenoid, 59, 205, 316
Carpel, 282
Carpellate cone, 499
Carpogonia, 98
Carpospores, 99
Carposporophyte, 98-99
Carrier hypothesis, 260
Carrot, 442
  family, 175, 442
  wild, 175
Carson, Rachel, 29
Carum carvi, 442
Carya cordiformis, 442
Carya illinoensis, 442
Caryophyllaceae, 437
Caryophyllales, 165, 437
Cascara buckthorn, 441
Cashew, 442
  family, 173, 442
Cassava, 173, 441
Cassia senna, 440
Castanea dentata, 437
Castor bean, 173
Castor-oil plant, 441
Catabolism, 190

Catalyst, 36
Category (categories), 5, 6
Catkins, 170
Cattail, 445
  family, 180, 445
Cattleya labiata, 446
Cauliflower, 170, 439
Ceanothus americanus, 441
Ceiba pentandra, 438
Ceiba tree, 438
Celastraceae, 441
Celastrales, 441
Celastrus scandens, 441
Celery, 175, 442
Cell, 3, 52, 55
  companion, 235, 267
  division, 297, 299, 321
  epidermal, 269
  guard, 267, 268, 270
  holdfast, 454
  lip, 492
  membrane, 53
  method, 54
  plate, 300
  prokaryotic, 5
  sex, 297
  size, 53, 54
  stone, 222
  theory, 53
  type, 54, 215
  wall, 4, 52, 55
    primary, 55
    secondary, 55
Cellulase, 305
Cellulose, 55
  synthesis, 318
Cenozoic era, 376
Central placenta, 285
Centrioles, 64
Centromere, 300
Century plant, 186, 187, 201, 445
Cereus giganteus, 438
Cereus thurberi, 438
Cercis canadensis, 440
Cesalpino, Andrea, 17
Chalazal, 505
Chamaesiphon, 84
Chaparral, 405
Chara, 88
  life cycle, 456, 457
  sp., 93
Characeae, 93
Charales, 88, 93
Chard, 165
Chemical environment, 389
Chemical reactions, 36
Chemistry, 33
Chemosynthetic, 76
Chenopodiaceae, 165, 438

Chenopodium album, 438
Cherry, 171
  sweet, 440
Chestnut, 164
  American, 437
    horse, 173, 442
      family, 442
    water, 172, 441
      family, 441
Chiasma, 322, 323
Chicle, 441
Chicory, 178, 444
Chilling, 318
Chinchona, 444
Chinese, 353
  date, 441
  lantern plant, 443
Chitin, 56
Chlamydomonas, 88
  life cycle, 450, 451
  sp., 90
Chlamydospore, 109
Chlorella, 88
  sp., 89, 90
Chlorenchyma, 221
Chlorine, 254
Chlorococcales, 88
Chlorophyll, 5, 59, 201, 204, 205
Chlorophyta, 88
Chlorophytes, 91
Chloroplast, 52, 53, 60, 201, 202, 269
Chlorosis, 254
Chocolate tree, 438
Cholera, 72
Cholla, 438
Chondrus, 99
Chromatid, 299, 300
Chromatin, 52, 64, 65
Chromatograph, 24
Chromatophores, 98
Chromoplast, 59
Chromosome, 64, 299, 321, 324, 341, 421
  mapping, 333
  variation, 337
Chroococcus, 84
Chrysanthemum, 315, 444
Chrysanthemum morrifolium, 444
Chrysophyceae, 93, 94
Chrysophyta, 93
Chytridiomycetes, 106
Cichorium endivia, 444
Cichorium intybus, 444
Cicuta douglasii, 426, 442
Cinchona, 178
Cinchona ledgeriana, 444
Cinnamomum camphora, 436
Cinnamomum zeylanicum, 436
Cinnamon, 161, 436
Circinate vernation, 141, 143

Circulation in plants, 263
Cirsium arvense, 444
Citric acid, 191, 195
  cycle, 194, 195
Citrullus vulgaris, 439
Citrus limon, 442
Citrus paradisi, 442
Citrus sinensis, 442
Citrus tree, 173
Civilization, 418
Cladophora, 88
Cladophorales, 88
Clamp connections, 117
Clamp connectors, 117
Classification, 5
  former, 7
Clausen, Keck, and Hiesey, 8
Claviceps purpurea, 112
Climax, 410
Clintonia uniflora, 446
Clone, 276
Closed system, 22
Clostridium, 390
Clove, 441
Clover, 172
  white, 329, 440
Coalescent, 283
Coca, 175
  cocaine plant, 442
  family, 442
Cocaine, 175
Coccus, 513
Cocklebur, 444
Cockspur, 440
Cocoa plant, 438
Coconut, 180, 183
Cocos nucifera, 183, 445
Codon, 344
Coefficient of coincidence, 334
Coelenterata, 4
Coenocyte, 91
Coenzyme, 43
  A, 191
  Q, 196
Coevolution, 363
  plants, 363
  animals, 363
Cofactor, 43
Coffea arabica, 444
Coffee, 178, 444
Cohesion, 264
  force, 270
  tension theory, 264
Cola, 438
Cola acuminata, 438
Colchicine, 340, 341
Colchicum autumnale, 340, 446
Cold hardiness, 311
Coleoptile, 218, 219

Collenchyma, 221
Colloid, 44
Colony, 278
 formation, 76
 motile, 90
Columbine, 437
Columella, 483
Commelinaceae, 179, 444
Commelinales, 179, 444
Commensals, 107, 399
Commercial regulators, 311, 312
Commiphora myrrha, 442
Common fat, 39
Common water lily, 436
Community (communities), 396, 410
 biotic, 401
Competition
 different kind, 391
 plants of same kind, 391
Complete flower, 279
Compositae, 444
Compound
 leaf, 248
 organic, 38
 pistil, 285
Comptonia peregrina, 437
Computers, 21
Conceptacles, 460
Cones, 152, 156, 495
Conidiophores, 109
Conidium, 109, 472
Coniferophyta, 150
Coniferous forest, 406
Conifers, 150
 classification, 436
Conium maculatum, 174, 442
Conjugation, 91
Conservation, plant, 413
Constitutive system, 350
Continental drift, 371, 372
 recent, 373
Controlled system, 22
Controlling genes, 348
Convallaria majalis, 446
Converter gene, 330
Convolvulaceae, 443
Copper, 254, 257
Corchorus capsularis, 438
Core, 289
Cork, 221
 cambium, 234
 tissue, 222
Corkwood family, 437
Corm, 243, 244
Corn, 445
 high-lysine, 424
Cornaceae, 172, 441
Cornales, 172, 441
Cornus florida, 172, 441

Corolla, 281
Cortex, 225
Corylus aveliana, 437
Cosmic theory, 33
Cosmopolitan, 360
Cosmos, yellow, 444
Cosmos sulphureus, 444
Cotton, 167, 315, 423, 438
Cottonwood, 170, 439
Cotyledons, 215
Covalent bond, 36
Cowslip, 164, 437
Crabgrass, 445
Cranberry, 170, 439
Crataegus crus-galli, 440
Crataegus intricata var. *straminea*, 440
Creator, 18
Creosote bush, 442
Cress seed, 311
Cretaceous period, 376
Cristae, 59
Crocus, 446
Crocus susianus, 446
Cro-Magnon, 17
Cross-pollination, 285
Crossing over, 323, 332, 333
Crowfoot, 162
 family, 436
Crown gall, 393
Cryptomonads, 96
Crystalloid, 44
Cucumber, 170, 315, 349
 squirting, 439
Cucumis melo, 439
Cucumis sativus, 439
Cucurbita argyrosperma, 439
Cucurbita maxima, 439
Cucurbita pepo, 439
Cucurbitaceae, 170, 439
Culpeper, Edward, 18
Cultivated plants, 419
 New World and Old World, 419
 origin, 419
Cupules, 480
Curare, 175, 442
Currant, 171, 440
Custard-apple family, 436
Cuticle, 245, 267, 269
Cutin, 56, 235
Cyanide in clover, 329
Cyanophage, 69
Cyanophyta, 81
Cycadophyta, 149
Cycads, 149
Cycas revoluta, 148, 149
Cyclosis, 267
Cydonia oblonga, 440
Cynara scolymus, 444
Cynodon dactylon, 445

Cynoglossum officinale, 443
Cyperaceae, 180, 445
Cyperales, 180, 445
Cyperus papyrus, 445
Cypress, 153
  cones, 153
  knees, 386
Cypripedium acaule, 446
Cypripedium calceolus, 187
Cystocarp, 464
Cytochromes, 196, 204
Cytokinins, 270, 301, 208, 317
Cytology, 14, 53, 54, 55
Cytoplasm, 53, 57
Cytoplasmic inheritance, 340
Cytosine, 42, 341, 345

D

Dahlia, 444
Dahlia pinnata, 444
Dandelion, 177, 178, 293, 315, 444
Danielli, 56
Dark Ages, 17
Dark reaction, 207
Darwin, Charles, 7, 353
  theory, 355
Dasycladales, 88
Data, 19
  analysis, 26
  kinds used, 21
  organization, 22
  processing, 21, 22
Dates, 180
Datura wrightii, 177
Daucus carota, 442
Dayflower, blue, 179
De Vries, Hugo, 8
Deciduous, 249
  forest, 408, 409
Defective virus, 71
Delphinium ajacis, 437
Democritus, 354
Deoxyribonucleic acid, 41
Deoxyribose, 40, 42
Desert, 405, 406
  Sonoran, 407
Design, experimental, 24
Desmid, 45
Desmodium tortuosum, 440
Determinate flower, 284
Deuteromycota, 120
Development, 297
Devonian, 374
Diakinesis, 323
Dianthus barbatus, 437
Diatomaceous earth, 95
Diatoms, 93, 94, 95
Dicentra spectabilis, 437
Dicotyledoneae, 161

Dicotyledones, classification, 436
Dictyosiphon, 98
  sp., 97
Dictyosome, 52, 63
Differentiation, 220, 297
Diffusion, 46
  deficit, 270
  passive, 258
  pressure, 46
Digitalis, 428
Digitalis purpurea, 444
Digitaria sanguinalis, 445
Dihybrid, 326, 327
  cross, 329
  modified, 328
Dikaryotic, 109
Dill, 175, 442
Dioecious, 160, 330
Dionaea muscipula, 250, 440
Dioscorea sativa, 446
Dioscorides, 17
Diospyros ebenum, 440
Diospyros virginiana, 440
Diphenylurea, 308
Diplanetism, 466
Diplococcus pneumoniae, 73
Diploid, 322
Diplotene, 323
Dipsacaceae, 178, 444
Dipsacales, 178, 444
Dipsacus sylvestris, 444
Disaccharide, 38
Disease, 27, 80
Dissociation, 37
Distribution, 359
  environmental factors, 397
  mechanisms, 399
Diversity, 2
DNA, 41, 64, 299, 341, 344
  replication, 345
Dobzhansky, Theodosius, 8
Dock, 438
Documents, 21
  source, 23
Dogbane, 443
  family, 175, 443
Dogtooth violet, 446
Dogwood, 172, 441
  family, 441
  flowering, 172
Dominant characteristic, 326
Donnan equilibrium, 258, 259
Dormancy, 317
  cycle, 317
Dormin, 310
Droseraceae, 171
Drosophila, 332
Drupe, 289, 290
Dryopteris austriaca, 134

Duckweed, 445
family, 180, 445
Dujardin, Félix, 52
Dyad, 323
Dynamic equilibrium, 46

E

Earth, 32
history, 366
Eaton agent, 71
Ebenaceae, 440
Ebenales, 440
Ebony, 440
family, 440
Ecballium elaterium, 439
Ecology, 13, 78
Ecosystems, 30, 380, 381
Ecotone, 402
Ectocarpus sp., 98
life cycle, 462, 463
Egg, 277
nuclei, 508
Eggplant, 177, 443
Egyptians, 353
Eichhornia crassipes, 445
Elaeagnaceae, 441
Elaeis guineensis, 183
Elaiosome, 400
Elater, 140, 489
Electrical energy, 33
Electromagnetic spectrum, 203
Electromagnetic waves, 204
Electron, 34
microscope, 54
transport, 195, 197
Electro-osmosis, 266
Element, 33
ancient, 16
Elm, 167, 438
family, 438
Elodea, 49
Embryo, 292, 297
sac, 292
Embryonic region, 223
Empedocles, 354
Emulsion, 45
Endergonic reaction, 189
Endive, 444
Endocarp, 289
Endodermis, 141, 225
Endoplasmic reticulum, 52, 55, 59
Endosperm, 216
Energy, 33, 189
cycle, 191
extraction, 191
free, 189
kinetic, 33
magnetic, 33

Energy—cont'd
potential, 33
radiant, 33
English ivy, 442
Entropy, 190
Environment, 381
abiotic, 381
biotic, 391
physical, 381
plant form and, 391
Enzyme, 36, 43, 44
Eocene, 377
Ephedra, 155
Epicotyl, 215
Epidemic, 80
Epidermis, 221
Epigynous, 160
Epiphytes, 180, 408
Epiphytotic, 516
Epistasis, 329, 341
Epizootic, 80
Equisetaceae, 141
Equisetales, 141
Equisetum sp., 141, 490
life cycle, 489
Ergotism, 112
Ericaceae, 439
Ericales, 170, 439
Erosion, 413
anchoring plants, 414, 415
Erysiphe cichoriacearum, 112
Erysiphe graminis, 112
Erythronium americanum, 446
Erythroxylaceae, 442
Erythroxylon coca, 442
Ethyl alcohol, 194
Ethylene, 301, 303, 309, 310
Etioplasts, 59
Eucalyptus, 172
globulus, 159
Eudorina, 88
Euglena sp., 95, 96
Euglenophyta, 95
Eukaryon, 53
Eukaryotic, 5, 53
Euphorbia antisyphilitica, 173, 441
Euphorbia pulcherrima, 441
Euphorbiaceae, 172, 441
Euphorbiales, 172, 441
Eusporangiopsida, 143
Evening primrose, 172, 441
family, 441
Evergreen forest
northern, 406
ponderosa pine, 408
southern, 407, 408
western, 408
Evolution, 7, 51, 301, 353

Evolution—cont'd
  anatomy and, 359
  data, 359
  development and, 359
  genetics and, 359
  physiology and, 359
  plant, 373
  process, 356
  theory, 353
Exergonic reaction, 189
Exine, 508
Exocarp, 289
Experimental design, 24
Experimentation, 19, 22

F

F_1, 324
F_2, 324
Fagaceae, 164, 437
Fagales, 164, 172, 437, 440
Fagopyrum esculentum, 438
Fagus grandifolia, 437
Fairy primrose, 440
Fairy ring, 119
Fascicular cambium, 238
Fat, 38
Fermentation, 189, 190, 192, 194
Fern, 135
  compared with lily, 503
  front, 143
  giant tropical, 144
  heterosporous water, 144
  leaf, 142
  stem, 142
  tree, 140
  whisk, 135
Ferrodoxin, 257
Ferrodoxin-reducing substance, 207
Fertilization
  double, 509
Fibrous root, 227
Ficus bengalensis, 168
Ficus carica, 438
Fig, 438
  strangling, 15
Filament, 278
Fire, 410
  exclusion, 410
Flagellum, 87, 460
Flavin adenine dinucleotide, 196
Flavin mononucleotide, 196
Flax, 173, 442
  family, 442
Fleming, Sir Alexander, 112
Florigen, 316
Flower, 279
  adaptation, 282
  cultivated, 420

Flower—cont'd
  imperfect, 282
  indeterminate, 284
  irregular, 283
  monocot and dicot comparison, 282
  naked, 283
  parts, 279, 280
  perfect, 282
  staminate, 283
Flowering, 303
  factor, 316
  hormone, 316
  plants, classification, 160, 436
  response, 314, 315, 316
Foliar fertilization, 261
Food, 421
  preservation, 312
Foolish seedling disease, 305
Forest
  temperate rain, 6
  tropical rain, 409, 410
Forget-me-not, 177, 443
Forsythia, 443
Forsythia suspensa, 443
Fossil, 1, 359
  formation, 367
Fouquieria splendens, 169, 439
Fouquieriaceae, 167, 439
Four-o'clocks, 165, 437
  family, 437
Foxglove, 444
Fragaria chiloensis var. *ananassa*, 440
Frangipani, 175, 176
Frankincense, 442
Fraxinum americana, 443
Frond, 141
Frugivores, 363
Fruit, 287, 303
  accessory, 289
  aggregate, 289, 291
  dehiscent, 291
  development, 288
  dispersal, 293, 294
  dry, 289
  fleshy, 289
  indehiscent, 291
  kinds, 289
  multiple, 289
  pineapple, 292
  simple, 289
Fucus
  life cycle, 460, 461
  sp. 97
Fuels, 39
Fungi, 311
  bird's nest, 119
  chytrids, 106
  classification, 435

Fungi—cont'd
club, 112, 113
conjugation, 107
ergot, 112
imperfect, 120
jelly, 118
oosphere, 106
parasitic, 121
sac, 109
saprophytic, 102, 121
stinkhorns, 119
Fungicides, 44, 429
Fungus, 102
bracket, 113, 114
tinderbox, 115
Funiculus, 285

G

Galápagos Islands, 352
Galilei, Galileo, 18
Galls, 119
Gametangium, 99
Gamete, 275, 321
Gametophores, 480
Gametophyte, 87, 277, 322
Gamopetalous, 283
Gamosepalous, 283
Gap, 357
Garcinia mangostana, 438
Garden, botanical, 25
Garlic, 187
wild, 445
Gaultheria procumbens, 439
Gause's principle, 397
Gaylussacia baccata, 439
Gazelles, 380
Gel, 45
Gelatin, 45
Gelidium sp., 99
Gemma, 276
Gene, 297, 321, 324
modifying, 330
multiple, 336
operator, 348
pool, 358
promoter, 348
regulator, 348
structural, 348
Generation
spontaneous, 51
three alternate, 99
Generative nucleus, 508
Generic name, 8, 9
Genetic code, 345
Genetic drift, 360
Genetics, 13
Genotype, 325, 363

Gentian
family, 175, 442
fringed, 442
Gentiana crinita, 442
Gentianaceae, 442
Gentianales, 175, 442
Geological record, 368
time, 368
time scale, 369
Geotropism, 216, 302, 303
Geraniaceae, 442
Geraniales, 173, 442
Geranium, 175
family, 442
Germ plasm, 324
Germ theory, 80
Germination, 216
Giant cactus, 438
Gibberella fujikuroi, 307
Gibberellic acid, 270, 307
effect, 307, 308
Gibberellins, 301, 305, 313, 317
in flowering, 316
Gill, 478
Ginkgo, 64, 149
Ginkgo biloba, 149, 150
Ginkgophyta, 149
Giraffe, 380
Gladiolus, 446
Gladiolus gandavensis, 446
Gleditsia triacanthos, 440
Globule, 456
Gloeocapsa, 84
Glucan, 104
Glucanase, 304
Glucoabscisin, 302
Glucose, 192
Glycerol, 39
Glycine max, 440
Glycolysis, 192, 193
Gnetes, 155
Gnetophyta, 155
Golden tuft, 439
Goldenrod, 315, 444
Golgi apparatus, 52, 55, 62, 63
Gonidia, 452
Gooseberry, 171, 440
Barbados, 438
family, 438
Gosypium arboreum, 438
Gourd, 439
family, 170, 439
Grafting, 430, 431
Graminae, 179
Grana, 52, 202
Grape, 441, 442
family, 173, 441
fruit, 442

Graphica scripta, 125
Grass, 178
  arrow, 427
  family, 179
  holy, 415, 445
  land, 405
  sweet, 445
  tree, 187
Greenbrier, 446
Grew, Nehemiah, 52
Ground meristem, 229
Growth, 297
  hormones, 272
  secondary, 238
Guanine, 42, 341, 345
Guava, 441
Gum
  sweet, 437
  Tasmanian blue, 158
Gum tree, 172
Guttation, 49, 261
Gymnocladus dioica, 440
Gymnodinium sp., 96
Gynoecium, 282

H

Habitats, plant, 395
Hamamelidaceae, 437
Hamamelidales, 164, 437
Hamamelis virginiana, 437
Haploid, 322
Hapteron, 120
Harvey, William, 51
Hashish, 167
Hatch-Slack pathway, 208, 210
Haustorium, 472
Hawthorn, 440
Hazel, 164
  nut, 437
Heath family, 170, 439
Hedera helix, 442
Helianthus annuus, 444
Hellebore, 164, 437
Helleborus niger, 437
Helmont, Jan van, 202
Heme, 196, 204
Hemlock, 175
  poison, 174, 442
Hemoglobin, 257
Hemp, 438
  Indian, 167, 443
  Manila, 445
  sisal, 445
Henbane, 316
Hepatophyta, 127
Heraclitus, 354
Herbaceous stem, 240
Herbalists, 17

Herbals, 18
Herbarium, 21
Herbicides, 429
Heredity, 14
Heterocysts, 449
Heterogamete, 91
Heterogamous, 466
Heteroploidy, 339
Heterosis, 335
Heterosporous, 145
Heterothallic, 454
Heterotrophs, 1
Heterozygous, 325
Hevea brasiliensis, 441
Hexaploid, 337
Hibiscus, 167, 438
  leaves, 321
Hibiscus syriacus, 438
Hickory, 173
  swamp, 442
Hierochloe alpina, 445
Hill, Robert, 206
Hilum, 292
Hippocastanacea, 442
Hippocrates, 16
Hippomane mancinella, 441
Histology, 14
Histone, 64
History, plant science, 16
Hobby, 431
Holly family, 441
Hollyhock, 167, 438
Holozoic, 96
Holy fire, 112
Holy grass, 415, 445
Homologous chromosomes, 321
Homologs, 321
Homology, 26, 357
Homozygous, 325
Honey locust, 440
Honeysuckle, 444
  family, 178, 444
Hooke, Robert, 51
Hop, 439
Hordeum vulgare, 445
Hormodendrum sp., 120
Hormogones, 449
Hormogonium, 449
Hormone, 297, 301
Hornbeam, 164, 437
Hornworts, 127
Horse nettle, 443
Horseradish, 170, 439
Horsetails, 139
  ancient, 375
Hot springs, 82
Hound's-tongue, 177, 443
Huckleberry, 170, 439

Humors, 16
Humulus americanus, 439
Hurricanes, 32
Hybrid, 362
    vigor, 335
Hybridization, 362
Hydathodes, 261
Hydra, 4
Hydra sp., 90
Hydrangea, 440
Hydrangea paniculata var. *grandiflora*, 440
Hydrodictyon, 88, 89
Hydrogen, 254
    ion, 37
Hydroponics, 253
Hydrostatic pressure, 265
Hydrotropism, 226, 398
Hydroxyl, 37
Hyella, 84
Hypanthium, 518
Hypericaceae, 438
Hypericum perforatum, 438
Hypertonic, 47, 49, 64
    solution, 49
Hypha(e), 103
Hypochytridiomycetes, 107
Hypochytrids, 107
Hypocotyl, 215
Hypogynous, 160
Hypothesis, 19
Hypotonic solution, 49
Hyssop, 443
Hyssopus officinalis, 443

I

Identification, 433
Idria columnaris, 439
Ilex opaca, 441
Imbibition, 47
Immunity, 80
Impatiens biflora, 442
Incomplete dominance, 326, 327
Independent assortment, 324
Indian jujube, 441
Indian pipe, 170, 171, 439
Indole ethanol, 302
Indoleacetaldehyde, 302
Indoleacetonitrile, 302
Inducer, 348
Indusium, 492
Inflorescence, 284
Information, 19
    retrieval system (IRS), 21
    sources, 20
Inhibitors, 310
Inorganic compound, 37
Inorganic molecules, 37
Insecticides, 44, 425
Insects, 425

Integuments, 508
Interfascicular cambium, 238
Interference, 334, 335
International Code of Botanical Nomenclature, 8
Internode, 93
    elongation, 312
Interphase, 299
Introgressive hybridization, 360, 361
Inversion, 339
Ion, 34
    exchange, 258, 259
Ipomoea batatas, 443
Ipomoea purpurea, 443
Iridaceae, 446
Iris, 446
    family, 446
Iris versicolor, 446
Iron, 254
    absorption, 257
Ironwood, 437
Isoetales, 139
Isoetes sp., 139
Isogamete, 451
Isolation mechanism, 359
Isotonic, 47, 49
    solution, 49
Isotope, 3, 35
    tracer, 24
Israelites, 354

J

Janssen, Zacharias, 18
Jellyfish, 4
Jimsonweed, 177
Johannsen, Wilhelm Ludwig, 8
Joshua tree, 186, 187
Juglandaceae, 173, 442
Juglandales, 173, 442
Juglans regia, 442
Juncaginaceae, 444
Juniper, alligator, 141
Juniperus deppeana, 151
Jurassic period, 376
Jute, 438

K

Kale, 170, 439
Kalmia latifolia, 436
Kapok, 438
Kelp, 97
Kentucky coffee tree, 440
Kew Gardens, 25
Kinetochore (centromere), 300
Kingdoms, 4
Koch, Robert, 80
    postulates, 80
Kohlrabi, 170, 439

Kornberg, Arthur, 344
Krebs cycle, 191, 192, 194

L

Labiatae, 443
Labyrinthulomycetes, 105
Lactose operon, 348, 349
Lactuca sativa, 444
Lacustrine, 389
Lady's slipper, 446
   yellow, 187
Lamarck, Jean Baptiste de, 7
Lamb's quarters, 438
Lamiales, 177, 443
Laminaria sp., 98
Laminarin, 87
Land bridges, 371
Land development, 415
Land, plants for, 413
Landslide, 389
Larkspur, 437
Larrea tridentata, 442
Lateral
   conjugation, 93
   roots, 226
   transport, 265
Lauraceae, 436
Laurales, 161, 436
   family, 436
Lavandula officinalis, 443
Lavender, 443
Layering, 431
Leaf, 269
   abscission, 321
   adaptation, 247
   axil, 242
   compound, 247, 249
   epidermal appendates, 246
   monocot and dicot compared, 245
   mosaic, 242
   scars, 243
   senescence, 316, 317
   simple, 247
   trichomes, 246
   venation, 248
Leek, 187, 445
Legume, 290
Leguminosae, 171, 440
Leitneria floridana, 437
Leitneriaceae, 164
Leitneriales, 164, 437
Lemaireocerus thurberi, 166
Lemna, 180
Lemnaceae, 180, 445
Lemon, 442
Lemon geranium, 442
Lenticel, 242, 243
Leprosy, 78
Leptosporangiopsida, 143

Leptotene, 323
Lethal gene, 336
Lettuce, 178, 315, 444
Leucosin, 87
Leukoplast, 59
Lewisia rediviva, 438
Lianas, 240
Lichen, 123, 124
   foliose, 124
   fruticose, 125
Life, origin, 1, 366
Light, 204, 381
   intensity, 272
   reaction, 206
Lignin, 222
   synthesis, 318
Ligule, 485
Ligustrum vulgare, 443
Lilac, 443
Liliaceae, 180, 445
Liliales, 180, 445
Lilium canadense, 446
   life cycle, 502
   sp., 504-509
Lily, 446
   compared with fern, 503
   family, 180, 445
   of the valley, 446
Linaceae, 442
Linden, 167, 438
   family, 438
Linen, 173, 442
Linkage, 331, 332
   group, 332
Linnaeus, Carolus, 8, 355
Linné, Carl von, 8
Linum usitatissimum, 442
Lipid, 38
   bilayer, 56
Liquidambar styraciflua, 437
Liriodendron tulipifera, 436
Literature, 21
Liverwort, 126, 311
Locoweed, 440
Locules, 282
Locus, 331
Loganberry, 171, 440
Logania family, 175, 442
Loganiaceae, 442
Lonicera sempervirens, 444
Loranthaceae, 172, 441
Lucerne, 440
Lumber, 423
   erosion, 424
Lycopersicon esculentum, 443
Lycophyta, 135, 375
Lycopodiales, 135
Lycopodium, 137
Lycopods, 135

Lyngbya, 84
Lysimachia nummularia, 440
Lysis, 72
Lysosome, 52, 55, 63

M

Mace, 436
Maclura pomifera, 439
Macronutrients, 253, 254
Macrosporangium, 460
Madder family, 177, 444
Magnesium, 254
  absorption, 256
Magnolia soulangeana, 161, 436
Magnolia virginiana, 436
Magnoliaceae, 161, 436
Magnoliales, 161, 436
Magnus, Albertus, 18
Mahogany family, 442
Malaria, 178
Mallow, 438
  family, 438
Malpighi, Marcello, 51
Malt, 445
Malus sylvestris, 440
Malva neglecta, 438
Malvaceae, 167, 438
Malvales, 167, 438
Manchineel, 428, 441
Manganese, 254, 257
Mangifera indica, 442
Mango, 442
Mangosteen, 438
Mangrove, 172, 386, 440
  family, 440
Manihot esculenta, 441
Manna, 125
Map unit, 334
Maple, 173
  family, 442
  silver, 174, 442
Mapping, 334
Marchantia sp., 126, 128, 481, 482
  life cycle, 480
Marigold, 178, 444
Marijuana, 438
Marjoram, 443
Marsh marigold, 163, 437
Marsilea sp., 144, 145
Marsileaceae, 145
Mass-flow, 265-267
Matrix, 57
Matter, 33
Maxwell, 204
May apple, 163, 437
Measurements, 21, 24
Medicago sativa, 440
Medullary rays, 231
Megaspore, 138

Megasporophyll, 485
Meiosis, 87, 321, 322
Meliaceae, 442
Membrane
  ATPase, 259
  plasma, 56
  unit, 56
Mendel, Gregor, 7, 323, 353
Mendelism, 323, 324
Mentha piperita, 443
Mentha spicata, 443
Meristem, 220
Mesembryanthemum family, 437
Mesocarp, 289
Mesophyll, 245, 269
  spongy, 269
Mesophytes, 398
Mesozoic era, 376
Mesquite, 252, 440
Messenger RNA, 346
Metabolism, 190
Metaphase, 298-300
Metaxylem, 226
Methionine, 309
Microbiology, 69
Microcystis, 84
Microgametophyte, 487
Micronutrients, 253, 254, 257
Micropyle, 282
Microscope, 18, 51
  Culpeper, 18
  electron, 19, 28
Microsphaera sp., 473
  life cycle, 472
Microsporangium, 460
Microspore, 138
Microsporophyll, 485
Microsurgery, 54
Microtechnique, 432
Middle Ages, 18
Middle lamella, 55, 303
Midrib, 247
Milkweed, 175, 443
Millet, 445
Mimosa pudica, 440
Mineral, 253
  deficiency, 254
Mint, 177
  family, 443
Miocene, 377
Mirabilis jalapa, 437
Mississippian period, 375
Missouri Botanical Garden, 26
Mistletoe, 173, 441
  family, 172, 441
Mitochondrion (mitochondria), 52, 53, 55, 58, 59
Mitosis, 296, 298, 321
Mohl, Hugo, 52
Mole, 36

Molecular biology, 28
Molecular weight, 36
Molecule, 33
Molucella laevis, 443
Molybdenum, 254, 257
Moneywort, 440
Monocotyledoneae, 178
Monocotyledones, 444
Monoecious, 160
Monohybrid, 325, 326
Monosaccharide, 38
Monosomic, 339
Monotropa uniflora, 171, 439
Moraceae, 168, 438
Morning glory, 177, 443
　family, 443
Morphogenesis, 297
Morphology, 14
Morus alba, 439
Moss, 130, 132
　Spanish, 180
Mountain laurel, 436
Mucor sp., 109
Mulberry, 167, 439
　family, 438
Mullein, 267, 444
Multicellular, 275
Mumps, 71
Murtales, 440
Musa paradisiaca, 445
　var. *sapienta*, 182, 445
Musa textilis, 445
Musaceae, 180, 445
Mushrooms, 113, 116
Mustard, 170, 439
　family, 170, 439
Mutagenic agent, 336
Mutation, 336
　rate, 336
Mycelium, 103
Mycobacteriophages, 72
Mycologist, 119
Mycorrhiza, 121
　pine, 122
Mycota, 2, 5
Myosotis, 177
　scorpioides, 443
Myrica gale, 437
Myrica pensylvanica, 437
Myricaceae, 437
Myricales, 437
Myristica fragans, 436
Myristicaceae, 161, 436
Myrrh, 173, 442
Myrtaceae, 172, 441
Myrtales, 172
Myrtle, 172, 441
　bog, 437
　family, 172, 441

Myrtus communis, 441
Myxomycetes, 105
Myxomycota, 104

N

n number, 322
NAD, 191
NADP, 191
NADPH, 190
Naiad, 178, 444
Najadaceae, 178, 444
Najadales, 178, 444
Najas, 178, 444
　family, 444
　marina, 444
Name
　generic, 9
　specific, 9
Narcissus, 446
Narcissus tazetta, 446
Neck canal, 482
Nectar, 281
Nectaries, 274, 281
Nectarine, 171, 440
Nemalion sp., 99, 465
　life cycle, 464
Nepenthaceae, 162, 436
Nepenthes pervillei, 162
Nettles, 439
　family, 439
Neurospora, 112
Neutral day plants, 315
Neutron, 34
New Jersey tea, 441
Newton, Isaac, 204
Niche, 397
Nicotiana tabacum, 443
Nicotinamide adenine dinucleotide, 196
　phosphate, 196
Nightshade, 177
　family, 443
Nitrate, 255
　reductase, 255
Nitrogen, 254
　absorption, 256
　cycle, 391
　fixation, 255
　fixing bacteria, 390
　reduction, 255
Node, 93
Nonhistone, 64
　protein, 64
Nostoc sp., 83, 94
　life cycle, 449
Nothofagus, 377
Nucelles, 292
Nuclear envelope, 65
Nucleolar organizing region, 66
Nucleolus, 52, 55, 66

Nucleotide, 40, 41, 44
Nucleus, 52, 53, 55, 64, 269
  tube, 508
Nucule, 456
Nullisomic, 339
Nutmeg, 161, 436
  family, 436
Nutrition, 253
Nux-vomica tree, 442
Nyctaginaceae, 165, 437
Nymphaea odorata, 162, 436
Nymphaeaceae, 162, 436
Nymphaeales, 162, 436
Nyssa aquatica, 441
Nyssa sylvatica, 441
Nyssaceae, 172, 441

O

Oak, 164
  cork, 437
  live, 437
  post, 437
  water, 437
  white, 437
Oat, 445
  field, 404
  root, 260
  wild, 444
Observation, 22
Ocean, 402
Ocimum basilicum, 443
Ocotillo, 167, 198, 439
  family, 439
Oedogoniales, 88
Oedogonium, 84, 88, 91
Oenothera biennis, 441
Oidium (oidia), 109
Oil, 39
Olea africana, 176
Olea europaea, 443
Oleaceae, 443
Oleaster family, 441
Oligocene, 377
Olive, 175, 443
  family, 175, 443
  small fruited, 176
Onagraceae, 172, 441
Onion, 187, 445
Oogonium, 88
Oomycetes, 106
Oomycota, 106
Oosphere, 453
Oospore, 453
Opaque-2, 425
Operculum, 483
Operon, 348
Opposite leaves, 241
Opuntia chlorotica, 166, 438
Opuntia versicolor, 438

Orange, 442
Orchid, 187, 446
Orchidaceae, 187
Orchidales, 187, 446
Ordovician, 374
Organ, 214, 215
  system, 215
Organelle, 53
Organism, 1
Origanum vulgare, 443
Oryza sativa, 445
Osage orange, 439
Oscillatoria, 83, 84
Osmosis, 46, 47
Osmotic pressure, 260
Ostrya virginiana, 437
Ovary, 282, 283
  type, 285
Overgrazing, effects of, 416
Ovule, 282
Oxaloacetate, 195, 210
Oxidation, 36
  -reduction, 36
Oxidative phosphorylation, 191, 192, 195-197
Oxygen, 254
  in atmosphere, 385
  in carbon dioxide cycle, 385
  cycle, 385
Oxytropis splendens, 440

P

P_{660}, 313
P_{730}, 313
Pachytene, 323
Paeonia suffruticosa, 437
Paleocene, 377
Paleomagnetism, 372
Paleozoic era, 373
Palisade layer, 246
Palm
  betel, 445
  coconut, 445
  date, 445
  family, 180, 445
  oil, 183
  sago, 148, 149
  traveler's, 180, 182, 445
Palmate, 248, 445
Palmetto, saw, 445
Palynology, 282
Pandana, 445
  family, 445
Pandanaceae, 180, 445
Pandanales, 180, 445
Pandanus, 180
  pacificus, 184, 445
Pandorina, 88, 90
Pangaea, 372
Pansy, 439

Papaver somniferum, 437
Papaveraceae, 164, 437
Papaverales, 164, 437
Papaw, 436
Papaya, 169, 170, 439
Paper, 180
   birch, 437
Papyrus, 180, 445
Paramecium, 4, 90
Paramylon, 87
Parasitic, 5
Parasitism, principle, 81
Parenchyma, 221
Parietal, 285
Pars amorpha, 66
Parsnip, 442
Parthenocarpy, 308
Parthenocissus quinquefolia, 441
Parthenocissus tricuspidata, 441
Parthenospore, 453
Pasteur, Louis, 51
Pasteurization, 51
Pastinaca sativa, 442
Pathogen, 80
Pea, 172, 325, 440
   family, 171, 440
   garden, 421
   sweet, 283
Peach, 171, 440
Peanut, 440
Pear, 440
Pecan, 173, 442
Pectin, 52, 55, 303
Pediastrum, 88, 89
Pedicel, 279
Peduncle, 279
Pelargonium crispum, 442
Penicillium, 103
   chrysogenum, 111
   sp., 109, 110
Pennisetum glaucum, 445
Pennsylvanian period, 375
Pentaploid, 337
Pentose-shunt pathway, 198, 199
Peony, 164, 437
Pepper, 162
   bell, 177, 443
   family, 436
   green, 177, 443
Peppermint, 443
Peptide, 40
   bond, 40
Perennial, 213, 339
Perianth, 281
Pericarp, 288
Pericycle, 141, 226
Periderm, 238
Peridinium sp., 96
Peridiole, 120

Perigynous, 285
Perithecium, 473
Periwinkle, 175, 443
Permeable, selectively, 47
Permeation, 47
Permian period, 376
Persea americana, 436
Persimmon, 440
Pesticides, 29
Petal, 281
Petiole, 247
Petunia hybrida, 443
pH, 37, 305
Phaeophyta, 96
Phage, 70, 71
Phallales, 119
Phanerogams, 147
Phaseolus limensis, 440
Phaseolus vulgaris, 440
Phelloderm, 238
Phellogen, 238
Phenotype, 325, 363
Phenylalanine ammonia-lyase (PAL), 313
Phloem, 261, 262
   translocation, 264
Phlox, 177
   annual, 443
   family, 443
   wild blue, 443
Phlox
   divaricata, 443
   drummondii, 443
Phoenix dactylifera, 445
Phoradendron juniperinum, 173, 441
Phosphoenolpyruvate carboxylase, 210
Phospholipid, 308
Phosphorus, 254
   absorption, 256
Photoactivation, 314
Photoelectric effect, 204
Photomicroscope, 20
Photomorphogenesis, 312
Photon, 204
Photoperiod, 272, 297
Photoperiodism, 312, 383
Photoreaction center, 206
Photoreceptors, 312
Photosynthesis, 4, 201, 209, 210
Photosynthetic, 69
Photosystem
   I, 206, 207
   II, 207
Phototropic, 301
Phototropism, 302
Phycobilins, 205
Phycocyanin, 82
Phycoerythrin, 98
Phycomyces sp., 108
Phylogenetic tree, 6, 365

Phylogeny, 364
Physalis alkekengi, 443
Physiology, 13, 76
Phytochrome, 313
    system, 313-315, 318
Phytophagous, 394
Phytophthora infestans, 107
Pickerelweed family, 445
Piereskia aculeata, 438
Pigment systems, 204
Pigweed, 438
Pileus, 478
Pimenta dioica, 441
Pimpinella anisum, 442
Pine
    New Caledonian, 146
    screw, 184
Pineapple family, 180, 445
Pink family, 437
Pinnae, 492
Pinnate, 248
Pinnately
    bicompound, 248
    compound, 248
Pinus
    aristata, 155
    life cycle, 495
    sp., 496-500
Piper nigrum, 436
Piperaceae, 162, 436
Piperales, 162, 436
Pistil, 282
    arrangement, 286
    simple, 285
Pistillate, 147
Pisum sativum, 324, 440
Pitcher plant, 13, 162, 250, 399, 437
    family, 164
    green, 165
    Old World family, 436
Pith, 231
    rays, 231
Placenta, 285
Placentation, 286
Plane tree, 164
    family, 437
Plankton, 394
Plant(s)
    aerial, 398
    body evolution, 278
        organization, 214
    carnivorous, 250
    changes, 421
        cultivation, 421
    differences, 4
    disease-producing, 428
    diseases, 429
    distinct from animals, 2

Plant(s)—cont'd
    features common to animals, 3
    fibers, 423
    food, 421
    geological history, 374
    life cycles, 447
    long-day, 313
    major groups, 435
    medicinal, 429
    new, 424
    parasitic, 399
    poisonous, 426
    primitive vascular classification, 436
    science subjects, 13
    short-day, 315
    source of carbohydrate, 422
    specimen preparation, 432
    vascular, classification, 136
    wild for food, 425
Plantae, 2, 5, 7
    classification, 435
    kingdom, 87
Plantaginales, 177, 443
Plantago family, 443
Plantago lanceolata, 443
Plantain, 443, 445
    fruiting, 392
Plantanaceae, 164, 437
Plaques, 72
Plasma, 56
    membrane, 52, 53, 55-57
Plasmodesmas, 52
Plasmodiophoromycetes, 105
Plasmodium, 104
Plasmolysis, 49
Plasmopara viticola, 107
Plastid, 55, 59
Plastocyanin, 207
Platanus occidentalis, 437
Plate tectonics, 370, 371
Pleiotrophic effect, 359, 522
Pleuropneumonia-like organism (PPLO), 71
Pliny the Elder, 17
Pliocene, 377
Plow, 418
Plum, 171, 440
Plumeria rubia, 176
Plumuli, 215
Plurilocular, 462
Pneumococcus, 73
Pneumonia, 73
Poales, 179, 444
Podophyllum peltatum, 163, 437
Poinsettia, 173, 315, 441
Poison ivy, 427, 442
Polar nuclei, 508
Polemoniaceae, 443
Polemoniales, 175, 443

Poles, 299
Pollen
    grains, 64, 281
    sacs, 281
    tube, 281, 508
Pollination, 27, 285, 287, 410
    air, 430
    self, 285
Pollinators, 274
Polygenes, 336
Polygonaceae, 438
Polygonales, 438
Polygonum cilinode, 438
Polymer, 39
Polymerase, 346
Polymorphic, 360
Polymorphism, 360
Polypeptide, 40
    primary structure, 40
    secondary structure, 40
Polyploidy, 337-339, 421
Polypodium sp., 493
    life cycle, 492
Polysaccharides, 38
Polysiphonia sp., 99
Polysomic, 339
Polytrichum, 32
    life cycle, 483
    sp., 448
Pome, 289
Pomegranate, 441
    family, 172, 441
Pondweed, 179, 444
Pontederiaceae, 445
Poplar, 170
    lombardy, 439
Poppy, 437
    family, 437
    prickly, 164, 165, 437
Population, 396
    explosion, 27
    genetics and, 360
Populus
    deltoides, 439
    nigra var. *italica*, 439
    tremuloides, 439
Porphyra, 99
Portulaca, 165, 438
Portulaca oleracea var. *sativa*, 438
Portulacaceae, 165, 438
Potassium, 254
    absorption, 256
Potamogeton distinctus, 309
Potamogeton natans, 444
Potamogetonaceae, 179
Potato, 177, 443
    sweet, 443
PPLO, 71

Prasiola, 88
Prickles, 243
Primrose family, 440
Primula malacoides, 440
Primulaceae, 440
Primulales, 440
Privet, 443
Procarp, 464
Proembryos, 501
Prokaryotae, 2, 5, 53
    classification, 435
    kingdom, 69
Promeristem, 229
Prophase, 298, 299
Prosopis juliflora, 440
Proteales, 441
Protein, 38, 39
    synthesis, 346, 347
Prothallium, 140
Prothallus, 487
Protochlorophyll, 314
Protococcus, 88
    viridis, 88
Proton, 34
Protonema, 93
Protoplasmic streaming, 267
Protostele, 226
Protoxylem, 226
Protozoa, 1
Prune, 171, 440
Prunus
    amygdalus, 440
    armeniaca, 440
    avium, 440
    domestica, 440
    persica, 440
        var. *nectarina*, 440
Psidium guajava, 441
Psilophyta, 135
Psilotum sp., 135, 137
Psittacosis, 71
Pteridophytes, 141
Pterophyta, 141
Puccinia graminis, 117, 475, 476
    life cycle, 474
Puffballs, 113, 115
Pumpkin, 170, 439
Punica granatum, 441
Punicaceae, 172, 441
Purines, 42
Purkinje, Johannes, 52
Purslane, 438
    family, 438
Pycnium, 477
Pyrimidines, 42
Pyrrophyta, 96
Pyrus communis, 440
Pyruvate, 192

Q

Q, 207
Quantum, 204
Quassia family, 442
Quaternary period, 377
  structure, 40
Queen Anne's lace, 175
Queen-cup, 446
Quercus
  alba, 437
  nigra, 437
  stellata, 437
  suber, 437
  virginiana, 437
Quillwort, 139
Quince, 440

R

Rachis, 248
Radiation
  environmental, 388
  of species, 360
Radicle, 215, 216
Radioautography, 55
Radish, 170, 439
Ragweed, 178, 315, 444
Ranunculaceae, 162, 163, 436
Ranunculales, 162, 436
Ranunculus acris, 437
Raphanus sativus, 439
Raphe, 292
Raspberry, 171, 440
Ravenala madagascariensis, 182, 445
Ray, John, 8
Ray, vascular, 227, 264
Realm, 401
  geographical, 402
Recent (epoch), 377
Receptacle, 279
Recessive characteristic, 326
Record, 21
Red clover, 171
Red laurel, 439
Redbud, 440
Redi, Francesco, 51
Redox potential, 196
Reduction, 36
Reef, 402
Refugia, 377
Region of elongation, 223
Region of maturation, 223
Relative humidity, 272
Repressor, 348
  protein, 348
Reproduction, 75, 84, 275
  asexual, 276
  basic principles, 275
  sexual, 276

Research, 20
  botanical, 13
  requirement, study box, 14
Respiration, 189, 190
Rhamnaceae, 441
Rhamnales, 173, 441
Rhamnus purshiana, 441
Rheum rhaponticum, 438
Rhinitis, 109, 120
Rhizobium, 390
Rhizoids, 104
Rhizome, 91
Rhizomorphs, 478
Rhizophoraceae, 172, 440
Rhizopus stolonifer, 107, 469
  life cycle, 468
Rhododendron arborescens, 439
  canadense, 439
  catawbiense, 439
Rhodophyta, 98
Rhodora, 439
Rhodymenia, 99
Rhubarb, 438
Rhus radicans, 442
Rhus typhina, 442
Ribes grossularia, 440
Ribes sativum, 440
Ribose, 40, 42
Ribosome, 53, 55, 59, 61, 63
Ribulose-1, 5-diphosphate carboxylase, 207
Rice, 445
Ricinus communis, 441
Rickettsiae, 78
Rickettsias, 71
Ringworm, 103, 120
RNA, 41, 341
  polymerase, 346
  synthetase, 347
  transfer, 346-348
Rockrose family, 439
Rocky Mountain spotted fever, 78
Root, 216
  adaptation, 227
  adventitious, 244
    on stems, 240
  buttercup, 224
  cap, 223
  fibrous, 227
  hair, 223, 260
    region, 223
  lateral, 226
  monocot and dicot compound, 224
  pressure, 260, 261
  primary, 216, 226
  secondary, 218
  systems, 228
  tip, 222
Rooting, 229

Rosa cathayensis, 440
Rosaceae, 170, 440
Rosales, 170, 440
Rose, 440
    family, 170, 440
Rose of Sharon, 438
Rosemary, 443
Rosette, 213
Rosmarinus officinalis, 443
Royal Botanic Gardens, 25
Rubber tree, 173, 441
Rubiaceae, 177, 444
Rubiales, 177, 444
Rubus occidentalis, 440
Rubus strigosus, 440
Rubus ursinus var. *loganobaccus*, 440
Rue family, 442
Rumex crispis, 438
Runner, 240, 276
Rust, 113
    wheat, 118
Rutabaga, 170, 439
Rutaceae, 442
Rye, 332, 445

S

Saccharum officinarum, 445
Sage, 443
Sagittaria graminea, 444
Sagittaria pygmaea, 309
Saguaro, 166, 438
Saint-John's wort, 438
    family, 438
Salicaceae, 439
Salicales, 170, 439
Salix discolor, 439
Salsify, 444
Salt, 37, 38
    plant growth, 390
Saltbush, 438
Salvia, 315
Salvia officinalis, 443
Samara, 291
Sandalwood, 172, 441
    family, 441
Sanguinaria canadensis, 164, 437
Santalaceae, 441
Santalales, 172, 441
Santalum, 441
Sap, 261, 262
Sapindales, 173, 442
Sapium biloculare, 441
Sapodilla, 441
Saprolegnia sp., 467
    life cycle, 466
Saprophytes, 76, 399
Saprophytic, 5
Sargassum sp., 98

Sarracenia
    minor, 13
    oreophila, 165
    purpurea, 250, 437
Sarraceniaceae, 437
Sarraceniales, 164, 437
Sassafras, 161, 436
Sassafras albidum, 436
Satureja hortensis, 443
Savory, 443
Saxifragaceae, 440
Saxifrage family, 171, 440
Scalariform conjugation, 93
Scales, 495
Scenedesmus, 88, 89
Schizogoniales, 88
Schizophyta, 73
    classification, 435
Schizosaccharomyces sp., 110
    life cycle, 471
Schleiden, Matthias, 52
Scholasticism, 18
Schwann, Theodor, 53
Science, 27
    dangers, 29
    future, 30
Scientific method, 16, 19
    name, 8, 9
    notation, study box, 22
Scientists, 30
Scion, 430
Sclerenchyma, 221
Scouring rushes, 140
Screw pine, 184
Scrophulariaceae, 443
Scrophulariales, 177, 443
Sebastiana pavoniana, 441
Secale cereale, 445
Sedge, 180, 445
    family, 445
Sedimentation-velocity coefficient, 63
Seed, 213, 292, 297
    coat, 292
    dicot, 216
    dispersal, 293, 294
    germination, 215, 217, 312
    kinds, 292
    monocot, 217, 218
    parts, 219
    peach, 219
Seed plants, classification, 148
Segregation, 324
Selaginella sp., 147, 486-488
    life cycle, 485
    oregona, 138
Self-pollination, 285
Semiconservative replication, 344
Semipermeable, 47
Senescence, 304, 308, 310, 311, 316

Senna, 440
Sensitive plant, 172, 440
Sepal, 281
Septate, 104
Sere, 410
Serenoa repens, 445
Sex determination, 330
Sexual, 276
Shepherdia argentea, 441
Shoot, 228
Shore, 404
Shrubs, 240
Sieve tubes, 235, 267
Silurian, 374
Simaroubaceae, 442
Singer, 57
Siphonales, 88
Siphonocladiales, 88
Skunk cabbage, 180, 445
Slime mold, 105
    cell-net, 105
    cellular, 105, 106
    endoparasitic, 105
    plasmodial, 105
Smilax herbacea, 446
Smuts, 113
    common, 114
Snapdragon, 177, 443
    family, 443
Society, science and, 27
Sodium chloride, 35
Soil, 253, 388
    water, 270
Sol, 45
Solanaceae, 177, 443
Solanum
    carolinense, 443
    melongena var. *esculentum*, 443
    nigrum, 443
    tuberosum, 443
Solar day, 312
Solidago speciosa, 444
Soredium, 124
Sorghum, 445
Sorghum vulgare, 445
Sorus, 98
Sour gum, 441
    family, 172, 441
Source documents, 23
Soybean, 440
Spanish moss, 445
Spearmint, 443
Speciation, 358
Species, 5, 13, 361
Species Plantarum, 9
Specific name, 8, 9
Sperm, 277
Spermatium, 477
Spermatophytes, 147

Sphagnum sp., 131
Spiderwort, 179, 444
    family, 444
Spinach, 165, 438
Spinacia oleracea, 438
Spindle, 299
Spine, 243, 524
Spiraea prunifolia, 440
Spirillum, 75
Spirochetes, 78
Spirodela polyrrhiza, 445
Spirogyra, 88, 92
Spongy layer, 246
Sporangiophores, 108, 470
Sporangium, 140
Spore, 104
    swarm, 91
Sporocarp, 143
Sporogenesis, 505
Sporophylls, 485
Sporophyte, 87, 277
Sporulation, 140
Spruce, 151
Spurge family, 172, 441
Squash, 170, 439
Staff-tree, 441
Staining, selective, 55
Stamen, 281
Staminate cone, 147
Starch sheath, 236
Statospores, 95
Stebbins, G. Ledyard, 8
Stele, 225
Stem, 228
    adaptation, 243
    aerial, 240
    growing, 231
    leaf arrangement, 241
    modification, 239
    monocot and dicot compared, 237
    regions of growth, 230
    tendrils, 244
    underground, 240
    woody, 236
    development, 233
Stentor sp., 90
Stercula family, 438
Sterculiaceae, 438
Sterigma, 479
Sterol, 39
Sticktight, 444
Stigma(s), 282
Stinkhorns, 113
Stipule, 247
Stock, 430
Stolon, 104, 240, 243
Stoma, 127, 268, 269
Stomatal closure, 311
Stomatal movement, 270-272

Stonewort, 93
Strawberry, 440
Streptomyces sp., 120
  rimosum, 111
Strelitzia nicolai, 181
Strelitziaceae, 181
Stripe, 478
Strobilus, 140
Strychnine, 175, 442
Strychnos, 175, 442
  toxifera, 442
Style, 282
Suberin, 234
Substrate, 43, 388
Succession, 410
Sugarcane, 445
Sulfur, 254
  absorption, 256
Sumac, 173
  staghorn, 442
Sunflower, 178, 309, 315, 444
  family, 178
Suppressors, 330
Sweet bay, 436
Sweet fern, 437
Sweet gale, 437
  family, 437
Sweet william, 437
Swietenia mahogani, 442
Sycamore, 437
Symbionts, 104
Symbiosis, 393
Symbiotically, 90
Sympatric, 361
Symplocarpus foetidus, 445
Synapsis, 323
Syncytium (syncytia), 279
Synecology, 396
Synergids, 508
Syngamy, 501
Syringa vulgaris, 443
System, 214, 525
  closed, 22
  controlled, 22
Systematics, 4, 7, 365
Systematists, 4
Syzygium aromaticum, 441

T

T-even phage, 440
Tagetes lucida, 444
Tapioca, 173
Taproot, 213, 227
Taraxacum officinale, 177, 444
Taxon, 3, 5
Taxonomy, 13
Tea, 438
  family, 438
  plant, 165

Teasel, 444
  family, 178, 444
Tectonic, plates, 370, 371
Teliospores, 474
Telium, 474
Telophase, 298-300
Temperature, environmental, 388
Tendrils, 243
Terramycin, 111
Tertiary period, 377
Tertiary structure, 41
Testcross, 326
Tetrad, 323, 332
Tetraploid, 337
Tetraspora, 88
Tetrasporales, 88
Tetraspores, 99
Tetrasporophyte, 99
Thallus, 87, 91, 287
Thea sinensis, 438
Theaceae, 165, 438
Theales, 165, 354, 438
Theobroma cacao, 438
Theophrastus, 17
Theory, 19
Thermodynamics, 89
Thigmomorphogenesis, 309
Thistle, 178, 444
Thorn, 243
Thorn apple, 177
Thylakoid, 201, 202, 207
  discs, 201
Thyme, 443
Thymine, 42, 341, 345
Thymus vulgaris, 443
Tilia americana, 438
Tiliaceae, 438
Tillandsia usneoides, 445
Tissue, 215
  complex, 220
  complex permanent, 222
  primary permanent, 227
  secondary permanent, 227
  simple, 220
  simple permanent, 221
  supporting, 221
  vascular, 87
TMV, 341
Toadstools, 119
Tobacco, 177, 443
  mosaic virus, 341, 343
Tomato, 177, 315, 443
  berries, 320
Tonoplast, 64
Touch-me-not, 173, 442
Toxins, 77
Trace elements, 257
Tracer isotope, 24
Tracheid, 234, 297

Trachoma agents, 71
Tradescantia sp., 301
  virginiana, 444
Tragopogon porrifolius, 444
Transformation, 341, 342
Transforming principle, 341
Translocation, 253, 261, 339
  passive, 270
Transpiration, 253, 261, 264, 267, 268, 270, 308, 311
Transport
  lateral, 265
  passive, 267
Trapa natans, 441
Trapaceae, 172, 441
Traveler's tree, 180, 182, 445
Tree, 240
Tree of heaven, 442
Treponema pallidum, 80
Trianthema portulacastrum, 438
Triassic period, 376
Trichogyne, 98
Trichomes, 245
Trifoliolate, 248
Trifolium pratense, 171
Trifolium repens, 440
Triglochin palustre, 444
Triploid, 337
Triticum aestivum, 445
Trumpet hyphae, 98
Tuber, 243
Tuberculosis, 78
Tulip, 446
  tree, 436
Tulipa gesneriana, 446
Tumbleweed, 438
Tundra, 404
Tupelo, swamp, 441
Turgid, 49
Turgor pressure, 49, 270
Turnip, 170, 439
Twig, winter, 241
Typha latifolia, 445
Typhaceae, 180, 445
Typhales, 180, 445
Typhus, 78

U

Ubiquinones, 196
Ulmaceae, 438
Ulmus americana, 438
Ulotrichales, 88
Ulothrix, 88
  sp., 454, 455
Ultrastructure, 13
Ulva, 88
Ulvales, 88
Umbellales, 175, 442
Umbelliferae, 175, 442
Unicellular, 275

Unilocular, 98
Uracil, 42
Uredinales, 118
Uredium, 474
Uredospore, 474
Urtica dioica, 439
Urticaceae, 439
Urticales, 167, 438

V

Vaccinium corymbosum, 439
Vaccinium macrocarpon, 439
Vacuoles, 52, 55, 64
Valence, 35
Valonia, 88
Vapor pressure, 270
Variation, 321, 362
Varro, 17
Vascular bundle, 141
Vascular cambium, 231, 238
Vascular system, 141, 232
Vasculum, 432
Vaucheria, 88
  life cycle, 458, 459
Vegetable oyster, 444
Vegetation, floating, 398
Vegetative body, 276
  growth, 312
  types, map, 403
Veins, 248
Venter, 484
Venus's-flytrap, 171, 250, 440
Verbascum thapsus, 444
Verbena, 177, 443
  family, 443
Verbena hybrida, 443
Verbenaceae, 443
Vergil, 17
Vernalin, 316
Vernalization, 314, 316
Vessel, 234, 235
Vetch, 443
Vicia sativa, 443
Vinca minor, 443
Vines, 240
Viola odorata, 168, 439
Viola tricolor var. *hortensis*, 439
Violaceae, 167, 439
Violales, 167, 439
Violet, 167, 168, 439
  family, 439
Virginia creeper, 173, 441
Virion, 72
Virulent, 341
Virulenta, 69
  classification, 435
Virus 69-70
  helper, 71
  life cycle, 448

Virus—cont'd
  nomenclature, 71
  poliomyelitis, 71
  tobacco necrosis, 71
Vitaceae, 173, 441
Vitamin, 43
  A, 205
Vitis vinifera, 441
Volvocales, 88
Volvox, 88
  aureus, 90
  life cycle, 452, 453
  sp., 90

W

Walnut, 442
  family, 173, 442
Water, 44, 383
  absorption, 261
  cycle, 383
  hemlock, 442, 426
  hyacinth, 384, 445
  lettuce, 384
  lily, 162
    family, 436
  plantain family, 444
  pot vine, 162
  use cycle, 382
Watermelon, 439
Wavelength, 203
Weeds, 429
Welwitschia mirabilis, 154, 155
Wheat, 445
  Wichita, 420
Whorls, 241
Willow, 170
  family, 439
  pussy, 170, 439
Wind and atmosphere, 38
Windbreaks, plants for, 414
Winter hardiness, 317
Winter rest, 318
Wintergreen, 170, 439
Wisteria floribunda, 440
Witch hazel, 164, 437
  family, 437
Wolffia punctata, 445

X

Xanthium, 316
  spinosum, 444
Xanthophyceae, 93
Xanthophylls, 88, 205, 311
Xanthorrhoea preisii, 185
Xanthoxin, 310, 311
Xerophyte, 395
Xerophytes, 398
X-ray diffraction, 54
Xylem, 141, 261, 262
  sap, 262
  translocation, 262

Y

Yam, 446
  family, 446
Yucca brevifolia, 186

Z

Z, 207
Zamina, 64
  floridana, 149
Zea mays, 445
Zeatin, 308
Zinc, 254, 257
Zingiberales, 180, 445
Zinnia, 178, 444
Zinnia elegans, 444
Zizyphus
  jujuba,. 441
  mauritania, 441
Zone of differentiation, 229
Zone of elongation, 229
Zoogametes, 94
Zoogeographical realms, 401
Zoology, 2
Zoospore, 88, 451
Zygnema sp., 93
Zygnematales, 88
Zygomycota, 107
Zygophyllaceae, 442
Zygospore, 470
Zygote, 87, 321
Zygotene, 322, 323